Modern Control Engineering

Modern Control Engineering

Editor: Anabelle Holmes

MURPHY & MOORE
www.murphy-moorepublishing.com

Published by Murphy & Moore Publishing
1 Rockefeller Plaza,
New York City, NY 10020, USA
www.murphy-moorepublishing.com

Modern Control Engineering
Edited by Anabelle Holmes

International Standard Book Number: 978-1-63987-369-2 (Hardback)

Cataloging-in-Publication Data

Modern control engineering / edited by Anabelle Holmes.
p. cm.
Includes bibliographical references and index.
ISBN 978-1-63987-369-2
1. Automatic control. 2. Control theory. 3. Engineering. I. Holmes, Anabelle.
TJ213 .M63 2022
629.8--dc23

Contents

Preface

The purpose of the book is to provide a glimpse into the dynamics and to present opinions and studies of some of the scientists engaged in the development of new ideas in the field from very different standpoints. This book will prove useful to students and researchers owing to its high content quality.

Control engineering is a field of engineering which applies automation to the design of systems with desirable behaviors in controlled settings. By using sensors and detectors, the output performance of the controlled process is measured. Such measurement can provide corrective feedback to achieve the desired performance. Control engineering can have an essential role in diverse control systems, from flight and propulsion systems used in commercial airliners to household washing machines. Automatic control systems such as cruise control in a car are designed to perform without requiring human input. Modern control engineering integrates computer-automated design for controller system optimization, system identification, etc. This book is compiled in such a manner, that it will provide in-depth knowledge about the theory and practice of control engineering. From theories to research to practical applications, case studies related to all contemporary topics of relevance to this field have been included herein. This book is a resource guide for experts as well as students.

At the end, I would like to appreciate all the efforts made by the authors in completing their chapters professionally. I express my deepest gratitude to all of them for contributing to this book by sharing their valuable works. A special thanks to my family and friends for their constant support in this journey.

Editor

1

Route-Based Signal Preemption Control of Emergency Vehicle

Haibo Mu, Yubo Song, and Linzhong Liu

School of Traffic and Transportation, Lanzhou Jiaotong University, Lanzhou, Gansu 730070, China

Correspondence should be addressed to Haibo Mu; mhbmmm@mail.lzjtu.cn

Academic Editor: Ai-Guo Wu

This paper focuses on the signal preemption control of *emergency vehicles* (EV). A signal preemption control method based on route is proposed to reduce time delay of EV at intersections. According to the time at which EV is detected and the current phase of each intersection on the travelling route of EV, the calculation methods of the earliest start time and the latest start time of green light at each intersection are given. Consequently, the effective time range of green light at each intersection is determined in theory. A multiobjective programming model, whose objectives are the minimal residence time of EV at all intersections and the maximal passing numbers of general society vehicles, is presented. Finally, a simulation calculation is carried out. Calculation results indicate that, by adopting the signal preemption method based on route, the delay of EV is reduced and the number of society vehicles passing through the whole system is increased. The signal preemption control method of EV based on route can reduce the time delay of EV and improve the evacuation efficiency of the system.

1. Introduction

Once an unexpected event has happened, an evacuation or a rescue must be carried out to help people in hazardous situations, and emergency vehicles (EV) play an important role in this process. To reduce the negative impact of unexpected events, EV is used to transferr people from dangerous areas to emergency shelters or medical assistance organizations as rapidly as possible. As is well-known to us all, the quality of the emergency service relies on the travel time that EV spends in the evacuation route [1, 2]. Therefore, EV has a greater priority than normal social vehicles, and signal preemption policy should be adopted to ensure the rapid transit of EV. However, under the condition of traffic jams or if there is no emergency lane, the queuing vehicles may prevent EV from passing through the intersection. Even if there is an emergency lane, to avoid an urgent spectacle, EV also has to wait if green traffic light for the opposite direction is going on.

To ensure that EV can pass through intersections safely and rapidly, some scholars proposed the signal preemption strategies. Paniati and Amoni [3] summarized the advantages of traffic signal preemption for EV in improving response speed, ensuring safety, saving cost, and other aspects and pointed out some of the key technologies used in traffic signal preemption system as well as some associated responsibility institutions. Mirchandani and Lucas [4] studied control strategies by integrating transit signal priority and rail/emergency preemption within a dynamic programming-based real-time traffic adaptive signal control system. Song [5] proposed traffic signal priority control method for multiple EVs based on multiagent and obtained signal priority control strategies through the coordination of phase agent and management agent. Taking the minimum delay of EV at single intersection as control target, three signal priority strategies, namely green time extension, red time shortening, and multiphase integrated control were proposed by Yang et al. [6], and Vissim was adopted to evaluate the proposed strategies. Ma and Cui [7] studied the signal priority problem of EV that was coming from different directions and had to pass the same intersection in the certain time period and presented a signal priority control system for multiple EV based on multi-agent. Wang et al. [8] proposed a degree-of-priority based control strategy for emergency vehicle preemption operation to decrease the impacts of emergency vehicles on normal traffic, and the performance of the proposed strategy was compared with the conventional local detection based method under the microscopic simulation model. To reduce the response time and minimize the impact of EV operation on general traffic, Qin [9] proposed two new control strategies for EV signal preemption, where the first strategy was developed to enable signal transition from normal operation to EV signal preemption and the second control strategy was used for the transition from EV signal preemption back to normal operation.

Most of the signal preemption methods adopted currently are based on isolated intersection, namely, taking single intersection as a basis, and signal preemption of this intersection will be activated after local detector has detected EV and thus form a signal priority sequence from intersection to intersection [10, 11]. This kind of signal preemption strategy, which is based on local detection and clears the intersection one by one, can start only after an EV has been detected and usually leads to inevitable intersection delay. Moreover, when multiple signal preemptions are implemented during peak period, the queue and delay of social vehicles in a given network will be significantly affected [12].

Studies have shown that implementing signal preemption from the perspective of entire route can reduce the response time of EV [13]. Although substantial progress has been made in signal preemption of isolated intersection, research of dynamic signal preemption based on entire route is rarely seen. Kwon and Kim [14] developed a route-based signal preemption strategy. They used Dijkstra's algorithm to obtain an optimal path from a given origin to a desired destination so as to ensure a minimal travel cost under current traffic conditions. The control algorithm chose a specific phase combination for each intersection on the selected route sequentially when EV arrived. But, the main purpose of this research was to ensure the safety of pedestrians passing through intersection, and real-time traffic conditions of various intersections were ignored. To reduce travel time of EV, Gedawy et al. [15] combined signal preemption with dynamic route planning and designed route of EV from the perspective of network. A signal preemption strategy, which could ensure the safe operation of an EV and maximize traffic flow passing through the intersection at the same time, was proposed. In order to provide green wave for EV, Kang et al. [16] proposed a coordinated signal control method and carried out traffic simulation taking eight intersections as example. All the methods mentioned above ensured the smooth running of EV by determining in advance the signal phase of each intersection after EV had been detected. Since green signal stayed at the designated phase, traffic flow from other signal phases lost right of way and they had to stay a long time at the intersection, which in turn increased the burden of traffic system and even caused traffic jams. Therefore, it is especially important to establish a route-based dynamic preemption strategy so as to provide efficient and safe operating environment for EV and minimize the interference on social vehicles.

So far, most of the developed preemption systems operate on a single intersection basis and require local detection of EV at each intersection to activate signal preemption sequence. Owing to the fact that these kinds of strategies depend on local detection and clear intersection one by one, and signal preemption procedure can not start until an EV is detected, inherent delays at intersections are unavoidable. The basic method of route-based signal preemption control of EV is that, when unexpected events occur, in consideration of dynamically changing traffic situations, the route that EV can use to reach the scene of the accident in the shortest time is calculated and recommended as the evacuation route. Once an evacuation route is selected, particular signal preemption control strategy is presented to determine the activation time for preemption at each intersection on the emergency route and realize the signal preemption control of EV ultimately.

Evacuation route planning is an important component of emergency management that seeks to minimize the loss of life or harm to the public during unexpected events [17]. The aim of evacuation route planning is to minimize evacuation time under certain constraints and minimize the possibility of depth hazards to people from accident areas. Nowadays, almost all of the intersections in urban traffic network are signalized intersections. In order to realize smooth traffic under real road condition, traffic management departments usually adopt regulation measures, including no-left-turn, no-right-turn, P-turn, and U-turn, to improve the safety and efficiency of traffic. For this reason, the road network that is connected ostensibly is not connected actually, which add difficulties to the shortest path algorithms and make some classical finding methods unavailable. Many scholars have intensively studied the shortest path problems of road network with traffic restrictions [18–23]. Various techniques have been proposed to eliminate certain turning and crossing maneuvers at intersections so as to obtain evacuation routes that can provide continuous traffic flow and reduce accidents [24–30]. There also has been a considerable amount of evacuation route works that consider delay and capacity of intersections [31, 32]. The majority of the evacuation route works in the literature can be divided into three categories, namely, network flow methods, which can be further divided into linear programming and dynamic minimum cost flow problem [24, 33–38], simulation methods [39, 40], and heuristic methods, which mainly includes the well-known Capacity Constrained Route Planner (CCRP) algorithm and its improved algorithms [41–47]. The research of evacuation route planning has made substantial progress, and the development of modern detecting equipment and information technology ensures the enforceability of real-time traffic flow detection and information exchange among different intersection controllers. These technologies make it possible to implement route-based signal preemption control of EV, but a safe and efficient manner to change the traffic signal timing plan, namely, control mechanism, has not been well studied.

In this paper, we will study the problem of dynamic signal preemption control of EV based on entire route. We limit the scope of this paper to signal preemption control of EVs under the assumption that the evacuation route has been determined, and the problem of evacuation route calculation will not be discussed further. The objectives of research reported here were to (1) propose a route-based signal preemption strategy to provide efficient and safe operating environment for EV and minimize the impact on social vehicles in traffic network as far as possible, (2) develop a multiobjective optimization model to minimize the residence time of EV at

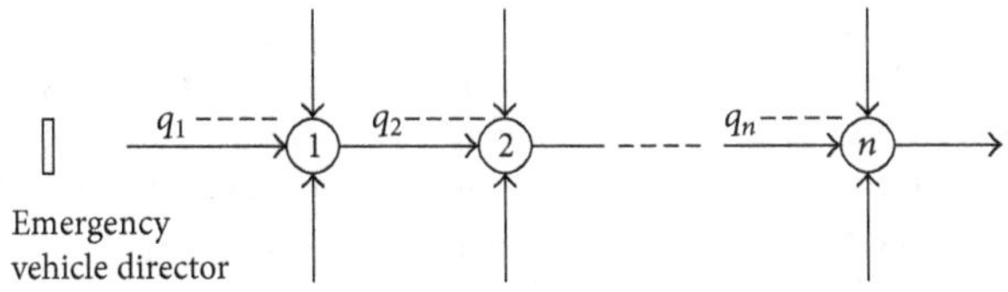

FIGURE 1: Diagram of intersection.

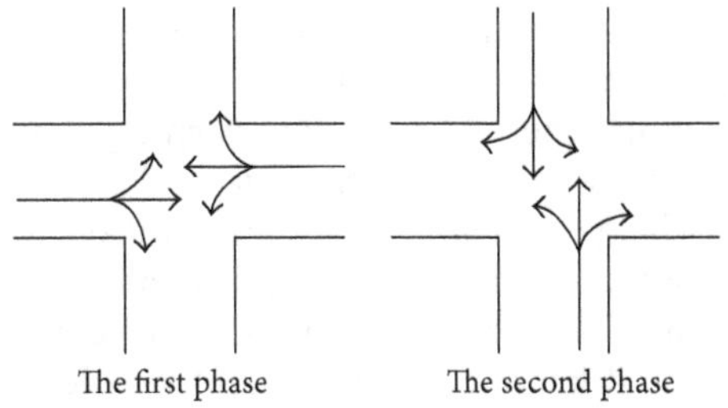

FIGURE 2: Phase diagram of traffic signal.

all intersections and maximize the number of social vehicles passing through all intersections, and (3) design a solving algorithm for the presented optimization model and verify the efficiency of the proposed signal preemption strategy.

First, on the premise of having selected a concrete evacuation route, taking isolated intersection as objective and considering the distance between EV detector and each intersection, the operating speed of EV and the number of queued vehicles at each intersection, the earliest-possible start time, and the latest-possible start time of green light at each intersection in the route are given. And then, in order to establish a real-time signal control strategy to ensure that EV can pass through intersection with operating speed or without stopping and minimize the impact of EV on social vehicles, a multiobjective programming model is presented and a particle swarm algorithm is designed to find the Pareto optimal solution set of this model. Finally, a simulation analysis is carried out.

2. Calculation of Intersection Time Parameters

After the optimal evacuation route of EV has been given, to ensure that the delay of EV is much smaller and the social vehicles passing through the whole system are much larger, we need to determine the most suitable time of opening the green light for EV of each intersection in the given route. Suppose that there are n intersections in the evacuation route except the given origin and the desired destination, and the evacuation direction is from west to east as shown in Figure 1. What needs to be stressed here is that we just take Figure 1 as an example to introduce the following concepts and calculation methods of intersection parameters. In fact, an EV may turn left or right at some of the intersections on the given evacuation route, and the following methods are also applicable to this kind of route. We assume that EV is detected at time t_0, and q_i, $i = 1, 2, \ldots, n$ denotes vehicles waiting to pass through intersection i in front of EV along the direction of evacuation at time t_0; q_{iN}, q_{iS}, q_{iW}, q_{iE}, $i = 1, 2, \ldots, n$ denotes vehicles waiting to pass through intersection i in north-southward direction, south-northward direction, west-eastward direction, and east-westward direction at time t_0, respectively. Suppose that all the intersections are two-phase control intersections and have fixed signal cycle, as shown in Figure 2. Each intersection i corresponds to a time window sequences $WL(i) = (ws_i, w_{i1}, w_{i2})$, where ws_i denotes the start time of the first time window, and w_{ik}, $k = 1, 2$ denotes the kth time window of intersection i.

The time parameters mentioned in this paper include the earliest-possible start time and the latest-possible start time of green light at each intersection.

Definition 1. The earliest-possible start time is defined as the earliest time that traffic light of EV direction can be changed to green from the time the EV is detected and on the premise of ensuring the efficiency of the whole system.

Definition 2. The latest-possible start time is defined as the latest time that traffic light of EV direction must be changed to green so as to ensure that EV can pass through the intersection without reducing speed or without stopping.

2.1. Calculation of Time Parameters of Intersection 1. We employ d_{ij} and t_0 to denote the duration time of phase j at intersection i and the time that EV is detected. If the starting time of the first time window of intersection 1, denoted by ws_1, is determined and the duration time of the two phases denoted by d_{11} and d_{12} is also determined, then the current phase of intersection 1 and the elapsed green time of this phase can be calculated correspondingly.

Provided that C_i denotes the cycle length of intersection i and $D_{i,j}$ denotes the elapsed green time of intersection i after j phases, the elapsed number of cycles in period $t_0 - ws_1$ is denoted as a, and the remaining time after a cycles in period $t_0 - ws_1$ is denoted as b. Then we have

$$D_{1,0} = 0,$$
$$D_{1,j} = \sum_{p=1}^{j} d_{1p}, \quad j = 1, 2. \tag{1}$$

Let

$$a = \frac{(t_0 - ws_1)}{C_i},$$
$$b = (t_0 - ws_1) \operatorname{mod} C_i; \tag{2}$$

then the current phase is i if it satisfies that $D_{u,i-1} \le b < D_{u,i}$, and the elapsed green time of phase i is $t_0 - ws_1 - a \times C_i - \sum_{p=0}^{j-1} D_{1,p}$.

Let $G^j_{\min}$, $G^j_{\max}$, G^j_{ela}, and G^j_{real} denote the minimal green time, the maximal green time, the elapsed green time, and the real green time of phase j, respectively; $L_{\mathrm{EV}-i}$ denotes the distance between EV detection and intersection i; v_{EV}, v_{reg} denote the speed of EV and mean speed of general traffic, respectively; t_{switch} denotes the time needed to switch green indication from one phase to another phase; t_{safe} is the safety time interval that must be kept between the last vehicle in the queue on the EV approach and the EV so as to avoid collision between EV and social vehicles. It is an important factor to ensure the safe operation of emergency vehicles and can be provided by an appropriate advance notice of the approaching of EV and a real-time traffic queue control strategy; B is the average effective length of each vehicle when parking, which is equal to the sum of the length of vehicle itself and the distance between adjacent vehicles. Let $t_E(i)$ and $t_L(i)$ $i = 1, 2, \ldots, n$ denote the earliest-possible start time and the latest-possible start time of phase i in EV approach; then we will illustrate the calculation method of $t_E(i)$ and $t_L(i)$ $i = 1, 2, \ldots, n$ beginning from the first intersection.

2.1.1. The Calculation of $t_E(1)$

(1) Phase 1 is the Current Phase of Intersection 1 at Time t_0. If the elapsed green time is G^1_{ela} and satisfies $L_{\text{EV}-1}/v_{\text{EV}} + G^1_{\text{ela}} \leq G^1_{\max}$, then $t_E(1) = t_0$ and the real start time of green signal at intersection 1 is t_0. Otherwise, if $L_{\text{EV}-1}/v_{\text{EV}} + G^1_{\text{ela}} > G^1_{\max}$, then we have to determine whether $G^1_{\text{ela}} \geq G^1_{\min}$ is met at time t_0. If it is met, the green indication of phase 1 is ended and starts again after the minimal green time of pahse 2 is finished. Before this, we have to check, after phase 1 regains the green indication, whether the queuing vehicles at the intersection can be cleared before EV arrives and can maintain a certain safety interval between EV and social vehicles. If we have

$$t_{\text{switch}} + G^2_{\min} + t_{\text{switch}} + \frac{q_1 \times B}{v_{\text{reg}}} + t_{\text{safe}} \leq \frac{L_{\text{EV}-1}}{v_{\text{EV}}} \quad (3)$$

then $t_E(1) = t_0 + 2t_{\text{switch}} + G^2_{\min}$. If $G^1_{\text{ela}} < G^1_{\min}$ is met, then the green indication of phase 1 will remain $G^1_{\min} - G^1_{\text{ela}}$ seconds. After $G^1_{\min} - G^1_{\text{ela}}$ seconds, the signal transformation mentioned above will be carried out on the basis of satisfying

$$G^1_{\min} - G^1_{\text{ela}} + t_{\text{switch}} + G^2_{\min} + t_{\text{switch}} + \frac{q_1 \times B}{v_{\text{reg}}} + t_{\text{safe}} \leq \frac{L_{\text{EV}-1}}{v_{\text{EV}}}. \quad (4)$$

It can be seen from (3) and (4) that, when $G^1_{\text{ela}} < G^1_{\min}$, if (4) is satisfied then (3) is sure to be satisfied. Let G_{low} denote the allowed lower bound of green light time that is smaller than $G^1_{\min}$, and G^j_{short} denote the shortened green light time of phase j. If (3) is satisfied while (4) is not satisfied, then G_{low} is given. If $G^1_{\text{ela}} \geq G_{\text{low}}$, then $t_E(1) = t_0 + G^2_{\min} + 2t_{\text{switch}}$. If (3) is also not satisfied, then the green light time of pahse 2 will be shortened to G^2_{short} to satisfy the following equation:

$$t_{\text{switch}} + G^2_{\text{short}} + t_{\text{switch}} + \frac{q_1 \times B}{v_{\text{reg}}} + t_{\text{safe}} = \frac{L_{\text{EV}-1}}{v_{\text{EV}}}. \quad (5)$$

Thus we have

$$G^2_{\text{short}} = \frac{L_{\text{EV}-1}}{v_{\text{EV}}} - \left(2t_{\text{switch}} + \frac{q_1 \times B}{v_{\text{reg}}} + t_{\text{safe}}\right). \quad (6)$$

If $G^2_{\text{short}} < G_{\text{low}}$, it shows that the green light time of phase 2 is very short and has little effect on alleviating traffic pressure and can lead to security problems. Therefore, $t_E(1) = t_0$ under this condition. Otherwise, the green light time of pahse 2 is equal to G^2_{short}, and $t_E(1)$ is equal to $t_0 + G^2_{\text{short}} + t_{\text{switch}}$. The real start time of green signal is $t_E(1)$.

(2) Phase 2 is the Current Phase of Intersection 1 at Time t_0. Calculate G^2_{ela} firstly. If $G^2_{\text{ela}} \leq G^2_{\min}$, it should be guaranteed generally that the green duration of phase 2 is $G^2_{\min}$ seconds and then changed to phase 1 after fixed yellow and red duration. However, it should be guaranteed that

$$G^2_{\min} - G^2_{\text{ela}} + t_{\text{switch}} + \frac{q_1 \times B}{v_{\text{reg}}} + t_{\text{safe}} \leq \frac{L_{\text{EV}-1}}{v_{\text{EV}}}. \quad (7)$$

Now, $t_E(1)$ is equal to $t_0 + G^2_{\min} - G^2_{\text{ela}} + t_{\text{switch}}$. Otherwise, we have to check whether the following equation is satisfied:

$$G_{\text{low}} - G^2_{\text{ela}} + t_{\text{switch}} + \frac{q_1 \times B}{v_{\text{reg}}} + t_{\text{safe}} \leq \frac{L_{\text{EV}-1}}{v_{\text{EV}}}. \quad (8)$$

If it is satisfied, then $t_E(1)$ is equal to $t_0 + G_{\text{low}} - G^2_{\text{ela}} + t_{\text{switch}}$.

If $G^2_{\text{ela}} > G^2_{\min}$, then determine whether the following equation is satisfied:

$$t_{\text{switch}} + \frac{q_1 \times B}{v_{\text{reg}}} + t_{\text{safe}} \leq \frac{L_{\text{EV}-1}}{v_{\text{EV}}}. \quad (9)$$

If it is satisfied, then $t_E(1)$ is equal to $t_0 + t_{\text{switch}}$.

2.1.2. The Calculation of $t_L(1)$. Assume that t^i_{arr} denotes the time that EV arrive at intersection i without reducing its speed; then we have

$$t^1_{\text{arr}} = t_0 + \frac{L_{\text{EV}-1}}{v_{\text{EV}}}. \quad (10)$$

Let t^i_{clear} denote the time needed to clear the queues at intersection i before EV arrives, then we have

$$t^1_{\text{clear}} = \frac{q_1 \times B}{v_{\text{reg}}}. \quad (11)$$

The value of $t_L(1)$ can be described by the following equation:

$$t_L(1) = t_0 + \frac{L_{\text{EV}-1}}{v_{\text{EV}}} - \frac{q_1 \times B}{v_{\text{reg}}} - t_{\text{safe}}. \quad (12)$$

If the distance between EV detector and intersection 1 is far enough and queued vehicles are not too much, $t_L(1)$ is sure greater than or equal to $t_E(1)$. However, if the traffic is rather heavy and clearing queued vehicles takes a long time, $t_L(1)$ may be less than $t_E(1)$ at phase 2. Under this condition, to ensure the smooth and safe passing of EV, the green time of phase 2 has to be sacrificed; namely, the green signal is converted to phase 1 after phase 2 has executed a short green time that is less than the minimum green time. Now, the value of $t_E(1)$ is $t_0 + t_{\text{switch}}$. If $t_L(1)$ is still less than $t_E(1)$, then let $t_L(1)$ and $t_E(1)$ have the same value, say, $t_0 + t_{\text{switch}}$. Now, it is unable to meet the requirement of EV's passing through the intersection without deceleration and stopping, and all that we can do is to minimize the waiting time of EV at intersection 1.

2.2. Calculation of Time Parameters at Other Intersections

2.2.1. The Calculation of $t_L(2)$. Let t_{pass} be the time EV needs to pass through the intersection, t_{arr}^2 the time EV needs to arrive at the parking line of intersection 2 after passing through intersection 1 without stopping, t_{start}^i the real start time of green signal of intersection i along the evacuation direction of EV, and Q_i and D_i the number of vehicles arriving at intersection $i+1$ from intersection i and vehicles departing from intersection i in t_0 to t_{start}^i time period, respectively. Then we have

$$t_{arr}^2 = t_0 + \frac{L_{EV-2}}{v_{EV}} + t_{pass}. \tag{13}$$

And the queuing vehicles q_2' at intersection 2 can be described as

$$q_2' = q_2 + Q_1 - D_2, \tag{14}$$

where Q_i and D_i are calculated according to the situation that the green light of intersection 1 starts at the latest-possible start time. Let t_{clear}^2 be the time needed for clearing q_2' before EV arrives; then we have

$$t_{clear}^2 = \frac{q_2' \times B}{v_{reg}} = \frac{(q_2 + Q_1 - D_2) \times B}{v_{reg}}. \tag{15}$$

We have the following equation to determine the latest start time of green signal at intersection 2:

$$\begin{aligned} t_L(2) &= t_{arr}^2 - t_{clear}^2 - t_{safe} \\ &= t_0 + \frac{L_{EV-2}}{v_{EV}} + t_{pass} - \frac{(q_2 + Q_1 - D_2) \times B}{v_{reg}} \\ &\quad - t_{safe}. \end{aligned} \tag{16}$$

2.2.2. The Calculation of $t_E(2)$

(1) Phase 1 is the Current Phase of Intersection 2 at Time t_0. If the elapsed green time of phase 1 is G_{ela}^1, $G_{ela}^1 + L_{EV-2}/v_{EV} + t_{pass}$ is the sum of the elapsed green time of phase 1 and the time EV spent before reaching intersection 2. We can discuss the calculation of $t_E(2)$ in the following three cases.

Case 1.

$$G_{ela}^1 + \frac{L_{EV-2}}{v_{EV}} + t_{pass} \le G_{real}^1. \tag{17}$$

Case 2.

$$G_{real}^1 < G_{ela}^1 + \frac{L_{EV-2}}{v_{EV}} + t_{pass} \le G_{max}^1. \tag{18}$$

Case 3.

$$G_{ela}^1 + \frac{L_{EV-2}}{v_{EV}} + t_{pass} > G_{max}^1. \tag{19}$$

In Case 1, there is no need to perform signal preemption at intersection 2, and both $t_E(2)$ and the real start time of green signal are equal to t_0.

In Case 2, after both phase 1 and phase 2 have executed the minimum green time and then green indication goes back to phase 1, we should check whether the following equation is satisfied:

$$\begin{aligned} &\left(G_{min}^1 - G_{ela}^1\right) + 2t_{switch} + G_{min}^2 + \frac{q_2' \times B}{v_{reg}} + t_{safe} \\ &\quad \le \frac{L_{EV-2}}{v_{EV}} + t_{pass}, \end{aligned} \tag{20}$$

where q_2' is the queue length of intersection 2 at time $(G_{min}^1 - G_{ela}^1) + 2t_{switch} + G_{min}^2$ and is calculated according to (14). In (14), Q_1 is calculated according to the situation that the green light of intersection 1 starts at the earliest-possible start time; D_2 is the number of vehicles passing through intersection 2 in the effective green duration of phase 1 in the period of t_0 to $t_E(2)$. If (20) is satisfied, then $t_E(2)$ is equal to $t_0 + (G_{min}^1 - G_{ela}^1) + 2t_{switch} + G_{min}^2$; otherwise, $t_E(2)$ is equal to t_0.

In Case 3, we need to consider the restriction of the maximum green time. Let both phase 1 and phase 2 execute the maximum green time, and then let green indication go back to phase 1. But we should check whether q_2' can be cleared before EV arrives, namely, whether the following equation is satisfied:

$$\begin{aligned} &\left(G_{max}^1 - G_{ela}^1\right) + 2t_{switch} + G_{max}^2 + \frac{q_2' \times B}{v_{reg}} + t_{safe} \\ &\quad \le \frac{L_{EV-2}}{v_{EV}} + t_{pass}. \end{aligned} \tag{21}$$

The calculation of q_2' is similar to Case 2. If (21) is satisfied, $t_E(2)$ is equal to $t_0 + (G_{max}^1 - G_{ela}^1) + 2t_{switch} + G_{max}^2$. Otherwise, let phase 1 and phase 2 execute the minimum green time and the maximum green time, respectively, and then let green indication go back to phase 1. But the following equation should be satisfied:

$$\begin{aligned} &\left(G_{min}^1 - G_{ela}^1\right) + 2t_{switch} + G_{max}^2 + \frac{q_2' \times B}{v_{reg}} + t_{safe} \\ &\quad \le \frac{L_{EV-2}}{v_{EV}} + t_{pass}. \end{aligned} \tag{22}$$

Now, $t_E(2)$ is equal to $t_0 + (G_{min}^1 - G_{ela}^1) + 2t_{switch} + G_{max}^2$.

If (22) is not satisfied, then both phase 1 and phase 2 will execute the minimum green time, and then green indication will go back to phase 1. Let the cycle length be C_{max} when all of the two phases execute the maximum green time and C_{min} when all of the two phases execute the minimum green time.

(2) Phase 2 Is the Current Phase of Intersection 2 at Time t_0. If the elapsed green time of phase 2 is G_{ela}^2, we can discuss the calculation of $t_E(2)$ in the following four cases.

Case 1.

$$G_{\text{real}}^2 - G_{\text{ela}}^2 + t_{\text{switch}} + t_{\text{clear}}^2 + t_{\text{safe}} \le \frac{L_{\text{EV}-2}}{v_{\text{EV}}} + t_{\text{pass}}. \tag{23}$$

Case 2.

$$G_{\min}^2 - G_{\text{ela}}^2 + t_{\text{switch}} + t_{\text{clear}}^2 + t_{\text{safe}} \le \frac{L_{\text{EV}-2}}{v_{\text{EV}}} + t_{\text{pass}} < G_{\text{real}}^2 - G_{\text{ela}}^2 + t_{\text{switch}} + t_{\text{clear}}^2 + t_{\text{safe}}. \tag{24}$$

Case 3.

$$G_{\text{low}} - G_{\text{ela}}^2 + t_{\text{switch}} + t_{\text{clear}}^2 + t_{\text{safe}} \le \frac{L_{\text{EV}-2}}{v_{\text{EV}}} + t_{\text{pass}} < G_{\min}^2 - G_{\text{ela}}^2 + t_{\text{switch}} + t_{\text{clear}}^2 + t_{\text{safe}}. \tag{25}$$

Case 4.

$$\frac{L_{\text{EV}-2}}{v_{\text{EV}}} + t_{\text{pass}} < G_{\text{low}} - G_{\text{ela}}^2 + t_{\text{switch}} + t_{\text{clear}}^2 + t_{\text{safe}}. \tag{26}$$

Case 1 corresponds to the situation that the time needed by green light of phase 1 turning green after the normal green duration of phase 2 and clearing the queue length of phase 1 is no more than the time spent by EV for arriving at intersection 2; then there is no need to perform signal preemption at intersection 2 and $t_E(2)$ is equal to $t_0 + G_{\text{real}}^2 - G_{\text{ela}}^2 + t_{\text{switch}}$. Case 2 corresponds to the situation that the time needed by phase 1 turning green after the minimum green duration of phase 2 and clearing the queue length of phase 1 is no more than the time spent by EV for arriving at intersection 2; then signal preemption is necessary and $t_E(2)$ is equal to $t_0 + G_{\min}^2 - G_{\text{ela}}^2 + t_{\text{switch}}$. In Case 3, green light of phase 1 turns green after phase 2 has passed G_{low} time, and now $t_E(2)$ is equal to $t_0 + G_{\text{low}} - G_{\text{ela}}^2 + t_{\text{switch}}$. In Case 4, signal preemption is necessary at intersection 2, and the green duration of phase 2 has to be sacrificed; namely, the green signal has to turn to phase 1 immediately after detecting EV, and now $t_E(2)$ is equal to $t_0 + G_{\text{ela}}^2 + t_{\text{switch}}$.

The time parameters of the other intersections on the line are similar to those of intersection 2, which is no longer introduced here.

3. Multi-Objective Optimization Model

Let n be the number of intersections on the evacuation route, m the number of entrance sections on the cross road of evacuation route; t_{dep}^i the time instant EV departing from intersection i; $g_{\text{eff}}^{i,j}$, $s_{i,j}^{y\text{-}z}$, and $S_{i,j}$ the effective green time of phase j at intersection i, the average departure rate of vehicles from y to z directions on the entrance sections, and the saturation flow rate of all lanes on the entrance sections, respectively; and $A_{\text{thr}}^{i,j}$, $A_{\text{left}}^{i,j}$, and $A_{\text{right}}^{i,j}$ the ratio of straight, left turn and right turn of phase j at intersection i, respectively. The multiobjective optimization model for EV signal preemption control based on route is

$$\min \quad f_1 = \sum_{i=1}^{n} \left(t_{\text{start}}^i + t_{\text{clear}}^i + t_{\text{safe}} - t_{\text{arr}}^i\right) \times \delta_i, \tag{27}$$

$$\max \quad f_2 = \sum_{i=1}^{n} g_{\text{eff}}^{i,2} \times \left[\left(s_{i,2}^{N\text{-}S} + s_{i,2}^{S\text{-}N}\right) \times A_{\text{thr}}^{i,2} + s_{i,2}^{N\text{-}S} A_{\text{left}}^{i,2} + s_{i,2}^{S\text{-}N} A_{\text{right}}^{i,2}\right], \tag{28}$$

$$\text{s.t.} \quad t_E(i) \le t_{\text{start}}^i \le t_L(i) \quad i = 1, 2, \ldots, n, \tag{29}$$

$$t_{\text{arr}}^i = \frac{L_{\text{EV}-i}}{v_{\text{EV}}} \quad i = 1, 2, \ldots, n, \tag{30}$$

$$t_{\text{clear}}^i = \frac{q_i' \times B}{v_{\text{reg}}} \quad i = 2, \ldots, n, \tag{31}$$

$$q_i' = q_i + \min\left\{q_{i-1}' \times A_{\text{thr}}^{i-1,1}, g_{\text{eff}}^{i-1,1} \times S_{i-1,j} \times A_{\text{thr}}^{i-1,1}\right\} + g_{\text{eff}}^{i-1,2}\left(s_{i,2}^{N\text{-}S} A_{\text{left}}^{i,2} + s_{i,2}^{S\text{-}N} A_{\text{right}}^{i,2}\right) - \sum_{e=1}^{l} \min\left\{q_e'', g_{\text{eff}}^{i,1} \times A_{\text{thr}}^{i,1}\right\}, \tag{32}$$

$$\delta_i = \begin{cases} 0 & t_{\text{start}}^i + t_{\text{clear}}^i + t_{\text{safe}} \le t_{\text{arr}}^i \\ 1 & \text{otherwise} \end{cases} \quad i = 2, \ldots, n, \tag{33}$$

$$t_{\text{safe}} \ge 2, \tag{34}$$

$$t_E(i) \ge t_0 \quad i = 1, 2, \ldots, n, \tag{35}$$

$$t_L(i) \ge t_0 \quad i = 1, 2, \ldots, n, \tag{36}$$

where (27) and (28) are objective functions and represent minimizing the residence time of EV at all intersections and maximizing the number of social vehicles passing through all intersections of the evacuation route, respectively. The number of social vehicles includes the vehicles entering in the system from the cross road of evacuation route and passing through intersections and the vehicles passing through intersections in the direction of the evacuation route of EV. Equation (29) ensures that the real start time of green signal of intersection i is between $t_E(i)$ and $t_L(i)$; (30) represents the arrival time of EV at intersection i; (31) represents the time needed to clear the queued vehicles before EV arrives; (32) denotes the calculation methd of the queued vehicles at intersection i at the moment the green signal is starting, and $q'', e = 1, 2, \ldots, l$ represents the number of queued vehicles of phase 1 at the moment the green signal is starting in each signal cycle of intersection i before t_{start}^i, and l denotes the number of passed signal cycles; (33) represents the 0-1 variable of intersection i, which means that the waiting time of EV at intersection i is 0 if the green signal is on at intersection i, and the queued vehicles have been cleared and security interval is satisfied, namely, $t_{\text{start}}^i + t_{\text{clear}}^i + t_{\text{safe}} \le t_{\text{arr}}^i$. Otherwise, the EV has to wait in line, and the waiting time is euqual to $t^i_{\text{start}} + t^i_{\text{clear}} + t_{\text{safe}} - t^i_{\text{arr}}$; (34) shows the lower limit of the value of t_{safe}; (35) and (36) represent the relationship between $t_E(i)$, $t_L(i)$, and t_0. Variable $g^{i,2}_{\text{eff}}$ will take different values according to different t_0 and t^i_{start}. The value of $g^{1,2}_{\text{eff}}$ and $g^{2,2}_{\text{eff}}$ can be calculated by (3)–(9) and (10)–(16) and (20)–(22), respectively, and the calculation method of $g^{i,2}_{\text{eff}}$, $i = 3, 4, \ldots, n$ is similar to $g^{2,2}_{\text{eff}}$.

4. Design of Solving Algorithm

At present, the evolutionary algorithm is the main algorithm for solving multiobjective programming problems. Owing to the advantages of fast convergence speed and easy implementation, the particle swarm algorithm (PSO) has been successfully applied in many optimization problems [48–52]. In this paper, PSO is adopted to solve the given model.

4.1. Encoding and Particle Swarm Generation. Generate particle swarm P with N_1 particles randomly, and each particle can be represented as $X = (x_1, x_2, \ldots, x_n)$, where x_i, $i = 1, 2, \ldots, n$ denotes t^i_{start} of intersection i and $x_i \in [t_E(i), t_L(i)]$. If rand( ) represents a random number between 0 and 1, then the specific method of producing x_i, $i = 1, 2, \ldots, n$ is defined as follows:

$$x_i = \text{int}\left[t_E(i) + (t_L(i) - t_E(i)) \times \text{rand}()\right]. \quad (37)$$

4.2. Density of Solutions. Since the density of nondominated solutions is inversely proportional to their diversity [48], to obtain uniformly distributed Pareto frontier, the diversity of solutions should be guaranteed; that is, the sparse solutions should be retained as many as possible. For this reason, the solution density is introduced in this paper. Suppose that there are p objectives and there are S solutions in dominant population E. If the solutions in E are sorted in ascending order according to their fitness values and $f_j(i)$, $j = 1, 2, \ldots, p$; $i = 1, 2, \ldots, S$ represents the ith solution when sorting by the jth objective, then the solution density is defined as follows:

$$\begin{aligned}&\sum_{j=1}^{p}\left[f_j(i+1) - f_j(i-1)\right] + \sum_{j=1}^{p}\left[f_j(i) - f_j(i-1)\right]\\&\quad = \sum_{j=1}^{p} f_j(i+1) + \sum_{j=1}^{p} f_j(i) - 2\sum_{j=1}^{l} f_j(i-1).\end{aligned} \quad (38)$$

4.3. Updating of Location and Velocity. In standard multiobjective PSO algorithm, location updating requires the inertia factor w and learning factors c_1 and c_2. Since a larger w is suitable for a wide range exploration of solution space, while a smaller w is suitable for detailed search on the target, the learning factors are adopted to balance the cognitive abilities of individuals and groups. To achieve better search and balance, some scholars put forward the idea that inertia factor and learning factors can change dynamically [49]. In this paper, inertia factor and learning factors are changed as follows:

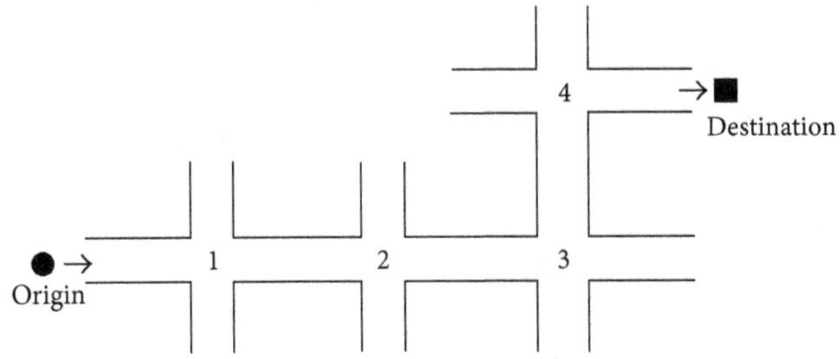

Figure 3: Intersections on evacutation route.

$$\begin{aligned}w(k) &= (w_f - w_s) \times \frac{\text{Max} - k}{\text{Max}} + w_s,\\c_1(k) &= (c_{1f} - c_{1s}) \times \frac{\text{Max} - k}{\text{Max}} + c_{1s},\\c_2(k) &= (c_{2f} - c_{2s}) \times \frac{\text{Max} - k}{\text{Max}} + c_{2s},\end{aligned} \quad (39)$$

where k, Max, w_s, and w_f represent current iterations, maximum iterations, the initial value of inertia weight, and the final value of inertia weight, respectively; c_{1s}, c_{2s} represent the initial value of c_1 and c_2, respectively; c_{1f}, c_{2f} represent the final value of c_1 and c_2, respectively.

4.4. Mutation. In order to prevent the algorithm from falling into local optimum, mutation operator is introduced. Let O be the mutation probability; if $k < \text{Max} \times O$, then $\text{Max} \times O - k$ particles are selected from P and mutate according to the following equation:

$$x(i) = x(i) + \text{rand}() \times \text{int}\left(\frac{t_L(i) - t_E(i)}{2} - \frac{x(i)}{2}\right). \quad (40)$$

This will not only ensure the diversification of particles but also ensure the feasibility of particles after mutation.

5. Simulation Analysis

Assume that the evacuation route is composed of 4 intersections, as shown in Figure 3. The evacuation direction of EV is from west to east. It will pass through intersection 1 and intersection 2 straight ahead, turn left at intersection 3, and turn right at intersection 4, and $L_{\text{EV-1}} = 400$ m, $L_{\text{EV-2}} = 850$ m, $L_{\text{EV-3}} = 1350$ m, $L_{\text{EV-4}} = 1850$ m, $v_{\text{EV}} = 45$ km/h, $v_{\text{reg}} = 45$ km/h; $d_{i1} = d_{i2} = 28$ s, $C_i = 56$ s, $i = 1, 2, 3, 4$, $C_{\max} = 120$ s, $C_{\min} = 30$ s; $ws_1 = 0$ s, $ws_2 = 30$ s, $ws_3 = 60$ s, $ws_4 = 90$ s; $t_{\text{switch}} = 3$ s, $t_{\text{pass}} = 2$ s, $B = 7$ m; $G^1_{\min} = G^2_{\min} = 15$ s, $G^1_{\max} = G^2_{\max} = 60$ s, $G^1_{\text{real}} = G^2_{\text{real}} = 25$ s, $G_{\text{low}} = 10$ s, $G^j_{\text{short}} = 10$ s.

In south-northward direction and west-eastward direction, the ratio of straight, right turn, and left turn is 0.7, 0.2, and 0.1, respectively. And it is 0.6, 0.3, and 0.1 respectively, in north-southward direction. In the particle swarm algorithm, $N_1 = 100$, Max $= 1000$, and $O = 0.1$. The size of external dominance population is 100, and $w_s = 0.4$, $w_f = 0.9$, $c_{1f} = 1$, $c_{2f} = 2$, $c_{1s} = 2$, $c_{2s} = 1$.

TABLE 1: The Pareto solution set of of EV when t_0 is 9 o'clock.

Number	The real start time of green signal(s)				f_1 (s)	f_2 (pcu)
	Intersection 1	Intersection 2	Intersection 3	Intersection 4		
(1)	10	33	53	81	0	112.1
(2)	10	34	54	81	2.3	113.6
(3)	10	34	55	82	2.4	114.8
(4)	11	34	55	83	3.9	116.6
(5)	12	34	56	83	4.7	118.7
(6)	13	34	58	84	4.8	120.4

TABLE 2: The Pareto solution set of EV when t_0 is 10 o'clock.

Number	The real start time of green signal(s)				f_1 (s)	f_2 (pcu)
	Intersection 1	Intersection 2	Intersection 3	Intersection 4		
(1)	10	31	56	82	0	119.7
(2)	11	33	58	83	2.9	119.9
(3)	11	34	57	83	3.2	122.4
(4)	12	34	58	83	4.9	122.7
(5)	13	34	57	83	5.5	123.5

Assume that the number of vehicles waiting to pass through each intersection at time t_0 is $q_1 = 10$, $q_2 = 5$, $q_3 = 6$, and $q_4 = 8$, respectively, and $q_{iN} = q_{iS} = 10$, $i = 1, 2$, $q_{3S} = q_{3E} = 10$, and $q_{4N} = q_{4W} = 10$. At each intersection, the arrival rate from south to north is 1080–1440 pcu/h and the average departure rate is 1440 pcu/h, the arrival rate from east to west is 1440–1800 pcu/h, and the saturation flow rate is 1800 pcu/h. Assume that t_0 is 9 o'clock and 10 o'clock, respectively. The Pareto solution set obtained by using C# for simulation calculation is shown in Tables 1 and 2. The data corresponding to each intersection indicates how many seconds after 9 o'clock or 10 o'clock will each intersection start the green signal; f_1 and f_2 are the residence time of EV at all the intersections and the number of vehicles passing through the system, respectively.

Assume that the number of vehicles waiting to pass through each intersection at time t_0 is $q_i = 20$, $i = 1, 2, 3, 4$, and $q_{iN} = q_{iS} = 10$, $i = 1, 2$, $q_{3S} = q_{3E} = 10$, $q_{4N} = q_{4W} = 10$. Traffic flow from north to south and from west to east is the key traffic flow. At each intersection, the arrival rate from north to south is 1400–1600 pcu/h and the average departure rate is 1440 pcu/h; the arrival rate from west to east is 1800–1900 pcu/h and the saturation flow rate is 1800 pcu/h; the arrival rate in other directions remain unchanged. If t_0 is 9 o'clock and 10 o'clock, respectively, the Pareto solution set is shown in Tables 3 and 4.

To verify the control effect of the signal preemption control method presented in this paper under the condition of much larger traffic volume, for example, in the moring peak or the evening peak period, we assume that t_0 is 8 o'clock, namely, in the moring peak period, and $q_i = 20$, $i = 1, 2, 3, 4$, $q_{iN} = q_{iS} = 15$, $i = 1, 2$, $q_{3S} = q_{3E} = 15$, $q_{4N} = q_{4W} = 15$. At each intersection, the arrival rate from north to south and from west to east is 1800–1900 pcu/h and 2200–2400 pcu/h, respectively. The saturation flow rate is 1800 pcu/h and the arrival rate in other directions remain unchanged. The Pareto optimal set is shown in Figure 4.

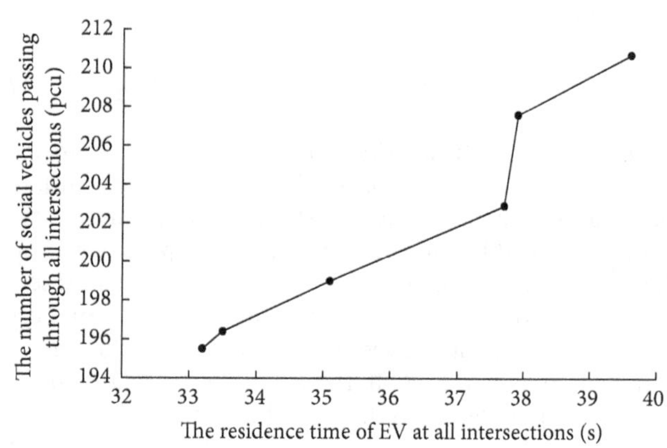

FIGURE 4: The Pareto optimal set.

We can see from the simulation results shown in Tables 1 and 2 that, for each solution in the Pareto solution set, the variation law of the two optimization objectives is that with the reduction of one objective; the other objective is reduced simultaneously, which implies that the decrease in the residence time of EV at intersections is at the expense of the reduction of the number of vehicles passing through the system. It also can be seen from the simulation results shown in Tables 3 and 4 that, at the moment the EV is detected, if there are many queued vehicles at each intersections and the traffic volume of key traffic flow exceeds the amount of traffic that the system can release, the control method can still guarantee the EV to pass through the intersection at a faster speed, and more social vehicles can pass through the system. The results shown in Figure 4 also indicate that the signal preemption control method is effective when the

TABLE 3: The Pareto solution set for an increasing arrival rate when t_0 is 9 o'clock.

Number	The real start time of green signal(s)				f_1 (s)	f_2 (pcu)
	Intersection 1	Intersection 2	Intersection 3	Intersection 4		
(1)	2	46	73	108	29.4	165.7
(2)	2	45	75	108	29.5	167.1
(3)	0	45	76	114	33.2	171.2
(4)	1	43	76	111	36.7	175.4
(5)	1	44	80	116	37.9	183.5
(6)	4	44	80	111	38.4	184.4
(7)	7	43	82	123	38.4	186.9
(8)	4	43	80	117	39.3	189.1
(9)	7	43	84	119	39.9	193.6

TABLE 4: The Pareto solution set for an increasing arrival rate when t_0 is 10 o'clock.

Number	The real start time of green signal(s)				f_1 (s)	f_2 (pcu)
	Intersection 1	Intersection 2	Intersection 3	Intersection 4		
(1)	11	43	79	110	27.2	169.9
(2)	10	45	80	108	29.4	174.7
(3)	4	33	76	106	29.7	180.2
(4)	2	34	73	111	30.1	183.8
(5)	6	46	79	116	31.8	185.7
(6)	4	35	71	107	36.6	187.0
(7)	12	43	77	108	38.3	189.2

arrival rate are much larger in the morning peak period. Simulation results under different traffic states show that the signal preemption control method presented in this paper can reduce the delay of EV at intersections and lessen the impact of EV on social vehicles.

When multiple EVs arrive at the same time, the signal priority control method proposed in this paper is also applicable. If the arrival interval of adjacent EV is relatively long, it should be regarded as two signal priority processes. Now, after the front EV has passed through the intersection, we must consider signal transiton from priority signal to normal signal firstly and then adopt the method mentioned above to control the following EV.

6. Conclusion

In order to overcome the disadvantages of signal preemption control method from intersection to intersection, we study the dynamic signal preemption control method based on route for EV in this paper. Firstly, a single intersection is taken as the research object to determine the earliest-possible start time and the latest-possible start time of each intersection in evacuation direction of EV on the selected route. Then, considering the delay of EV at intersections and the effect of EV to social vehicles comprehensively, to ensure that EV can pass through the intersections without reducing speed and without stopping as far as possible and the impact on social vehicles can be minimal, a multiobjective programming model is established and the solving algorithm is designed. Simulation calculations, considering EV is detected at different times and different traffic states, are carried out, and the corresponding Pareto sets are given. The result shows that, by adopting this method, we can get more reasonable start time of green light at each intersection, reduce the delay of EV at intersections, and improve the efficiency of evacuation. Determine how the signal should be converted to the normal signal when signal preemption has been finished according to the signal timing, the traffic flow characteristics, and other parameters; namely, the signal transition strategy is the next focus of the study.

Conflicts of Interest

The authors declare that they have no conflicts of interest.

Acknowledgments

This research is supported by National Nature Science Foundation of China (nos. 61563029, 71671079, 71361018, and 71571090).

References

[1] Y.-S. Huang, Y.-S. Weng, and M.-C. Zhou, "Design of traffic safety control systems for emergency vehicle preemption using timed petri nets," *IEEE Transactions on Intelligent Transportation Systems*, vol. 16, no. 4, pp. 2113–2120, 2015.

[2] M. B. Younes and A. Boukerche, "An efficient dynamic traffic light scheduling algorithm considering emergency vehicles for intelligent transportation systems," *Wireless Networks*, pp. 1–13, 2017.

[3] J. F. Paniati and M. Amoni, *Traffic signal preemption for emergency vehicle: a cross-cutting study*, US Federal Highway Administration, 2006.

[4] P. B. Mirchandani and D. E. Lucas, "Integrated transit priority and rail/emergency preemption in real-time traffic adaptive signal control," *Journal of Intelligent Transportation Systems: Technology, Planning, and Operations*, vol. 8, no. 2, pp. 101–115, 2004.

[5] H. H. Song, *Study on multiple emergency vehicles signal priority control based on multi-agent approach , Harbin: Harbin Institute of Technology*, Harbin Institute of Technology, Harbin, 2008.

[6] Z.-S. Yang, X.-M. Sun, and P.-C. Sun, "Signal priority control strategy and implementation for emergency vehicle at single intersection under traffic accidents," *Jilin Daxue Xuebao (Gongxueban)/Journal of Jilin University (Engineering and Technology Edition)*, vol. 41, no. 3, pp. 640–644, 2011.

[7] C. Ma and J. X. Cui, "Multiple emergency vehicles signal priority control based on multi-agent approach," *Journal of Transportation Systems Engineering and Information Technology*, vol. 13, no. 1, pp. 57–62, 2013.

[8] J. Wang, W. Ma, and X. Yang, "Development of degree-of-priority based control strategy for emergency vehicle preemption operation," *Discrete Dynamics in Nature & Society*, vol. 2013, no. 1, pp. 215–227, 2013.

[9] X. Qin, *Real time control and optimization of traffic signal timing transition for emergency vehicle preemption*, University of Ottawa, Ottawa, Canada, 2005.

[10] F. A. Marcianò, G. Musolino, and A. Vitetta, "Signal setting optimization on urban road transport networks: The case of emergency evacuation," *Safety Science*, vol. 72, pp. 209–220, 2015.

[11] T. J. Kimpel, J. Strathman, R. L. Bertini, and S. Callas, "Analysis of transit signal priority using archived TriMet Bus Dispatch System data," *Transportation Research Record*, no. 1925, pp. 156–166, 2005.

[12] X. Qin and A. M. Khan, "Control strategies of traffic signal timing transition for emergency vehicle preemption," *Transportation Research Part C: Emerging Technologies*, vol. 25, pp. 1–17, 2012.

[13] S. Shibuya, T. Yoshida, Z. Yamashiro, and M. Miyawaki, "Fast emergency vehicle preemption systems," *Transportation Research Record Journal of the Transportation Research Board*, vol. 1739, pp. 44–50, 2000.

[14] E. Kwon and S. Kim, *Development of dynamic route clearance strategies for emergency vehicle operations, Phase I*, University of Minnesota: Center for Transportation Studies, Center for Transportation Studies, 2003.

[15] H. K. Gedawy, M. B. Dias, and K. Harras, *Dynamic path planning and traffic light coordination for emergency vehicle routing*, Computer Science Department Carnegie Mellon University Pittsburgh, Pennsylvania, 2008.

[16] W. Kang, G. Xiong, Y. Lv, X. Dong, F. Zhu, and Q. Kong, "Traffic signal coordination for emergency vehicles," in *Proceedings of the 2014 17th IEEE International Conference on Intelligent Transportation Systems, ITSC 2014*, pp. 157–161, China, October 2014.

[17] S. Shekhar, K. S. Yang, V. M. V. Gunturi et al., "Experiences with evacuation route planning algorithms," *International Journal of Geographical Information Science*, vol. 26, no. 12, pp. 2253–2265, 2012.

[18] R. F. Kirby and R. B. Potts, "The minimum route problem for networks with turn penalties and prohibitions," *Transportation Research*, vol. 3, no. 3, pp. 397–408, 1969.

[19] E. H. E. Singer, "A method of finding cost-minimizing routes incorporating turn costs," *Traffic, Engineering and Control*, vol. 15, pp. 771-772, 1974.

[20] S. Namkoong and H. S. Mahmassani, "A note on least time path computation considering delays and prohibitions for intersection movements," *Transportation Research Part B: Methodological*, vol. 30, no. 5, pp. 359–367, 1996.

[21] X. D. Zou, S. F. Zheng, X. G. Ban, and X. M. Lian, "Optimal path algorithm for road Net with traffic capacity limits," *Journal of Highway and Transportation Research and Development*, vol. 19, pp. 82–84, 2002.

[22] Y. Z. Li and R. C. He, "Modeling for shortest path in traffic network with node limits," *Operations Research and Management Science*, vol. 14, pp. 40–43, 2005.

[23] F.-Y. Wang, F.-Q. Pan, L.-X. Zhang, and X.-D. Zou, "Optimal path algorithm of road network with traffic restriction," *Journal of Traffic and Transportation Engineering*, vol. 5, no. 1, pp. 92–95, 2005.

[24] T. J. Cova and J. P. Johnson, "A network flow model for lane-based evacuation routing," *Transportation Research Part A: Policy and Practice*, vol. 37, no. 7, pp. 579–604, 2003.

[25] R. Song, H. E. Shiwei, and L. Zhang, "Optimum transit operations during the emergency evacuations," *Journal of Transportation Systems Engineering and Information Technology*, vol. 9, no. 6, pp. 154–160, 2009.

[26] X. Chen and F. B. Zhan, "Agent-based modeling and simulation of urban evacuation: relative effectiveness of simultaneous and staged evacuation strategies," *Agent-Based Modeling and Simulation*, pp. 78–96, 2014.

[27] P. Murray-Tuite and B. Wolshon, "Evacuation transportation modeling: an overview of research, development, and practice," *Transportation Research Part C: Emerging Technologies*, vol. 27, pp. 25–45, 2013.

[28] E. Urbina and B. Wolshon, "National review of hurricane evacuation plans and policies: a comparison and contrast of state practices," *Transportation Research Part A: Policy and Practice*, vol. 37, no. 3, pp. 257–275, 2003.

[29] S. A. Parr and E. Kaisar, "Critical intersection signal optimization during urban evacuation utilizing dynamic programming," *Journal of Transportation Safety & Security*, vol. 3, no. 1, pp. 59–76, 2011.

[30] V. Campos, R. Bandeira, and A. Bandeira, "A Method for Evacuation Route Planning in Disaster Situations," *Procedia - Social and Behavioral Sciences*, vol. 54, pp. 503–512, 2012.

[31] M. Gao and G. He, "Optimization of evacuation routes considering movements at intersections," *Tumu Gongcheng Xuebao/China Civil Engineering Journal*, vol. 40, no. 6, pp. 80–83, 2007.

[32] H. B. Mu, J. N. Yu, and L. Z. Liu, "Evacuation Routes Optimization with Effects of Traffic Light at Intersections," *Journal of Transportation Systems Engineering & Information Technology*, vol. 11, no. 3, pp. 76–82, 2011.

[33] L. G. Chalmet, R. L. Francis, and P. B. Saunders, "NETWORK MODELS FOR BUILDING EVACUATION.," *Management Science*, vol. 28, no. 1, pp. 86–105, 1982.

[34] W. L. Wilkinson, "An algorithm for universal maximal dynamic flows in a network," *Operations Research*, vol. 19, pp. 1602–1612, 1971.

[35] E. Minieka, "Maximal, lexicographic, and dynamic network flows," *Operations Research*, vol. 21, pp. 517–527, 1973.

[36] A. Stepanov and J. M. Smith, "Multi-objective evacuation routing in transportation networks," *European Journal of Operational Research*, vol. 198, no. 2, pp. 435–446, 2009.

[37] V. Bayram, B. Ç. Tansel, and H. Yaman, "Compromising system and user interests in shelter location and evacuation planning," *Transportation Research Part B: Methodological*, vol. 72, pp. 146–163, 2015.

[38] N. Nassir, M. Hickman, H. Zheng, and Y.-C. Chiu, "Network flow solution method for optimal evacuation traffic routing and signal control with nonuniform threat," *Transportation Research Record*, vol. 2459, pp. 54–62, 2014.

[39] R. L. Church and R. M. Sexton, *Modeling small area evacuation: Can existing transportation infrastructure impede public safety? Caltrans Testbed Center for Interoperability Task Order 3021 Report*, Vehicle Intelligence and Transportation Analysis Laboratory, University of California, Santa Barbara, CA, USA, 2002.

[40] M. Ben-Akiva, D. Mcfadden, K. Train, J. Walker, C. Bhat, and M. Bierlaire, *Development of a deployable real-time dynamic traffic assignment system: Dynamit and Dynamit-p users guide*, Massachusetts Institute of Technology, Intelligent Transportation Systems Program, Cambridge, 2002.

[41] Q. Lu, Y. Huang, and S. Shekhar, "Evacuation Planning: A Capacity Constrained Routing Approach," in *Intelligence and Security Informatics*, vol. 2665 of *Lecture Notes in Computer Science*, pp. 111–125, Springer Berlin Heidelberg, Berlin, Heidelberg, 2003.

[42] Q. Lu, B. George, and S. Shekhar, "Capacity Constrained Routing Algorithms for Evacuation Planning: A Summary of Results," in *Advances in Spatial and Temporal Databases*, vol. 3633 of *Lecture Notes in Computer Science*, pp. 291–307, Springer Berlin Heidelberg, Berlin, Heidelberg, 2005.

[43] Q. Lu, B. George, and S. Shekhar, "Evacuation route planning: a case study in semantic computing," *International Journal of Semantic Computing*, vol. 1, no. 2, pp. 249–303, 2007.

[44] K. Shahabi and J. P. Wilson, "CASPER: Intelligent capacity-aware evacuation routing," *Computers, Environment and Urban Systems*, vol. 46, pp. 12–24, 2014.

[45] E. Pourrahmani, M. R. Delavar, P. Pahlavani, and M. A. Mostafavi, "Dynamic evacuation routing plan after an earthquake," *Natural Hazards Review*, vol. 16, no. 4, Article ID 04015006, 2015.

[46] K. Shahabi and J. P. Wilson, "Scalable evacuation routing in a dynamic environment," *Computers Environment Urban Systems*, vol. 67, pp. 29–40, 2018.

[47] M. Zeng and C. Wang, "Evacuation route planning algorithm: Longer route preferential," *Lecture Notes in Computer Science (including subseries Lecture Notes in Artificial Intelligence and Lecture Notes in Bioinformatics): Preface*, vol. 5551, no. 1, pp. 1062–1071, 2009.

[48] D. F. Wang, L. Meng, and W. J. Zhao, "Improved bare bones particle swarm optimization with adaptive search center," *Chinese Journal of Computers. Jisuanji Xuebao*, vol. 39, no. 12, pp. 2652–2667, 2016.

[49] C. A. C. Coello, G. T. Pulido, and M. S. Lechuga, "Handling multiple objectives with particle swarm optimization," *IEEE Transactions on Evolutionary Computation*, vol. 8, no. 3, pp. 256–279, 2004.

[50] B. Wei, Y. X. Li, X. Xu, and D. C. Shen, "Partical swarm optimization algorithm based on stable strategy," *Computer Science*, vol. 38, no. 12, pp. 221–223, 2011.

[51] H. Wang and F. Qian, "Improved PSO-based multi-objective optimization by crowding with mutation and particle swarm optimization dynamic changing," *Kongzhi yu Juece/Control and Decision*, vol. 23, no. 11, pp. 1238–1248, 2008.

[52] A. Ratnaweera, S. K. Halgamuge, and H. C. Watson, "Self-organizing hierarchical particle swarm optimizer with time-varying acceleration coefficients," *IEEE Transactions on Evolutionary Computation*, vol. 8, no. 3, pp. 240–255, 2004.

2

Robust H_∞ Output-Feedback Yaw Control for Vehicles with Differential Steering

Paul Oke,[1] **Sing Kiong Nguang**,[1] **and Wentai Qu**[2]

[1]*Department of Electrical and Computer Engineering, The University of Auckland, Auckland 1010, New Zealand*
[2]*School of Information Science and Engineering, Ningbo Institute of Technology, Zhejiang University, Ningbo Higher Educational Zone, China*

Correspondence should be addressed to Sing Kiong Nguang; sk.nguang@auckland.ac.nz

Academic Editor: Darong Huang

This paper examines the problem of designing a robust H_∞ output-feedback yaw controller with both input and output constraints for four-wheel independently driven in-wheel electric vehicles (EVs) with differential steering. Specifically, the controller aims are to ensure the stability and improve the performance of the EV despite variations in the road adhesion coefficient, longitudinal velocity, and external disturbance. Based on the linear matrix inequalities approach, sufficient conditions for the existence of an H_∞ output-feedback controller for linear systems with polytopic uncertainties, and input and control output constraints, are derived. Then those sufficient conditions are utilized to design an H_∞ output-feedback yaw controller that guarantees the robust performance and stability of an EV over a wider range of road conditions. Finally, the capability of the developed controller is simulated on a vehicle model with uncertain road conditions and longitudinal velocities.

1. Introduction

Active vehicle control (AVC) is a system that continuously observes the vehicle's dynamic responses and implements a corrective control action when necessary to improve drivability and stability of the vehicle. These subtle corrective actions can be a blend of regulating the steering angle and optimizing the transmitted tyre forces to the road [1–4]. Examples of AVC systems are active four-wheel steering (4WS), direct yaw moment control (DYC), active roll-over protection, anti-lock brake system (ABS), electronic stability control (ESC), active front-wheel steering (AFS), and active suspension system [2, 5]. The main aim of AVC is to improve vehicle safety, vehicle handling, and ride comfort and reduce the driving stress on poor road conditions.

In recent years, studies addressing active yaw moment control for physically steerable wheels to maintain the vehicle stability are enormous. In [4], stability control strategies of the vehicle via DYC are studied. Braking forces and traction controls are used to establish the stability threshold of the vehicle. Similarly, robust controller design [6–9] and optimal distribution of tyre forces [10–13] have been developed for steerable wheels. However, there are currently no studies, as far as we know, addressing the robust control of EVs without physically steerable wheels.

Vehicles are subject to uncertainties such as the crosswind disturbances, payload, vehicle's longitudinal velocity, and, most importantly, friction differences in the tyre-ground interaction among many other factors. Several authors have adopted robust controller design for yaw stability control in an independently driven electric vehicle. In [14], a μ-synthesis robust controller design improves the yaw motion and chassis sideslip. In a similar study, in [15], the authors designed a controller with the aim of achieving both a neutral steer and a reduction in the sideslip of a four-wheel EV. These studies do not consider the tire force saturation.

In [16, 17], the authors consider the tire force saturation, parametric uncertainties, and external disturbances in solving the lateral-plane motion and yaw stability problem of an independently driven electric vehicle. Reference [18] investigated an active steering failure mechanism in a differential assisted steering for an autonomous system. The

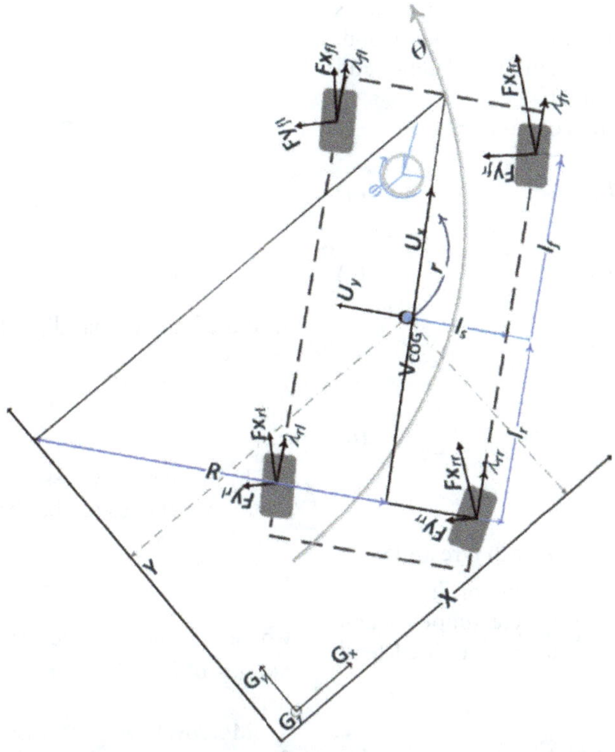

FIGURE 1: Differential Speed Steer EV maneuvering a curve.

authors consider the tyre force saturation and the use of a robust multiple-disturbances observer-based controller for a path following control.

In [19], a dynamic output-feedback controller is designed to realize a differential speed steering control strategy. However, in the previous design, the controller design was performed by linearizing the vehicle under a single operating condition tested for robustness. As one knows, the vehicle needs to operate over a wide range of guaranteed operating conditions, especially at different vehicle speed and road conditions. Based on the LMI approach, this paper designs a robust H_∞ dynamic output-feedback controller for four-wheel independently driven electric vehicles (FWIDEV) under bounded uncertainties and external environmental disturbance. The bounded uncertainties under consideration are the road coefficients of friction, vehicle's longitudinal speed, and the presence of lateral wind disturbances.

The main contributions of this paper are (1) the derivation of sufficient conditions for the existence of an H_∞ output-feedback controller for linear systems with polytopic uncertainties regarding input and output constraints and (2) the design of an H_∞ robust yaw controller that can achieve both robust stability and performance for EVs in the presence of uncertainties considering both input and output constraints.

The rest of the paper is organized as follows. Section 2 provides the mathematical model and the uncertain linear system of a four-wheel independently driven in-wheel electric vehicle. Section 3 presents the control problem and the robust controller design. Simulation results and discussion are given in Section 4. Concluding remarks are given in Section 5.

2. Modelling of a Four-Wheel Independently Driven In-Wheel Electric Vehicle

The four-wheel independently driven in-wheel electric vehicle under consideration is shown in Figure 1. Each wheel in the EV is independently driven. Schematically, the longitudinal, lateral, and yaw equations of motion are as follows:

$$\dot{U}_x = U_y r + \left(\frac{1}{m}\right)\left(F_{xfl} + F_{xfr} + F_{xrl} + F_{xrr}\right) \tag{1}$$

$$\dot{U}_y = -U_x r + \left(\frac{1}{m}\right)\left(F_{yfl} + F_{yfr} + F_{yrl} + F_{yrr}\right) \tag{2}$$

$$\begin{aligned} J\dot{r} &= l_f\left(F_{yfl} + F_{yfr}\right) - l_r\left(F_{yrl} + F_{yrr}\right) \\ &\quad + l_s\left[F_{xfl} + F_{xrl} - F_{xfr} - F_{xrr}\right] + M_d \end{aligned} \tag{3}$$

where U_x, U_y, and r are the vehicle's velocities along the longitudinal motion, lateral motion, and the yaw motion, respectively. m is the vehicle mass, l_f is the distance between the centre of gravity (COG) and the centre point of the front wheel, l_r is the distance between the COG and the rear wheel, l_s is half of the vehicle width, J is the yaw moment of inertia, and M_d is an external disturbance.

The tyre forces, F_{xi} and F_{yi} with $i = fl, fr, rl, rr$, are the result of the pneumatic deflections of the tire properties due to the weight of the vehicle, tyre pressure distribution, and also the tyre-ground interaction. To determine the forces generated by the tyres as a result of the tyre-road interaction, the tyre nonlinear model for the combined lateral and longitudinal forces will be used as in [20–23].

The longitudinal and lateral forces components are

$$F_{xi} = \frac{\sigma_{xi}}{\sigma_i} F_{\mathrm{T}i}, \quad (4)$$

$$F_{yi} = \frac{\sigma_{yi}}{\sigma_i} F_{\mathrm{T}i}, \quad (5)$$

where

$$F_{\mathrm{T}i} = \begin{cases} \mu F_{zi} \left\{ 3\theta_c \sigma_i - \frac{1}{3} (3\theta_c \sigma_i)^2 + \frac{1}{27} (3\theta_c \sigma_i)^3 \right\} & \text{if } \sigma_i \le \sigma_m \\ \text{or } \mu F_{zi} & \text{if } \sigma_i > \sigma_m, \end{cases} \quad (6)$$

where $\sigma_i = \sqrt{\alpha_{xi}^2 + \lambda_{yi}^2}$ is the total slip for each tyre and $\sigma_m = 1/\theta_c$ and $\theta_c = C_y / 3\mu F_{zi}$ are the limiting values of slip.

Parameters C_y and μ are, respectively, tyre lumped lateral stiffness and the coefficient of tyre-road friction. The lateral tyre slip, σ_{yi}, is

$$\sigma_{yi} = \frac{U_y}{\overline{w}_i r_w} \tan(\lambda_i) \quad (7)$$

and the longitudinal tyre slip, σ_{xi}, is

$$\sigma_{xi} = \begin{cases} \dfrac{\overline{w}_i r_w - U_x}{\overline{w}_i r_w} & \text{during acceleration} \\ \text{or } \dfrac{\overline{w}_i r_w - U_x}{U_x} & \text{during braking.} \end{cases} \quad (8)$$

r_w is the effective wheel radius, $\overline{w}_i$ is the angular velocity of each wheel, and λ_i is the tyre side slip angle. The normal load, F_{zi}, is based on the vehicle's geometry as

$$F_{zfl} = F_{zfr} = \frac{mgl_r}{2(l_f + l_r)}$$
$$F_{zfl} = F_{zfr} = \frac{mgl_f}{2(l_f + l_r)}. \quad (9)$$

Here a linear form of the combined tyre force modelled by (4) and (5) is used to describe the tyre frictional forces on each wheel [24]. We assume the vertical load of the vehicle is evenly distributed on each wheel. So that when the tyre side slip angle is small, by [21, 25], the tyre forces can be approximated as

$$F_{xi} = \mu C_{xi} \alpha_i, \quad (10)$$

$$F_{yi} = \mu C_{yi} \lambda_i, \quad (11)$$

where C_x is the longitudinal tyre stiffness; the tyre side slip angles, λ_i, are approximated as

$$\lambda_{fl} = -\left[\frac{U_y + rl_f}{U_x}\right],$$
$$\lambda_{fr} = -\left[\frac{U_y + rl_f}{U_x}\right],$$
$$\lambda_{rl} = -\left[\frac{U_y - rl_r}{U_x}\right],$$
$$\lambda_{rr} = -\left[\frac{U_y - rl_r}{U_x}\right], \quad (12)$$

And the longitudinal slip ratio, α_i, is

$$\alpha_i = \frac{r_w \overline{w}_i - U_x}{U_x}, \quad (13)$$

In this paper, we assume the longitudinal velocity, U_x, is an unknown constant but bounded as

$$\underline{U_x} < U_x < \overline{U_x}, \quad (14)$$

where $\underline{U_x}$ and $\overline{U_x}$, are, respectively, the lower and upper bounds of U_x. U_x is a constant; hence (1) is not considered here.

Moreover, we also assume that the coefficient of road friction, μ, is an unknown constant but bounded as

$$\underline{\mu} < \mu < \overline{\mu}, \quad (15)$$

where $\underline{\mu}$ and $\overline{\mu}$ are, respectively, the lower and upper bounds of μ.

We model the uncertainties in U_x and μ as polytopic uncertainties; hence (2) and (3) can be expressed as follows:

$$\begin{aligned} \dot{x}(t) &= A(\theta_u) x(t) + B_1 w(t) + B_2(\theta_u) u(t) \\ y(t) &= C_2 x(t) + D_2 w(t) \end{aligned} \quad (16)$$

where $x(t) = [U_y, r]^T$, $u(t) = \overline{w_r} - \overline{w_l}$, $w(t) = [M_d\ n_d]^T$, and n_d is the measurement noise. The uncertain matrices $A(\theta_u)$ and $B_2(\theta_u)$ belong to the polytopic uncertain domain

$$\Omega_p = \{A(\theta_u), B_2(\theta_u)\} = \sum_{i=1}^{4} \theta_{ui} \{(A_i, B_{2i}\}, \quad (17)$$

where $\theta_{ui} \ge 0\ \forall i$ and $\sum_{i=1}^{m} \theta_{ui} = 1$, and

$$A_1 = \begin{bmatrix} -\dfrac{C_y \underline{\mu}}{0.25m\underline{U_x}} & -\dfrac{C_y \underline{\mu}(l_f - l_r)}{0.5m\underline{U_x}} - \underline{U_x} \\ \dfrac{C_y \underline{\mu}(l_r - l_f)}{0.5J\underline{U_x}} & -\dfrac{C_y \underline{\mu}(l_f^2 + l_r^2)}{0.5J\underline{U_x}} \end{bmatrix},$$

$$A_2 = \begin{bmatrix} -\dfrac{C_y \underline{\mu}}{0.25m\overline{U_x}} & -\dfrac{C_y \underline{\mu}(l_f - l_r)}{0.5m\overline{U_x}} - \overline{U_x} \\ \dfrac{C_y \underline{\mu}(l_r - l_f)}{0.5J\overline{U_x}} & -\dfrac{C_y \underline{\mu}(l_f^2 + l_r^2)}{0.5J\overline{U_x}} \end{bmatrix},$$

$$A_3 = \begin{bmatrix} -\frac{C_y\bar{\mu}}{0.25m\underline{U_x}} & -\frac{C_y\bar{\mu}\left(l_f - l_r\right)}{0.5m\underline{U_x}} - \underline{U_x} \\ \frac{C_y\bar{\mu}\left(l_r - l_f\right)}{0.5J\underline{U_x}} & -\frac{C_y\bar{\mu}\left(l_f^2 + l_r^2\right)}{0.5J\underline{U_x}} \end{bmatrix},$$

$$A_4 = \begin{bmatrix} -\frac{C_y\bar{\mu}}{0.25m\overline{U_x}} & -\frac{C_y\bar{\mu}\left(l_f - l_r\right)}{0.5m\overline{U_x}} - \overline{U_x} \\ \frac{C_y\bar{\mu}\left(l_r - l_f\right)}{0.5J\overline{U_x}} & -\frac{C_y\bar{\mu}\left(l_f^2 + l_r^2\right)}{0.5J\overline{U_x}} \end{bmatrix}.$$

$$B_{2_1} = \begin{bmatrix} 0 \\ \frac{r_w l_s C_x \underline{\mu}}{0.5J\underline{U_x}} \end{bmatrix},$$

$$B_{2_2} = \begin{bmatrix} 0 \\ \frac{r_w l_s C_x \underline{\mu}}{0.5J\overline{U_x}} \end{bmatrix},$$

$$B_{2_3} = \begin{bmatrix} 0 \\ \frac{r_w l_s C_x \overline{\mu}}{0.5J\underline{U_x}} \end{bmatrix},$$

$$B_{2_4} = \begin{bmatrix} 0 \\ \frac{r_w l_s C_x \overline{\mu}}{0.5J\overline{U_x}} \end{bmatrix}.$$

$$B_1 = \begin{bmatrix} 0 & 0 \\ \frac{1}{J} & 0 \end{bmatrix},$$

$$D_2 = \begin{bmatrix} 0 & 1 \end{bmatrix}.$$

$$C_2 = \begin{bmatrix} 0 & 1 \end{bmatrix} \tag{18}$$

In the next section, a robust H_∞ dynamic output-feedback controller design technique will be developed for general linear uncertain systems with polytopic uncertainties.

3. Robust Controller Design

Consider the following uncertain system:

$$\begin{aligned} \dot{x}(t) &= A(\theta_u)x(t) + B_1 w(t) + B_2(\theta_u)u(t) \\ z(t) &= C_1 x(t) + D_1 u(t) \\ z_s(t) &= C_s(\theta_u)x(t) + D_{1s}w(t) + D_{2s}(\theta_u)u(t) \\ y(t) &= C_2 x(t) + D_2 w(t) \end{aligned} \tag{19}$$

where $x(t)$ is the state vector, $u(t)$ is the control input, $w(t)$ is the disturbance, $z(t)$ is the performance output, $z_s(t)$ is the control output, $y(t)$ is the measured output, and θ_u exists in a convex hull. The uncertain matrices $A(\theta_u)$, $B_2(\theta_u)$, $C_s(\theta_u)$, and $D_{2s}(\theta_u)$ belong to the polytopic uncertain domain

$$\begin{aligned} \Omega_p &= \{A(\theta_u), B_2(\theta_u), C_s(\theta_u), D_{2s}(\theta_u)\} \\ &= \sum_{i=1}^{m}\theta_{ui}\{(A_i, B_{2i}, C_{si}, D_{2si})\}, \end{aligned} \tag{20}$$

where $\theta_{ui} \geq 0\ \forall i$ and $\sum_{i=1}^{m}\theta_{ui} = 1$. m is the number of vertices in the polytopic system. The matrices A_i, B_{2i}, C_{si}, and D_{2si} build the $i-th$ vertex of the polytope.

Without loss of generality, we assume

$$\begin{aligned} &\int_0^\infty w^T(t)w(t)\,dt \leq \rho \\ &\text{and } \|w(t)\|^2 \leq w_{\max}, \\ &\qquad \forall t > 0. \end{aligned} \tag{21}$$

Since the sensors required to measure the lateral velocity are costly, in this paper, we will consider the dynamic output-feedback controller of the form

$$\begin{aligned} \dot{\hat{x}}_c(t) &= A_c\hat{x}_c(t) + B_c y(t) \\ u(t) &= C_c\hat{x}_c(t) \end{aligned} \tag{22}$$

where A_c, B_c, and C_c are the parameters of the controller.

Robust H_∞ Controller Formulation. Given γ, find a controller of the form (22) such that

(1) system (19) with (22) is stable and satisfying the following inequality

(2) under zero initial conditions, the following inequality is satisfied:

$$\int_0^\infty z^T(t)z(t)\,dt < \gamma\int_0^\infty w^T(t)w(t)\,dt \tag{23}$$

(3) the following control output and input constraints are met:

$$\begin{aligned} &\left\|z_{sj}(t)\right\| \leq z_{s,\max,j} \\ &\text{and } \|u(t)\| \leq u_{\max}, \\ &\qquad \forall t > 0 \end{aligned} \tag{24}$$

where $j = 1, 2, \ldots, n_r$, and n_r is the number of rows in $z_s(t)$ and $z_{s,\max,j}^2 > (D_{1sj}^2 + D_{1sj})w_{\max}$.

The closed-loop system of (19) with (22) is

$$\begin{aligned} \dot{\tilde{x}}(t) &= A_{cl}\tilde{x}(t) + B_{cl}w(t) \\ z(t) &= C_{cl}\tilde{x}(t) \\ z_s(t) &= C_{scl}\tilde{x}(t) + D_{1s}w(t), \end{aligned} \tag{25}$$

where

$$\tilde{x}(t) = \begin{bmatrix} x(t) \\ \hat{x}(t) \end{bmatrix},$$

$$A_{cl} = \begin{bmatrix} A(\theta_u) & B_2(\theta_u) C_c \\ B_c C_2 & A_c \end{bmatrix},$$

$$B_{cl} = \begin{bmatrix} B_1 \\ B_c D_2 \end{bmatrix}, \tag{26}$$

$$C_{cl} = [C_1 \;\; D_1 C_c],$$

$$C_{scl} = [C_s(\theta_u) \;\; D_{2s}(\theta_u) C_c].$$

Theorem 1. *Given γ, δ_1, and δ_2, suppose that there exist symmetric matrices W_{11}^i, W_{22}^i, Q_{1i}, Q_{2i}, and Y and matrices W_{12}^i, G, A_f, X, K, and L such that for $i = 1,\ldots,m$ and $j = 1,\ldots,n_r$ the following inequalities conditions hold:*

$$W^i > 0, \tag{27}$$

$$\begin{bmatrix} -\dfrac{u_{max}^2}{\gamma\rho} W^i & [0 \;\; L]^T \\ * & -I \end{bmatrix} < 0, \tag{28}$$

$$\begin{bmatrix} -\dfrac{\rho_{sj}}{\gamma\rho} W^i & [C_{sij} Y \;\; D_{2sij} L]^T \\ * & -\left(1 + \left\|D_{1sj}\right\|\right)^{-1} I \end{bmatrix} < 0, \tag{29}$$

$$\begin{bmatrix} -Q_{1i} & YA_i^T + L^T B_{2i}^T - G^T & 0 \\ * & -\delta_2 I & A_i Y + B_{2i} L - G \\ * & * & -Q_{2i} \end{bmatrix} < 0, \tag{30}$$

$$\begin{bmatrix} \phi_{11i} & \phi_{12i} & \phi_{13} & \phi_{14} & \phi_{15} \\ * & \phi_{22i} & \phi_{23} & 0 & 0 \\ * & * & -\gamma I & 0 & 0 \\ * & * & * & -I & 0 \\ * & * & * & * & -\delta_2 I \end{bmatrix} < 0, \tag{31}$$

where $\rho_{sj} = z_{s,\max,j}^2 - (\|D_{1sj}\|^2 + \|D_{1sj}\|) w_{\max}$,

$$W^i = \begin{bmatrix} W_{11}^i & W_{12}^i \\ * & W_{22}^i \end{bmatrix},$$

$$\phi_{11i} = \begin{bmatrix} YA_i^T + L^T B_{2i}^T + A_i Y + B_{2i} L + Q_{1i} & A_i + A_f \\ * & A_i^T X + C_2^T K^T + XA + KC_2 \end{bmatrix}$$

$$\phi_{12i} = \begin{bmatrix} \delta_1 (A_i Y + B_{2i} L) - Y + W_{11}^i & \delta_1 A_i - I + W_{12}^i \\ -\delta_1 A_f^T - I + W_{12}^i & \delta_1 (X A_i + K C_2) - X + W_{22}^i \end{bmatrix}$$

$$\phi_{13} = \begin{bmatrix} B_1 \\ XB_1 + KD_2 \end{bmatrix}$$

$$\phi_{14} = \begin{bmatrix} YC_1^T + L^T D_1^T \\ C_1^T \end{bmatrix}$$

$$\phi_{15} = \begin{bmatrix} 0 \\ \delta_2 X \end{bmatrix}$$

$$\phi_{22i} = \begin{bmatrix} -2Y + Q_{2i} & -2I \\ -2I & -X - X^T \end{bmatrix}$$

$$\phi_{23} = \delta_1 \begin{bmatrix} B_1 \\ XB_1 + KD_2 \end{bmatrix}. \tag{32}$$

Then, system (19) with (22) is stable and the H_∞ performance (23) and the input and output constraints (24) hold. Moreover, the controller parameters are given as follows:

$$C_c = LY^{-1},$$

$$B_c = \left(Y^{-1} - X\right)^{-1} K, \tag{33}$$

$$A_c = \left(Y^{-1} - X\right)^{-1} \left\{A_f - \left(Y^{-1} - X\right) B_c C_2 Y - XG\right\} Y^{-1}.$$

Proof. Refer to the Appendix for the proof. □

4. Simulation Results

In this section, by using the parameters given in Table 1, the proposed control design is verified by stabilizing the EV along a straight path. The uncertainties in the coefficient of road friction and the longitudinal velocity are varying parameters from $0.2 \le \mu \le 1$ and $20 \le U_x \le 120$km/hr, respectively. For comparison, the nominal operating points of 50km/hr and $\mu = 0.8$ are selected for the H_∞ design. The uncertainty in μ and U_x, respectively, represents −75% to +25% and −60% to +140% of the nominal points.

In this simulation, the maximum speed of each wheel is assumed to be 219rad/s or 260km/hr in relation to vehicle's speed. Since the maximum operating speed of the vehicle is 120km/hr (101rad/s), the input constraint is given as

$$\|u(t)\| <= u_{max} = 118\text{rad/s}. \tag{34}$$

For the comfort of the driver and passengers, the output constraint for the yaw acceleration is chosen as $0.4g/l_c$; that is,

$$|\dot{r}(t)| <= Z_{s,max} = \frac{(0.4g)}{l_c}, \tag{35}$$

where $l_c = 1.36$m is the centroid of the vehicle.

The disturbance term M_d used in the simulation is shown in Figure 2. The yaw rate measurement white noise's power density is 0.02 with a maximum amplitude of 0.1.

TABLE 1: Vehicle parameters.

Symbol	Parameters	Values
m	vehicle mass	1450 kg
l_f	distance from COG to front axle	1.013 m
l_r	distance from COG to rear axle	1.3 m
l_s	half of track width	0.7180 m
r_w	tire effective radius	0.33 m
g	acceleration of gravity	9.81 ms^{-2}
l_c	Midpoints of diagonals	1.36 m
J	yaw moment of inertia	2300 $kg.m^2$
μ	coefficient of friction	0.2 - 0.9
C_x	longitudinal tire stiffness	50000 N/rad
C_y	lateral tire stiffness	25000 N/rad

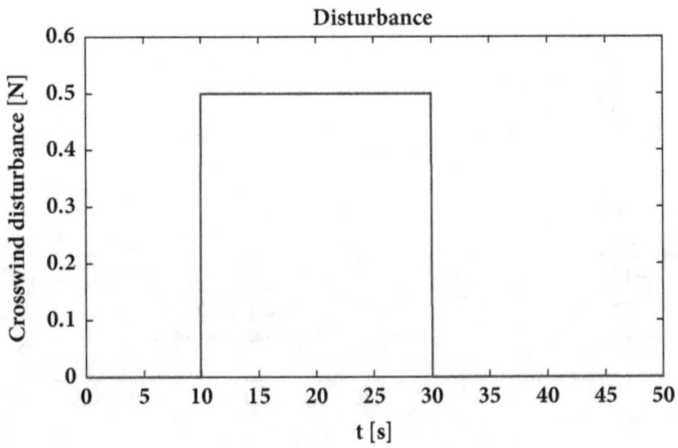

FIGURE 2: The disturbance.

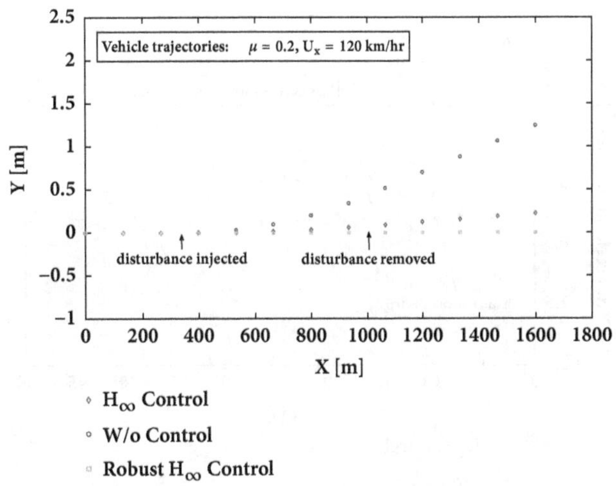

FIGURE 3: Vehicle trajectories: $\mu = 0.2$; vehicle speed: 120km/hr.

With this disturbance and the measurement noise, $\|w(t)\|^2 \leq 0.26$ and $\rho = 5$.

By Theorem 1, with $\delta_1 = 0.3$, $\delta_2 = 200$, and $\gamma = 1.2$, the following robust control parameters are obtained by solving the LMIs in (27), (28), (29), (30), and (31) using YALMIP toolbox (MOSEK solver) [26].

$$A_c = \begin{bmatrix} -26.5443 & 4878.64 \\ 1.0912 & -218.41 \end{bmatrix},$$
$$B_c = \begin{bmatrix} -4684.78 \\ 187.12 \end{bmatrix}, \quad (36)$$
$$C_c = \begin{bmatrix} 1.7951 & -33.0317 \end{bmatrix}.$$

In comparison, we design an H_∞ with $\delta_1 = 0.5$, $\delta_2 = 8$, and $\gamma = 3.55$; the control parameters are

$$A_c = \begin{bmatrix} -0.1668 & -4.09 \\ 4.2732 & -124.47 \end{bmatrix},$$
$$B_c = \begin{bmatrix} -94.77 \\ 87.24 \end{bmatrix}, \quad (37)$$
$$C_c = \begin{bmatrix} 0.2111 & -2.0492 \end{bmatrix}.$$

To test the effectiveness of the robust controller design and control strategy, different simulations are performed under multiple vehicle operating conditions. The nonlinear vehicle dynamic model in Section 2 is used in the final simulations. Each simulation ran for 50 seconds and generated 505 data points per second. The total displayed distance for the vehicle trajectory depends on the road condition (μ) and the vehicle's longitudinal velocity (U_x).

Figure 3 shows the trajectories of the EV with its longitudinal speeds at 120km/hr and coefficient of road friction at 0.2. The simulation starts with all the three vehicles operating without the crosswind disturbance, Figure 2; the trajectories of the vehicles can be seen to be on track. At the situation when the disturbance is injected from 10 sec to 30 sec, the vehicle without a controller begins to swerve off the desired path while the vehicles with a controller stay the course. The vehicle with the H_∞ controller shows a slight lateral deviation of less than 2 cm. After the vehicles exit the disturbance, the vehicles with a controller are still able to continue on the trajectory, but the vehicle with the robust controller shows an improved performance by maintaining a straight path.

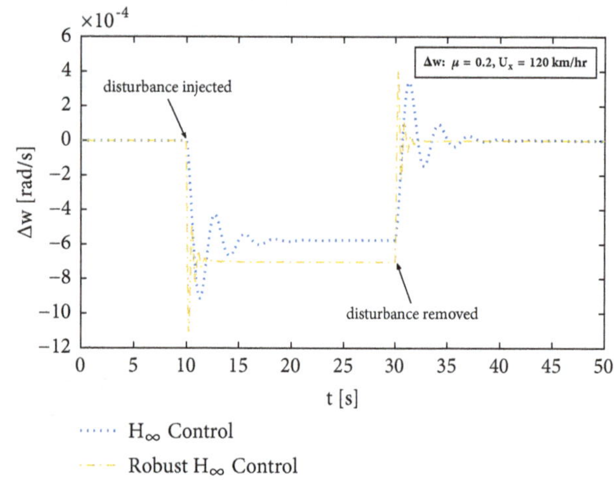

FIGURE 4: Δw: $\mu = 0.2$; vehicle speed: 120km/hr.

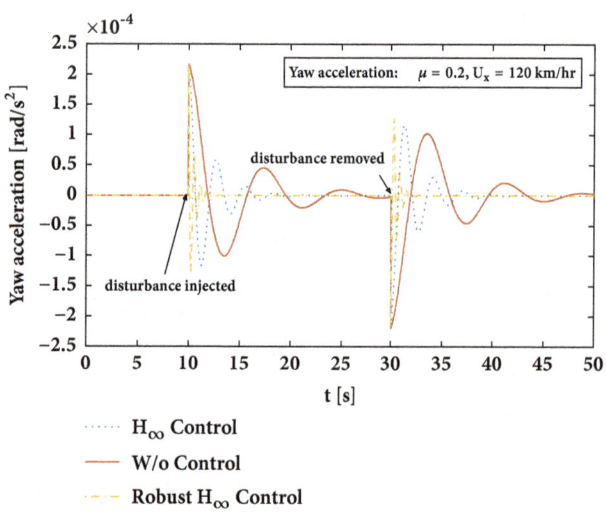

FIGURE 5: Yaw acceleration: $\mu = 0.2$; vehicle speed: 120 km/hr.

The controller output, Δw, and $\dot{r}$ are shown in Figures 4 and 5, respectively. The robust controller shows a fast and good damping of the disturbance while still being within the constraints.

Figure 6 shows the vehicle's trajectories at the operating condition, $\mu = 0.8$ and $U_x = 20$km/hr. Both of the vehicles with a controller indicated a strong performance, but an improvement can be seen with the robust controller as shown in the input, Figure 7, and control output, Figure 8. We can observe that the robust controller can quickly attenuate the disturbance upon entry and exit of the disturbance without violating the constraints. Figure 7 shows an improved performance of the robust controller. It is interesting to note the fast dynamics of the robust controller and its corresponding output constraint in Figure 8, for a straight-ahead driving, Figure 6. Figure 8 shows the vehicle's yaw acceleration at $\mu = 0.8$ and an entry speed of 20km/hr. The controllers are still able to stabilize the vehicle on the straight path with minimal lateral deviation.

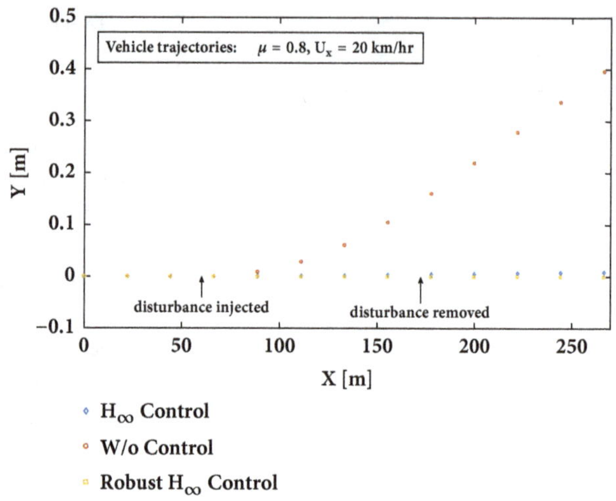

FIGURE 6: Vehicle trajectories: $\mu = 0.8$ and 20km/hr.

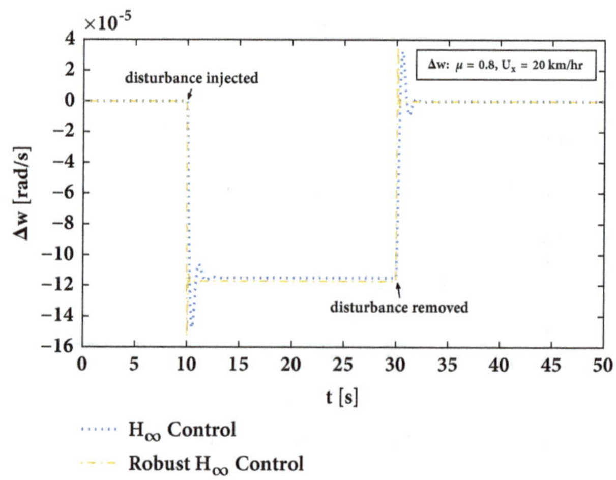

FIGURE 7: Control input: $\mu = 0.8$; vehicle speed: 20km/hr.

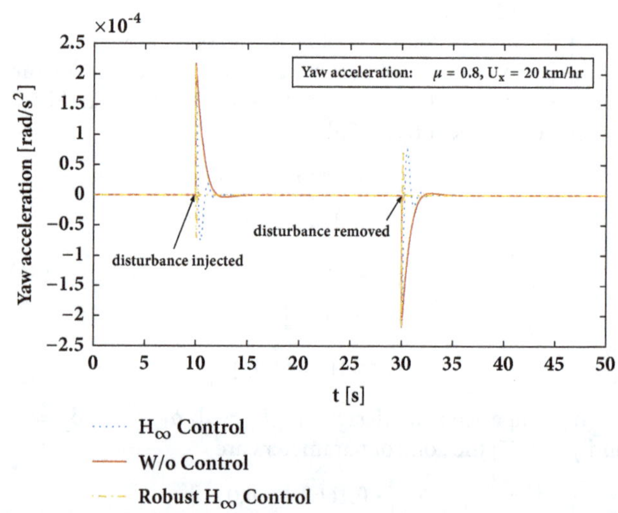

FIGURE 8: Yaw acceleration: $\mu = 0.8$; 20km/hr.

From the simulations, we observe that the vehicle with the robust controller is able to attenuates disturbance while satisfying both the input and output constraints.

5. Conclusion

This paper has designed a robust H_∞ dynamic output-feedback controller that attenuates external disturbances of four-wheel independently driven in-wheel electric vehicles. The electric vehicle has been modelled as an uncertain polytopic system with bounded uncertainties and external disturbance. We have developed sufficient conditions for the existence of a robust controller with input and output constraints that ensures the vehicle stays on course despite different road conditions and different longitudinal speeds. The performance of the robust controller has been tested on a complete nonlinear vehicle dynamic model. The simulation results have shown that despite significant differences in the operating conditions, the proposed robust controller ensures that the vehicle's trajectory stays on the desired path and satisfies both the input and control output constraints.

Appendix

Proof of Theorem 1. For the closed-loop system (25), consider the following parameter Lyapunov function:

$$V = \tilde{x}^T(t)\overline{P}(\theta_u)\tilde{x}(t), \tag{A.1}$$

where

$$\overline{P}(\theta_u) = \sum_{i=1}^{m}\theta_{ui}(t)\overline{P}_i. \tag{A.2}$$

Taking the time derivative of (A.1) along with (25) leads to

$$\dot{V} = \dot{\tilde{x}}^T(t)\overline{P}(\theta_u)\tilde{x}(t) + \tilde{x}^T(t)\overline{P}(\theta_u)\dot{\tilde{x}}(t) \tag{A.3}$$

Adding

$$2\left[\tilde{x}^T(t)P_1 + \dot{\tilde{x}}^T(t)P_2\right]\left[A_{cl}\tilde{x}(t) - \dot{\tilde{x}}(t) + B_{cl}w(t)\right] = 0 \tag{A.4}$$

to (A.3) by [27], for robust stability and performance, yields

$$\dot{V} = \varphi^T(t)Q\varphi(t) \tag{A.5}$$

where $\varphi^T(t) = \left[\tilde{x}^T(t)\ \ \dot{\tilde{x}}^T(t)\ \ w^T(t)\right]$ and

$$Q = \begin{bmatrix} A_{cl}^T P_1^T + P_1 A_{cl} & \overline{P}(\theta) - P_1 + P_2 A_{cl} & P_1 B_{cl} \\ * & -P_2 - P_2^T & P_2 B_{cl} \\ * & * & 0 \end{bmatrix}. \tag{A.6}$$

Adding and subtracting $(z^T(t)z(t) - \gamma w^T(t)w(t))$ to and from (A.5) results in

$$\dot{V} = \varphi^T\overline{Q}\varphi - z^T(t)z(t) + \gamma w^T(t)w(t), \tag{A.7}$$

where

$$\overline{Q} = \begin{bmatrix} A_{cl}^T P_1^T + P_1 A_{cl} + C_{cl}^T C_{cl} & \overline{P}(\theta) - P_1 + P_2 A_{cl} & P_1 B_{cl} \\ * & -P_2 - P_2^T & P_2 B_{cl} \\ * & * & -\gamma I \end{bmatrix}. \tag{A.8}$$

Suppose the conditions in Theorem 1 hold; then $\phi_{22} < 0$. This condition implies that Y and X are nonsingular matrices. Without the loss of generality, let partition P_1 be

$$P_1 = \begin{bmatrix} X & Y^{-1} - X \\ Y^{-1} - X & X - Y^{-1} \end{bmatrix}. \tag{A.9}$$

We assume $P_2 = \delta_1 P_1$. Define

$$J = \begin{bmatrix} Y & I \\ Y & 0 \end{bmatrix}. \tag{A.10}$$

Multiplying the left-hand side of (A.8) by $\text{diag}\{J^T, J^T, I\}$ and its right-hand side by $\text{diag}\{J, J, I\}$ reads

$$\text{diag}\left\{J^T, J^T, I\right\}\overline{Q}\,\text{diag}\{J, J, I\} = \begin{bmatrix} \Lambda_{11} & \Lambda_{12} & \Lambda_{13} \\ * & \Lambda_{22} & \Lambda_{23} \\ * & * & -\gamma I \end{bmatrix}, \tag{A.11}$$

where

$$\Lambda_{11} = \begin{bmatrix} YA^T(\theta_u) + YC_c^T B_2^T(\theta_u) + A(\theta_u)Y + B_2(\theta_u)C_c Y & \Gamma_1(\theta_u) \\ * & \Gamma_2(\theta_u) \end{bmatrix} + \phi_{14}\phi_{14}^T, \tag{A.12}$$

with

$$\begin{aligned} &\Gamma_1(\theta_u) \\ &= A(\theta_u) + YA^T(\theta_u)X^T + YC_2^T B_c^T\left(Y^{-1} - X^T\right) \\ &\quad + YC_c^T B_2(\theta_u)^T X^T + YA_c^T\left(Y^{-1} - X^T\right), \\ &\Gamma_2(\theta_u) \\ &= A^T(\theta_u)X^T - C_2^T B_c^T\left(Y^{-1} - X^T\right) + XA(\theta_u) \\ &\quad + \left(Y^{-1} - X\right)B_c C_2 \end{aligned} \tag{A.13}$$

$$\Lambda_{12} = \begin{bmatrix} \delta_1\left(A(\theta_u)Y + B_2(\theta_u)C_c Y\right) - Y + W_{11}(\theta) & \delta_1 A(\theta_u) - I + W_{12}(\theta_u) \\ \Gamma_3(\theta_u) & \delta_1\left(XA(\theta_u) + \left(Y^{-1} - X\right)B_c C_2\right) \end{bmatrix}$$

with

$$\begin{aligned} \Gamma_3(\theta_u) &= \delta_1\left(XA(\theta_u)Y + \left(Y^{-1} - X\right)B_c C_2 Y\right. \\ &\quad \left. + XB_2(\theta_u)C_c Y + \left(Y^{-1} - X\right)A_c Y\right) - I \\ &\quad + W_{12}(\theta_u), \end{aligned}$$

$$\begin{bmatrix} W_{11}(\theta_u) & W_{12}(\theta_u) \\ W_{21}(\theta_u) & W_{22}(\theta_u) \end{bmatrix} = J^T\overline{P}(\theta_u)J.$$

$$\Lambda_{13} = \begin{bmatrix} B_1 \\ XB_1 + \left(Y^{-1} - X\right)B_c D_2 \end{bmatrix},$$

$$\Lambda_{23} = \delta_1 \begin{bmatrix} B_1 \\ XB_1 + \left(Y^{-1} - X\right) B_c D_2 \end{bmatrix},$$

$$\Lambda_{22} = -\begin{bmatrix} 2Y & 2I \\ 2I & X + X^T \end{bmatrix}. \tag{A.14}$$

With the notations given in Theorem 1, Λ_{11} can be rewritten as

$$\Lambda_{11} = \Lambda_{11}^1 + \Lambda_{11}^2 \tag{A.15}$$

where

$$\Lambda_{11}^1 = \sum_{i=1}^{m} \theta_i \phi_{11i} + \phi_{14}\phi_{14}^T \tag{A.16}$$

and

$$\Lambda_{11}^2 = \begin{bmatrix} -Q_1(\theta_u) & YA^T(\theta_u)X^T + \left(YC_c^T B_2^T(\theta_u)X^T\right) - G^T X^T \\ * & -\delta_2 XX^T \end{bmatrix}. \tag{A.17}$$

Also rewrite Λ_{12} as

$$\Lambda_{12} = \sum_{i=1}^{m} \theta_{ui}\phi_{12i} + \begin{bmatrix} 0 & 0 \\ XA(\theta_u)Y + XB_2(\theta_u)C_c Y - GX & 0 \end{bmatrix} \tag{A.18}$$

$$= \Delta_{12}^1 + \Delta_{12}^2$$

and Λ_{22} as

$$\Lambda_{22} = \sum_{i=1}^{m} \theta_{ui}\phi_{22i} + \begin{bmatrix} -Q_2(\theta_u) & 0 \\ 0 & 0 \end{bmatrix} = \Delta_{22}^1 + \Delta_{22}^2. \tag{A.19}$$

Then

$$\Lambda = \begin{bmatrix} \Lambda_{11}^1 & \Lambda_{12}^1 & \Lambda_{13} \\ * & \Lambda_{22}^1 & \Lambda_{23} \\ * & * & -\gamma I \end{bmatrix} + \begin{bmatrix} \Lambda_{11}^2 & \Lambda_{12}^2 & 0 \\ * & \Lambda_{22}^2 & 0 \\ 0 & 0 & 0 \end{bmatrix} \tag{A.20}$$

$$= \Lambda^1 + \Lambda^2$$

Multiplying diag$\{I, X^{-1}, I\}$ to the left of Δ^2 and diag$\{I, X^{-T}, I\}$ to the right of Δ^2, we obtain

$$\Lambda^2 = \begin{bmatrix} -Q_1(\theta_u) & S_{c12} & 0 & 0 & 0 \\ * & -\delta_2 I & S_{c22} & 0 & 0 \\ * & * & -Q_2(\theta_u) & 0 & 0 \\ * & * & * & 0 & 0 \\ * & * & * & * & 0 \end{bmatrix}, \tag{A.21}$$

with

$$S_{c12} = YA^T(\theta_u) + L^T B_2^T(\theta_u) - G^T.$$
$$S_{c22} = A(\theta_u)Y + B_2(\theta_u)L - G \tag{A.22}$$

The condition in (30) implies $\Lambda^2 \leq 0$. Applying Schur complement with respect to the last two rows of (31), it can be shown that $\Lambda^1 \leq 0$. Therefore,

$$\dot{V} < -z^T(t)z(t) + \gamma w^T(t)w(t). \tag{A.23}$$

Integrating both sides and using the fact that $x(0) = \widehat{x}(0) = 0$, we have

$$\int_0^{\infty} z^T(t)z(t)\,dt < \gamma \int_0^{\infty} w^T(t)w(t)\,dt. \tag{A.24}$$

When $w(t)$ is zero,

$$\dot{V} < -z^T(t)z(t) < 0 \tag{A.25}$$

This implies that (25) is stable.

For the control output and input constraints,

$$V(t) - V(0) < \gamma \int_0^{t} w^T(t)w(t)\,dt < \gamma\rho \tag{A.26}$$

With the zero initial condition,

$$V(t) = \widetilde{x}^T(t)\overline{P}(\theta_u)\widetilde{x}(t) < \gamma\rho \tag{A.27}$$

Note that $u(t) = C_c\widehat{x}_c(t) = \widetilde{C}_c\widetilde{x}$, where $\widetilde{C}_c = [0 \ \ C_c]$. Hence,

$$\|u(t)\|^2 = \left\|\widetilde{C}_c\widetilde{x}(t)\right\|^2$$
$$\leq \widetilde{x}^T\overline{P}(\theta_u)\widetilde{x}\overline{P}^{-1/2}(\theta_u)\widetilde{C}_c^T\widetilde{C}_c\overline{P}^{-1/2}(\theta_u). \tag{A.28}$$

Using (A.27),

$$\|u(t)\|^2 \leq \gamma\rho\overline{P}^{-1/2}(\theta_u)\widetilde{C}_c^T\widetilde{C}_c\overline{P}^{1/2}(\theta_u). \tag{A.29}$$

From (A.29), $\|u(t)\| \leq u_{\max}$ if

$$\gamma\rho\overline{P}^{-1/2}(\theta_u)\widetilde{C}_c^T\widetilde{C}_c\overline{P}^{1/2}(\theta_u) < u_{max}^2. \tag{A.30}$$

Applying the Schur complement on (A.30) gives

$$\begin{bmatrix} -\dfrac{u_{max}^2}{\gamma\rho} & \overline{P}^{-1/2}(\theta_u)\widetilde{C}_c^T \\ * & -I \end{bmatrix} < 0. \tag{A.31}$$

Multiplying left and right of (A.31) by $\begin{bmatrix} \overline{P}^{1/2}(\theta_u) & 0 \\ 0 & I \end{bmatrix}$ gives

$$\begin{bmatrix} -\dfrac{u_{max}^2}{\gamma\rho}\overline{P}(\theta_u) & \widetilde{C}_c^T \\ 0 & -I \end{bmatrix} < 0. \tag{A.32}$$

Then multiplying the left and right of (A.32) by

$$\begin{bmatrix} J^T & 0 \\ 0 & I \end{bmatrix}$$

$$\text{and} \begin{bmatrix} J & 0 \\ 0 & I \end{bmatrix}, \tag{A.33}$$

respectively, gives (28).

Note that $z_{sj} = C_{sclj}\tilde{x}(t) + D_{1sj}w(t)$, where $j = 1, \ldots, n_r$. Therefore,

$$\begin{aligned}\left\|z_{sj}\right\|^2 &= \left\|C_{sclj}\tilde{x}(t) + D_{1sj}w(t)\right\|^2 \\ &\leq \left(1 + \left\|D_{1sj}\right\|\right)\left\|C_{sclj}\tilde{x}(t)\right\|^2 \\ &\quad + \left(\left\|D_{1sj}\right\|^2 + \left\|D_{1sj}\right\|\right)\|w(t)\|^2 \\ &\leq \left(1 + \left\|D_{1sj}\right\|\right)\left\|C_{sclj}\tilde{x}(t)\right\|^2 \\ &\quad + \left(\left\|D_{1sj}\right\|^2 + \left\|D_{1sj}\right\|\right) w_{\max}\end{aligned} \tag{A.34}$$

Using (A.27),

$$\begin{aligned}\left\|z_s(t)\right\| &\leq \left(1 + \left\|D_{1sj}\right\|\right)\gamma\rho\left\|C_{sclj}\tilde{x}(t)\overline{P}^{-1/2}(\theta_u)\right\|^2 \\ &\quad + \left(\left\|D_{1sj}\right\|^2 + \left\|D_{1sj}\right\|\right) w_{\max}\end{aligned} \tag{A.35}$$

From (A.35), $\|z_s(t)\| \leq z^2_{s,\max,j}$ if

$$\left(1 + \left\|D_{1sj}\right\|\right)\left\|C_{sclj}\tilde{x}(t)\overline{P}^{-1/2}(\theta_u)\right\|^2 \leq \frac{\rho_{sj}}{\gamma\rho} \tag{A.36}$$

Applying the Schur complement on (A.36) gives

$$\begin{bmatrix} -\dfrac{\rho_{sj}}{\gamma\rho} & \overline{P}^{-1/2}(\theta_u)C_{sclj} \\ * & -\left(1 + \left\|D_{1sj}\right\|\right)^{-1} I \end{bmatrix} < 0. \tag{A.37}$$

Multiplying left and right of (A.37) by $\left[\begin{smallmatrix} \overline{P}^{1/2}(\theta_u) & 0 \\ 0 & I \end{smallmatrix}\right]$ gives

$$\begin{bmatrix} -\dfrac{\rho_{sj}}{\gamma\rho}\overline{P}(\theta_u) & C_{sclj} \\ 0 & -\left(1 + \left\|D_{1sj}\right\|\right)^{-1} I \end{bmatrix} < 0. \tag{A.38}$$

Then multiplying the left and right of (A.38) by

$$\begin{bmatrix} J^T & 0 \\ 0 & I \end{bmatrix} \quad \text{and} \quad \begin{bmatrix} J & 0 \\ 0 & I \end{bmatrix}, \tag{A.39}$$

respectively, gives (29). This completes the proof. □

Conflicts of Interest

The authors declare that they have no conflicts of interest.

Acknowledgments

This work was supported by the National Natural Science Foundation of China under Grant 61573315.

References

[1] F. Yu, D.-F. Li, and D. A. Crolla, "Integrated vehicle dynamics control-state-of-the art review," in *Proceedings of the IEEE Vehicle Power and Propulsion Conference (VPPC '08)*, Harbin, China, September 2008.

[2] Y. Furukawa and M. Abe, "Advanced chassis control systems for vehicle handling and active safety," *Vehicle System Dynamics*, vol. 28, no. 2-3, pp. 59–86, 1997.

[3] P. Yih and J. C. Gerdes, "Modification of vehicle handling characteristics via steer-by-wire," *IEEE Transactions on Control Systems Technology*, vol. 13, no. 6, pp. 965–976, 2005.

[4] N. Ding and S. Taheri, "An adaptive integrated algorithm for active front steering and direct yaw moment control based on direct Lyapunov method," *Vehicle System Dynamics*, vol. 48, no. 10, pp. 1193–1213, 2010.

[5] A. Sigl and H. Demel, "ASR-traction control, state of the art and some prospects," *SAE Technical Papers*, 1990.

[6] H. Du, N. Zhang, and G. Dong, "Stabilizing vehicle lateral dynamics with considerations of parameter uncertainties and control saturation through robust yaw control," *IEEE Transactions on Vehicular Technology*, vol. 59, no. 5, pp. 2593–2597, 2010.

[7] A. Nasiri, S. Kiong Nguang, and A. Swain, "Adaptive sliding mode control for a class of MIMO nonlinear systems with uncertainties," *Journal of The Franklin Institute*, vol. 351, no. 4, pp. 2048–2061, 2014.

[8] J. Zhang, P. Shi, J. Qiu, and S. K. Nguang, "A novel observer-based output feedback controller design for discrete-time fuzzy systems," *IEEE Transactions on Fuzzy Systems*, vol. 23, no. 1, pp. 223–229, 2015.

[9] M. Shino and N. Masao, "Yaw-moment control of electric vehicle for improving handling and stability," *JSAE Review*, vol. 22, no. 4, pp. 473–480.

[10] Y. Suzuki, Y. Kano, and M. Abe, "A study on tyre force distribution controls for full drive-by-wire electric vehicle," *Vehicle System Dynamics*, vol. 52, no. 1, pp. 235–250, 2014.

[11] C.-C. Wang, S.-Y. Cheng, T. Hsiao, and W.-Y. Chou, "Application of optimum tire force distribution to vehicle motion control," in *Proceedings of the 2012 IEEE/ASME International Conference on Advanced Intelligent Mechatronics, AIM 2012*, pp. 538–543, Taiwan, July 2012.

[12] E. Ono, Y. Hattori, Y. Muragishi, and K. Koibuchi, "Vehicle dynamics integrated control for four-wheel-distributed steering and four-wheel-distributed traction/braking systems," *Vehicle System Dynamics*, vol. 44, no. 2, pp. 139–151, 2006.

[13] B. Li, H. Du, and W. Li, "Optimal distribution control of non-linear tire force of electric vehicles with in-wheel motors," *Asian Journal of Control*, vol. 18, no. 1, pp. 69–88, 2016.

[14] G. Yin, R. Wang, and J. Wang, "Robust control for four wheel independently-actuated electric ground vehicles by external yaw-moment generation," *International Journal of Automotive Technology*, vol. 16, no. 5, pp. 839–847, 2015.

[15] G. Zhai, M. Naka, T. Kobayashi, and J. Imae, "Towards neutral steer and sideslip reduction for four-wheeled electric vehicles," *Frontiers of Mechanical Engineering*, vol. 7, no. 1, pp. 16–22, 2012.

[16] R. Wang, H. Zhang, J. Wang, F. Yan, and N. Chen, "Robust lateral motion control of four-wheel independently actuated electric vehicles with tire force saturation consideration," *Journal of The Franklin Institute*, vol. 352, no. 2, pp. 645–668, 2015.

[17] J.-S. Hu, Y. Wang, H. Fujimoto, and Y. Hori, "Robust yaw stability control for in-wheel motor electric vehicles," *IEEE/ASME Transactions on Mechatronics*, vol. 22, no. 3, pp. 1360–1370, 2017.

[18] C. Hu, R. Wang, F. Yan, and H. R. Karimi, "Robust Composite Nonlinear Feedback Path-Following Control for Independently Actuated Autonomous Vehicles With Differential Steering," *IEEE Transactions on Transportation Electrification*, vol. 2, no. 3, pp. 312–321, 2016.

[19] P. Oke, S. K. Nguang, and J. Wen, "H∞ dynamic output feedback control for independently driven four-wheel electric vehicles with differential speed steering," in *Proceedings of the 2016 IEEE International Conference on Information and Automation, IEEE ICIA 2016*, pp. 567–572, China, August 2016.

[20] J. Svendenius and M. Gäfvert, "A brush-model based semi-empirical tire-model for combined slips," *SAE Technical Paper*, vol. 01, no. 1064, 2004.

[21] H. B. Pacejka, *Tire and Vehicle Dynamics*, SAE, USA, 2nd edition, 2005.

[22] B. J. Olson, S. W. Shaw, and G. Stépán, "Nonlinear dynamics of vehicle traction," *Vehicle System Dynamics*, vol. 40, no. 6, pp. 377–399, 2003.

[23] R. Rajamani, *Vehicle Dynamics and Control*, Springer, New York, NY, USA, 2012.

[24] E. Bakker, L. Nyborg, and H. B. Pacejka, "Tyre modelling for use in vehicle dynamics studies," *SAE Technical Papers*, Article ID 870421, pp. 1–15, 1987.

[25] G. Baffet, A. Charara, and J. Stephant, "Sideslip angle, lateral tire force and road friction estimation in simulations and experiments," in *Proceedings of the 2006 IEEE International Conference on Control Applications, CCA 2006*, pp. 903–908, Germany, October 2006.

[26] J. Löfberg, "Automatic robust convex programming," *Optimization Methods & Software*, vol. 27, no. 1, pp. 115–129, 2012.

[27] D. S. Bernstein and W. M. Haddad, "Robust stability and performance analysis for state-space systems via quadratic Lyapunov bounds," *SIAM Journal on Matrix Analysis and Applications*, vol. 11, no. 2, pp. 239–271, 1990.

3

Models and Algorithms of Conflict Detection and Scheduling Optimization for High-Speed Train Operations Based on MPC

Zhihui Wang,[1,2] Yonghua Zhou,[1] and Deng Liu[1,3]

[1] *School of Electronic and Information Engineering, Beijing Jiaotong University, Beijing, China*
[2] *Beijing Research and Development Center, Zhongxing Telecommunication Equipment Corporation, Shenzhen, China*
[3] *China Railway Communication & Signaling (CRCS) Survey and Design Co., Ltd., Beijing, China*

Correspondence should be addressed to Yonghua Zhou; yhzhou@bjtu.edu.cn

Academic Editor: Juan-Albino Méndez-Pérez

The high-speed train with high transport speed and punctuality rate has become a leading tool of daily traveling. Compared with the general-speed railway transportation, the high-speed railway transportation requires a timely and automatic adjustment capability in the centralized traffic control (CTC) system. In order to improve the capability, this paper mainly explores the models and algorithms of conflict detection and scheduling optimization of high-speed train operations. They are implemented under the mechanism of model predictive control (MPC) over the rolling prediction horizon. A train movement prediction model with conflict resolution is established. The mechanism of conflict detection and scheduling optimization is implemented through a case study. The simulation results demonstrate the effectiveness of the proposed mechanism.

1. Introduction

High-speed train operations may be inevitably disturbed by unexpected events such as bad weather or equipment failure. In these cases, the delays of some trains in a railway network will cause operation conflicts to take place, and actual train operations will deviate from preset timetables. If there is no timely adjustment to train operation plans, transportation efficiency will be decreased and, even, operation safety may be endangered. Therefore, the centralized traffic control (CTC) system requires timely detection of potential operation conflicts and engendering optimal scheduling strategies in advance.

Some literature has dealt with train scheduling and rescheduling problems. Szpigel [1] studied the problem of train scheduling on a single track line and modeled it as a job shop scheduling problem which is solved by the branch and bound algorithm. After that, train scheduling problem had aroused the interests to the scholars from various countries. Cordeau et al. [2] conducted a comprehensive survey on train scheduling and routing problems. Chiang et al. [3] put up a knowledge-based scheduling approach, employing global scheduling to deal with main plans and local scheduling to handle conflicts. Caprara et al. [4] put forward a solution method of a timetabling problem, utilizing Lagrangian relaxation to deal with constraints. D'Ariano et al. [5] proposed an alternative graph to describe scheduling strategies and adopted the branch and bound algorithm to solve the rescheduling problem under disturbances. D'Ariano et al. [6] proposed a branch and bound algorithm for the train scheduling problem in railway networks. Salido et al. [7] utilized graph decomposition technique to break a train scheduling problem into several subproblems. Zhou and Zhong [8] proposed a Lagrangian relaxation approach to provide a lower bound to the branch and bound algorithm which is utilized to solve the single-line timetabling problem. Cacchiani et al. [9] proposed an exact and heuristic algorithm to solve the train timetabling problem based on linear programming (LP) relaxation of an integer linear programming (ILP) formulation. Khan and Zhou [10] established a stochastic optimization model of timetabling problem and proposed a heuristic sequence solution method. Mu and Dessouky [11] presented an optimization scheduling approach using genetic algorithm for freight trains.

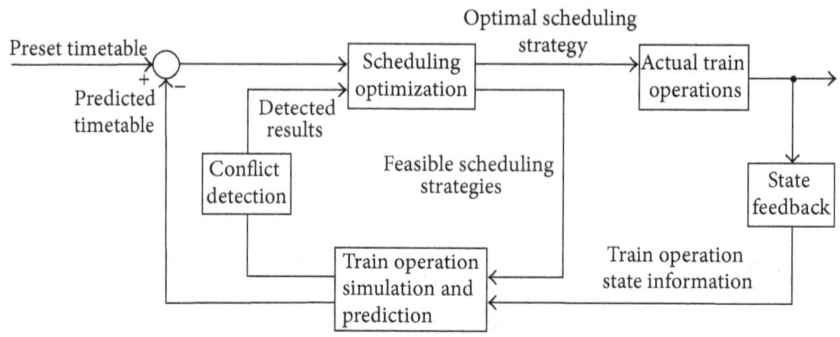

FIGURE 1: The framework of conflict detection and scheduling optimization of train operations based on model predictive control.

Yang et al. [12] explored the fuzzy and stochastic programming for freight transportation and utilized genetic algorithm to solve the chance constraint programming model. Yalçinkaya and Bayhan [13] put up a random simulation approach to produce feasible timetables. Krasemann [14] proposed a greedy algorithm using depth-first searching and evaluation function ordering to find the satisfactory solution of a train scheduling problem. Chevrier et al. [15] explored the energy-efficient timetabling problem using multiobjective evolution approach. Espinosa-Aranda and García-Ródenas [16] put up a heuristic algorithm using alternative arcs to minimize the traveling delay time. Li et al. [17] proposed the greedy algorithm to solve the train rescheduling problem with random delay. Kang et al. [18] presented the model and genetic solution algorithm of train rescheduling in urban railway transit networks. Zhan et al. [19] established a two-stage algorithm for the mixed programming model of high-speed train rescheduling problem to decide the wait stations and leave orders of scheduled trains. Šemrov et al. [20] proposed a reinforcement learning approach for the train rescheduling problem and compared its superiority with the rescheduling methods without learning. Kersbergen et al. [21] presented a dynamic model predictive control for railway traffic management which can improve calculation speeds and achieve good performances. Bettinelli et al. [22] proposed a real-time processing algorithm to avoid potential conflicts for railway traffic management. Eaton et al. [23] explored the ability of ant colony optimization to engender nondominated solutions for the dynamic multiobjective train rescheduling problem.

Statistics pointed out that 16% of the working hours for train dispatchers are to modify and implement train operation plans, only 4% of the time is to develop conflict resolution strategies, and the remaining 80% of the time is used to identify train operation conflicts. It can be seen that the discovery and settlement of train operation conflicts due to random disturbances are the most important work of dispatchers. How to utilize the advanced computer technology to timely and accurately detect train operation conflicts and optimize scheduling strategies has great practical significance to reduce the workload of dispatchers.

Model predictive control (MPC) has been successfully applied into industrial control, such as chemical, petroleum, metallurgy, and electric power control [24, 25]. The mechanism of prediction, feedback, and optimization in MPC is also applicable to the conflict resolution and scheduling optimization. This paper attempts to establish the framework of conflict resolution and scheduling optimization under the mechanism of MPC, focusing on the global framework elucidation, conflict detection approaches, and prediction model construction with conflict resolution [26, 27], which are different from the current literature.

The remainder of this paper is organized as follows. In Section 2, the framework is outlined. Section 3 addresses the conflict detection approaches and resolution strategies. Section 4 establishes the train movement prediction model taking conflict resolution into account. Section 5 elucidates the optimization scheduling model and algorithm. Section 6 demonstrates the case study. Finally, the concluding remarks are given in Section 7 to summarize the contribution of this paper.

2. Conflict Detection and Scheduling Optimization Framework

MPC is a computer-based optimization control methodology in industry. Based on the current states of a system and the prediction model, the future states of the system are predicted in the prediction horizon under the presumed control input sequences. The control inputs are optimized such that the predicted states approach the preset ones. Such a process repeats at a regular period or is driven by an event, and the prediction horizon rolls forward. MPC has various implementation ways in industrial control systems but is composed of three basic parts: prediction model, rolling optimization, and feedback adjustment.

In this paper, the mechanism of MPC is introduced into the conflict detection and scheduling optimization for high-speed railway traffic. The basic framework is shown in Figure 1. Train safe and punctual operations are regarded as the control objectives. In accordance with the MPC mechanism, the conflict detection and scheduling optimization are implemented over the rolling prediction horizon so that the predicted timetable approaches the preset one, and, moreover, train operation safety is guaranteed.

The main implementation procedure is described as follows.

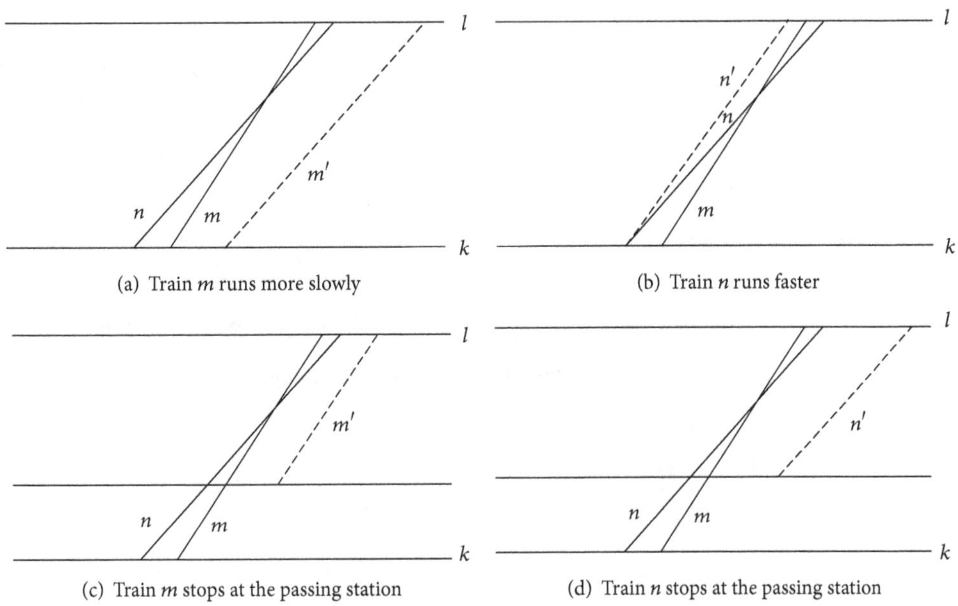

FIGURE 2: Section conflict and resolution.

Step 1. At current instant, sample the train operation state information such as positions and speeds.

Step 2. According to train operation models, predict train operation states in the prediction horizon on the basis of acquired position and speed information.

Step 3. Detect if there exist operation conflicts in the prediction horizon. If conflicts occur, produce feasible scheduling strategies. Scheduling strategies are optimized so that no conflicts occur and predicted timetable approaches the preset one.

Step 4. The prediction horizon moves forward a specified period. Go to *Step 1.*

3. Conflict Detection and Resolution

According to tempo-spatial constraints of high-speed train operations, operation conflicts can be categorized into the main four types, that is, section conflict, departure time conflict, arrival time conflict, and track number conflict, which are critical to train operation safety and often are paid great attention in practice. The section conflict means the operation time (or space) interval between two adjacent trains does not satisfy the safe constraint. The departure time conflict denotes the fact that the departure time interval from a station between two trains does not meet the minimum departure interval constraint, and the arrival time conflict represents the arrival time interval at a station between two trains not meeting the minimum arrival interval constraint. The track number conflict refers to the fact that the number of trains to dwell at or pass through a station has exceeded the number capacity of station tracks for trains' arrival and departure. Notations section (The Symbols' Definition of Conflict Detection and Scheduling Optimization) defines the symbols related to conflict description and conflict resolution.

3.1. Section Conflict and Resolution. When two adjacent trains run between stations, the distance between them is less than the minimum safety interval, and the running speed of the rear train is higher than that of the front one, the section conflict occurs, as shown in Figure 2. Figure 2 demonstrates the running trajectories of position (vertical) versus time (horizontal) of two trains. Train m runs faster than train n. The intersection point between two running lines is the collision place, which locates between stations.

The section conflict between two adjacent trains running in the same direction can be formulated as

$$\left(t^D_{m,k} - t^D_{n,k}\right)\left(t^A_{m,l} - t^A_{n,l}\right) < 0. \quad (1)$$

Equation (1) signifies the fact that the orders that trains m and n depart from station k and arrive at station l are different.

Two main approaches can be utilized to resolve the section conflict.

(1) When there is no passing station between stations k and l, the conflict can be prevented by driving the rear train m towards a smaller maximum running speed as shown by the dashed line (m') in Figure 2(a). The cost is the delay of train m. Otherwise, it can be achieved through driving the front train n towards a bigger maximum running speed as shown by the dashed line (n') in Figure 2(b) without delaying train m.

(2) When there is a passing station between stations k and l, let train m or n stop at the passing station for a while to resolve the conflict, as shown by the dashed line (m' or n') in Figure 2(c) or 2(d), respectively. The cost is the delay of train m or n.

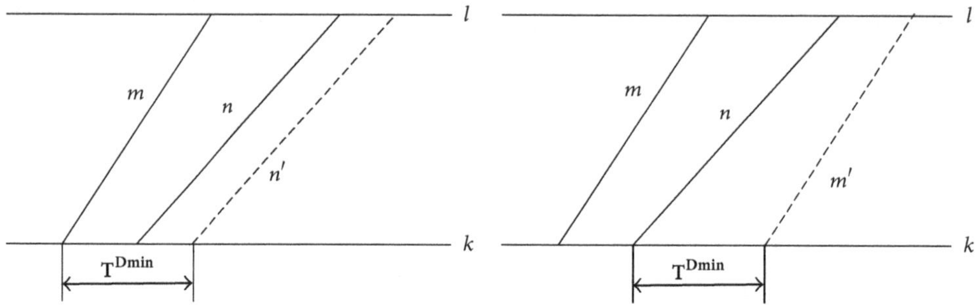

(a) Delay the departure of train n

(b) Delay the departure of train m

FIGURE 3: Departure time conflict and resolution.

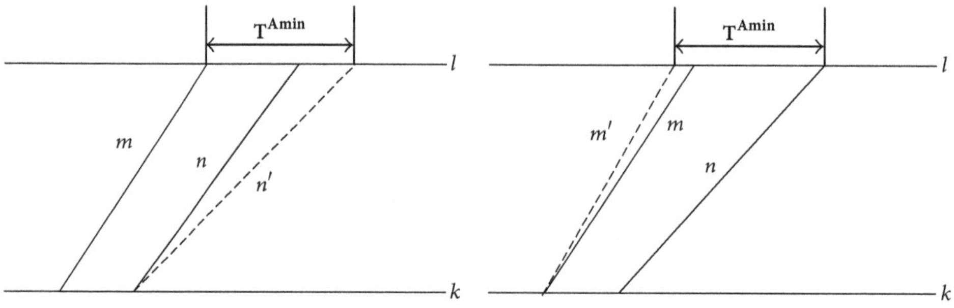

(a) Slow down train n

(b) Speed up train m

FIGURE 4: Arrival time conflict and resolution.

3.2. Departure Time Conflict and Resolution. When two adjacent trains depart from the same station, the relevant technical standards should be satisfied; that is, the departure time interval cannot be less than the configured minimum value, or the conflict will arise as shown in Figure 3.

According to Figure 3, the departure time conflict is expressed as

$$\left|t^D_{m,k} - t^D_{n,k}\right| < T^{D\min} \quad (m, n \in T,\ k \in S). \tag{2}$$

Equation (2) denotes the departure time interval between trains m and n being less than the limit of minimum departure time interval.

The possible methods to resolve the conflict are as follows.

(1) Transform the Operation Line of Train n. The departure time of train n is delayed so that the departure time interval between trains m and n is greater than or equal to the limit of minimum departure time interval, as shown by the dashed line (n') in Figure 3(a).

(2) Transform the Operation Line of Train m. The departure time of train m is delayed to meet the technical standard of departure time intervals, as shown by the dashed line (m') in Figure 3(b).

3.3. Arrival Time Conflict and Resolution. When two adjacent trains arrive at the same station, the arrival time interval at the same station should be greater than the configured minimum value; otherwise the conflict disobeying the technical standards will occur as shown in Figure 4.

As Figure 4 indicates, the arrival time conflict at a station is identified as

$$\left|t^A_{m,l} - t^A_{n,l}\right| < T^{A\min} \quad (m, n \in T,\ l \in S). \tag{3}$$

Equation (3) reveals the case dissatisfying the technical standard of arrival time constraints.

To resolve the arrival time conflict, the following measures can be taken.

(1) Slow Down Train n. As shown by the dashed line (n') in Figure 4(a), when the maximum operation speed of train n decreases, the requirement of arrival time interval at a station can be satisfied.

(2) Speed Up Train m. The dashed line (m') in Figure 4(b) demonstrates that increasing the operation speed of train m is a possible measure to make the arrival time interval at a station between two trains greater than the minimum value specified by the technical standard.

3.4. Track Number Conflict and Resolution. The total number of trains to arrive at or pass through a station cannot exceed the capacity that the station can provide. If a train is approaching or directly passing through a station, but there is no surplus track at the station to accommodate the train, the track number conflict will be brought about. The judgement rule of track number conflict is represented as

$$\sum_{m \in T} \alpha_{m,k} > M_k \quad (k \in S). \tag{4}$$

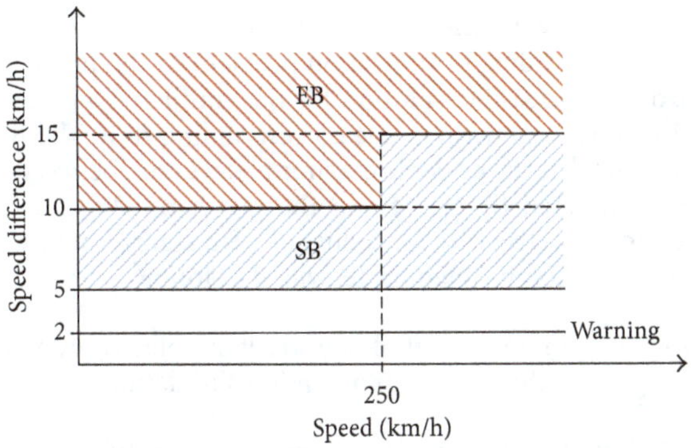

FIGURE 5: The application ranges of service braking and emergency braking.

When a track number conflict takes place at a station, the alternative strategy to resolve it is to command the related train to slow down towards the station or even stop outside the station.

4. Train Operation Simulation and Prediction

4.1. Train Operation Simulation. In the moving-block train control system, a train runs according to the preset logical processes. At first, the equipment of automatic train protection (ATP) constantly judges whether the current train speed exceeds the speed limit along a railway line. If an overspeed phenomenon occurs, according to the overspeed degree, the corresponding braking deceleration will be implemented. The more the speed exceeds a limit, the bigger deceleration will be adopted. Figure 5 demonstrates the application ranges of service braking (SB) and emergency braking (EB). On the other hand, the onboard control equipment of a train determines if braking should be applied according to the braking reference distance of a train to its front adjacent train. If braking is not applied, the train can accelerate towards the maximum speed or hold its current speed. Otherwise, the train will decelerate towards the target speed at the target point.

The symbols for train movements are defined in Notations section (The Symbols' Definition of Train Movement Model). The train movement model for simulation and prediction is described as follows, where the speed, acceleration, and deceleration are in the time unit, so the time unit is omitted in the rules of speed and position update.

(i) Speed Update

(1) IF ($v(i) \le 250$ km/h AND $v(i) - v_{\lim}(x(i)) > 5$ km/h AND $v(i) - v_{\lim}(x(i)) \le 10$ km/h) OR

(2) ($v(i) > 250$ km/h AND $v(i) - v_{\lim}(x(i)) > 5$ km/h AND $v(i) - v_{\lim}(x(i)) \le 15$ km/h)

(3) $v(i+1) = \max(v(i) - b_s, 0)$

(4) ENDIF

(5) IF ($v(i) \le 250$ km/h AND $v(i) - v_{\lim}(x(i)) > 10$ km/h) OR

(6) ($v(i) > 250$ km/h AND $v(i) - v_{\lim}(x(i)) > 15$ km/h)

(7) $v(i+1) = \max(v(i) - b_e, 0)$

(8) ENDIF

(9) IF $d_t(i) > d_r$, $v(i+1) = \min(v(i) + a, v_{\max})$

(10) ELSEIF $d_t(i) = d_r$, $v(i+1) = v(i)$

(11) ELSE

(12) $v(i+1) = \max(v(i) - b_s, v_t)$

(13) ENDIF

(ii) Position update

(14) $x(i+1) = x(i) + v(i)$

In the train movement model, sentences (1) to (4) indicate the service braking application case according to Figure 5, while sentences (5) to (8) represent the emergency braking application case. Sentence (9) implies that if the distance of a train to its operation target point is greater than the braking reference distance, the train will accelerate towards the maximum speed; sentence (10) shows the case that the train holds its current speed; sentence (12) denotes that the target speed for deceleration is v_t. Sentence (14) demonstrates the position update within the unit time.

4.2. Conflict Resolution in the Prediction Horizon. According to the discussion in Section 2, the conflict resolution algorithm in the prediction horizon is described as follows:

(1) IF $(t^D_{m,k} - t^D_{n,k})(t^A_{m,l} - t^A_{n,l}) < 0$ after prediction

(2) $v_m(i+1) = \min(v(i) + a, v^s_{\max})$ if train m accelerates,

(3) or $v_n(i+1) = \min(v(i) + a, v^b_{\max})$ if train n accelerates,

(4) or if there is a passing station, stop at the passing station for train m or n.

(5) ENDIF

(6) IF $|t^D_{m,k} - t^D_{n,k}| < T^{D\min}$

(7) Delay the departure time of train m or n.

(8) ENDIF

(9) IF $|t^A_{m,l} - t^A_{n,l}| < T^{A\min}$

(10) $v_m(i+1) = \min(v(i) + a, v^b_{\max})$ if train m accelerates,

(11) or $v_n(i+1) = \min(v(i) + a, v^s_{\max})$ if train n accelerates.

(12) ENDIF

(13) IF $\sum_{m \in T} \alpha_{m,k} > M_k$

(14) Train m decelerates towards or even stops outside the arrival station.

(15) END

In the above conflict resolution algorithm, sentences (1) to (5) deal with the section conflict, sentences (6) to (8) the departure time conflict, sentences (9) to (12) the arrival time conflict, and sentences (13) to (15) the track number conflict, which are based on Section 3. In sentences (9) to (12), train m or n runs at different maximum running speeds.

5. Scheduling Optimization

The objective of scheduling optimization is to generate new scheduling strategies if there exist potential operation conflicts or the deviation between the practical timetable and the preset timetable is too large. The new scheduling strategies must meet the following conditions:

(1) They resolve the existing operation conflicts

(2) There is no new conflict in the newly generated timetable

(3) The difference between the new timetable and the preset timetable is minimized.

Therefore, the objective function and constraint conditions are described as

$$\min \quad \sum_{j\in S}\sum_{m\in T}\left(w_m\left|t_{m,j}-t'_{m,j}\right|\right) \tag{5}$$

$$\text{s.t.} \quad \left|t_{m,k}-t_{n,k}\right| \geq T^{\min} \quad (m,n\in T,\ k\in B) \tag{6}$$

$$\left|t^{D}_{m,k}-t^{A}_{m,k}\right| \geq \mathrm{DT}^{\min}_{k} \quad (m\in T,\ k\in S) \tag{7}$$

$$\left|t^{D}_{m,k}-t^{D}_{n,k}\right| \geq T^{D\min} \quad (m,n\in T,\ k\in S) \tag{8}$$

$$\left|t^{A}_{m,k}-t^{A}_{n,k}\right| \geq T^{A\min} \quad (m,n\in T,\ k\in S) \tag{9}$$

$$\sum_{m\in T}\alpha_{m,k} \leq M_k \quad (k\in S), \tag{10}$$

where w_m is the weighting coefficient of train m. The higher the priority of train m is, the bigger the weight w_m is. Equation (6) represents the constraint of safe time interval between two adjacent trains arriving at a section to avoid section conflicts. Equation (7) implies that the dwelling time at a station cannot be less than the specified minimum value. Equation (8) points out the limitation condition of departure time interval between two trains from a station. Equation (9) shows the restraint of arrival time interval between two trains at a station. Equation (10) demonstrates the fact that the number of trains arriving at a station cannot exceed the capacity limit of the station.

Branch and Bound (B&B) algorithm is an important method to solve the combinatorial optimization problem. It searches the solution of a problem in the solution space tree. The core idea of the B&B algorithm is to expand the whole feasible solution space for the optimization problem with constraints and gradually decide the optimal solution satisfying constraint conditions.

The two core steps of the B&B algorithm are bounding and pruning. Bounding is to determine a reasonable range of objective function values, and the reasonable reference boundary is further provided to the pruning operation. For the maximization problem, bounding is to determine the lower bound of the objective function values. For the minimization problem, bounding is to find out the upper bound of the objective function values. The main function of pruning is to remove the branches that cannot produce the optimal solution so as to improve the search efficiency of the algorithm.

When applying the B&B algorithm to the conflict detection and scheduling optimization problem in the prediction horizon, each feasible solution for the conflict resolution in Section 3 has a new timetable corresponding to it. The B&B algorithm takes the possible maximum deviation between the newly generated timetable and the preset timetable as the upper bound (UB). The procedure of B&B algorithm to solve the optimization problem is described as follows.

Step 1. Determine the initial boundary of objective function values [0, UB] through train operation simulation.

Step 2. Initialize the list of live nodes to be empty.

Step 3. Take the place where the conflict occurs as a potential extension node to implement branch operation.

Step 4. Find the node with the smallest objective function value Val in the list of live nodes with regard to the newly generated tree.

Step 4.1. Calculate Val of a child node x in the list of live nodes according to train operation simulation.

Step 4.2. If Val $\leq$ UB, add the child node x to the list of live nodes.

Step 4.3. If the node x is the leaf node and its Val is the smallest in the list of live nodes, the corresponding scheduling strategy is optimal. Output the scheduling strategy, and the algorithm is finished.

Step 4.4. If node x *is* a leaf node but its Val is not the smallest in the list of live nodes, then let UB = Val, and, at the same time, delete the nodes whose Vals are greater than UB.

Step 4.5. Return to *Step 4.1.*

6. Case Study

The simulation is performed around a node, Changsha South Station in the China's "four vertical and four horizontal" high-speed railway network, as shown in Figure 6. Two railway lines are involved in the simulation, that is, Beijing-Guangzhou line (in red) and Shanghai-Kunming line (in green).

The layout of Changsha South Station is shown in Figure 7. At the entrance of 20 km away from the Changsha South Station, two railway lines are merged. Two platforms are provided at Changsha South Station. At the exit of 20 km away from the Changsha South Station, there exist two forks towards different terminal stations. The maximum speed of the trains on the Shanghai-Kunming line is 80 m/s and the dwelling time at Changsha South Station is 150 s. The maximum speed of the trains on the Beijing-Guangzhou line is 100 m/s and the dwelling time is 50 s. The minimum departure and arrival time intervals are 50 s. The minimum safe time interval is also 50 s. The prediction horizon is set to be 600 s, and the control horizon is equal to the prediction horizon. The sample period is configured to be equal to the simulation period.

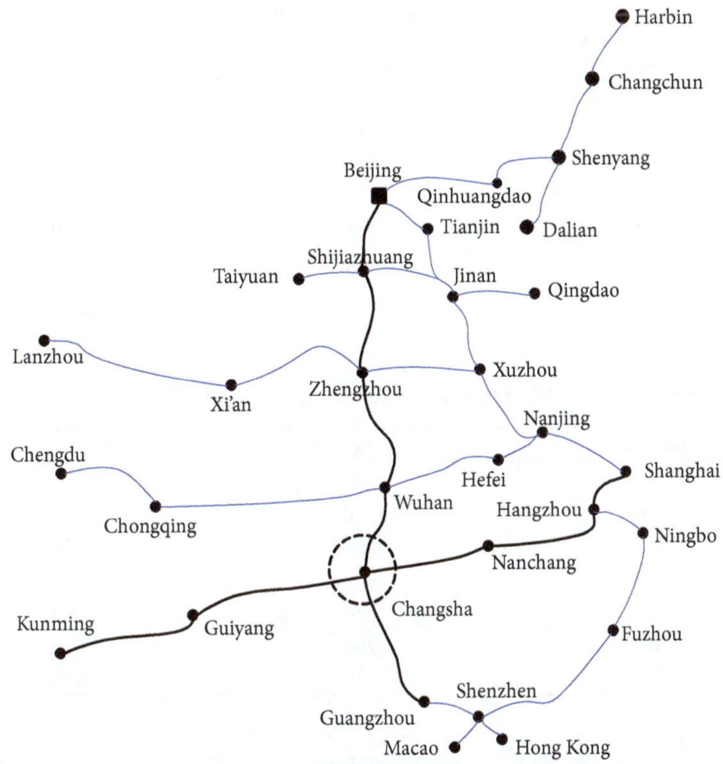

FIGURE 6: High-speed railway network.

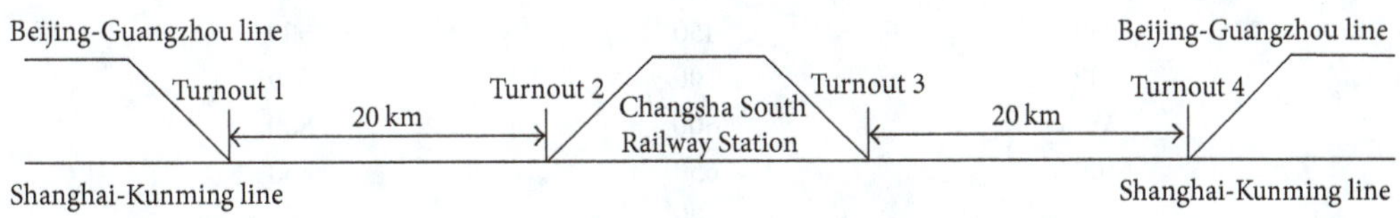

FIGURE 7: Changsha South Station layout.

In the simulation test, we set that there are four trains on the Beijing-Guangzhou line, and train numbers are 001, 003, 005, and 007. There are three trains on the Shanghai-Kunming line, and train numbers are 002, 004, and 006. The preset timetables of the trains are shown in Table 1.

The global implementation procedure of the MPC mechanism for conflict detection and scheduling optimization can refer to Section 2. An unexpected disturbance is configured, which causes a delay of 200 s for train 001 at turnout 1, based on the prediction of train operations considering current feedback information of train positions and speeds. According to the preset timetable, it should be train 004 which passes through turnout 1 at 200 s. However, due to the delay propagation in the whole network, train operation conflicts are gradually detected and resolved through the models and algorithms in Sections 3 and 4. Consequently, the new timetables are produced through the common First Come First Served (FCFS) and the B&B scheduling strategies represented in Section 5, which are listed in Tables 2 and 3, respectively. It can be calculated that the total delays of the trains are 4350 s and 4150 s in the timetables generated by FCFS and B&B strategies, respectively. The B&B strategy can achieve less delay than FCFS.

In order to further verify the validity and superiority of B&B algorithm used in the conflict detection and scheduling optimization, the delay times vary from 50 s to 350 s. The objective function and bound values for the B&B strategy are demonstrated in Table 4 and Figure 8. It can be seen that with the increase of the overall delay time, the B&B scheduling strategy will gradually demonstrate its superiority over the FCFS algorithm.

7. Conclusions

In this paper, the framework of conflict detection and scheduling optimization is proposed under the mechanism of MPC. Centering on the framework, conflict detection approaches and resolution strategies are outlined, a prediction model is established with conflict resolution, and the optimization model and algorithm are developed to unify the conflict resolution and scheduling optimization in the prediction horizon. A numerical experiment is carried out

Table 1: The preset timetables of the trains.

Train number	Turnout 1	Turnout 2	Turnout 3	Turnout 4
001	0	200	250	450
002	50	300	450	700
003	150	350	400	600
004	200	450	600	850
005	300	500	550	750
006	350	600	750	1000
007	450	650	700	900

Table 2: The obtained timetable through FCFS algorithm in case of delay.

Train number	Turnout 1	Turnout 2	Turnout 3	Turnout 4
001	200	400	450	650
002	250	500	650	900
003	300	550	600	800
004	350	600	750	1000
005	400	650	700	950
006	450	700	850	1100
007	500	750	800	1050

Table 3: The newly generated timetable through B&B algorithm in case of delay.

Train number	Turnout 1	Turnout 2	Turnout 3	Turnout 4
001	200	400	450	650
003	250	450	500	700
005	300	500	550	750
002	350	600	800	1050
004	400	650	850	1100
007	450	700	750	950
006	500	750	900	1150

Table 4: Objective function and bound values in case of different delays.

Delay time	50	100	150	200	250	300	350
FCFS	550	1600	2900	4350	5750	7150	8550
B&B	550	1600	2900	4150	5000	6600	8000
Bound value	1400	2800	4200	5600	7000	8400	9800

to justify the effectiveness of the proposed framework. The simulation results demonstrate that the proposed framework has a favorable performance.

Notations

The Symbols' Definition of Conflict Detection and Scheduling Optimization

T: Set of trains
S: Set of stations
B: Set of block sections in a railway line
m, n: Train numbers
j: Section number
k, l: Two adjacent station numbers
$T^{D\min}$: The minimum time interval between two trains departing from a station
$T^{A\min}$: The minimum time interval between two trains arriving at a station
$t^{D}_{m,k}$: The time of train m to depart from station k
$t^{A}_{m,k}$: The time of train m to arrive at station k
$\mathrm{DT}^{\min}_{k}$: The minimum dwelling time of a train at station k

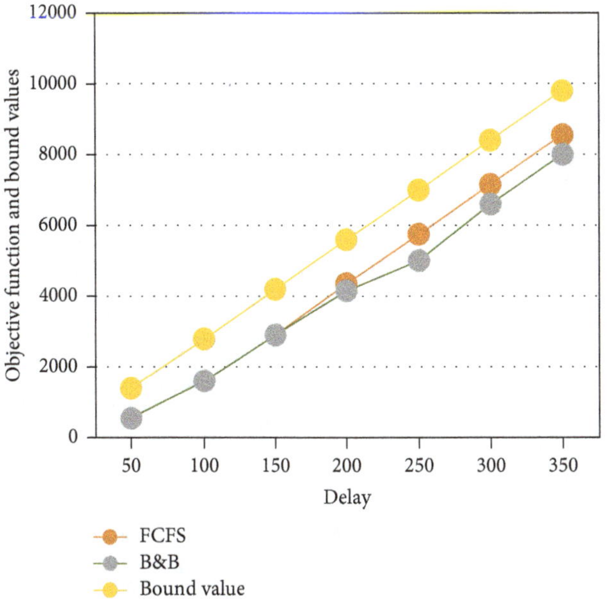

FIGURE 8: Simulation results in case of different delays.

M_k: The number of arrival-departure tracks at station k

$\alpha_{m,k}$: = 1, representing train m is now at station k; = 0, representing train m is now not at station k

$t_{m,j}$: The time that train m arrives at segment j in the new timetable

$t'_{m,j}$: The time that train m arrives at segment j in the preset timetable

$T^{\min}$: The minimum time interval of two adjacent trains arriving at the same section to meet the constraint of safe time interval.

The Symbols' Definition of Train Movement Model

$v(i)$: Train speed at instant i

$x(i)$: Train position at instant i

b_s: Deceleration of service braking

b_e: Deceleration of emergency braking

a: Acceleration of a train

$d_t(i)$: Distance of a train to its operation target point, that is, the front adjacent train or station, at instant i

v_t: Speed at operation target point

d_r: Braking reference distance

$v_{\max}$: The general maximum operation speed of a train

$v^b_{\max}$: The bigger maximum operation speed of a train

$v^s_{\max}$: The smaller maximum operation speed of a train.

Conflicts of Interest

All authors declare they have no conflicts of interest.

Acknowledgments

This work was financially supported by the National Natural Science Foundation of China (Grants no. 61673049 and no. 61074138) and the Fundamental Research Funds for the Central Universities of China (Grant no. 2017JBM302).

References

[1] B. Szpigel, "Optimal train scheduling on a single track railway," in *Operational Research '72*, M. Ross, Ed., pp. 343–352, Amsterdam, The Netherlands, 1973.

[2] J. Cordeau, P. Toth, and D. Vigo, "A survey of optimization models for train routing and scheduling," *Transportation Science*, vol. 32, no. 4, pp. 380–420, 1998.

[3] T.-W. Chiang, H.-Y. Hau, H.-M. Chiang, S.-Y. Ko, and C.-H. Hsieh, "Knowledge-based system for railway scheduling," *Data Knowledge Engineering*, vol. 27, no. 3, pp. 289–312, 1998.

[4] A. Caprara, M. Monaci, P. Toth, and P. L. Guida, "A Lagrangian heuristic algorithm for a real-world train timetabling problem," *Discrete Applied Mathematics*, vol. 154, no. 5, pp. 738–753, 2006.

[5] A. D'Ariano, M. Pranzo, and I. A. Hansen, "Conflict resolution and train speed coordination for solving real-time timetable perturbations," *IEEE Transactions on Intelligent Transportation Systems*, vol. 8, no. 2, pp. 208–222, 2007.

[6] A. D'Ariano, D. Pacciarelli, and M. Pranzo, "A branch and bound algorithm for scheduling trains in a railway network," *European Journal of Operational Research*, vol. 183, no. 2, pp. 643–657, 2007.

[7] M. A. Salido, M. Abril, F. Barber, L. Ingolotti, P. Tormos, and A. Lova, "Domain-dependent distributed models for railway scheduling," *Knowledge-Based Systems*, vol. 20, no. 2, pp. 186–194, 2007.

[8] X. Zhou and M. Zhong, "Single-track train timetabling with guaranteed optimality: branch-and-bound algorithms with enhanced lower bounds," *Transportation Research Part B: Methodological*, vol. 41, no. 3, pp. 320–341, 2007.

[9] V. Cacchiani, A. Caprara, and P. Toth, "A column generation approach to train timetabling on a corridor," *4OR*, vol. 6, no. 2, pp. 125–142, 2008.

[10] M. B. Khan and X. Zhou, "Stochastic optimization model and solution algorithm for robust double-track train-timetabling problem," *IEEE Transactions on Intelligent Transportation Systems*, vol. 11, no. 1, pp. 81–89, 2010.

[11] S. Mu and M. Dessouky, "Scheduling freight trains traveling on complex networks," *Transportation Research Part B: Methodological*, vol. 45, no. 7, pp. 1103–1123, 2011.

[12] L. Yang, Z. Gao, and K. Li, "Railway freight transportation planning with mixed uncertainty of randomness and fuzziness," *Applied Soft Computing*, vol. 11, no. 1, pp. 778–792, 2011.

[13] Ö. Yalçinkaya and G. M. Bayhan, "A feasible timetable generator simulation modelling framework for train scheduling problem," *Simulation Modelling Practice and Theory*, vol. 20, no. 1, pp. 124–141, 2012.

[14] J. T. Krasemann, "Design of an effective algorithm for fast response to the re-scheduling of railway traffic during disturbances," *Transportation Research Part C: Emerging Technologies*, vol. 20, no. 1, pp. 62–78, 2012.

[15] R. Chevrier, P. Pellegrini, and J. Rodriguez, "Energy saving in railway timetabling: a bi-objective evolutionary approach for computing alternative running times," *Transportation Research Part C: Emerging Technologies*, vol. 37, pp. 20–41, 2013.

[16] J. L. Espinosa-Aranda and R. García-Ródenas, "A demand-based weighted train delay approach for rescheduling railway networks in real time," *Journal of Rail Transport Planning and Management*, vol. 3, no. 1-2, pp. 1–13, 2013.

[17] X. Li, B. Shou, and D. Ralescu, "Train rescheduling with stochastic recovery time: A new track-backup approach," *IEEE Transactions on Systems, Man, and Cybernetics: Systems*, vol. 44, no. 9, pp. 1216–1233, 2014.

[18] L. Kang, J. Wu, H. Sun, X. Zhu, and B. Wang, "A practical model for last train rescheduling with train delay in urban railway transit networks," *Omega* , vol. 50, pp. 29–42, 2015.

[19] S. Zhan, L. G. Kroon, L. P. Veelenturf, and J. C. Wagenaar, "Real-time high-speed train rescheduling in case of a complete blockage," *Transportation Research Part B: Methodological*, vol. 78, pp. 182–201, 2015.

[20] D. Šemrov, R. Marsetič, M. Žura, L. Todorovski, and A. Srdic, "Reinforcement learning approach for train rescheduling on a single-track railway," *Transportation Research Part B: Methodological*, vol. 86, pp. 250–267, 2016.

[21] B. Kersbergen, T. van den Boom, and B. De Schutter, "Distributed model predictive control for railway traffic management," *Transportation Research Part C: Emerging Technologies*, vol. 68, pp. 462–489, 2016.

[22] A. Bettinelli, A. Santini, and D. Vigo, "A real-time conflict solution algorithm for the train rescheduling problem," *Transportation Research Part B: Methodological*, vol. 106, pp. 237–265, 2017.

[23] J. Eaton, S. Yang, and M. Gongora, "Ant colony optimization for simulated dynamic multi-objective railway junction rescheduling," *IEEE Transactions on Intelligent Transportation Systems*, vol. 18, no. 11, pp. 2980–2992, 2017.

[24] A. W. Hermansson and S. Syafiie, "Model predictive control of pH neutralization processes: A review," *Control Engineering Practice*, vol. 45, pp. 98–109, 2015.

[25] E. Camacho and C. Bordons, *Model Predictive Control in The Process Industry*, Spinger Press, London, UK, 1995.

[26] D. Liu, *Study on High-Speed Train Operation Conflict Detection and Scheduling Optimization [Master, Thesis]*, Beijing Jiaotong University Master Thesis, 2015.

[27] Z. Wang, *Study on the Model, Algorithm And Simulation Platform of Conflict Detection And Real-Time Scheduling of High-Speed Railway Train Operation [Master, Thesis]*, Beijing Jiaotong University Master Thesis, 2017.

4

Backstepping Control with Disturbance Observer for Permanent Magnet Synchronous Motor

Yong-Hong Lan and Lei-Zhou

College of Information Engineering, Xiangtan University, Xiangtan, Hunan 411105, China

Correspondence should be addressed to Yong-Hong Lan; 961501563@qq.com

Academic Editor: Darong Huang

For the speed tracking control problem of Permanent Magnet Synchronous Motor (PMSM), a disturbance observer-based (DOB) backstepping speed tracking control method is presented in this paper. First, to reconstruct the load disturbance, a slowly time-varying disturbance observer for PMSM is constructed. Next, by stability theory, the norm of the observation error and disturbance observer design method are obtained. On the basis of these, through the designing of the virtual control input that include the reconstruction disturbances, and using backstepping control strategy, the DOB controller of the closed-loop system is proposed. The obtained controller can achieve high precision speed tracking and disturbance rejection. Finally, some results of evaluative experiments verified the effectiveness of the proposed method for high-performance control and disturbance rejection for the PMSM drive.

1. Introduction

Permanent magnet synchronous motor (PMSM) is receiving increasing attention in high-performance industrial servo applications due to its advantages such as high torque to current ratio, super power density, and fast response, as well as low noise [1, 2]. However, PMSM is a typical high nonlinear, multivariable coupled system. It is sensitive to load disturbance, unmodeled dynamics, nonlinear uncertainties, and parameter uncertainties.

Over the last decades, various design methods have been developed for PMSM control systems, such as adaptive control [3], robust control [4], neural network control [5], predictive control [6], and so on. Recently, to improve the control performances in PMSM systems, much attention has been given to disturbance reject-based control method [7], which are insensitive to parameter variations and external disturbances. In [8, 9], the fractional order sliding-mode control (SMC) scheme has been proposed. The proposed control system not only obtained high control performance but also was robust against to external load disturbance and parameter variations.

To further improve the disturbance rejection performance of SMC, extended sliding-mode disturbance observer was proposed in [10, 11], and the estimated system disturbance is considered as the feed forward compensation part to compensate sliding-mode speed controller. In [12], an improved SMC scheme to solve time-varying parameters and disturbances for PMSM drive system was proposed. The new speed controller was designed by the nonsingular terminal SMC strategy with the disturbance observer. In [13], a new speed controller is designed by the nonsingular terminal SMC strategy with disturbance observer. The controller can make the motor speed reach the reference value in finite time, accompanied with a faster convergence and a better tracking precision. It is worth noting that one obvious disadvantage of SMC method is the chattering phenomenon, which is caused by discontinuous control law and frequent switching action near sliding surface. Besides, as the upper bound of lumped disturbances is not easy to be determined in advance, which could cause a large amount of chattering of SMC strategy in PMSM system.

Recently, disturbance observer-based (DOB) [14] control methods have been applied to PMSM system for better robustness against system disturbance. In [15], a DOB state feedback controller was designed for PMSM system. By using the same disturbance observer, a sensorless control method for PMSM drive was developed in [16]. The proposed DOBC method involved the use of a back electromotive force observer and a torque observer to estimate rotor position and compensate for load torque disturbance, respectively. For the mismatched disturbance, in [17], a DOB integral sliding-mode control approach for linear systems with mismatched disturbances was presented. The disturbance observer is proposed to generate the disturbance estimate, which can be incorporated in the controller to counteract the disturbance.

Backstepping is a well-known recursive and systematic design methodology for the feedback control of uncertain nonlinear system with matched uncertainties [18]. The key point is to use the virtual control variable to make the original high-order system simple; thus the final controller can be derived through Lyapunov stable theorem. In [19], an adaptive backstepping speed controller was proposed for the speed control of PMSM. The controller is robust against stator resistance, viscous friction uncertainties, and load torque disturbance. However, this approach uses the feedback linearization. By means of a nonlinear and adaptive backstepping design method, a speed and current control scheme for PMSM was presented in [20], in which all the parameters in both PMSM and load dynamics were considered unknown. In [21], a new nonlinear and full adaptive backstepping speed tracking control scheme was developed for an uncertain PMSM. Except for the number of pole pairs, all the other parameters in both PMSM and load dynamics were assumed unknown. Taking into account the unobservable of the systems states, a backstepping control method for speed sensorless PMSM based on slide model observer was proposed in [22], in which the slide model observer was designed by using slide model control technique and phase loop lock (PLL) method. In [23], a backstepping control algorithm based on disturbance observer was proposed. The minimum-order observer was established to observe the disturbance value of load inertia. Furthermore, the estimated disturbance value was used to identify the load inertia. Based on the identification results, the backstepping controller was designed. However, the above observer and controller are designed separately.

Motivated by the discussions above, in this paper, we mainly investigate backstepping speed control for PMSM based on disturbance observer. A nonlinear disturbance observer is first constructed to estimate the external slowly time-varying disturbance. Then, based on the backstepping control theory, the PMSM rotor speed and current tracking backstepping controllers are designed. Meanwhile, global asymptotic stability is guaranteed by Lyapunov stability analysis.

The rest of this paper is structured as follows. In Section 2, the mathematic model of PMSM and problem formulation are presented. The nonlinear disturbance observer design and stability analysis are derived in Section 3. In Section 4, the DOB backstepping controller design method is obtained. Section 5 presents the numerical simulation and experimental results. Finally, some conclusions are drawn in Section 6.

2. Mathematic Model of PMSM and Problem Formulation

Assume that magnetic circuit is unsaturated and hysteresis and eddy current loss are ignored. With above standard assumptions, the mathematical model of a conventional surface mounted PMSM with mismatched external disturbances can be given in the $d-q$ frame as follows [20, 22]:

$$\begin{aligned}\frac{d\omega}{dt} &= \frac{3p\phi_f}{2J}i_q - \frac{B}{J}\omega - \frac{T_L}{J},\\ \frac{di_q}{dt} &= -\frac{R}{L}i_q - p\omega i_d - \frac{p\phi_f}{L}\omega + \frac{1}{L}u_q + \frac{1}{L}d_1,\\ \frac{di_d}{dt} &= -\frac{R}{L}i_d + p\omega i_q + \frac{1}{L}u_d + \frac{1}{L}d_2,\end{aligned} \tag{1}$$

where ω is the rotor speed, i_d and i_q are the $d-q$ axis currents, u_d and u_q are the $d-q$ axis voltages, and d_i $(i = 1, 2)$ are external disturbances. R and L denote the stator resistance and inductance per phase, respectively, p is the number of pole pairs, ϕ_f is the permanent magnet flux, J is the rotor moment of inertia, B is the viscous friction factor, and T_L also represents the applied load torque disturbance.

To formulate the design problem, according to system (1), the state space model of the PMSM can be rewritten as the following nonlinear system:

$$\begin{aligned}\dot{x}(t) &= Ax(t) + f(x(t)) + Bu(t) + \eta + Bd(t),\\ y(t) &= Cx(t),\end{aligned} \tag{2}$$

where $x(t) = [\omega\ i_q\ i_d]^T$, $u = [u_q\ u_d]^T$ and

$$A = \begin{bmatrix} -\frac{B}{J} & \frac{3p\phi_f}{2J} & 0 \\ -\frac{p\phi_f}{L} & -\frac{R}{L} & 0 \\ 0 & 0 & -\frac{R}{L} \end{bmatrix},$$

$$B = \begin{bmatrix} 0 & 0 \\ \frac{1}{L} & 0 \\ 0 & \frac{1}{L} \end{bmatrix},$$

$$C = [1, 0, 0],$$

$$f(x(t)) = \begin{bmatrix} 0 \\ -p\omega i_d \\ -p\omega i_d \end{bmatrix},$$

$$\eta = \begin{bmatrix} -\frac{T_L}{J} \\ 0 \\ 0 \end{bmatrix}$$

$$d(t) = \begin{bmatrix} 0 \\ d_1 \\ d_1 \end{bmatrix}. \quad (3)$$

For the PMSM control system (2), we give the following assumptions.

Assumption 1. The external disturbance $d(t)$ is assumed that $d(t) \le \alpha, \dot{d}(t) = 0$, where α is a positive constant.

Assumption 2. For all $x_1, x_2 \in \Omega_1$ the nonlinear term $f(x(t))$ in system (2) satisfies

$$\|f(x_1) - f(x_2)\| \le \gamma \|x_1 - x_2\|, \quad (4)$$

where γ is Lipschitz constant.

The main control objective is to design a DOB backstepping controller to keep all the signals in the closed-loop system bounded and ensures global asymptotic convergence of the desired speed and current tracking errors to zero eventually.

3. Nonlinear Disturbance Observer and Stability Analysis

Motivated by the disturbance observers in [14–16], the following disturbance observer for system (2) can be employed:

$$\begin{aligned} \dot{v}(t) &= \Lambda \hat{d}(t) + \Lambda B^{\dagger} (Ax(t) + f(x(t)) + Bu(t) + \eta), \\ \hat{d}(t) &= v(t) - \Lambda B^{\dagger} x(t), \end{aligned} \quad (5)$$

where $B^{\dagger} = (B^T B)^{-1} B^T$, $\hat{d}(t)$ is the estimation of $d(t)$, $v(t)$ is the internal variable vector of the observer, and Λ is a Hurwitz matrix chosen by designer. For simplicity, we assume $\hat{d}(0) = 0$. Define the disturbance estimation error $\tilde{d}(t)$ as

$$\tilde{d}(t) = d(t) - \hat{d}(t). \quad (6)$$

To show that the disturbance can be observed asymptotically by observer (5), we first give the following lemma.

Lemma 3. *For nonlinear system (2), suppose that the disturbance observer holds form (5); then the disturbance can be observed asymptotically.*

Proof. It can be obtained that

$$\dot{v}(t) = \Lambda \hat{d}(t) + \Lambda B^{\dagger} \dot{x}(t) - \Lambda d(t), \quad (7)$$

which implies that $\dot{\hat{d}}(t) = -\Lambda \tilde{d}(t)$. Therefore, we have

$$\dot{\tilde{d}}(t) = \Lambda \tilde{d}(t). \quad (8)$$

Define a monochromatic Lyapunov function as

$$V_o(t) = \frac{1}{2} \tilde{d}^T(t) \tilde{d}(t). \quad (9)$$

Clearly,

$$\dot{V}_o(t) \le \tilde{d}^T(t) \left(\Lambda + \Lambda^T \right) \tilde{d}(t). \quad (10)$$

Since Λ is Hurwitz, then

$$\dot{V}_o(t) \le 2\lambda_{max}(\Lambda) \|\tilde{d}(t)\|^2 = -\beta \|\tilde{d}(t)\|^2, \quad \beta > 0, \quad (11)$$

which implies the error dynamics (8) is asymptotically stable. This completes the proof. □

Remark 4. It follows from Lemma 3 that parameter Λ in the disturbance observer (5) can change the rate of convergence of the observer and should be selected as large enough.

4. DOB Backstepping Controller Design

In the backstepping procedure, a virtual control state is firstly defined and then it is forced to become a stabilizing function. By defining a corresponding error variable and using Lyapunov stability theory, the related control input can be obtained to stabilize the error dynamics [21]. In this paper, the overall control design can be established by three steps in the following order.

Step 1. In the first step of the backstepping control design, a fictitious control input for the rotor speed ω has to be determined. Let ω_r^* be the desired trajectory. Furthermore, the trajectory of ω_r^* is sketched to be smooth and $\dot{\omega}_r^* = 0$. Define the speed tracking error $e_\omega = \omega_r^* - \omega$. The speed tracking error dynamics can be obtained as

$$\dot{e_\omega} = \dot{\omega}_r^* - \dot{\omega} = \frac{1}{J} \left(B\omega + T_L - \frac{3p\phi_f}{2} i_q \right). \quad (12)$$

The first positive definite Lyapunov function can be defined as

$$V_1(t) = \frac{1}{2} e_\omega^2 + V_o(t). \quad (13)$$

Differentiating $V_1(t)$ with respect to time and using the results of (11) and (12), the time derivative of $V_1(t)$ is given as

$$\begin{aligned} \dot{V}_1(t) &= e_\omega \dot{e_\omega} + \dot{V}_o(t) \\ &\le \frac{e_\omega}{J} \left(B\omega + T_L - \frac{3p\phi_f}{2} i_q \right) - \beta \|\tilde{d}(t)\|^2. \end{aligned} \quad (14)$$

In order to stabilize the speed tracking error dynamics, define the stabilizing virtual control input:

$$i_q = \frac{2}{3p\phi_f} (B\omega + T_L + k_1 J e_\omega), \quad (15)$$

where $k_1 > 0$ is a real constant. Therefore, this definition leads to

$$\dot{V}_1(t) \le -k_1 e_\omega^2 - \beta \|\tilde{d}(t)\|^2 \le 0, \quad (16)$$

which guarantees that the tracking error for rotor speed will converge asymptotically.

Step 2. In order to realize the complete decoupling and speed tracking of PMSM, the virtual input current can be chosen as

$$i_q^* = \frac{2}{3p\phi_f}\left(B\omega + T_L + k_1 J e_\omega\right), \quad i_d^* = 0. \tag{17}$$

Define the tracking error of q axis current as $e_q = i_q^* - i_q$. The time derivative of e_q is

$$\begin{aligned} \dot{e}_q &= \frac{i_q^*}{dt} - \frac{di_q}{dt} = \frac{2}{3p\phi_f}\left(B\frac{d\omega}{dt} + k_1 J\frac{de_\omega}{dt}\right) - \frac{di_q}{dt} \\ &= \frac{2\left(B - k_1 J\right)}{3P\phi_f J}\left(\frac{3p\phi_f}{2}i_q - B\omega - T_L\right) + \frac{Ri_q}{L} + p\omega i_d \\ &\quad - \frac{u_q}{L} + \frac{p\phi_f}{L}\omega - \frac{\widetilde{d_1}}{L} - \frac{\widehat{d_1}}{L}, \end{aligned} \tag{18}$$

where $\widetilde{d_1} + \widehat{d_1} = d_1$.

Choose the second Lyapunov function to stabilize q axis current tracking error dynamics as

$$V_2(t) = V_1(t) + \frac{1}{2}e_q^2. \tag{19}$$

By some mathematical manipulation, the time derivative of $V_2(t)$ is given by

$$\begin{aligned} \dot{V}_2(t) &= \dot{V}_1(t) + e_q\dot{e}_q \le -k_1 e_\omega^2 \\ &\quad + e_q\left[\frac{2\left(B - k_1 J\right)}{3p\phi_f J}\left(\frac{3p\phi_f J}{2}i_q - B\omega - T_L\right) + \frac{Ri_q}{L}\right. \\ &\quad \left. + p\omega i_d - \frac{u_q}{L} + \frac{p\phi_f}{L}\omega - \frac{\widehat{d_1}}{L}\right] - \frac{\widetilde{d_1}}{L}e_q - \beta\left\|\tilde{d}(t)\right\|^2. \end{aligned} \tag{20}$$

Setting

$$\begin{aligned} &\frac{2\left(B - k_1 J\right)}{3p\phi_f J}\left(\frac{3p\phi_f J}{2}i_q - B\omega - T_L\right) + \frac{Ri_q}{L} + p\omega i_d \\ &\quad - \frac{u_q}{L} + \frac{p\phi_f}{L}\omega - \frac{\widehat{d_1}}{L} = -k_2 e_q, \end{aligned} \tag{21}$$

where $k_2 > 0$ and using the generic inequality $\pm ab \le \varepsilon_1 a^2 + (1/4\varepsilon_1)b^2(\varepsilon_1 > 0)$, it yields

$$\begin{aligned} \dot{V}_2(t) &\le -k_1 e_\omega^2 - k_2 e_q^2 - \frac{\widetilde{d_1}}{L}e_q - \beta\left\|\tilde{d}(t)\right\|^2 \\ &\le -k_1 e_\omega^2 - k_2 e_q^2 + \frac{1}{L}\left(\varepsilon_1 e_q^2 + \frac{1}{4L\varepsilon_1}\left\|\tilde{d}_1(t)\right\|^2\right) \\ &\quad - \beta\left\|\tilde{d}(t)\right\|^2 = -k_1 e_\omega^2 - C_1 e_q^2 - C_2\left\|\tilde{d}(t)\right\|^2, \end{aligned} \tag{22}$$

where

$$\begin{aligned} C_1 &= k_2 - \frac{\varepsilon_1}{L}, \\ C_2 &= \beta - \frac{1}{4L\varepsilon_1}. \end{aligned} \tag{23}$$

If parameters k_2 and ε_1 are properly selected such that $C_1 > 0$ and $C_2 > 0$, then $\dot{V}_2(t) < 0$, which indicates the tracking error for q-axis stator current will converge asymptotically to zero.

On the other hand, from (21), the stabilizing control law u_q can be designed to stabilize q axis current tracking error dynamics as follows:

$$\begin{aligned} u_q^* &= L\left[\left(\frac{B}{J} + \frac{R}{L}\right)i_q - \frac{2B^2}{3p\phi_f J}\omega + \frac{p\phi_f}{L}\omega - \frac{2BT_L}{3p\phi_f J}\right. \\ &\quad \left. + p\omega i_d - \frac{2k_1^2 J}{3p\phi_f}e_\omega + k_2 e_q - \frac{\widehat{d_1}}{L}\right). \end{aligned} \tag{24}$$

Step 3. As to design of command input for u_d, define the tracking error

$$e_d = i_d^* - i_d \tag{25}$$

with $\widehat{i}_d^* = 0$ being the desired stator current of d axis. The time derivative of e_d is

$$\dot{e}_d = \frac{di_d^*}{dt} - \frac{di_d}{dt} = \frac{R}{L}i_d - p\omega i_d - \frac{1}{L}u_d - \frac{\widetilde{d_2}}{L} - \frac{\widehat{d_2}}{L}, \tag{26}$$

where $\widetilde{d_2} + \widehat{d_2} = d_2$.

Choose the last Lyapunov function candidate as

$$V_3(t) = V_2(t) + \frac{1}{2}e_d^2, \tag{27}$$

which results in

$$\begin{aligned} \dot{V}_3(t) &= \dot{V}_2(t) + e_d\dot{e}_d \\ &\le -k_1 e_\omega^2 - C_1 e_q^2 - C_2\left\|\tilde{d}(t)\right\|^2 \\ &\quad + e_d\left(\frac{R}{L}i_d - p\omega i_d - \frac{1}{L}u_d - \frac{\widehat{d_2}}{L}\right) - \frac{\widetilde{d_2}}{L}e_d. \end{aligned} \tag{28}$$

If the stabilizing control law for u_d is defined as

$$u_d^* = Ri_d - pL\omega i_q + Lk_3 e_d - \frac{\widehat{d_2}}{L}, \quad k_3 > 0, \tag{29}$$

then the following result is obtained:

$$\dot{V}_3(t) \le -k_1 e_\omega^2 - C_1 e_q^2 - C_2\left\|\tilde{d}(t)\right\|^2 - k_3 e_d^2 - \frac{\widetilde{d_2}}{L}e_d. \tag{30}$$

Again, based on the generic inequality, inequality (30) can be rewritten as follows:

$$\begin{aligned} \dot{V}_3(t) &\le -k_1 e_\omega^2 - C_1 e_q^2 - C_2\left\|\tilde{d}(t)\right\|^2 - k_3 e_d^2 + \frac{\varepsilon_2}{L}e_d^2 \\ &\quad + \frac{1}{4L\varepsilon_2}\left\|\tilde{d}_2(t)\right\|^2 \\ &= -k_1 e_\omega^2 - C_1 e_q^2 - \left(k_3 - \frac{\varepsilon_2}{L}\right)e_d^2 \\ &\quad - \left(C_2 - \frac{1}{4L\varepsilon_2}\right)\left\|\tilde{d}(t)\right\|^2. \end{aligned} \tag{31}$$

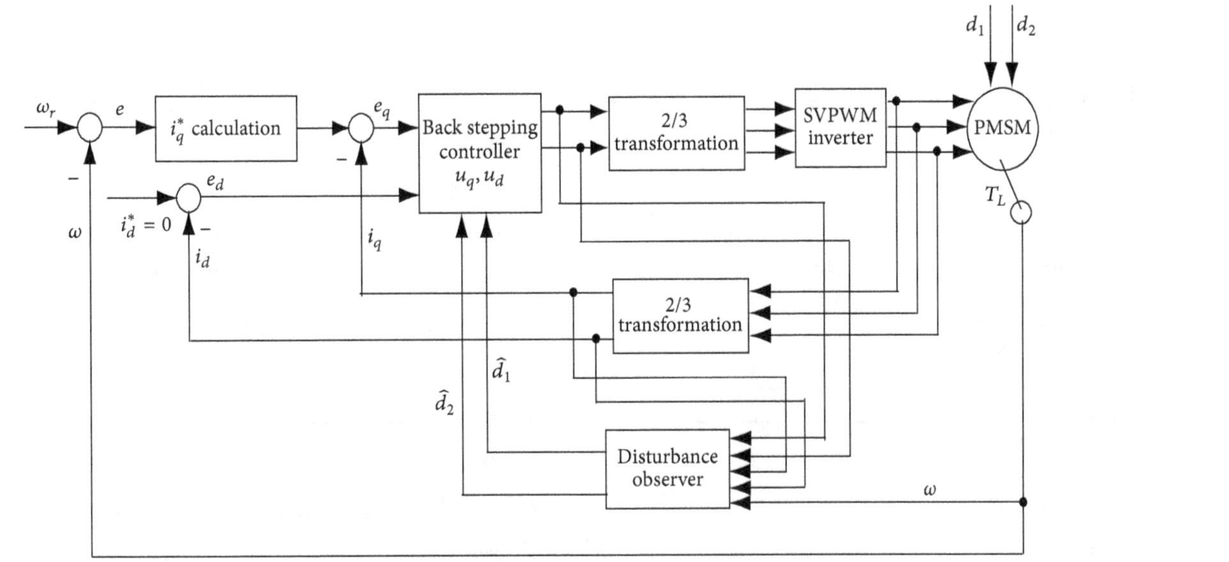

FIGURE 1: Block diagram of the proposed DOB backstepping PMSM control system.

If parameters k_1, k_2, k_3 and $\varepsilon_1, \varepsilon_2$ are properly chosen to make the following conditions hold:

$$\begin{aligned} & k_1 > 0, \\ & \left(k_2 - \frac{\varepsilon_1}{L}\right) = C_1 > 0, \\ & \left(\beta - \frac{1}{4L\varepsilon_1}\right) = C_2 > 0, \\ & \left(k_3 - \frac{\varepsilon_2}{L}\right) > 0, \\ & \left(C_2 - \frac{1}{4L\varepsilon_2}\right) > 0, \end{aligned} \tag{32}$$

then $\dot{V}_3(t) < 0$. Consequently, the tracking error for d-axis stator current will tend to zero eventually. The objective of tracking control of PMSM is completed.

5. Numerical Simulation and Experimental Results

In this section, the numerical example and experimental results are presented to demonstrate the validity of the proposed method.

(1) Numerical Simulation Results. The motor parameters used are listed in Table 1. The MATLAB/Simulink model of the proposed DOB backstepping PMSM control system is shown in Figure 1. The initial rotation speed of the motor is 1200 r/min, and the rotation speed is 1000 r/min at 0.6s. The initial load torque of the motor is 0N·m and the load disturbance torque is 10N·m at 0.8s. The parameters of the DOB-backstopping controller are selected as $\Lambda = -180I_2$, $\varepsilon_1 = \varepsilon_2 = 0.153$, $k_1 = 250$, $k_2 = 500$, $k_3 = 160$.

In this simulation, to illustrate the effectiveness of the proposed method, the phase traces by the conventional backstepping (BS) method [22] and the proposed DOB backstepping (DOB-BS) method are simulated and compared. The numerical simulation results are shown in Figures 2–4. Figure 2(a) indicates the actual rotation speed of the motor with BS method and the proposed DOB-BS method in the presence of the above disturbances load, respectively. Figure 2(b) shows the rotation speed response between $[0.008s, 0.012s]$, during which the PMSM motor just started. Figure 2(c) shows the rotation speed response between $[0.585s, 0.625s]$. In this time period, the reference rotation speed has a sudden change at $t = 0.6s$. Figure 2(d) shows rotation speed response between $[0.785s, 0.825s]$, in which the external disturbance load is added at $t = 0.8s$. From Figure 2, it can be seen that the rotation speed of the proposed DOB-BS method can rapidly track the reference rotation speed with smaller stability error, faster response, and smaller overshoot than that of the conventional BS method.

Figure 3 shows the three-phase current of the stator. The current amplitude is proportional to the rotation torque and changes rapidly as the load torque varies. The current frequency is inversely proportional to the rotation speed. Figure 4 shows the variation of electromagnetic torque as the load torque changes. Obviously, the DOB-BS method gives less load torque fluctuations.

TABLE 1: The parameters of PMSM.

Parameter	numerical value
Pole Pairs p	3
Friction factor B $(N \cdot m \cdot s/rad$	0.001
Stator Inductance L (H)	0.0153
Rotor moment of inertia J $(kg \cdot m^2)$	0.0021
Permanent magnet flux ϕ_f (wb)	0.82
Stator Resistance R (Ω)	0.56

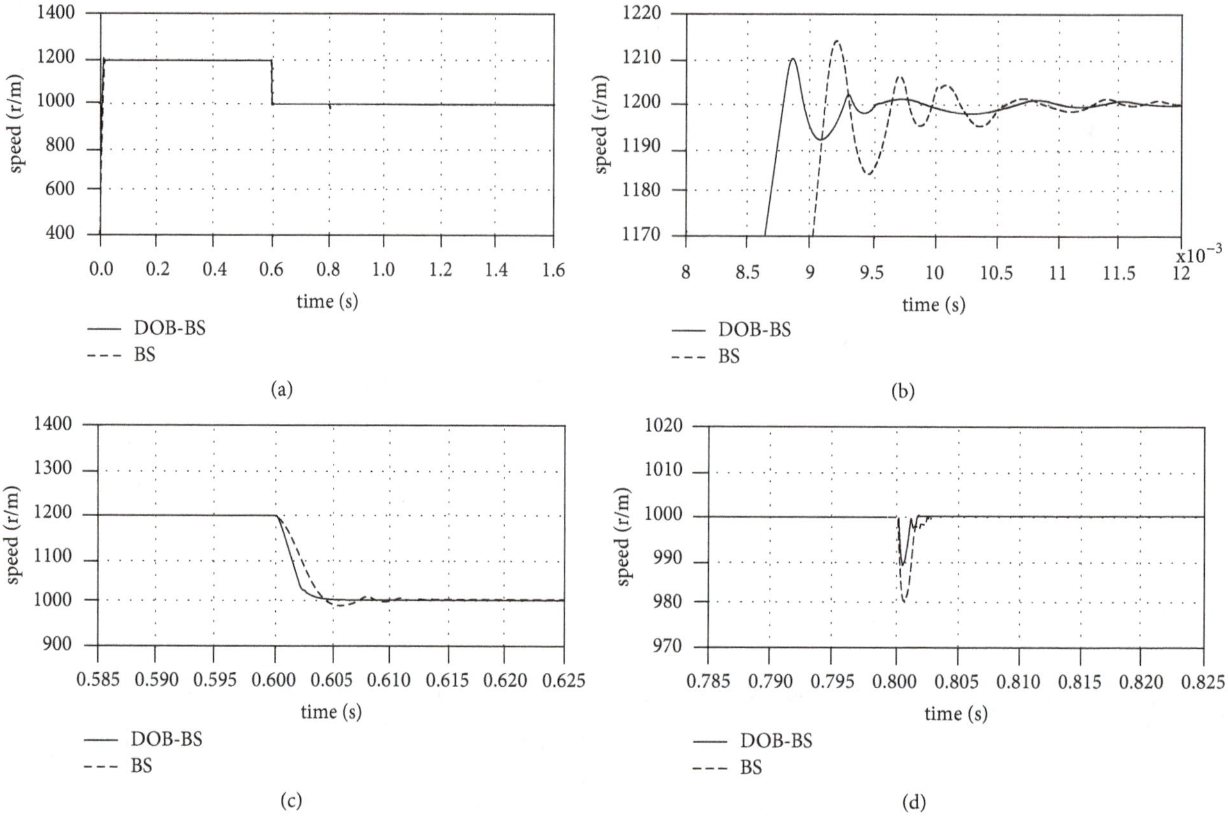

Figure 2: Rotation speed responses.

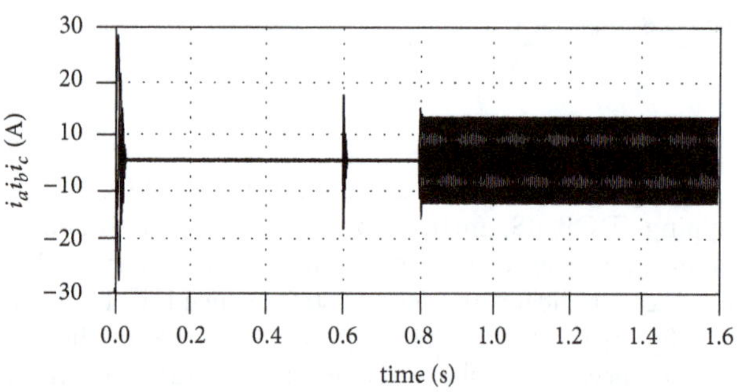

Figure 3: Three-phase stator current.

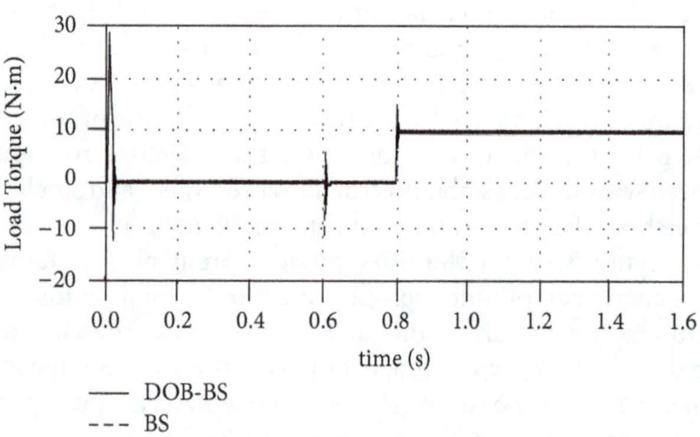

Figure 4: Load torque responses.

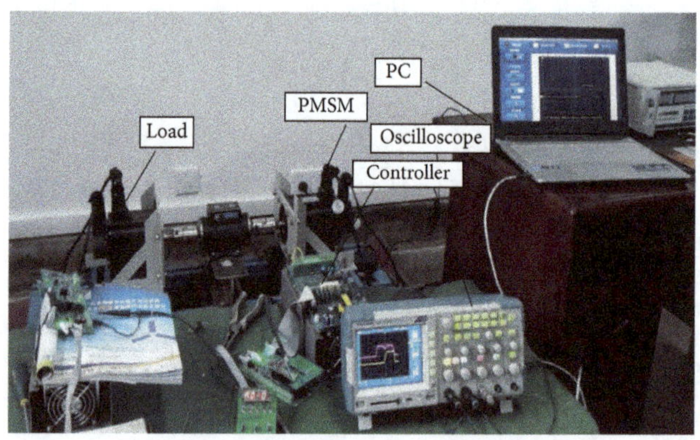

Figure 5: Load torque responses.

(2) Experimental Results. To evaluate the performance of the proposed method, a three-phase PMSM control system is set up. The experimental platform configuration is shown in Figure 5. The main chip of the inverter adopts the TMS320F28335 digital signal processor (DSP). The initial rotation speed of the motor is 1600 r/min, and the rotation speed is 600 r/min at 0.6*s*. The initial load torque of the motor is 0N·m and the load disturbance torque is 5N·m at 0.8*s*.

The results of the experiment are shown in Figures 6–8. Figure 6(a) shows the speed response of the closed-loop system under the BS and DOB-BS schemes. It can be seen

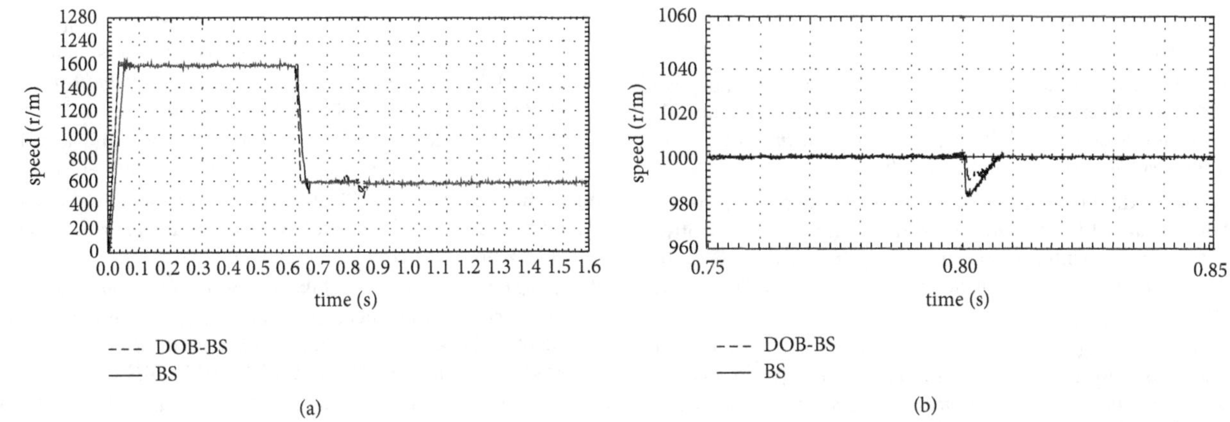

FIGURE 6: Rotation speed responses.

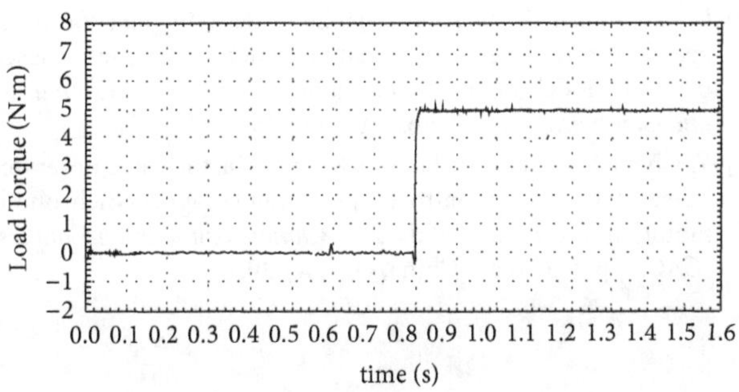

FIGURE 7: Load torque response.

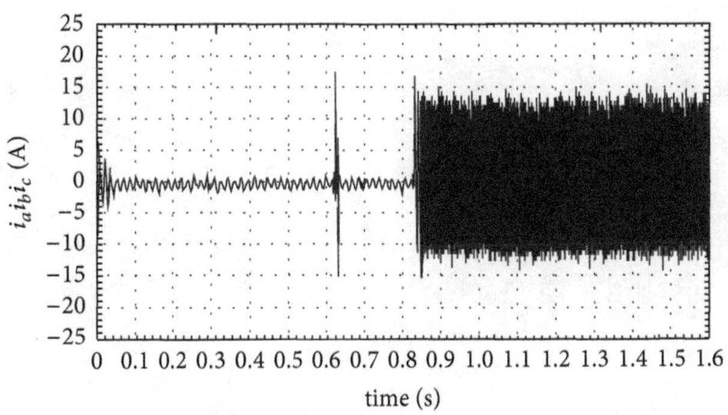

FIGURE 8: Three-phase stator current.

that the settling time of the DOB-BS method is shorter than that of the BS method. The experimental results of antiload disturbance of the two control schemes are shown in Figure 6(b). After the speed output of PMSM system (without load) is in the steady state, a step disturbance load, i.e., a rated torque of 10N.m at 0.8s, is added suddenly. Clearly, the DOB-BS method has a less speed drop, i.e., a better disturbance rejection capacity when the disturbance load is added.

The load torque and current responses of DOB-BS method is shown in Figures 7 and 8, which illustrates the quick response and low ripple.

From these results, it can be observed that the proposed DOB-BS method is effective under different operating conditions and has a better performance at most conditions than that of BS method.

6. Conclusion

In this paper, a disturbance observer-based (DOB) backstepping speed tracking control method has been presented for the speed tracking control PMSM. Through disturbance estimation, the DOB backstepping control strategy can achieve high precision speed tracking and disturbance rejection performance. Both simulation and experimental results have shown the effectiveness of the proposed method.

Conflicts of Interest

The authors declare that there are no conflicts of interest regarding the publication of this paper. There have no financial and personal relationships with other people or organizations that will influence our work, there is no professional or other personal interest of any nature or kind in any product, service, and/or company that could be construct as influencing the position presented in or the review of the manuscript.

Acknowledgments

This work was supported in part by the National Natural Science Foundation of China (61573298), Scientific Research Fund of Hunan Provincial Education Department (15B238), and the Key Laboratory of Intelligent Computing and Information Processing of Ministry of Education.

References

[1] S. Bolognani, R. Oboe, and M. Zigliotto, "Sensorless full-digital pmsm drive with ekf estimation of speed and rotor position,"

IEEE Transactions on Industrial Electronics, vol. 46, no. 1, pp. 184–191, 1999.

[2] Z. Chen, M. Tomita, S. Doki, and S. Okuma, "New adaptive sliding observers for position-and velocity-sensorless controls of brushless dc motors," *IEEE Transactions on Industrial Electronics*, vol. 47, no. 3, pp. 582–591, 2000.

[3] Y. Yi, D. M. Vilathgamuwa, and M. A. Rahman, "Implementation of an artificial-neural-network-based real-time adaptive controller for an interior permanent-magnet motor drive," *IEEE Transactions on Industry Applications*, vol. 39, no. 1, pp. 96–104, 2003.

[4] G. Lei, T. Wang, J. Zhu, Y. Guo, and S. Wang, "System-level design optimization method for electrical drive systems—robust approach," *IEEE Transactions on Industrial Electronics*, vol. 62, no. 8, pp. 4702–4713, 2015.

[5] M.-S. Wang, I. N. Syamsiana, and F.-C. Lin, "Sensorless speed control of permanent magnet synchronous motors by neural network algorithm," *Mathematical Problems in Engineering*, vol. 2014, pp. 1–7, 2014.

[6] X. Wu, H. Wang, X. Yuan, S. Huang, and D. Luo, "Design and implementation of recursive model predictive control for permanent magnet synchronous motor drives," *Mathematical Problems in Engineering*, vol. 2015, pp. 1–10, 2015.

[7] H. Wang, R. Zhao, F. Cheng, and H. Yang, "Active disturbance rejection control for PMSM servo system applied in industrial sewing machines," in *Proceedings of the International Conference on Electrical and Control Engineering, ICECE 2010*, pp. 3249–3252, China, May 2010.

[8] B. Zhang, Y. Pi, and Y. Luo, "Fractional order sliding-mode control based on parameters auto-tuning for velocity control of permanent magnet synchronous motor," *Isa Transactions*, vol. 51, no. 5, pp. 649–656, 2012.

[9] J. Huang, H. Li, Y. Chen, and Q. Xu, "Robust position control of PMSM using fractional-order sliding mode controller," *Abstract and Applied Analysis*, vol. 2012, no. 4, pp. 473–505, 2014.

[10] J. Yang, S. Li, and X. Yu, "Sliding-mode control for systems with mismatched uncertainties via a disturbance observer," *IEEE Transactions on Industrial Electronics*, vol. 60, no. 1, pp. 160–169, 2013.

[11] D. Ginoya, P. D. Shendge, and S. B. Phadke, "Sliding mode control for mismatched uncertain systems using an extended disturbance observer," *IEEE Transactions on Industrial Electronics*, vol. 61, no. 4, pp. 1983–1992, 2014.

[12] X. G. Zhang, L. Z. Sun, K. Zhao, and L. Sun, "Nonlinear speed control for PMSM system using sliding-mode control and disturbance compensation techniques," *IEEE Transactions on Power Electronics*, vol. 28, no. 3, pp. 1358–1365, 2013.

[13] W. Xu, Y. J. Jiang, and C. X. Mu, "Novel composite sliding mode control for PMSM system based on disturbance observer," *IEEE Transactions on Applied Superconductivity*, vol. 26, no. 7, pp. 12905–12910, 2016.

[14] W.-H. Chen, "Disturbance observer based control for nonlinear systems," *IEEE/ASME Transactions on Mechatronics*, vol. 9, no. 4, pp. 706–710, 2004.

[15] A. A. Apte, V. A. Joshi, R. A. Walambe, and A. A. Godbole, "Speed control of PMSM using disturbance observer," *IFAC-Papers OnLine*, vol. 49, no. 1, pp. 308–313, 2016.

[16] L. Xiaoquan, L. Heyun, and H. Junlin, "Load disturbance observer-based control method for sensorless PMSM drive," *IET Electric Power Applications*, vol. 10, no. 8, pp. 735–743, 2016.

[17] J. Zhang, X. Liu, Y. Xia, Z. Zuo, and Y. Wang, "Disturbance observer-based integral sliding-mode control for systems with mismatched disturbances," *IEEE Transactions on Industrial Electronics*, vol. 63, no. 11, pp. 7040–7048, 2016.

[18] P. V. Kokotovie, "The joy of feedback: nonlinear and adaptive," *IEEE Control Systems Magazine*, vol. 12, no. 3, pp. 7–17, 1992.

[19] J. Zhou and Y. Wang, "Real-time nonlinear adaptive backstepping speed control for a PMSM," *Control Engineering Practice*, vol. 13, no. 4, pp. 1259–1269, 2005.

[20] M. Karabacak and H. I. Eskikurt, "Speed and current regulation of a permanent magnet synchronous motor via nonlinear and adaptive backstepping control," *Mathematical and Computer Modelling*, vol. 53, no. 9-10, pp. 2015–2030, 2011.

[21] M. Karabacak and H. I. Eskikurt, "Design, modelling and simulation of a new nonlinear and full adaptive backstepping speed tracking controller for uncertain PMSM," *Applied Mathematical Modelling: Simulation and Computation for Engineering and Environmental Systems*, vol. 36, no. 11, pp. 5199–5213, 2012.

[22] C.-X. Chen, Y.-X. Xie, and Y.-H. Lan, "Backstepping control of speed sensorless permanent magnet synchronous motor based on slide model observer," *International Journal of Automation and Computing*, vol. 12, no. 2, pp. 149–155, 2015.

[23] Y. Chen, X. Wang, M. Li, and X. Liu, "Disturbance observer-based backstepping control for permanent magnet synchronous motor," in *Proceedings of the 2017 Chinese Automation Congress (CAC)*, pp. 1–5, Jinan, China, October 2017.

5

Improvement of Low-Frequency Radiated Emission in Electric Vehicle by Numerical Analysis

Feng Gao,[1] Chengkai Ye,[1] Zilong Wang,[2] and Xu Li[3]

[1]*School of Automotive Engineering, Chongqing University, Chongqing 400044, China*
[2]*China Automotive Technology and Research Center, Tianjin 300300, China*
[3]*Chongqing Chang'an Automobile Corporation Limited, Chongqing 401120, China*

Correspondence should be addressed to Feng Gao; gaofengl@cqu.edu.cn

Academic Editor: Daniela Proto

In this paper, a methodology for improving vehicle-level radiated electromagnetic interference (EMI) in electric vehicle (EV) is proposed. This methodology predicts vehicle-level radiated EMI by using the multiport network theory, based on analyzing the contribution from each electronic component to find out the main EMI source. To validate its effectiveness, the proposed methodology is applied in an actual EV for low-frequency radiated emissions. Simulation in a commercially available electromagnetics software and measurement in the EV are combined to predict the vehicle-level emissions, and then the electronic component with the greatest EMI that causes failure to meet the EMC standard is identified. After improving this component, the vehicle-level radiated emission is reduced to comply with the EMC standard, proving that the presented numerical method is effective. The proposed methodology can also be used in other EMC issues, regardless of the amount of interference sources and sensitive equipment.

1. Introduction

In recent years, the problems of environmental pollution and energy crisis have aroused global concern. Compared to internal combustion engine vehicles, electric vehicles (EVs) have been rapidly developed because of such advantages as no pollution, no direct emission, and using renewable energy sources. However, high-voltage and current levels in EVs bring much more serious vehicle-level electromagnetic emission problems especially at low frequencies [1]. The electric traction system in EV is the source of high-voltage and high-current electromagnetic interference (EMI). It mainly consists of a high-voltage battery, a power distribution unit (PDU), a drive inverter, a direct-current/direct-current converter (DC/DC), an electric motor, and cables connecting the aforementioned components. The usage of power electronic devices in the drive inverter and the DC/DC causes large amounts of EMI because of high frequency switching [2]. Moreover, the cables connecting the components in the electric traction system carry the EMI and serve as the transmitting antenna to radiate the EMI to the environment [3].

To avoid disturbing other equipment, the EMI generated by EV is restricted to a certain level. EVs are obliged to pass such standards as CISPR 12 and J551-5 before they are put on the market. The former is for the high frequency band and the latter is for the low-frequency band [4]. The compliance of EV to these standards only can be verified at a very late stage in the vehicle development process, because the electromagnetic compatibility (EMC) is greatly affected by the design changes in electronic/electrical components, electrical systems, and metal structures. The EMC problem discovered lately requires more time and cost to resolve. Moreover, when an EMC problem is found, it is difficult and time consuming to identify the actual source of EMI because (1) there are many sources causing radiated EMI in EV; (2) the parasitic parameters among structures cause many hidden transmission paths or antennas radiating/transmitting EMI. Compared to the experimental tests, the prediction technique using numerical simulation is an effective way to analyze and improve the EMC of EV. Benefitting from the development of computer technologies and numerical solution algorithms for three-dimensional electromagnetic field with the coupling of cables, some commercial software packages for analysis of EMC have been presented, such as FEKO, CST Studio, and EMC Studio.

To improve low-frequency radiated emission in EV, automotive manufacturers usually troubleshoot the component with the greatest EMI and then improve the component-level EMC. Therefore, the present efforts in the literature on the improvement of radiated emission focus on component-level EMC. Di Piazza et al. developed a novel common-mode EMI active filter for automotive motor drives to mitigate the leakage high frequency common-mode currents, so as to reduce the radiated EMI [5]. Marczok et al. reworked the interference mechanisms of spark plugs and then improved its EMI properties so as to improve vehicle-level EMC [6]. Chun et al. proposed several solutions of switching noise reduction of DC/DC converter in HEV and EV for improvement of vehicle-level EMC [7]. These aforementioned works improve radiated emission through reduction of component-level emissions without consideration of the relationship between vehicle-level EMC and component-level EMC, so the degree of improvement is difficult to control. Numerical methods are efficient tools to analyze vehicle-level EMC issues. For analysis of vehicle-level radiated EMC issues using numerical methods, one widely used method is to establish the transfer function between EMI and the conducted interference current, which is measured in component test platforms or actual vehicles. Ala et al. predicted the radiated EMI generated by a 42V/14V DC/DC step-down converter [8]. The transfer function from the time-domain interference current to the near electromagnetic field strength was calculated by finite-difference time-domain method. Chen et al. also anticipated the vehicle-level radiated EMI by characterizing the transfer function between the interference current carried in wire harnesses and the voltage at receiving antenna terminal [9]. Zeng et al. established the vehicle-level simulation model by using a three-dimensional electromagnetic analysis software package to calculate the conversion factor from the conductor current at the outputs of inverter to the voltage of antenna terminal [10]. Dong et al. obtained the coefficient between the electric field strength and the conducted current by analyzing the generation mechanism of radiated emissions [11]. Summing up the aforementioned studies, the transfer function should be obtained under the condition which is the same as the actual one, because the transfer function is greatly influenced by the impedance of source and victim. This poses great restrictions on its application in practice.

To improve the low-frequency radiated emission in an EV with less number of experiments and cost, a new numerical analysis method is presented to find out the main source, which has the following advantages: (a) the vehicle-level model of EMI is described by algebraic equations in frequency domain, by which the cost of computation is reduced; (b) the interactions of impedance among sources, paths, and victims are decoupled to integrate more approaches to set up the vehicle-level model of EMI; (c) it can be applied at the whole vehicle development stage by integrating different modeling approaches, test data, etc. This paper is organized as follows. Section 2 describes the studied EMC problem. Section 3 introduces the proposed numerical method used to model and analyze the problem. The application effect of this methodology on an EV is shown in Section 4. Finally, Section 5 concludes the paper.

2. Problem Description

SAE J551-5 limits the level of magnetic and electric emissions level from EVs over the frequency range from 150 kHz to 30 MHz. Figure 1 shows the test results of a four-seat sedan EV according to SAE J551-5. From Figure 1(b), it is found that the margin is sufficient at the low-frequency range, whereas it is narrow at the relative high frequency range and the magnetic field exceeds the limit about 10 dBμA/m at 13.8 MHz.

Vehicle-level EMC is affected by many factors, among which the performance of components is very important. Furthermore, at this stage, compared with redesigning the electrical systems or arranging the components, it costs less to improve the component-level EMC. Automotive manufacturers usually attempt to find out the component with the greatest EMI by experimental tests. The widely used way of troubleshooting is to do standard tests after turning off different suspected electronic/electrical components, until the vehicle-level EMC gets better. Troubleshooting with this method costs more than the standard EMC test due to the high cost of long-hours EMC test facility and the need for hours of testing. Additionally, this troubleshooting method cannot quantify the contribution of the components which cannot be turned off during the experimental test, such as the motor and motor controller of an EV.

To improve the vehicle-level radiated emission more efficiently, a numerical method will be introduced to model and find out the main emission sources in the EV. This methodology combines numerical simulation using electromagnetic software and experimental modeling approaches to predict vehicle-level radiated EMI and analyze the contribution from each electronic/electrical component. In other words, the greatest source of radiated EMI can be found by this numerical method without multiple experimental tests.

3. Radiated EMI Modeling

The low-frequency radiated emission is mainly caused by high-voltage cables which carry common-mode conducted interference current. To improve the vehicle-level radiated EMI by numerical method, a mathematical model to describe the EMC problem should be obtained first. Considering the fact that the electrically small component can be equivalent to a circuit, which is more effective to be solved numerically, a new strategy to model the EMC problem is presented as shown in Figure 2. The electronic/electrical components, transmission paths, and the measuring instruments constitute a multiport network system, in which each electrically small component or measuring instrument acts as a terminal. The electrically large parts including high-voltage cables, vehicle body, and other large structures are considered as a network which serves as a black box to describe the

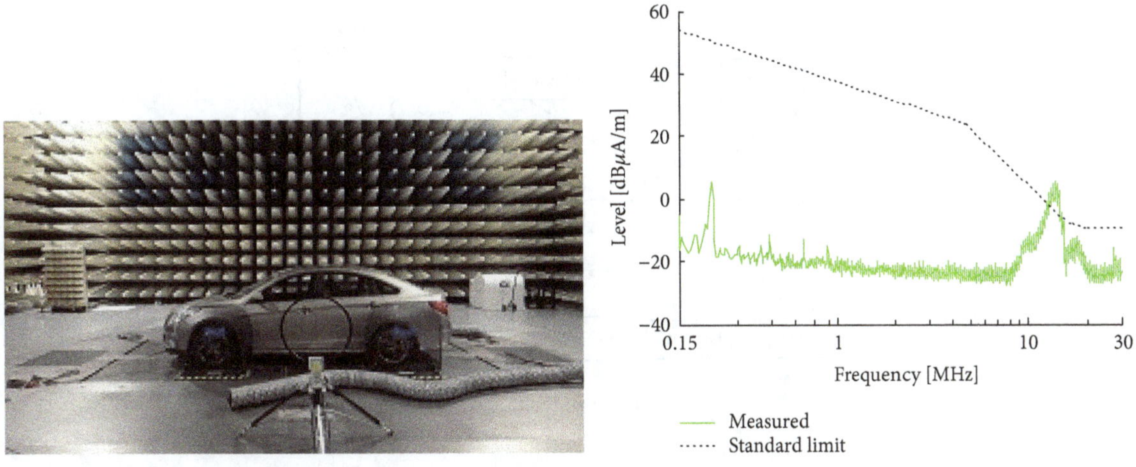

(a) (b)

FIGURE 1: Experimental test of J551-5: (a) test setup; (b) test results.

transmission paths. The characteristics of the network can be calculated by existing commercial software or measured by network analyzer. Each terminal is characterized by Thevenin's equivalent circuit, whose circuit parameters can be measured in the EV.

The vehicle-level radiated emission is evaluated by the electromagnetic field strength measured by an antenna. Port 1 in Figure 2 denotes the measuring antenna while others (Indexed by $i = 2, \cdots$) denote the on-board electronic components. Their Thevenin's equivalent circuits are described by the equivalent voltage source $V_{S,i}$ and Thevenin impedance $Z_{S,i}$. The port voltage and current are denoted by U_i and I_i, respectively. Note that the equivalent voltage sources of the devices without electromagnetic interference, including the measuring antenna, are equal to zero.

The open-circuit impedance parameter called **Z**-parameter is used to characterize the electromagnetic coupling relationship among the ports:

$$\begin{bmatrix} U_1 \\ U_2 \\ \vdots \\ U_n \end{bmatrix} = \left[\begin{array}{c|ccc} Z_{11} & Z_{12} & \cdots & Z_{1n} \\ \hline Z_{21} & Z_{22} & \cdots & Z_{2n} \\ \vdots & \vdots & \ddots & \vdots \\ Z_{n1} & Z_{n2} & \cdots & Z_{nn} \end{array}\right] \begin{bmatrix} I_1 \\ I_2 \\ \vdots \\ I_n \end{bmatrix} \quad (1)$$

The vehicle-level emission is measured by the terminal voltage of the measuring antenna, and to distinguish the port of antenna from other on-board electronic/electrical components, (1) is rewritten using subscripts *ant* and *ele*:

$$\begin{bmatrix} U_{ant} \\ \mathbf{U}_{ele} \end{bmatrix} = \begin{bmatrix} Z_{ant-ant} & \mathbf{Z}_{ant-ele} \\ \mathbf{Z}_{ele-ant} & \mathbf{Z}_{ele-ele} \end{bmatrix} \begin{bmatrix} I_{ant} \\ \mathbf{I}_{ele} \end{bmatrix} \quad (2)$$

Combining all Thevenin's equivalent circuits of the terminals together yields

$$\begin{bmatrix} U_{ant} \\ \mathbf{U}_{ele} \end{bmatrix} = \begin{bmatrix} V_{S,ant} \\ \mathbf{V}_{S,ele} \end{bmatrix} - \begin{bmatrix} Z_{S,ant} & \\ & \mathbf{Z}_{S,ele} \end{bmatrix} \begin{bmatrix} I_{ant} \\ \mathbf{I}_{ele} \end{bmatrix} \quad (3)$$

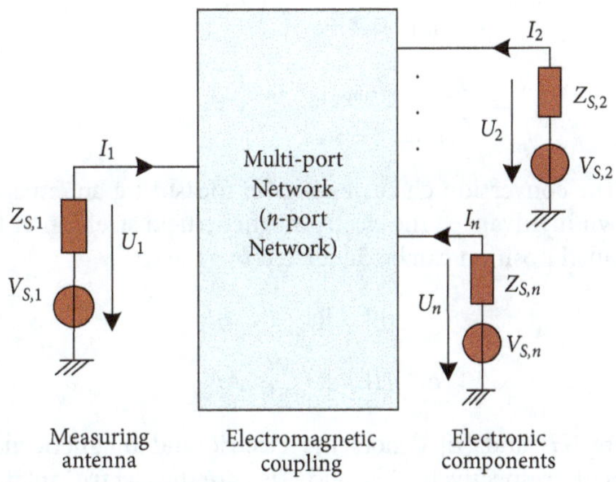

FIGURE 2: Model for pre-vehicle-level radiated EMI.

where $V_{S,ant} = 0$ and $Z_{S,ant} = Z_{S,1}$ are the equivalent voltage source and output impedances of measuring antenna. And $\mathbf{V}_{S,ele} = [V_{S,2}, \cdots, V_{S,n}]^{\mathrm{T}}$ and $\mathbf{Z}_{S,ele} = \mathrm{diag}(Z_{S,2}, \cdots, Z_{S,n})$ denote the equivalent voltage sources and output impedances of the on-board electronic components connected to the ports. Equations (2) and (3), respectively, describe the electromagnetic coupling relationship between the multiple ports and the equivalent characteristics of on-board electronic components.

Based on (2) and (3), the voltages at each port can be derived by

$$\begin{bmatrix} U_{ant} \\ \mathbf{U}_{ele} \end{bmatrix} = \begin{bmatrix} Z_{ant-ant} & \mathbf{Z}_{ant-ele} \\ \mathbf{Z}_{ele-ant} & \mathbf{Z}_{ele-ele} \end{bmatrix} \cdot \begin{bmatrix} Z_{ant-ant} + Z_{S,ant} & \mathbf{Z}_{ant-ele} \\ \mathbf{Z}_{ele-ant} & \mathbf{Z}_{ele-ele} + \mathbf{Z}_{S,ele} \end{bmatrix}^{-1} \begin{bmatrix} V_{S,ant} \\ \mathbf{V}_{S,ele} \end{bmatrix} \quad (4)$$

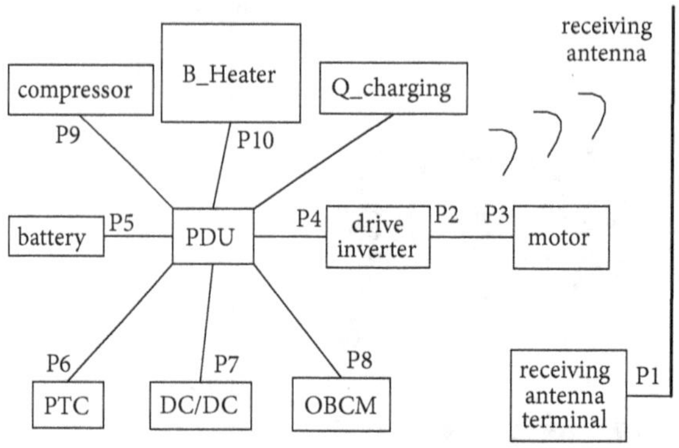

FIGURE 3: Multiport network of the EV.

Substituting the condition that $V_{S,ant} = 0$ to (4), the terminal voltage of measuring antenna is

$$U_{ant} = \frac{Z_{S,ant}\mathbf{Z}_{ant-ele}}{Z_{ant-ant} + Z_{S,ant}} \left(\mathbf{Z}_{S,ele} + \mathbf{Z}_{ele-ele} - \frac{\mathbf{Z}_{ele-ant}\mathbf{Z}_{ant-ele}}{Z_{ant-ant} + Z_{S,ant}} \right)^{-1} \bullet \mathbf{V}_{S,ele} \quad (5)$$

The conversion characteristics of measuring antenna are known in advance. The electromagnetic field strength at the specified position can be calculated by

$$|E| = |U_{ant}| \bullet AF_E \quad \text{or} \quad |H| = |U_{ant}| \bullet AF_H \quad (6)$$

where $|H|$ and $|E|$ denote the electric and magnetic field strength, respectively. AF_E and AF_H are the electric antenna factor and magnetic antenna factor, respectively. From (5) and (6), the electromagnetic field strength, which measures the vehicle-level radiated EMI, can be figured out by three parts: **Z**-parameter of the entire network, the characteristics of electronic components, and measuring antenna. The above three parts can be obtained independently.

4. Prediction Model Validation

In this section, the predictive model is validated by applying it to predict the vehicle-level radiated EMI of the studied sedan EV. Figure 3 shows its equivalent multiport network for EMI prediction, which is a 10-port network. The main components relating to EMI emission are determined from the previous standard and troubleshooting test results. Note that it is concluded from the previous results that the low-frequency radiated EMI originates from the power electronic devices in the drive inverter and the DC/DC, so, in Figure 3, P2, P4, and P7 act as interference source.

4.1. Modeling of Port Coupling. As described in Section 3, **Z**-parameter is used to characterize the port coupling of the 10-port network. However, at the high frequency range, **S**-parameter representation is more precise and easily obtained. In this paper, the **S**-parameter of the network is calculated by using FEKO, which is a commercial software package for numerical solving of three-dimensional electromagnetic field.

CAE preprocessing is necessary for EM simulation. It is widely accepted that the larger the EM energy is, the more detailed the EM model needs to be. Because the EMI source comes from drive inverter and the DC/DC and they are located at engine cabin, the EM energy mostly concentrates on engine cabin. In addition, the minor EM energy comes into the vehicle body because of the shielding effect of segregation board. So the EM model of engine cabin is detailed whereas the vehicle body can be highly simplified. The result of CAE preprocessing is shown in Figure 4(a). Figure 4(b) shows the calculation model in FEKO, which includes the main parts of EV and the measuring antenna arranged in the position required by SAE J551-5.

After calculating the **S**-parameter of the network, the required **Z**-parameter can be calculated by [9].

$$\begin{bmatrix} Z_{ant-ant} & \mathbf{Z}_{ant-ele} \\ \mathbf{Z}_{ele-ant} & \mathbf{Z}_{ele-ele} \end{bmatrix} = Z_0 \left(\begin{bmatrix} 1 & \\ & \mathbf{E}_{ele} \end{bmatrix} - \begin{bmatrix} S_{ant-ant} & \mathbf{S}_{ant-ele} \\ \mathbf{S}_{ele-ant} & \mathbf{S}_{ele-ele} \end{bmatrix} \right)^{-1} \cdot \left(\begin{bmatrix} 1 & \\ & \mathbf{E}_{ele} \end{bmatrix} + \begin{bmatrix} S_{ant-ant} & \mathbf{S}_{ant-ele} \\ \mathbf{S}_{ele-ant} & \mathbf{S}_{ele-ele} \end{bmatrix} \right) \quad (7)$$

where Z_0 denotes port reference impedance when calculating the **S**-parameter, which is usually set to 50Ω. And $\mathbf{E}_{ele}$ denotes the 9 × 9 identity matrix. Furthermore each part of **Z**-parameter can be obtained:

$$Z_{ant-ant} = Z_0 \frac{1 + S_{ant-ant}}{1 - S_{ant-ant}} + 2Z_0 \bullet \mathbf{S}_{ant-ele} \left[(1 - S_{ant-ant}) (\mathbf{E}_{ele} - \mathbf{S}_{ele-ele}) - \mathbf{S}_{ele-ant}\mathbf{S}_{ant-ele} \right]^{-1} \bullet \mathbf{S}_{ele-ant}$$

$$\mathbf{Z}_{ant-ele} = 2Z_0 \bullet \mathbf{S}_{ant-ele} \left[(1 - S_{ant-ant}) (\mathbf{E}_{ele} - \mathbf{S}_{ele-ele}) - \mathbf{S}_{ele-ant}\mathbf{S}_{ant-ele} \right]^{-1}$$

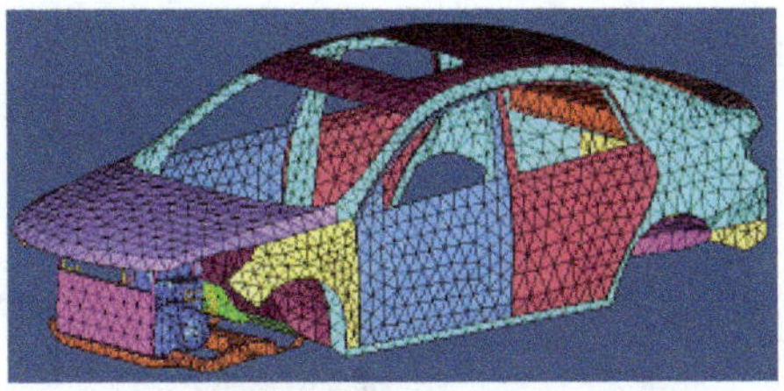

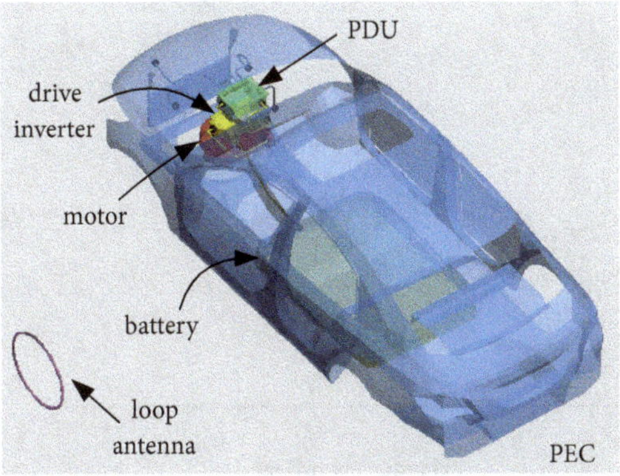

(a) (b)

FIGURE 4: Simulation model: (a) CAE preprocessing result; (b) EM model in FEKO.

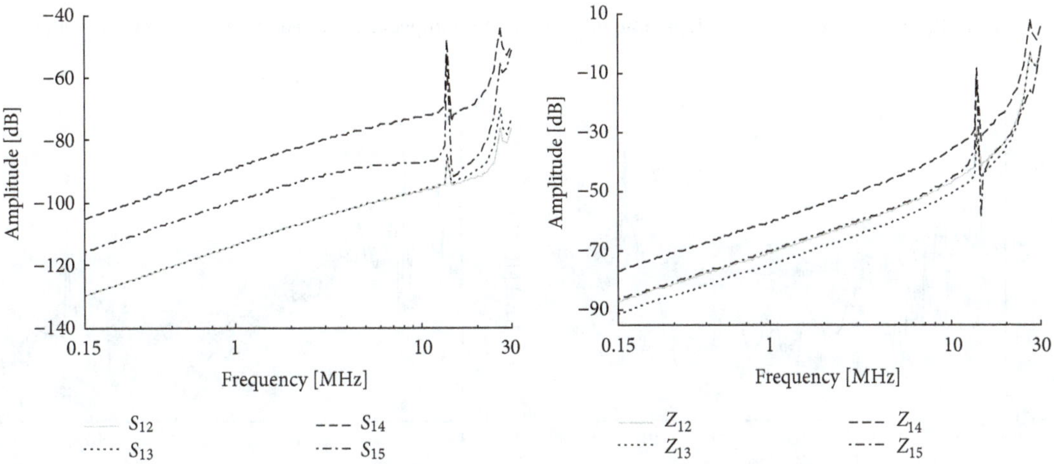

FIGURE 5: Network parameters: (a) S-parameter; (b) Z-parameter.

$$\begin{aligned}
\mathbf{Z}_{ele-ant} &= 2Z_0\left[\left(1 - S_{ant-ant}\right)\left(\mathbf{E}_{ele} - \mathbf{S}_{ele-ele}\right)\right. \\
&\quad \left. - \mathbf{S}_{ele-ant}\mathbf{S}_{ant-ele}\right]^{-1} \bullet \mathbf{S}_{ele-ant} \\
\mathbf{Z}_{ele-ele} &= Z_0\left[\left(1 - S_{ant-ant}\right)\left(\mathbf{E}_{ele} - \mathbf{S}_{ele-ele}\right)\right. \\
&\quad \left. - \mathbf{S}_{ele-ant}\mathbf{S}_{ant-ele}\right]^{-1} \bullet \left[\left(1 - S_{ant-ant}\right)\left(\mathbf{E}_{ele} + \mathbf{S}_{ele-ele}\right)\right. \\
&\quad \left. + \mathbf{S}_{ele-ant}\mathbf{S}_{ant-ele}\right]
\end{aligned} \tag{8}$$

Equation (8) is essentially the conversion formula from **S**-parameter to **Z**-parameter for vehicle-level radiated EMI prediction. Figure 5 shows the **S**-parameter and **Z**-parameter curves which describe the relationship between the measuring antenna (Port 1) and some of the component ports (Port 2 to 5) as an example.

4.2. Equivalent Model of Components. The parameters of Thevenin's equivalent circuits for electronic/electrical components are measured in the actual EV, which is explained in Figure 6. Note that it is concluded from previous results that the radiated EMI is mainly caused by common-mode interferences. And so the positive and negative lines are equivalent to one.

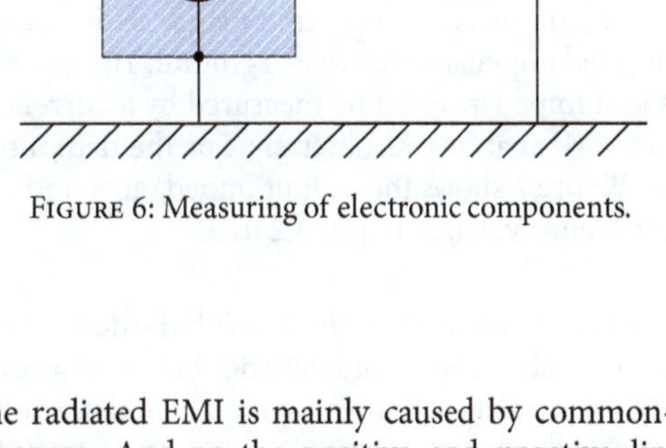

FIGURE 6: Measuring of electronic components.

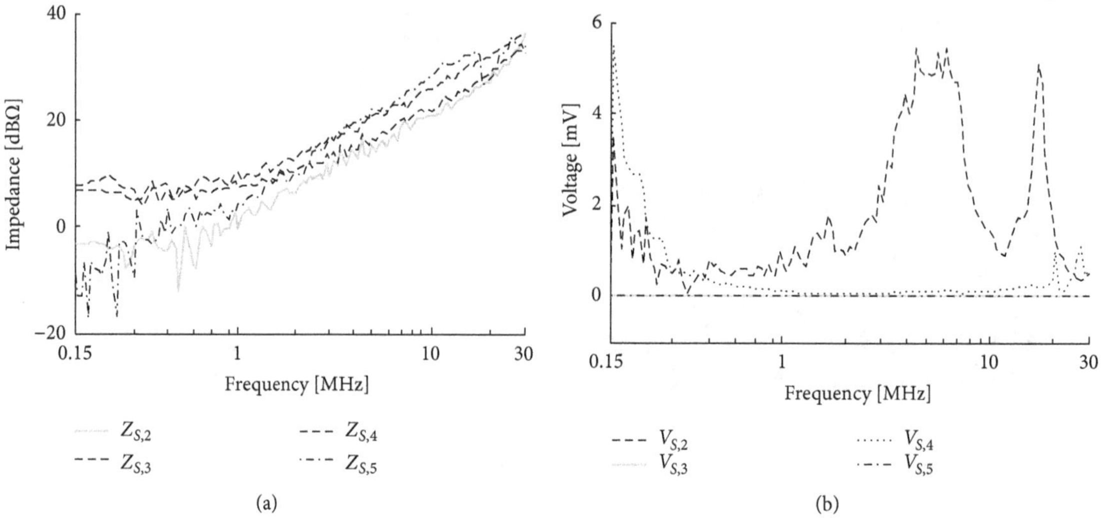

FIGURE 7: Thevenin's equivalent circuit parameters: (a) Thevenin impedance; (b) equivalent interference voltage.

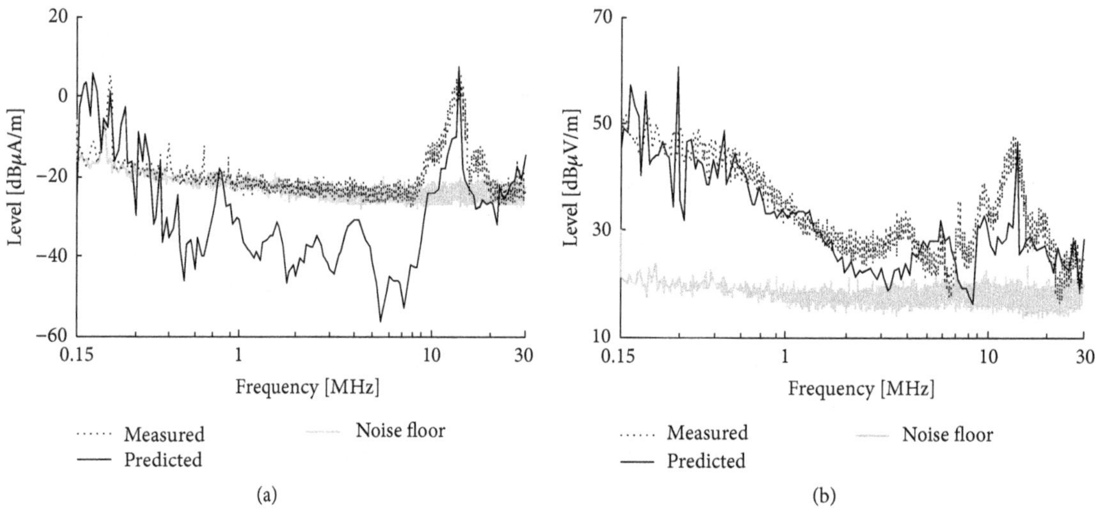

FIGURE 8: Predicted and measured EMI: (a) magnetic field; (b) electric field.

According to Figure 6, the parameters of Thevenin's equivalent circuit and the external load are formulated by

$$V_{S,i} = I_{C,i} \bullet (Z_{S,i} + Z_{L,i}) \tag{9}$$

where $I_{C,i}$ is the common-mode interference current, $V_{S,i}$ and $Z_{S,i}$ are the parameters of Thevenin's equivalent circuit, and $Z_{L,i}$ is the load impedance relative to ground. The interference current and impedance can be measured by a current clamp and a network analyzer, respectively. For the demonstration purpose, Figure 7 shows the output impedances and equivalent interference voltages of ports 2 to 5.

4.3. Prediction Results and Analysis. With the derived parameters of the coupling characteristic between ports and Thevenin's equivalent circuit for components, the voltage at the measuring antenna terminal is calculated by (5). Furthermore, after calibrating the antenna property in FEKO, the radiated EMI is obtained using (6), which is compared with the measurement results as shown in Figure 8. The trend of the predicted and measured results agrees with each other. Moreover, the peak point of magnetic field at around 18.83 MHz, which exceeds the limit, is predicted precisely. In the frequency range 0.3 kHz-8 MHz of the magnetic field, the predicted EMI is much smaller than the measured value on the whole. Comparing the measured value with the noise floor of EMC laboratory, we find that in the inaccurate frequency range the antenna only measures the noise floor, which implies that the emission generated by EV is much smaller.

5. Main EMI Analysis and Improvement

For the EV failing the requirements of SAE J551-5, a numerical model has been set up to predict its radiated low-frequency EMI and validated by comparison with the

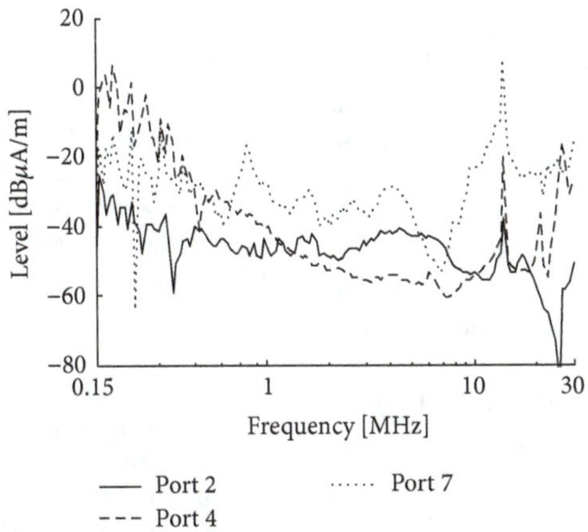

FIGURE 9: EMI from the three maximum components.

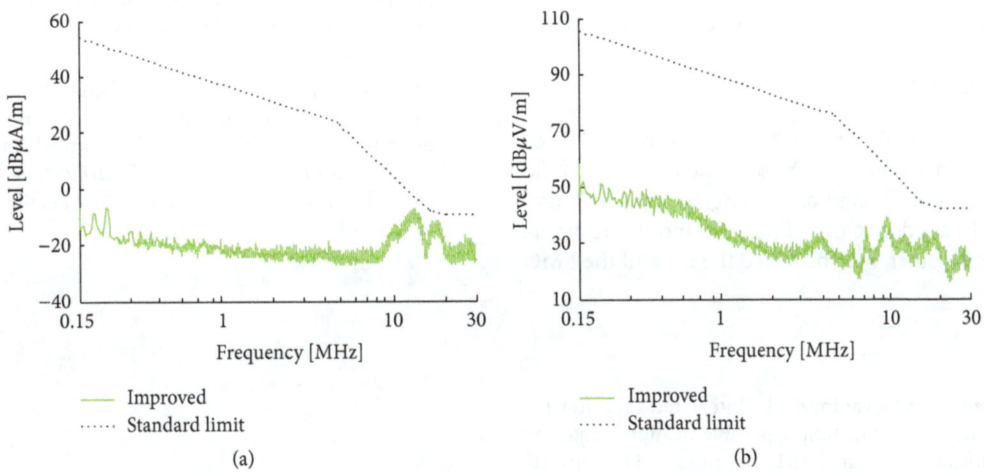

FIGURE 10: Test result after improvement: (a) magnetic field; (b) electric field.

measurement results. Instead of identifying the main source of EMI by expensive experimental testing, this predictive model is used to analyze the contribution to vehicle-level radiated EMI from each electronic component. Equation (5) shows that the vehicle-level radiated EMI is a sum of the induced interferences from every source. Then the generated interference magnetic field of each component can be derived, which is shown in Figure 9. Note that only the three maximum EMI sources are plotted to show clearly.

It is found that at the lower frequency range the greatest EMI is the drive inverter, whereas the EMI generated by the DC/DC is much higher than others especially at around the frequency where the EMI exceeds the limit. Therefore, this predictive model indicates that the DC/DC is the main source leading to excess vehicle-level magnetic field emission.

Although the radiated emission are also influenced by the coupling paths, from the standpoint of automotive enterprises, the easiest way is to improve EMC of the DC/DC converter because the vehicle may be required to be rearranged to reduce the coupling between components. This is because **S**-parameter and **Z**-parameter are affected by many factors like the structure of vehicle body, the layout of cables, and so on, and it is difficult to control the variation of them. Note that improvement of EMC is at the last stage of vehicle development process, so we should consider the easiest method first. If it failed, redesigning the electric system or vehicle body is the last resort.

After redesigning the DC/DC, the new test results of vehicle-level low-frequency emission according to SAE J551-5 are shown in Figure 10, which show that the EV can pass SAE J551-5. This implies that the troubleshooting result obtained by the presented numerical method is correct and effective. Compared with the original experimental method, the cost of both time and money is greatly reduced.

6. Conclusion

This paper proposes a methodology for improving vehicle-level radiated emission by numerical analysis. This methodology combines both simulation in electromagnetic numerical solving software and measurement in actual vehicle to predict the vehicle-level radiated emission. Then this model is applied to find out the main EMI source. The experimental application results show that the radiated emission of the EV is improved successfully with reduced cost.

This methodology can be used not only in the improvement of vehicle-level radiated EMI, but also in other kinds of EMC issues, which can be equivalent to a multiport network. However, there are some open problems that need to be further studied:

(1) The characteristics of electronic/electrical components may vary with working conditions. Theoretically, the parameters of the equivalent circuit should be obtained under the condition, which is the same with the studied EMC problem.

(2) The transmission path of the presented application is approximated to be linear. How to model and solve the nonlinear interactions is to be studied.

Conflicts of Interest

The authors declare that they have no conflicts of interest.

Acknowledgments

This research was supported the National Key R&D Program of China under grant 2017YFB0102504 and Scientific Technological Plans of Chongqing under grant cstc2017zdcy-zdzx0042. Thanks are due to the China Automotive Technology and Research Center, who provided the EV and the EMC laboratory.

References

[1] D. Hamza and M. Pahlevaninezhad, "Integrated capacitor for common-mode EMI mitigation applicable to high frequency planar transformers used in electric vehicles DC/DC converters," in *Proceedings of the 2014 IEEE Energy Conversion Congress and Exposition (ECCE' 14)*, pp. 4809–4814, Pittsburgh, Pa, USA, 2014.

[2] S. Jeschke, H. Hirsch, M. Trautmann et al., "EMI measurement on electric vehicle drive inverters using a passive motor impedance network," in *Proceedings of the 7th Asia-Pacific International Symposium on Electromagnetic Compatibility, APEMC '16*, pp. 292–294, Shenzhen, China, 2016.

[3] H. Shim, H. Kim, Y. Kwack et al., "Inverter modeling including non-ideal IGBT characteristics in Hybrid Electric Vehicle for accurate EMI noise prediction," in *Proceedings of the 2015 IEEE International Symposium on Electromagnetic Compatibility - EMC '15*, pp. 691–695, Dresden, Germany, 2015.

[4] A. R. Ruddle and R. Armstrong, "Review of current EMC standards in relation to vehicles with electric powertrains," in *Proceedings of the 2013 International Symposium on Electromagnetic Compatibility, EMC Europe '13*, pp. 298–303, Brugge, Belgium, September 2013.

[5] M. C. Di Piazza, M. Luna, A. Ragusa, and G. Vitale, "An improved common mode active filter for EMI reduction in vehicular motor drives," in *Proceedings of the 7th IEEE Vehicle Power and Propulsion Conference, VPPC '11*, pp. 1–8, Chicago, Ill, USA, 2011.

[6] C. Marczok, U. Maaß, E. Hoene, I. Ndip, K.-D. Lang, and D. Hasselberg, "Analysis and improvement of a spark plug for less radiated electromagnetic emissions," in *Proceedings of the 2014 International Symposium on Electromagnetic Compatibility, EMC Europe '14*, pp. 385–390, Gothenburg, Sweden, 2014.

[7] H. Chun et al., "Converter Switching Noise Reduction for Enhancing EMC Performance in HEV and EV," in *Proceedings of the PCIM Europe 2016; International Exhibition and Conference for Power Electronics, Intelligent Motion, Renewable Energy and Energy Management*, pp. 1–8, Nuremberg, Germany, 2016.

[8] G. Ala, M. C. Di Piazza, G. Tinè, F. Viola, and G. Vitale, "Evaluation of radiated EMI in 42-V vehicle electrical systems by FDTD simulation," *IEEE Transactions on Vehicular Technology*, vol. 56, no. 4 I, pp. 1477–1484, 2007.

[9] C. Chen, "Predicting vehicle-level radiated EMI emissions using module-level conducted EMIs and harness radiation efficiencies," in *Proceedings of the 2001 International Symposium on Electromagnetic Compatibility (EMC '01)*, pp. 1146–1151, Montreal, Canada, 2001.

[10] H. Zeng, I. Hoda, W. Ivan et al., "Vehicle-Level EMC Modeling for HEV/EV Applications," *SAE International Journal of Passenger Cars - Electronic and Electrical Systems*, vol. 8, no. 2, pp. 278–281, 2015.

[11] M. Dong, L. Zhai, R. Gao, and X. Zhang, "Research on radiated electromagnetic interference (EMI) from power cables of a three-phase inverter for electric vehicles," in *Proceedings of the 2014 IEEE Transportation Electrification Conference and Expo Transportation Electrification Asia-Pacific*, pp. 1–5, Beijing, China, 2014.

6

Suppression of Switched Reluctance Motor Vibration of In-Wheel Motor Electric Vehicle

Funing Yang and Yanyang Wang

The Key Laboratory of Automotive Engineering, Xihua University, Chengdu 610039, China

Correspondence should be addressed to Yanyang Wang; yywang@mail.xhu.edu.cn

Academic Editor: Manuel Pineda-Sanchez

Switched reluctance motor (SRM) has got great attention in in-wheel motor electric vehicle (IWM-EV), but SRM vertical force, the vertical component of SRM unbalanced radial force, yields SRM vertical vibration and does harm to dynamic performance of IWM-EV. In order to reduce the SRM vertical vibration, electromagnetic active suspension and a linear quadratic Gaussian (LQG) controller were used to suppress the unbalanced radial force in this paper. All the models and the controller were constructed in Matlab/Simulink R2015b. The controller considers five performance indexes: vehicle body acceleration, SRM airgap eccentricity, SRM stator acceleration, suspension dynamic deflections, and tyre deformation. Analytic Hierarchy Process (AHP) was used to calculate the weighted coefficients of performance indexes. Simulations indicate that this electromagnetic active suspension can reduce SRM vertical vibration obviously and improve dynamic performance of IWM-EV.

1. Introduction

Switched reluctance motor (SRM) has achieved good performance due to its remarkable advantages: high starting torque, wide operating speed range, and high efficiency. These advantages endow SRM with great potential on in-wheel motor electric vehicles (IWM-EV). However, asymmetrical magnetic pull caused by airgap eccentricity yields unbalanced radial force; this unbalanced radial force is recognized as one of the main reasons for the SRM vibration [1–5]. To reduce the vibration, scholars have focused on exploring the structure and control strategies of SRM [6–8].

On SRM structure, integrated design and the optimization of SRM components are main methods to reduce SRM vibration, such as multiobjective optimization design method [9], new stator tooth [10], new rotor tooth [11], and new poles ratio [12]. On SRM control strategies, varieties of methods have been proposed, such as torque distribution controller [13], multilevel systematic design method [14], and new motor torque controller [15]. The above studies do reduce the SRM vibration; however, these studies focus only on SRM itself; the negative influence of SRM on vehicle dynamic performance has not been fully considered. As the main component that flexibly connects the wheel and the vehicle body, suspension transmits the force acting between the wheel and the vehicle body. As a kind of suspension system, electromagnetic active suspension can produce controllable active force, which enables the suspension system to dynamically adjust the supporting force according to the vibration of wheel and vehicle body. In the previous work, we analyzed the influence of the SRM unbalanced radial force on comfort and stability of IWM-EV. The conclusion [16, 17] shows SRM vertical force is highly coupled with road excitation and SRM airgap eccentricity, and this coupling yields SRM vertical vibration that does harm to comfort and stability. This study used electromagnetic active suspension and linear quadratic Gaussian (LQG) controller to reduce the SRM vertical vibration and the negative influence on vehicle dynamic performance.

In order to conduct this study, a full IWM-EV model, SRM vertical force model, and control diagram for the electromagnetic active suspension are built in Matlab/Simulink R2015b. Then the weighted coefficients are calculated and LQG controller is completed. Finally, the frequency responses

Table 1: The SRM parameter values.

Definition	Symbol	Units	Value
Airgap	g	m	0.00025
Minimum inductance	L_q	H	0.0019
Maximum inductance	L_d	H	0.0318
Saturated inductance	L_{dsat}	H	0.0013
Maximum flux linkage	ψ_m	Wb	0.9

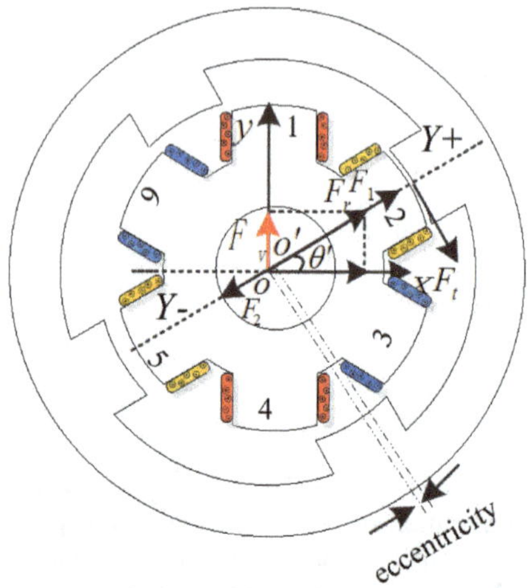

Figure 1: SRM vertical force.

on stochastic roads and time responses on representative roads are analyzed. Conclusion shows that the electromagnetic active suspension can effectively reduce SRM vibration and the negative influence on vehicle dynamic performance.

2. IWM-EV Modeling

Two models were constructed for studying the effects of electromagnetic active suspension. One is SRM vertical force model that reflects the SRM vertical force exerting on the wheel; another is full IWM-EV model that reflects the influence of electromagnetic active suspension force and unbalanced vertical force on vehicle.

2.1. SRM Vertical Force Model. Because of the geometrically balanced motor structure, the SRM radial force is always considered zero. But the vehicle load and road excitation [17] will yield SRM airgap eccentricity, due to which, the radial force is always not zero. In this study, The primary objective is to reduce the vertical vibration of SRM and the negative influence on vehicle dynamic performance, the well-known 6/4 outside-rotor SRM, like [18, 19], is shown in Figure 1. The SRM parameter values are listed in Table 1. According to [20], the difference magnetic pull between each pair of poles yields the unbalanced radial force, and the radial force of opposite stator poles can be described as follows.

$$F_1 = -\frac{\sin(\theta_0)}{g_m - \Delta g}T = -\frac{r\sin(\theta_0)}{g_m - \Delta g}F_t \tag{1}$$

$$F_2 = -\frac{\sin(\theta_0)}{g_m + \Delta g}T = -\frac{r\sin(\theta_0)}{g_m + \Delta g}F_t \tag{2}$$

The unbalanced radial force is follows.

$$F_r = F_1 - F_2 \tag{3}$$

The vertical force that is the vertical component of the unbalanced radial force can be described as

$$F_v = F_r \sin(\theta') \tag{4}$$

where θ_0 is the overlap angle of stator and rotor, F_t is tyre tangential force, g_m is airgap length of SRM, Δg is the airgap eccentricity, and θ' is the angle between stator and wheel longitudinal axis. Each pair of opposite stator with eccentricity will yield unbalanced radial force, for the convenience of presentation, only stators 1 and 4 are picked out to investigate the influence of SRM vertical force on SRM vibration and vehicle dynamic performance. When the θ' is 90°, SRM unbalanced radial force is equal to SRM vertical force.

2.2. Vehicle Model. To simulate the effects of electromagnetic active suspension force on IWM-EV, the required vehicle dynamics model needs to reflect contributions of both electromagnetic active suspension force and SRM vertical force. A vehicle model which has been proved reasonably well in previous work [16, 17] is developed as shown in Figure 2, and the governing equations of the vehicle motions can be described as

$$M_t\ddot{x} = -(F_{xfl} + F_{xfr})\cos\delta - (F_{yfl} + F_{yfr})\sin\delta - F_{xrl} - F_{xrr} \tag{5}$$

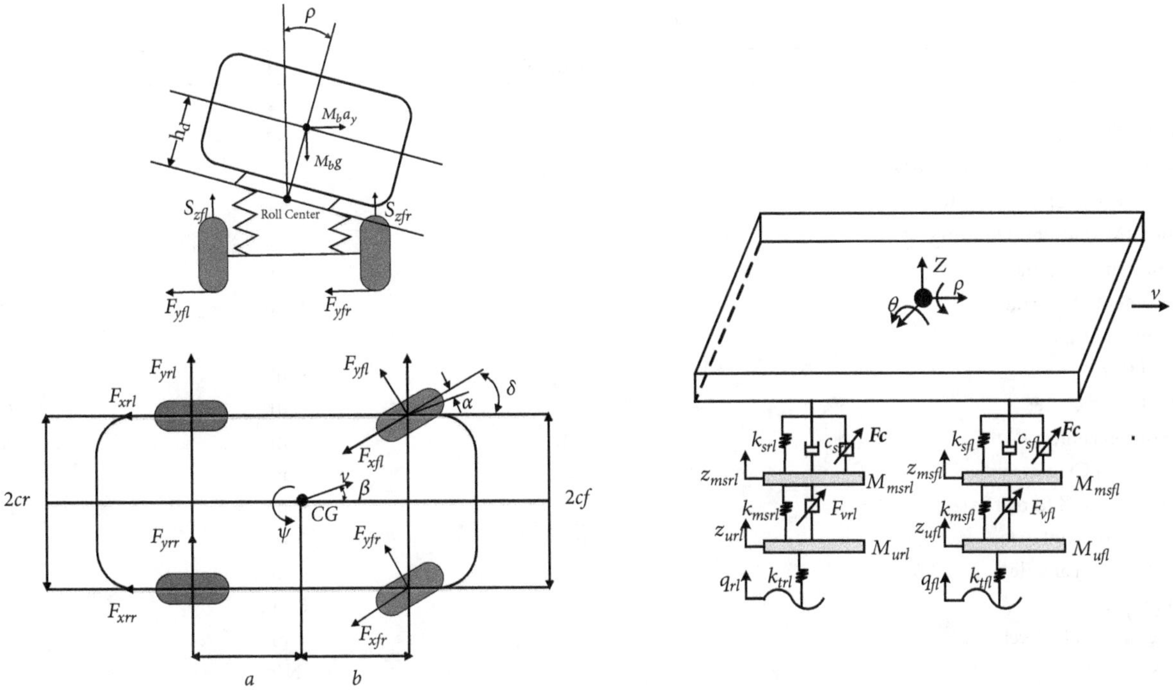

FIGURE 2: Schematic diagram of the vehicle model.

$$M_t \ddot{y} = -\left(F_{xfl} + F_{xfr}\right)\sin\delta + \left(F_{yfl} + F_{yfr}\right)\cos\delta + F_{yrl} + F_{yrr} \tag{6}$$

$$M_b \ddot{z} = S_{zfl} + S_{zfr} + S_{zrl} + S_{zrr} - Fc_{fl} - Fc_{fr} - Fc_{rl} - Fc_{rr} \tag{7}$$

$$\begin{aligned} Ix\ddot{p} = {} & c_f\left(S_{zfl} - S_{zfr}\right) + c_r\left(S_{zrl} - S_{zrr}\right) \\ & + h_c\left[-\left(F_{xfl} + F_{xfr}\right)\sin\delta + \left(F_{yfl} + F_{yfr}\right)\cos\delta \right. \\ & \left. + F_{yrl} + F_{yrr}\right] + M_b g\left(h_{cg} - h_c\right)\sin\rho \end{aligned} \tag{8}$$

$$\begin{aligned} Iy\ddot{\theta} = {} & -a\left(S_{zfl} + S_{zfr}\right) + b\left(S_{zrl} + S_{zrr}\right) \\ & + h_p\left[\left(F_{xfl} + F_{xfr}\right)\cos\delta + \left(F_{yfl} + F_{yfr}\right)\sin\delta + F_{xrl} \right. \\ & \left. + F_{xrr}\right] + M_b g\left(h_{cg} - h_p\right)\sin\theta \end{aligned} \tag{9}$$

$$\begin{aligned} Iz\ddot{\psi} = {} & a\left[-\left(F_{xfl} + F_{xfr}\right)\sin\delta + \left(F_{yfl} + F_{yfr}\right)\cos\delta\right] \\ & - b\left(F_{yrl} + F_{yrr}\right) + c_f\left[\left(F_{xfl} - F_{xfr}\right)\cos\delta \right. \\ & \left. + \left(F_{yfl} - F_{yfr}\right)\sin\delta\right] + c_r\left(F_{xrl} - F_{xrr}\right) \end{aligned} \tag{10}$$

where S_{zij}, the suspension vertical forces at each corner, is as follows.

$$\begin{aligned} S_{zfl} &= -k_{sfl}\left(z + c_f\rho - a\theta - z_{msfl}\right) - c_{sfl}\left(\dot{z} + c_f\dot{\rho} - a\dot{\theta} - \dot{z}_{msfl}\right) \\ S_{zfr} &= -k_{sfr}\left(z - c_f\rho - a\theta - z_{msfr}\right) - c_{sfr}\left(\dot{z} - c_f\dot{\rho} - a\dot{\theta} - \dot{z}_{msfr}\right) \\ S_{zrl} &= -k_{srl}\left(z + c_r\rho + b\theta - z_{msrl}\right) - c_{srl}\left(\dot{z} + c_r\dot{\rho} + b\dot{\theta} - \dot{z}_{msrl}\right) \\ S_{zrr} &= -k_{srr}\left(z - c_r\rho + b\theta - z_{msrr}\right) - c_{srr}\left(\dot{z} - c_r\dot{\rho} + b\dot{\theta} - \dot{z}_{msrr}\right) \end{aligned} \tag{11}$$

Equations of the vertical motion of the aggregate mass of the tyres, the rims, and the SRM rotors at each corner are as follows.

$$\begin{aligned} M_{ufl}\ddot{z}_{ufl} &= k_{msfl}\left(z_{msfl} - z_{ufl}\right) - k_{tfl}\left(z_{ufl} - q_{fl}\right) - F_{vfl} \\ M_{ufr}\ddot{z}_{ufr} &= k_{msfr}\left(z_{msfr} - z_{ufr}\right) - k_{tfr}\left(z_{ufr} - q_{fr}\right) - F_{vfr} \\ M_{url}\ddot{z}_{url} &= k_{msrl}\left(z_{msrl} - z_{url}\right) - k_{trl}\left(z_{url} - q_{rl}\right) - F_{vrl} \\ M_{urr}\ddot{z}_{urr} &= k_{msrr}\left(z_{msrr} - z_{urr}\right) - k_{trr}\left(z_{urr} - q_{rr}\right) - F_{vrr} \end{aligned} \tag{12}$$

Equations of the vertical motion of the aggregate mass of the SRM stator and housing are

$$M_{msfl}\ddot{z}_{msfl} = k_{sfl}\left(z + c_f\rho - a\theta - z_{msfl}\right) + c_{sfl}\left(z + c_f\rho - a\theta - z_{msfl}\right)$$

Table 2: The vehicle parameter values.

Definition	Symbol	Units	Value
Vehicle total mass	M_t	kg	1770
Sprung mass	M_b	kg	1350
Mass of tyre, rim and SRM rotor of front wheel	M_{ufj}	kg	70
Mass of tyre, rim and SRM rotor of rear wheel	M_{urj}	kg	65
Mass of SRM stator and housing	M_{msij}	kg	37. 5
Roll moment of inertia	I_x	Kg m^2	480
Pitch moment of inertia	I_y	Kg m^2	1895
Yaw moment of inertia	I_z	Kg m^2	1875
Height of CG	h_{cg}	m	0. 504
Height of pitch center	h_p	m	0. 4
Height of roll center	h_c	m	0. 25
Distance from CG to front axle	a	m	1. 08
Distance from CG to rear axle	b	m	1. 62
Track width of front axle	c_f	m	1. 4
Track width of rear axle	c_r	m	1. 385
Wheel inertia	I_{wi}	Kg m^2	1. 8
Effective radius of wheel	R	m	0. 269
Stiffness of front suspension	k_{sfj}	N/m	22500
Stiffness of rear suspension	k_{srj}	N/m	19600
Damping of front/rear suspension	c_{sij}	Ns/m	1695
Sum of SRM and hub bearing stiffness	k_{msij}	N/m	6500000
Stiffness of tyre	k_{tij}	N/m	150000

$$
\begin{aligned}
&\quad - k_{msfl}\left(z_{msfl} - z_{ufl}\right) + Fv_{fl} + Fc_{fl} \\
M_{msfr}\ddot{z}_{msfr} &= k_{sfr}\left(z - c_f\rho - a\theta - z_{msfr}\right) \\
&\quad + c_{sfr}\left(z - c_f\rho - a\theta - z_{msfr}\right) \\
&\quad - k_{msfr}\left(z_{msfr} - z_{ufr}\right) + Fv_{fr} + Fc_{fr} \\
M_{msrl}\ddot{z}_{msrl} &= k_{srl}\left(z + c_r\rho + b\theta - z_{msrl}\right) \\
&\quad + c_{srl}\left(z + c_r\rho + b\theta - z_{msrl}\right) \\
&\quad - k_{msrl}\left(z_{msrl} - z_{url}\right) + Fv_{rl} + Fc_{rl} \\
M_{msrr}\ddot{z}_{msrr} &= k_{srr}\left(z - c_r\rho + b\theta - z_{msrr}\right) \\
&\quad + c_{srr}\left(z - c_r\rho + b\theta - z_{msrr}\right) \\
&\quad - k_{msrr}\left(z_{msrr} - z_{urr}\right) + Fv_{rr} + Fc_{rr}
\end{aligned} \tag{13}
$$

where x, y, and z are the longitudinal displacement, the lateral displacement, and the vertical displacement, respectively, of the sprung mass of the vehicle; ρ, θ, and ψ are the vertical, pitch, and roll motion of the sprung mass of the vehicle; δ is the steering angle; Fc is the electromagnetic active suspension force; F_{xij} and F_{yij} are the longitudinal tyre force and the lateral tyre force at each wheel; q_{ij} is road displacement excitation at each wheel; Fv_{ij} is the unbalanced vertical force of each SRM; M_{msij} is the aggregate mass of tyre, hub, and SRM rotor; M_{uij} is the aggregate mass of SRM stator and housing; and k_{msij} is the bearing stiffness. The subscripts i in the equations refer to the front (f) or rear (r), and the subscripts j in the equations refers to the left (l) side or the right (r) side. The vehicle parameter values used for this study are listed in Table 2.

The Pacejka nonlinear tyre model [21] is adopted in this paper to simulate the nonlinearity tyre force. The equations of the Pacejka nonlinear tyre mode are

$$
\begin{aligned}
F_{xij}(\lambda) &= D_x \\
&\quad \cdot \sin\left(C_x \arctan\left\{B_y\lambda - E_x\left[B_x\lambda - \arctan\left(B_x\lambda\right)\right]\right\}\right)
\end{aligned} \tag{14}
$$

$$
\begin{aligned}
F_{yij}(\alpha) &= D_y \\
&\quad \cdot \sin\left(C_y \arctan\left\{B_y\alpha - E_y\left[B_x\alpha - \arctan\left(B_x\alpha\right)\right]\right\}\right)
\end{aligned} \tag{15}
$$

$$
\begin{aligned}
B_x &= \frac{a_3 \sin\left[2\arctan\left(F_z/a_4\right)\right]}{C_x D_x} \\
C_x &= b_0 \\
D_x &= \frac{a_1 F_z^2 + a_2 F_z}{1000} \\
E_x &= a_5 F_z + a_6 \\
B_y &= \frac{b_3 \sin\left[2\arctan\left(F_z/b_4\right)\right]}{C_y D_y} \\
C_y &= b_0
\end{aligned}
$$

$$D_y = \frac{b_1 F_z^2 + b_2 F_z}{1000}$$

$$E_y = b_5 F_z + b_6 \tag{16}$$

where λ is the slip ratio and B_x, C_x, D_x, E_x, B_y, C_y, D_y, and E_y are empirical parameters. F_{xij}, F_{yij} are the longitudinal force and lateral force.

3. LQG Controller of Electromagnetic Active Suspension

In this paper, several performance indexes of IWM-EV are considered by linear quadratic Gaussian (LQG) controller, such us vehicle body acceleration, airgap eccentricity, SRM stator acceleration, suspension dynamic deflections, and tyre deformation. It is easy to use the methodology of Analytic Hierarchy Process (AHP) to select the weighted coefficients of performance indexes [22]. This LQG controller is based on optimal control theory and the controller can propose different objective functions according to design requirements and improve the vehicle performance by comprehensively considering various performance indexes.

3.1. State Equation. 24 state variables and 14 output variables are chosen to constitute the vehicle system for controller, and the state equation can be described as

$$\dot{X} = AX + Bu + GQ \tag{17}$$

$$Y = CX + Du \tag{18}$$

where

$$X = \left[\dot{z}\ \dot{\theta}\ z\ \theta\ \dot{z}_{msfl}\ \dot{z}_{msfr}\ \dot{z}_{msrl}\ \dot{z}_{msrr}\ z_{msfl}\ z_{msfr}\ z_{msrl}\ z_{msrr}\ \dot{z}_{ufl}\ \dot{z}_{ufr}\ \dot{z}_{url}\ \dot{z}_{urr}\ z_{ufl}\ z_{ufr}\ z_{url}\ z_{urr}\ q_{fl}\ q_{fr}\ q_{rl}\ q_{rr}\right]^T \tag{19}$$

$$Y = \left[z\ \theta\ Z_{msfl}\ Z_{msfr}\ Z_{msrl}\ Z_{msrr}\ Z_{ufl}\ Z_{ufr}\ Z_{url}\ Z_{urr}\ q_{fl}\ q_{fr}\ q_{rl}\ q_{rr}\right]^T \tag{20}$$

$$u = F_c \tag{21}$$

where X is the state variables, Y is the output variables, u is the control vector, and A is the system matrix. B is control matrix. G is disturbance matrix. C is the output matrix. D is transfer matrix.

3.2. Selection of Performance Indexes' Weighted Coefficients for LQG Controller. AHP is a decision-making method. It decomposes elements that are always related to decision-making into objectives, criteria, and programs and then makes qualitative and quantitative analysis on this basis. It was used to select weighted coefficients of performance indexes for LQG controller [23].

3.2.1. Quantitating Scale Factor. The root mean square (RMS) of dynamic characteristics of the passive suspension vehicle can be produced by simulating the passive suspension vehicle model under the corresponding conditions. According to the literature, [22, 23], scale factors of vehicle dynamic characteristics can be described by the following equations:

$$\begin{aligned}\sigma_{BA}^2 \times 1 &= \sigma_{PA}^2 \times \beta_{PA} = \sigma_{AEij}^2 \times \beta_{AEij} = \sigma_{TDij}^2 \times \beta_{TDij} \\ &= \sigma_{SDDij}^2 \times \beta_{SDDij}\end{aligned} \tag{22}$$

where σ_{BA}, σ_{PA}, σ_{AEij}, σ_{TDij}, σ_{SDDij} are RMS of vehicle body vertical acceleration (BVA), pitch angle (PA), airgap eccentricity (AE), tyre deformation (TD), and suspension dynamic deflections (SDD). β_{PA}, β_{AEij}, β_{TDij}, β_{SDDij} are pitch angle scale factor, airgap eccentricity scale factor, tyre deformation scale factor, and suspension dynamic deflections scale factor; vehicle body vertical acceleration (BVA) scale factor is 1.

3.2.2. Subjective Weighted Coefficient

(1) Making Judgment Matrix H. h_{ij} (no unit) is the relative importance ratio of the index i and j. Table 3 is a comparison table of the relative importance of each index. According to Table 3, judgment matrix H shown in (23) can be constructed.

$$H = \left(h_{ij}\right)_{nxn} = \begin{matrix} 1 & h_{12} & h_{13} & \cdots & \cdots & h_{1n} \\ \frac{1}{h_{12}} & 1 & h_{23} & h_{24} & \cdots & h_{2n} \\ \frac{1}{h_{13}} & \frac{1}{h_{23}} & 1 & h_{34} & \cdots & h_{3n} \\ \vdots & \frac{1}{h_{24}} & \frac{1}{h_{34}} & \cdots & \cdots & \cdots \\ \vdots & \vdots & \vdots & & & \vdots \\ \frac{1}{h_{1n}} & \frac{1}{h_{2n}} & \frac{1}{h_{3n}} & \cdots & \cdots & 1 \end{matrix} \tag{23}$$

Calculate multiplying vector of every row.

$$\begin{aligned} M &= \left[M_1, M_2, M_3, \ldots, M_n\right]^T \\ M_i &= \prod_{j=1}^{n} hij, \quad (i, j = 1, 2, 3, \ldots, n)\end{aligned} \tag{24}$$

TABLE 3: Comparison table.

i/j	Equal	Moderately	Strongly	Very	Extremely
h_{ij}	1	3	5	7	9

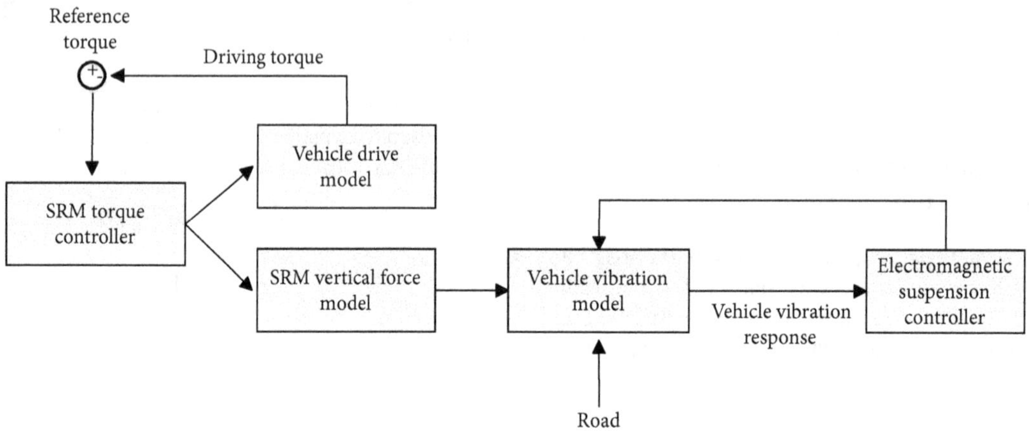

FIGURE 3: Control diagram for the electromagnetic active suspension.

Calculate $\sqrt[n]{W}$ and W

$$\begin{aligned} \overline{W} &= \left[\overline{W}_1, \overline{W}_2, \overline{W}_3, \ldots, \overline{W}_n\right]^T \\ \overline{W}_i &= \sqrt[n]{M_i}, \quad (i = 1, 2, 3, \ldots, n) \\ W &= \frac{\overline{W}}{\sum_{i=1}^{n} \overline{W}_i}, \quad (i = 1, 2, 3, \ldots, n) \end{aligned} \tag{25}$$

where W is the subjective weighted coefficient.

(2) Maximum Eigenvalue and Consistency Checking of Matrix H

$$\lambda_{max} = \sum_{i=1}^{n} \frac{(HW)_i}{nW_i}, \quad (i = 1, 2, 3, \ldots, n) \tag{26}$$

where λ_{max} is the maximum eigenvalue of H.

If every element in matrix H satisfies the equations $h_{ij} = 1/h_{ij}$ and $h_{ij} = h_{ik} \times h_{kj}$, the matrix H is the consistency matrix. The method of consistency checking is comparing random consistency ratio CR.

$$CR = \frac{\lambda_{max} - n}{RI\,(n-1)}, \quad (i = 1, 2, 3, \ldots, n) \tag{27}$$

RI is the random consistency index. When n is 14, RI is 1. 59. If CR is less than 1, H passes the consistency checking. If CR is more than 1, H needs to be revised [24].

3.2.3. Weighted Coefficients. Subjective weighted coefficients of pitch angle, airgap eccentricity, tyre deformation, and suspension dynamic deflections can be calculated as follows.

$$W_1 = \frac{W_i}{\gamma_i} \tag{28}$$

$$\begin{aligned} q_1 &= 1, \\ q_2 &= \beta_{PA}\gamma_2, \\ q_3 &= \beta_{AEfl}\gamma_3, \\ q_4 &= \beta_{AEfr}\gamma_4, \\ q_5 &= \beta_{AErl}\gamma_5, \\ q_6 &= \beta_{AErr}\gamma_6 \\ q_7 &= \beta_{TDfl}\gamma_7, \\ q_8 &= \beta_{TDfr}\gamma_8, \\ q_9 &= \beta_{TDrl}\gamma_9, \\ q_{10} &= \beta_{TDrr}\gamma_{10}, \\ q_{11} &= \beta_{SDDfl}\gamma_{11} \\ q_{12} &= \beta_{SDDfr}\gamma_{12}, \\ q_{13} &= \beta_{SDDrl}\gamma_{13}, \\ q_{14} &= \beta_{SDDrr}\gamma_{14} \end{aligned} \tag{29}$$

3.3. LQG Controller Design. The control diagram for the electromagnetic active suspension is shown in Figure 3. Vehicle body acceleration and pitch angle are chosen to be the performance indexes for improving ride comfort and handling stability. Airgap eccentricity, tyre deformation, and suspension dynamic deflections of each wheel are chosen to be the performance indexes for reducing the vertical vibration of SRM.

The performance function is defined as follows:

$$\begin{aligned} J = \frac{1}{2}\int_0^{\infty} & q_1\ddot{z}_b^2 + q_2\theta^2 + q_3\left(z_{msfl} - z_{ufl}\right)^2 \\ & + q_4\left(z_{msfr} - z_{ufr}\right)^2 + q_5\left(z_{msrl} - z_{url}\right)^2 \\ & + q_6\left(z_{msrr} - z_{urr}\right)^2 + q_7\left(z_{ufl} - q_{fl}\right)^2 \\ & + q_8\left(z_{ufr} - q_{fr}\right)^2 + q_9\left(z_{url} - q_{rl}\right)^2 \\ & + q_{10}\left(z_{urr} - q_{rr}\right) + q_{11}\left(z + c_f\rho - a\theta - z_{msfl}\right)^2 \\ & + q_{12}\left(z - c_f\rho - a\theta - z_{msfr}\right)^2 \\ & + q_{13}\left(z + c_r\rho - a\theta - z_{msrl}\right)^2 \\ & + q_{14}\left(z + c_r\rho - a\theta - z_{msrr}\right)^2 + Ru^2\Big]\, dt = \frac{1}{2} \\ & \cdot \int_0^{\infty} Y^TQY + u^TRu\Big)\, dt \end{aligned} \tag{30}$$

where q_1 is the weighted coefficient of body vertical acceleration; q_2 is the weighted coefficient of pitch angle; q_3, q_4, q_5, and q_6 are the weighted coefficients of airgap eccentricity; q_7, q_8, q_9, and q_{10} are the weighted coefficients of tyre deformation; q_{11}, q_{12}, q_{13}, and q_{14} are the weighted coefficients of suspension dynamic deflections; Q is the weight matrix of the state; u is the control vector of Fc; and R is the weighted coefficient of u.

Rewrite (30) as a standard form:

$$J = \frac{1}{2}\int_0^{\infty}\left[X^TQ'X + 2X^TNu + u^TR'u\right] dt \tag{31}$$

where

$$\begin{aligned} Q' &= C^TQC, \\ N &= C^TQD, \\ R' &= R + D^TQD \end{aligned} \tag{32}$$

$$u = -KX \tag{33}$$

and the optimal control feedback matrix K can be given by the Riccati equation.

$$\begin{aligned} & PA + A^TP + Q - (PB + N)\,R^{-1}\left(B^TP + N^T\right) = 0 \\ & K = B^TP + N^T \end{aligned} \tag{34}$$

It is easy to calculate feedback matrix K by using Matlab function as follows.

$$[K, S, E] = \mathrm{lqr}\,(A, B, Q, R, N) \tag{35}$$

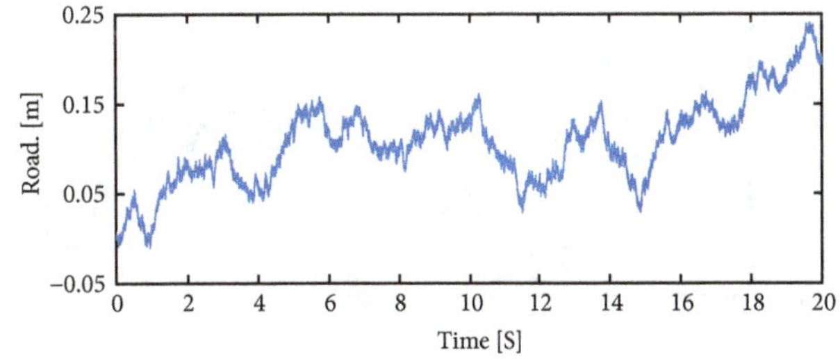

FIGURE 4: Effective road excitation.

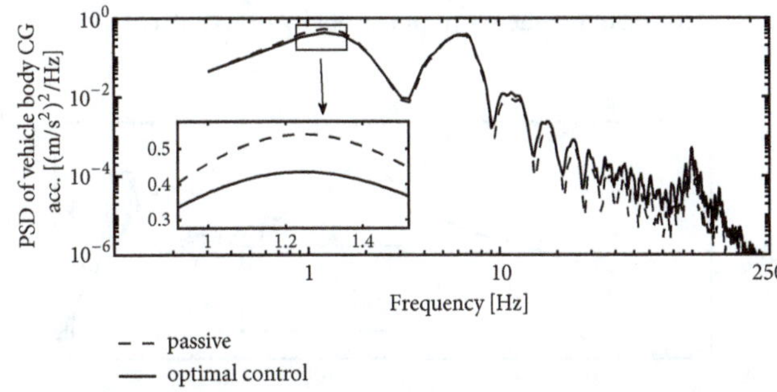

FIGURE 5: Frequency response of vehicle body acceleration.

4. Suppression of SRM Vertical Force and Promotion of Vehicle Dynamic Performance

SRM vibration and IWM-EV body movement influence each other, so it is difficult to improve the dynamic performance of the vehicle considering only the SRM structure and controller design. Therefore, this study used electromagnetic active suspension, connecting body and wheels, to reduce the SRM vibration and the negative influence on vehicle body.

4.1. Frequency Response on Stochastic Roads. For this simulation, the road excitations at four tyres are assumed to be the same, only the front left in-wheel SRM response is presented in this paper. In order to study the effects of electromagnetic active suspension under urban condition and high-speed condition, vehicle responses are simulated at 60 km/h vehicle speed on class C road and 100 km/h vehicle speed on class C road.

4.1.1. Response under Urban Condition. Vehicle responses are simulated at 60 km/h vehicle speed on class C road as shown in Figure 4.

Responses of vehicle body acceleration, airgap eccentricity, SRM stator acceleration, suspension dynamic deflections, and tyre deformation at 60 km/h are shown in Figures 5–9

It can be seen that the electromagnetic active suspension can reduce the power spectral density (PSD) values of vehicle body acceleration, airgap eccentricity, SRM stator acceleration, suspension dynamic deflections, and tyre deformation at their resonance frequency. Specifically, Figure 5 shows that the resonance frequency of vehicle body acceleration is 1.221 Hz, and the peak value of vehicle body acceleration decreases from 0.5395 $(m/s^2)^2$/Hz to 0.4343 $(m/s^2)^2$/Hz and

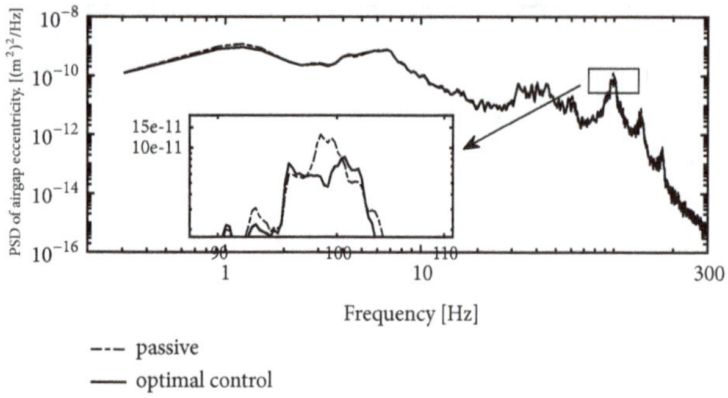

FIGURE 6: Frequency response of airgap eccentricity.

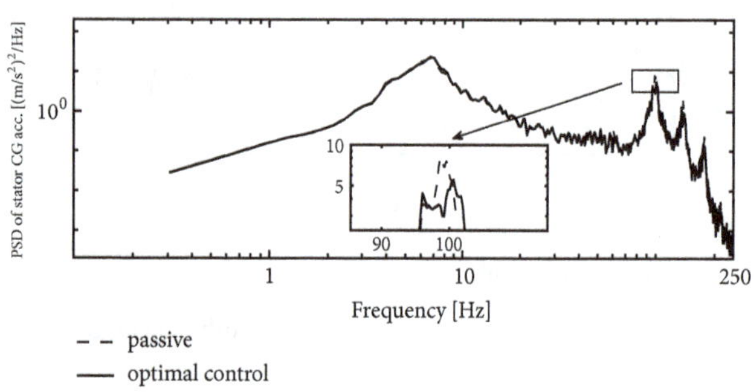

FIGURE 7: Frequency response of stator acceleration.

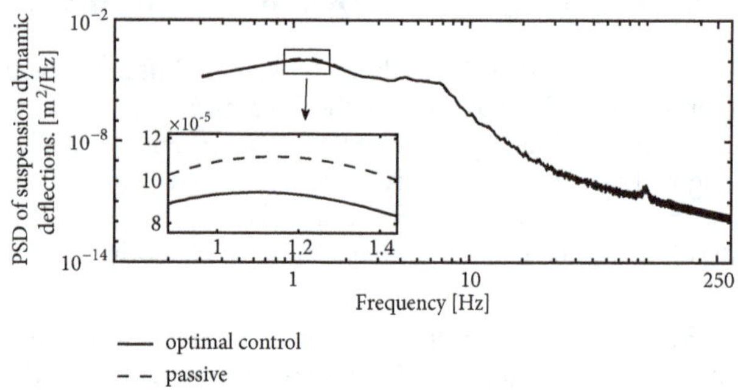

FIGURE 8: Frequency response of suspension dynamic deflections.

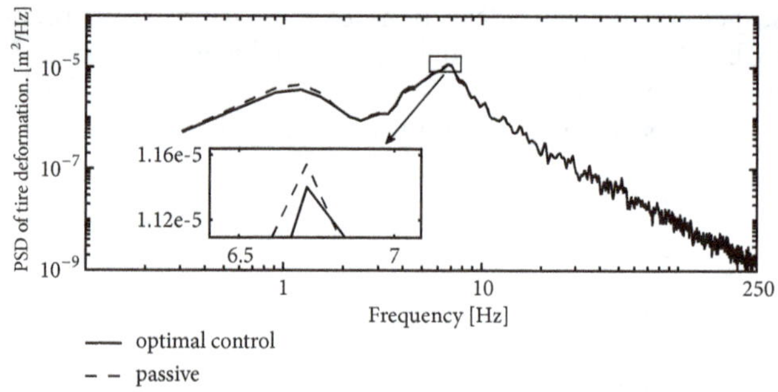

FIGURE 9: Frequency response of tyre deformation.

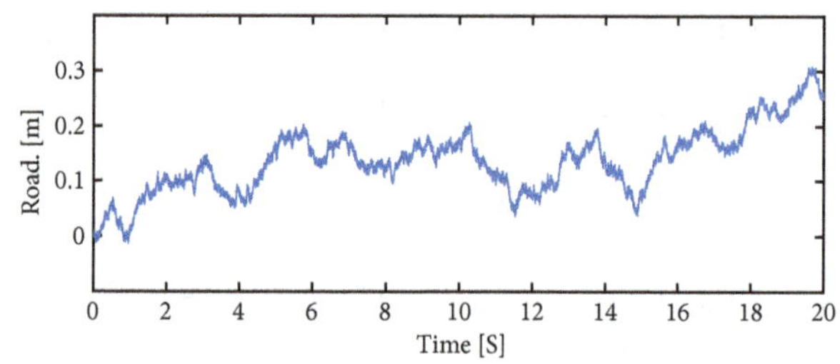

FIGURE 10: Effective road excitation.

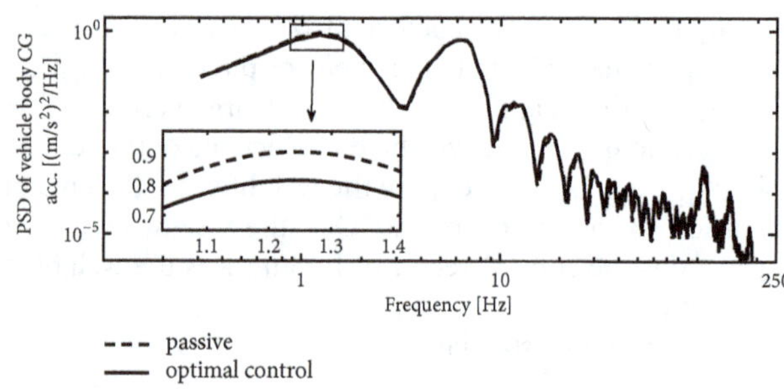

FIGURE 11: Frequency response of vehicle body acceleration.

reduced by 19.5%. This indicates that the electromagnetic active suspension can reduce the force acting between the wheel and the vehicle body, and ride comfort improved. Figure 6 shows that the resonance frequency of airgap eccentricity is 98.57 Hz, and the peak value of airgap eccentricity decreases from $1.307e^{-10}\,m^2/Hz$ to $8.339e^{-11}\,m^2/Hz$ and reduced by 36.2%. Figure 7 shows that the resonance frequency of stator acceleration is 98.57 Hz, and the peak value of stator acceleration decreases from 8.147 $(m/s^2)^2/Hz$ to 5.520 $(m/s^2)^2/Hz$ and reduced by 32.2%. Figures 6 and 7 mean that electromagnetic active suspension conspicuously suppressed SRM vertical vibration, and operating stability of SRM improved. Furthermore, SRM's negative influence on the vehicle reduced. Figure 8 shows that the resonance frequency of suspension dynamic deflections is 1.221 Hz, and the peak value of suspension dynamic deflections decreases from $1.30e^{-4}\,m^2/Hz$ to $1.06e^{-4}\,m^2/Hz$ and reduced by 18.4%. Figure 9 shows that the resonance frequency of tyre deformation is 6.714 Hz, and the peak value of tyre deformation decreases from $1.154e^{-5}\,m^2/Hz$ to $1.140e^{-5}\,m^2/Hz$ and reduced by 1.21%. Because driving stability is related to suspension dynamic deflections and tyre deformation, decrease of them implies that electromagnetic active suspension can improve vehicle safety. Therefore, compared with passive suspension, electromagnetic active suspension can availably reduce values of body acceleration, airgap eccentricity, SRM stator acceleration, and so on. Ride comfort and safety under urban condition improved.

4.1.2. Response under High-Speed Condition. Vehicle responses are simulated at 100 km/h vehicle speed on class C road as shown in Figure 10. Vehicle body acceleration, airgap eccentricity, SRM stator acceleration, tyre deformation, and suspension dynamic deflections responses at 100 km/h are shown in Figures 11–15.

Under high-speed condition, the power spectral density (PSD) values of vehicle body acceleration, airgap eccentricity,

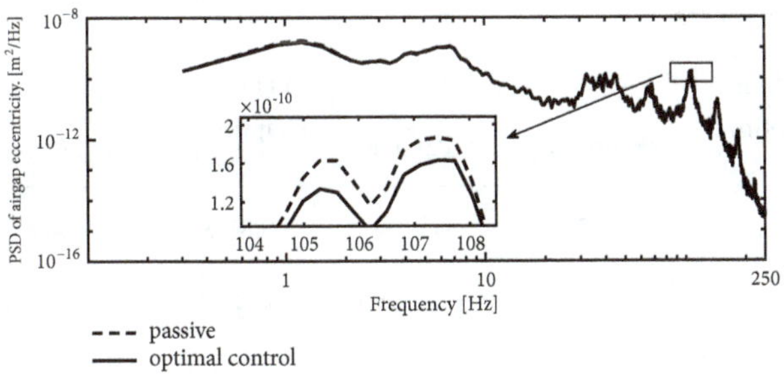

Figure 12: Frequency response of airgap eccentricity.

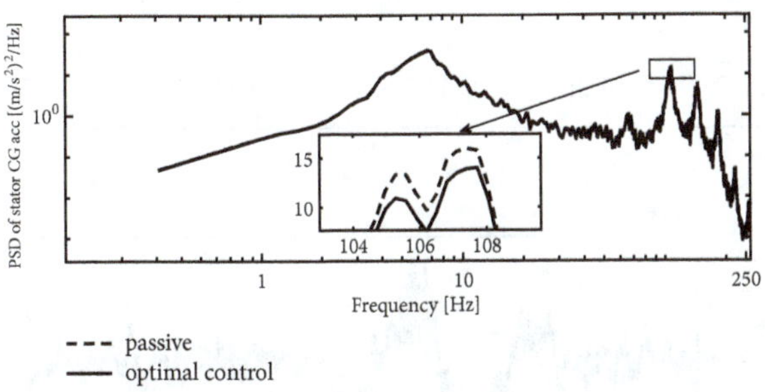

Figure 13: Frequency response of stator acceleration.

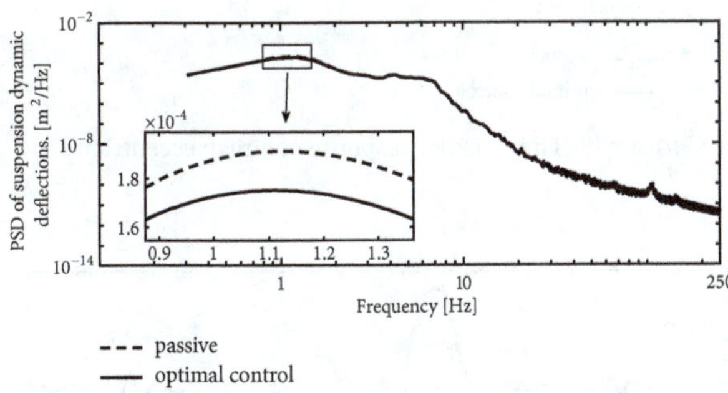

Figure 14: Frequency response of suspension dynamic deflections.

SRM stator acceleration, suspension dynamic deflections, and tyre deformation can be reduced at their resonance frequency. Figure 11 shows that the peak value of vehicle body acceleration decreases from 0.9104 $(m/s^2)^2$/Hz to 0.8158 $(m/s^2)^2$/Hz and reduced by 10. 4%. This means that, even under high-speed conditions, the electromagnetic active suspension can still significantly reduce the vibration of the vehicle body, and the ride comfort improved obviously. As can be seen from Figures 12 and 13, the peak value of airgap eccentricity decreases from $1.867e^{-10}$ m^2/Hz to $1.625e^{-10}$ m^2/Hz and reduced by 13.0%; the peak value of stator acceleration decreases from 15.95 $(m/s^2)^2$/Hz to 13.91 $(m/s^2)^2$/Hz and reduced by 12.8%. This indicates that the vibration of SRM under high-speed condition was reduced effectively. Figure 14 shows that the peak value of suspension dynamic deflections decreases from $2.24e^{-4}$ m^2/Hz to $2.02e^{-4}$ m^2/Hz and reduced by 9.8%. Figure 15 shows that the peak value of tyre deformation decreases from $1.902e^{-5}$ m^2/Hz to $1.894e^{-5}$ m^2/Hz and reduced by 0.42%.

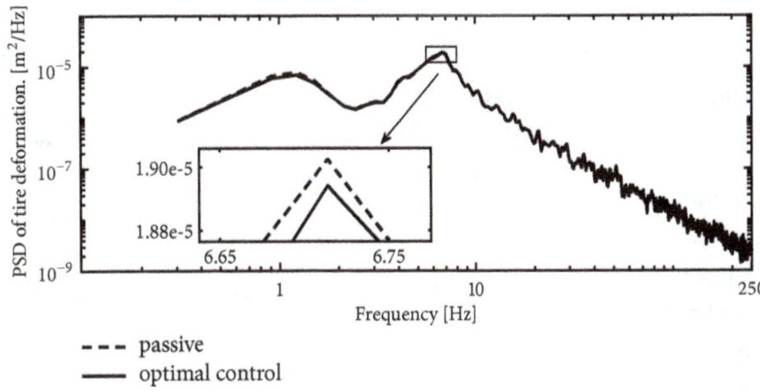

Figure 15: Frequency response of tyre deformation.

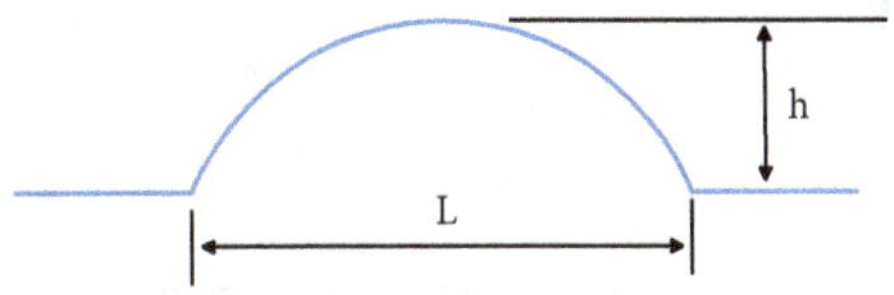

Figure 16: Bump road model.

Therefore, electromagnetic active suspension can improve ride comfort and reduce SRM vibration under high-speed condition. Because decreases of suspension dynamic deflections and tyre deformation can greatly enhance the driving stability under high-speed condition, the vehicle safety improved.

The effects of electromagnetic active suspension under high-speed condition are weaker than effects under urban condition, because the weighted coefficients of LQG controller are more suitable for urban condition. All the vehicle responses at both 60 km/h and 100 km/h were shown in Table 4.

4.2. Time Response on Representative Roads

4.2.1. Response on Bump Road. In order to simulate the speed bump, bump road model is shown in Figure 16. The simplified model can be described as (36), where L, h are the width and height of the bump and V is the vehicle speed. Setting different vehicle speed or changing the width of the bump can easily adjust road excitation frequency. To simulate the ordinary low frequency speed hump excitation, L=0.8m, h=0.07m, and V is 0.8 m/s. Vehicle body acceleration, pitch angle, airgap eccentricity, suspension dynamic deflections, and tyre deformation responses on class C road are shown in Figures 17–21.

Figure 17 shows the vehicle body acceleration responses. When t=1s, the vehicle body starts to vibrate under the excitation of bump road. When t=2.16s, the vehicle body acceleration reaches peak value, and the peak value decreases by 6.11%. Figure 18 shows the response of pitch angle. When t=1s, the bump road excitation starts to affect the vehicle. When t=3s, the vehicle system tends to be stable. The peak value of pitch angle decreases by 3.16%. Figures 17 and 18 mean that the vertical vibration of the body was reduced and the ride comfort improved, because the vertical motion of the unsprung mass and its impact on vehicle body were reduced.

TABLE 4: Vehicle responses.

Frequency response	Resonance frequency (Hz)	Units	60 km/h Passive	60 km/h Optimal control	60 km/h reduce by	100 km/h Passive	100 km/h Optimal control	100 km/h reduce by
Vehicle body acceleration	1.221	$(m/s^2)^2/Hz$	0.5395	0.4343	19.5%	0.9104	0.8158	10. 4%
Airgap eccentricity	98.57	m^2/Hz	$1.307e^{-10}$	$8.339e^{-11}$	36.2%	$1.867e^{-10}$	$1.625e^{-10}$	13.0%
Stator acceleration	98.57	$(m/s^2)^2/Hz$	8.147	5.520	32.2%	15.95	13.91	12.8%
Suspension dynamic deflections	1.221	m^2/Hz	$1.30e^{-4}$	$1.06e^{-4}$	18.4%	$2.24e^{-4}$	$2.02e^{-4}$	9.8%
Tyre deformation	6.714	m^2/Hz	$1.154e^{-5}$	$1.140e^{-5}$	1.21%	$1.902e^{-5}$	$1.894e^{-5}$	0.42%

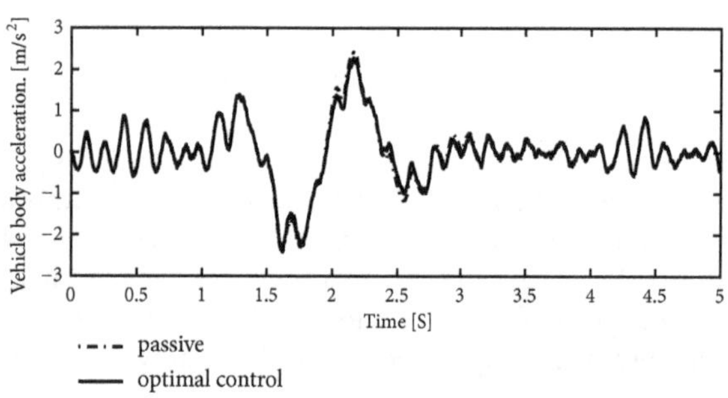

FIGURE 17: Time history response of vehicle body acceleration.

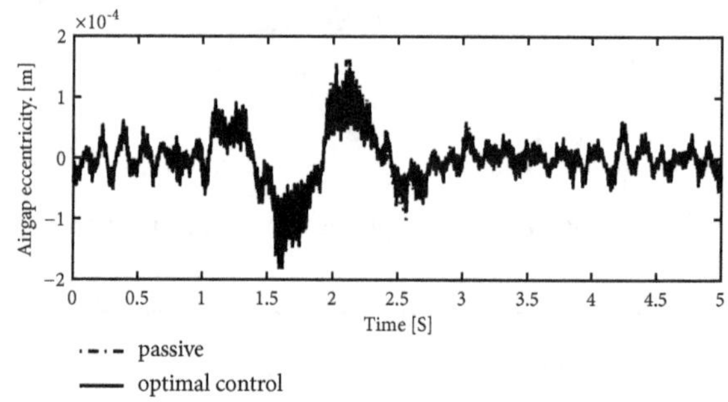

FIGURE 19: Time history response of airgap eccentricity.

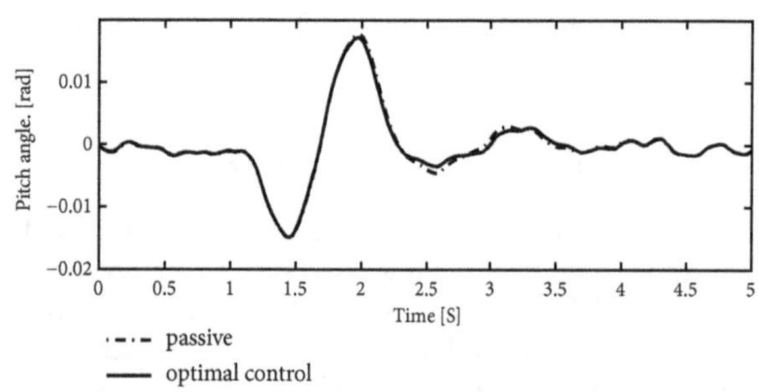

FIGURE 18: Time history response of pitch angle.

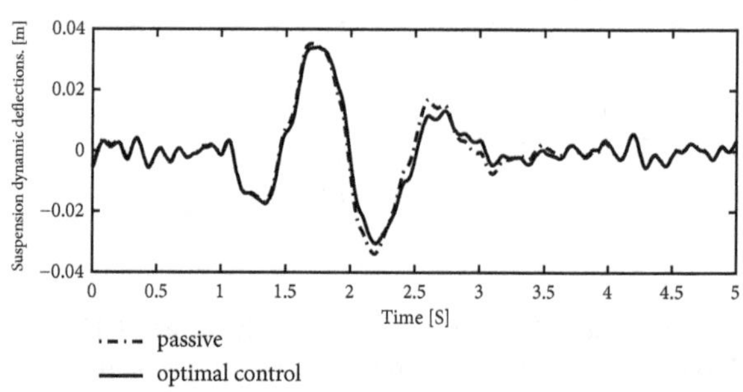

FIGURE 20: Time history response of suspension dynamic deflections.

As can be seen from Figures 19 and 20, the trend of airgap eccentricity is similar to vehicle body acceleration, and the peak value of airgap eccentricity decreases by 8.89%. When t=1s, the suspension system starts to shrink. When t=3s, the electromagnetic active suspension system approaches steady state that is faster than passive suspension system at t=3.7s. The peak value of suspension dynamic deflection decreases by 3.64%. Figure 21 shows that the peak value of tyre deformation decreases by 7.44%. This implies that the electromagnetic active suspension can effectively reduce the airgap eccentricity, suspension dynamic deflection, and tyre deformation on bump road. The unsprung mass had less

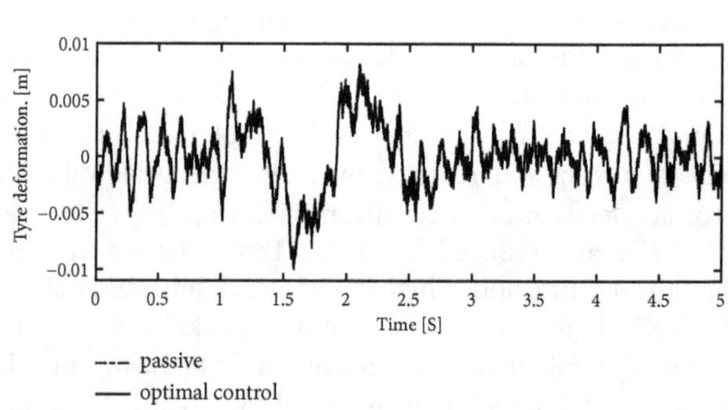

FIGURE 21: Time history response of tyre deformation.

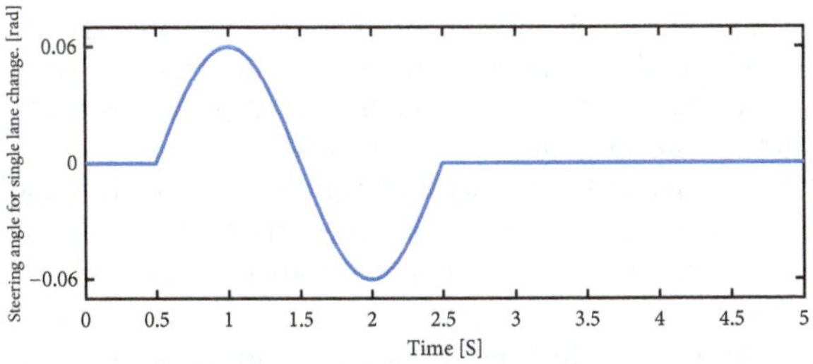

FIGURE 22: Single lane change.

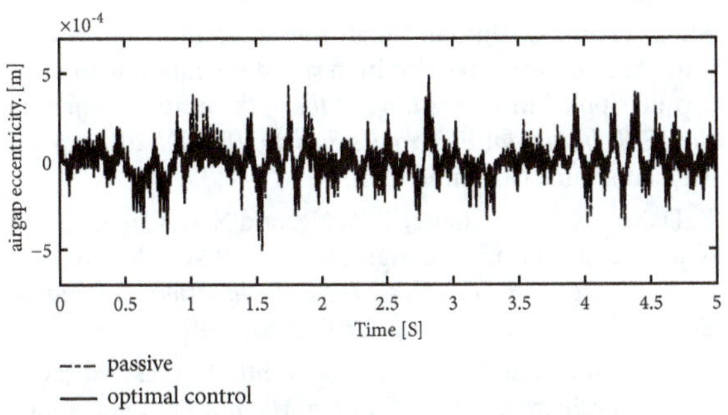

FIGURE 23: Time history response of airgap eccentricity.

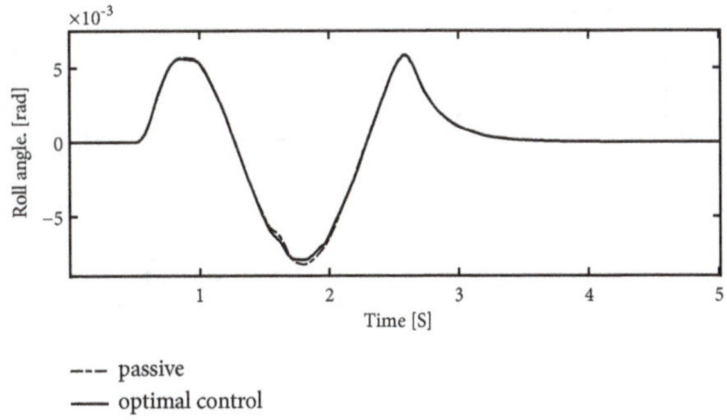

FIGURE 24: Time history response of roll angle.

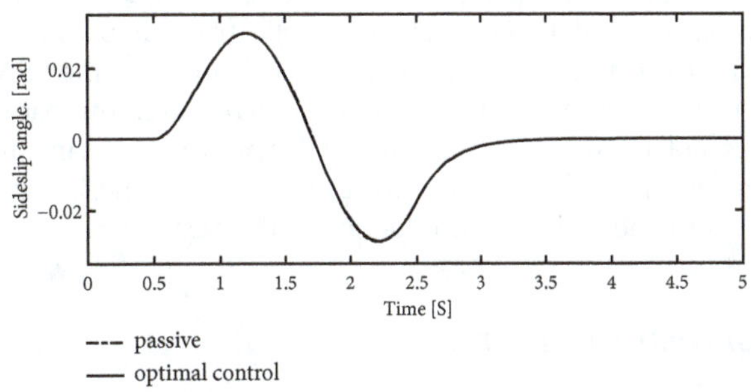

FIGURE 25: Sideslip angle.

influence on the body and can provide higher stability for the vehicle.

$$z_r = \frac{h(1 - \cos 2\pi f t)}{2r}$$
$$f = \frac{V}{L} \tag{36}$$

4.2.2. Response on Single Lane Change. In order to study the effects of electromagnetic active suspension on handling stability, vehicle responses are simulated under single lane change condition. The steering angle model is shown in Figure 22. Airgap eccentricity, roll angle, sideslip angle, and yaw rate responses at 100 km/h on class C road are shown in Figures 23–26.

Figure 23 shows that the airgap eccentricity response does not have much relevance to steering angle, but the electromagnetic active suspension efficiently reduces the airgap eccentricity. Figure 24 shows that the vehicle starts to roll at 0.5s. The roll angle reaches peak value at 1.8s and decreases by 3.66%. Figures 25 and 26 show that the trends of sideslip angle and yaw rate are similar to steering angle. Although the contributions of the electromagnetic active suspension to sideslip angle and yaw rate under this condition are not very significant, the electromagnetic active suspension system does have positive influences on the lateral dynamic performance.

5. Conclusions

In order to reduce SRM vertical vibration and improve IWM-EV dynamic performance, this paper used electromagnetic active suspension and LQG controller to suppress the SRM unbalanced radial force. The controller based on optimal control theory can efficiently match electromagnetic active suspension. In theory, compared with the passive suspension, this electromagnetic active suspension can reduce SRM vertical vibration and provide better dynamic performance to the IWM-VE. Based on the analyzed simulation results, the following 3 conclusions can be summarize:

(1) This electromagnetic active suspension could effectively improve the ride comfort. Vehicle body acceleration was reduced on urban condition, high-speed condition, and representative roads. For urban condition, the vehicle body acceleration reduced by 19.5% near the vehicle body's resonance frequency 1.22 Hz. For high-speed condition, the vehicle body acceleration reduced by 10.4% near the vehicle body's resonance frequency 1.22 Hz.

(2) This electromagnetic active suspension could effectively reduce the SRM vertical vibration. SRM airgap eccentricity and stator acceleration were both reduced. Specifically, when the frequencies of the SRM airgap eccentricity and stator acceleration are near the stator's resonance frequency of 107.4 Hz, the SRM vertical vibration shows a most obvious drop.

(3) This electromagnetic active suspension could improve the safety. Under both urban condition and high-speed condition, suspension dynamic deflection and tyre deformation were reduced. On representative roads, electromagnetic active suspension reduced roll angle, sideslip angle, and yaw rate. It indicated that the electromagnetic active suspension can improve lateral dynamic performance and reduce SRM vertical vibration at the same time.

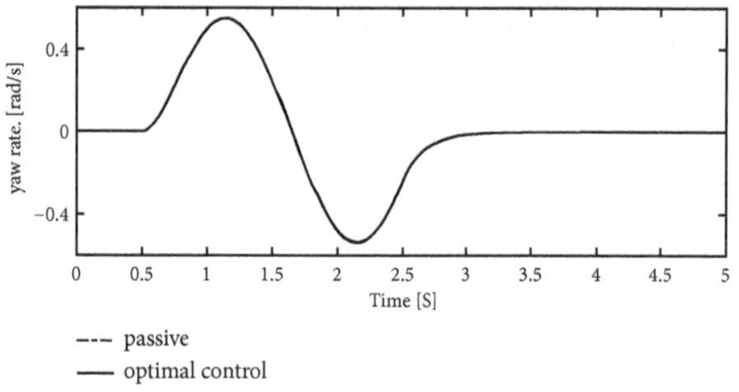

FIGURE 26: Yaw rate.

In summary, this paper used electromagnetic active suspension and LQG controller to effectively improve the IWM-EV ride comfort and safety and also reduce the SRM vertical vibration. Due to limited research resource, this paper does not construct an experimental platform to verify the simulation results. But we are preparing an experimental platform and will give introduction in the next work.

Conflicts of Interest

The authors declare that they have no conflicts of interest.

Acknowledgments

This research is supported by the National Natural Science Foundation of China (Grant No. 51605390), Foundation of Educational Commission of Sichuan Province (Grant No. 16ZA0161), the Open Research Subject of Key Laboratory of Sichuan Provincial Automotive Engineering (Grant No. szjj2015-048), Foundation of the Science and Technology Research and Development Program of Sichuan Province (Grant No. 2017GZ0103).

References

[1] M. Divandari and A. Dadpour, "Radial force and torque ripple optimization for acoustic noise reduction of SRM drives via fuzzy logic control," in *Proceedings of the 2010 9th IEEE/IAS International Conference on Industry Applications, INDUSCON 2010*, Brazil, November 2010.

[2] F. L. dos Santos, J. Anthonis, and H. Van der Auweraer, "Multiphysics thermal and NVH modeling: Integrated simulation of a switched reluctance motor drivetrain for an electric vehicle," in *Proceedings of the 2012 IEEE International Electric Vehicle Conference (IEVC)*, pp. 1–7, Greenville, SC, USA, March 2012.

[3] I. Husain, A. Radun, and J. Nairus, "Unbalanced force calculation in switched-reluctance machines," *IEEE Transactions on Magnetics*, vol. 36, pp. 330–338, 2000.

[4] N. K. Sheth and K. R. Rajagopal, "Variations in overall developed torque of a switched reluctance motor with airgap nonuniformity," *IEEE Transactions on Magnetics*, vol. 41, no. 10, pp. 3973–3975, 2005.

[5] B. M. Ebrahimi, J. Faiz, and M. J. Roshtkhari, "Static-, dynamic-, and mixed-eccentricity fault diagnoses in permanent-magnet synchronous motors," *IEEE Transactions on Industrial Electronics*, vol. 56, no. 11, pp. 4727–4739, 2009.

[6] N. K. Sheth and K. R. Rajagopal, "Effects of nonuniform airgap on the torque characteristics of a switched reluctance motor," *IEEE Transactions on Magnetics*, vol. 40, no. 4, pp. 2032–2034, 2004.

[7] E. Afjei and H. Torkaman, "Airgap eccentricity fault diagnosis in switched reluctance motor," in *Proceedings of the 1st Power Electronics and Drives, Systems and Technologies Conference, PEDSTC 2010*, pp. 290–294, Iran, February 2010.

[8] M. C. Ta and C. Dufour, "Real-time simulation and control of reluctance motor drives for high speed operation with reduced torque ripple," in *Proceedings of the 37th Annual Conference of the IEEE Industrial Electronics Society, IECON 2011*, pp. 4176–4181, Australia, November 2011.

[9] X. D. Xue, K. W. E. Cheng, T. W. Ng, and N. C. Cheung, "Multi-objective optimization design of in-wheel switched reluctance motors in electric vehicles," *IEEE Transactions on Industrial Electronics*, vol. 57, no. 9, pp. 2980–2987, 2010.

[10] H. Chaozhi, "A Stator novel structure to decrease vibration of switched reluctance motor," *China Micro Motor*, vol. 45, no. 9, pp. 37–40, 2017.

[11] X. Zhang, X. Wang, Y. Yang, and B. Wei, "Vibration reduction of a switched reluctance motor using new rotor tooth with slot on each side," *Zhongguo Dianji Gongcheng Xuebao/Proceedings of the Chinese Society of Electrical Engineering*, vol. 35, no. 6, pp. 1508–1515, 2015.

[12] P. C. Desai, M. Krishnamurthy, N. Schofield, and A. Emadi, "Novel switched reluctance machine configuration with higher number of rotor poles than stator poles: Concept to implementation," *IEEE Transactions on Industrial Electronics*, vol. 57, no. 2, pp. 649–659, 2010.

[13] J. Ye, B. Bilgin, and A. Emadi, "An extended-speed low-ripple torque control of switched reluctance motor drives," *IEEE Transactions on Power Electronics*, vol. 30, no. 3, pp. 1457–1470, 2015.

[14] P. Song, M. Tomizuka, and C. Zong, "A novel integrated chassis controller for full drive-by-wire vehicles," *Vehicle System Dynamics*, vol. 53, no. 2, pp. 215–236, 2015.

[15] S. Ko, J. Ko, S. Lee, J. Cheon, and H. Kim, "A study on the road friction coefficient estimation and motor torque control for an in-wheel electric vehicle," *Proceedings of the Institution of Mechanical Engineers, Part D: Journal of Automobile Engineering*, vol. 229, no. 5, pp. 611–623, 2015.

[16] Y. Wang, P. Li, and G. Ren, "Electric vehicles with in-wheel switched reluctance motors: Coupling effects between road excitation and the unbalanced radial force," *Journal of Sound and Vibration*, vol. 372, pp. 69–81, 2016.

[17] Y.-Y. Wang, Y.-N. Li, W. Sun, and L. Zheng, "Effect of the unbalanced vertical force of a switched reluctance motor on the stability and the comfort of an in-wheel motor electric vehicle," *Proceedings of the Institution of Mechanical Engineers, Part D: Journal of Automobile Engineering*, vol. 229, no. 12, pp. 1569–1584, 2015.

[18] J. Lin, K. W. E. Cheng, Z. Zhang, and X. Xue, "Experimental investigation of in-wheel switched reluctance motor driving system for future electric vehicles," in *Proceedings of the 2009*

3rd International Conference on Power Electronics Systems and Applications, PESA 2009, China, May 2009.

[19] A. Labak and N. C. Kar, "Outer rotor switched reluctance motor design for in-wheel drive of electric bus applications," in *Proceedings of the 2012 XXth International Conference on Electrical Machines (ICEM)*, pp. 418–423, Marseille, France, September 2012.

[20] R. Krishnan, *Switched Reluctance Motor Drives-Modeling, Simulation, Analysis, Design and Applications*, CRC Press, Boca Raton, Florida, 2001.

[21] E. Bakker, L. Nyborg, and H. B. Pacejka, "Tyre modelling for use in vehicle dynamics studies," *SAE Technical Papers*, Article ID 870421, pp. 1–15, 1987.

[22] L. Chai and T. Sun, "The design of LQG controller for active suspension based on analytic hierarchy process," *Mathematical Problems in Engineering*, vol. 2010, Article ID 701951, 19 pages, 2010.

[23] S.-A. Chen, F. Qiu, R. He, and S.-L. Lu, "Method for choosing weights in a suspension LQG control," *Journal of Vibration and Shock*, vol. 27, no. 2, pp. 65–68, 2008.

[24] T. Xiao and Y. Shihua, "Method of comparison matrix consistency adjustment based on AHP," *Ordnance Industry Automation*, vol. 27, no. 4, 2008.

7

Study on Load Parameters Estimation during AMT Startup

Yunxia Li, Yufeng Lu, Huilai Sun, and Peng Yan

School of Mechanical & Automotive Engineering, Qilu University of Technology (Shandong Academy of Sciences), Jinan, China

Correspondence should be addressed to Yunxia Li; ldyffr@126.com

Academic Editor: James Lam

The identification problem of the load parameters is the basis of researching the clutch control strategy and the shift schedule for heavy-duty AMT (Automated Mechanical Transmission). AMT is proposed as the soft starter for the belt conveyor which expands AMT application areas. In order to accomplish the goal of AMT as the soft starter, the load parameters estimation methods including the load resistance torque and the load inertia are firstly studied to accommodate the actual control for AMT startup as the soft starter for belt conveyor. In the paper, the estimation models of the load resistance torque and the load inertia are established during AMT startup for belt conveyor. On the basis of the recursion least squares method, the on-line estimations for the load parameters are recognized according to the test data during AMT startup. The estimation results of these two parameters prove that the load parameters estimation models and the estimation methods during AMT startup are valid.

1. Introduction

Heavy-duty AMT is a type of transmission with complicated working condition and wider load range. The load parameters identification has significant influences on control method and control law for AMT. The researchers have studied the transmitting torque estimation of the clutch during the shifting process in terms of smaller vehicles for decades. Zhao et al. studied the transmitting torque estimation of the clutch during the shifting process on dry double-clutch transmission [1–3]. Pettersson et al. studied the output torque of the vehicle drive shaft during the shifting process based on the transmission model with elastic drive shaft [4, 5]. Baumann et al. designed a Luenberger observer to estimate the actual output torque of the drive shaft [6]. Yi et al. designed a self-adapt slid mode observer to estimate the transmitting torque of the automatic transmission [7]. The identification accuracies of these studies should be verified on real vehicles to demonstrate the theoretical validity. The recursive least squares method is widely used to identify the load parameters. The researchers have studied the least squares parameter estimation methods to for nonlinear system. Wang et al. studied highly efficient identification methods for dual-rate Hammerstein systems [8]. Zhang et al. studied separate block-based parameter estimation method for Hammerstein systems [9].

Belt faults such as breakage and slipping are often caused on account of excessive belt acceleration during the starting process of belt conveyors [10, 11]. Generally, the belt acceleration value is under 0.3 meter per square seconds in order to accelerate the belt conveyor slowly [12–15]. Li et al. studied the starting process and the shifting process for AMT softly starting the belt conveyor based on known load parameters [16, 17]. In application of AMT as the soft starter for belt conveyors, the unknown load parameters should be identified so that the starting process and the shifting process can be better controlled.

The half engagement point of the clutch is a critical point during the starting process which is related to the driving resistance torque on the transmission input shaft. As is known, the half engagement point of the clutch is related to the load resistance torque and the gear position. As a result, it is different if the load resistance torque or the gear position is different. In consequence, the identification for the half engagement of the clutch is the premise to recognize the load parameters. In this paper, the characteristics of the half engagement of the clutch will be used to establish the estimation model of the load resistance torque. Then the estimation model of the load inertia is proposed on the basis of the information about the output shaft speed and the clutch location. The recursion least squares method is applied to the load inertia estimation during AMT startup.

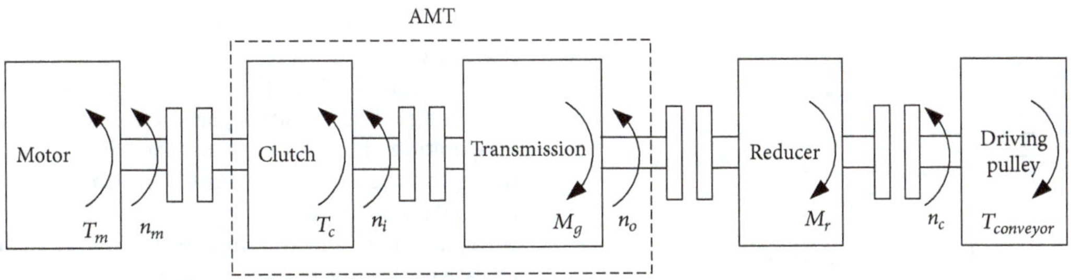

FIGURE 1: Drive line of AMT softly starting the belt conveyor.

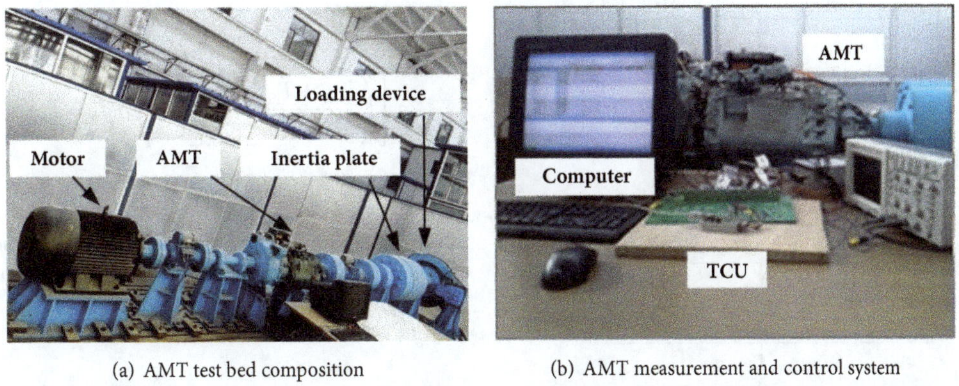

(a) AMT test bed composition (b) AMT measurement and control system

FIGURE 2: AMT test bed system.

As is shown in Figure 1, the drive line of AMT softly starting the belt conveyor mainly composes of the three-phase induction motor, AMT, the reducer, and the driving pulley of the belt conveyor.

The symbolic meanings are as follows: T_m is the output torque of the three-phase induction motor, T_c is the transmitting torque of the clutch, $T_{conveyor}$ is the resistance torque at the driving pulley of the belt conveyor, M_g is the friction loss of the transmission, M_r is the friction loss of the reducer, n_m is the motor speed, n_i is the input speed of the transmission, n_o and n_c are the output speed of the transmission and the driving pulley speed separately.

2. Structure of AMT Test Bed

AMT test bed consisting of the three-phase induction motor with 55 kilowatts, the efficiency instrument, HW19710 type pneumatic AMT manufactured by China Sinotruck, the inertia plate which is used to simulate the load inertia, the loading device which is used to simulate the load resistance torque, the hydraulic station which is used to provide the hydraulic pressure for the loading device to change the resistance torque conveniently, the air compressor which is used to provide the compressed air for AMT actuator, the AMT measurement and control system which is used to provide the command controls, and the sensors signal collections for communicates between AMT and the computer. The reduce gearbox is omitted for the limited space in the laboratory. The test bed system is shown in Figure 2.

3. Identification Models of the Load Parameters

3.1. Clutch Torque Characteristics. As is known, the torque characteristics of the dry clutch are related to the wear loss. Based on the clutch displacement sensor and the input shaft speed sensor, the estimation of the clutch wear loss can be measured which the diagram of the wear testing program is shown in Figure 3.

The dry clutch torque characteristics can be expressed by the displacement of the clutch actuator with a cubic displacement polynomial in

$$T_c(x_c) = \mu' \left[K_1(x_c - x_{ce} - \Delta x_{ce}) + K_2(x_c - x_{ce} - \Delta x_{ce})^2 + K_3(x_c - x_{ce} - \Delta x_{ce})^3\right] \quad (1)$$

where K_1, K_2, and K_3 are first-order coefficient, second-order coefficient, and third-order coefficient respectively, μ' is the dynamic friction coefficient between the driving disc and the fiction disc, x_c is the displacement of the clutch actuator, x_{ce} is the corresponding displacement of the clutch actuator related to the critical point of the new clutch empty running, and Δx_{ce} is the increase of the clutch actuator empty running displacement caused by the clutch wear.

3.2. Estimation Model of the Load Resistance Torque. The load resistance torque can be estimated on the basis of the relationship between the half engagement joint and the clutch

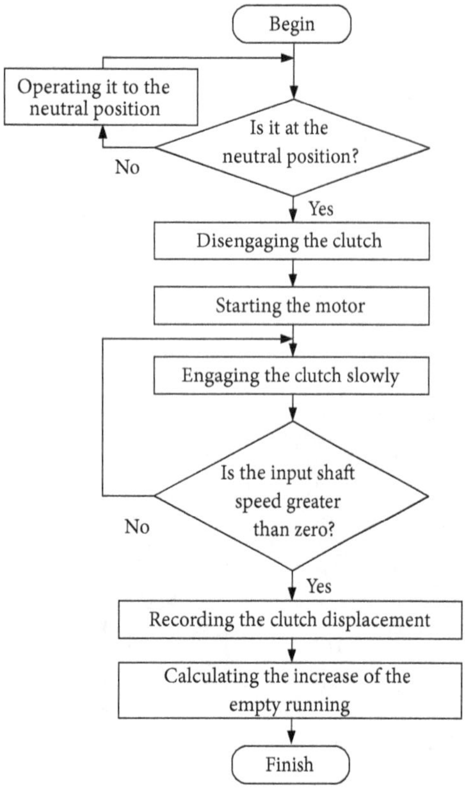

FIGURE 3: The diagram of the clutch wear testing program.

transmitted torque. The signal of the transmission output shaft speed sensor can be collected after the signal gear turns the first tooth. The speed of the transmission output shaft can be measured as long as the output shaft turns. The load resistance torque can be calculated until the transmission control unit collects the first output shaft speed signal during the starting process.

The estimation model for the equivalent load resistance torque on the transmission output shaft can be expressed in

$$\hat{T}_{load} = T_c(x_{hc})\eta_g i_g \quad (2)$$

where $\hat{T}_{load}$ is the load equivalent resistance torque, $T_c(x_{hc})$ is the clutch transmitted torque at the half engagement point, x_{hc} is the actuator corresponding position for the clutch engagement joint, and η_g and i_g are the transmission efficiency and ratio separately.

3.3. Estimation Model for the Load Inertia. We suppose that the load resistance torque is invariable during the starting process. The load inertia can be estimated by the use of the momentum theorem on the basis of the clutch transmitted torque and the output shaft speed. The time when the transmission output speed is more than zero is considered as the zero time for the load inertia estimation. Δt_s is the interval time which is ten millisecond in this paper. The sampling times are zero, t_1, t_2, ..., and t_n orderly. The clutch actuator corresponding positions are x_{c0}, x_{c1}, x_{c2}, ..., and x_{cn} orderly. The clutch transmitted torques are T_{c0}, T_{c1}, T_{c2}, ..., and T_{cn} successively. The transmission output shaft speeds are n_{o0}, n_{o1}, n_{o2}, ..., and n_{on} in sequence.

We suppose that the data updates n times from the beginning of the half engagement joint to the finishing acceleration of the output shaft which it reaches the top speed under the selected gear position. The time of the output shaft speed data updates m-1 times (m is less than or equal to n) is defined as t_k and the time of the output shaft speed data update m times is considered as t_l. The parameter of Δn_m is ordered as n_{ol} minus n_{ok}. The load angular momentum on the transmission output shaft can be calculated in (3) during the period of the time of t_k to that of t_l.

$$M_{loadm} = \left[\left(\frac{T_{ck}+T_{cl}}{2} + \sum_{x=k+1}^{l-1} T_{cx}\right)\eta_g i_g - \hat{T}_{load}(l-k)\right]\Delta t_s \times 10^{-3} \quad (3)$$

Consequently, the observation equation between the load angular momentum on the transmission output shaft and the unknown load inertia during the period of the time of t_k to that of t_l can be expressed in

$$M_{loadm} = \frac{\pi \Delta n_m}{30} J_{load} + v_m \quad (4)$$

where J_{load} is the load equivalent inertia on the transmission output shaft.

The observation matrix of the load equivalent inertia on the transmission output shaft can be described in (5) if the load angular momentum on the transmission output shaft can be taken as the observation matrix and the output shaft speed variation can be treated as the gain matrix.

$$M_{loadi} = H_i J_{load} + V_i \quad (5)$$

where i is equal to 1,2, ..., m, M_{loadi} is the observation matrix which is equal to the transpose of $[M_{load1}\ M_{load2}\ \cdots\ M_{loadm}]$, H_i is the gain matrix which is equal to the transpose of $[\Delta n_1\ \Delta n_2\ \cdots\ \Delta n_m]$, and V_i is the measurement error matrix which is equal to the transpose of $[v_1\ v_2\ \cdots\ v_m]$.

3.4. Recursion Least Squares Estimation Method of the Load Inertia. $\hat{J}_{load}$ is the estimation value of J_{load} which is the load inertia. The quadratic index of the least squares estimation for the load inertia can be shown in

$$J_W(\hat{J}_{load}) = (M_{loadm} - H_m\hat{J}_{load})^T W_m (M_{loadm} - H_m\hat{J}_{load}) \quad (6)$$

where $(M_{loadm} - H_m\hat{J}_{load})^T$ is the transposed matrix of $(M_{loadm} - H_m\hat{J}_{load})$ and W_m is a positive weighting matrix.

If the mean value of the observation errors is zero and the variance matrix is Q_m which is equal to the mathematical expectation of the product between V_m and its transpose, the optimal weighting least squares estimation for the load inertia

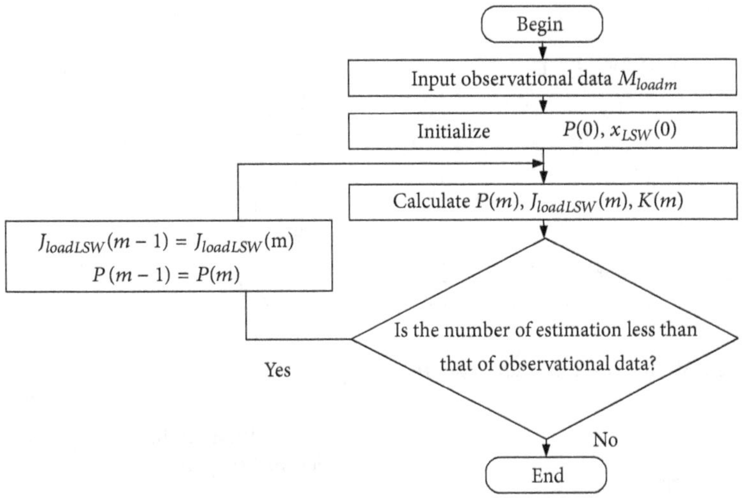

FIGURE 4: The flow block diagram of the recursive least squares estimation algorithm.

with m sets of observation data can be expressed in (7) under the condition that the weighting matrix of W_m is equal to the inverse of the variance matrix of Q_m.

$$\hat{J}_{loadLSW}(m) = \left[H_m{}^T Q_m{}^{-1} H_m\right]^{-1} H_m{}^T Q_m{}^{-1} M_{loadm} \quad (7)$$

where Q_m^{-1} is the inverse matrix of Q_m, H_m^T is the transpose matrix of H_m, and $[H_m{}^T Q_m{}^{-1} H_m]^{-1}$ is the inverse matrix of $[H_m{}^T Q_m{}^{-1} H_m]$.

In the same way, the optimal weighting least squares estimation for the load inertia with one plus m sets of observation data can be expressed in

$$\hat{J}_{loadLSW}(m+1) = \left[H_{m+1}{}^T Q_{m+1}{}^{-1} H_{m+1}\right]^{-1} H_{m+1}{}^T Q_{m+1}{}^{-1} M_{load(m+1)} \quad (8)$$

where Q_{m+1} is equal to $\left[\begin{smallmatrix} Q_m & 0 \\ 0 & q_{m+1} \end{smallmatrix}\right]$ and q_{m+1} is the measurement error variance of number m+1 times.

If P_m and P_{m+1} are $H_m{}^T Q_m{}^{-1} H_m$ and $H_{m+1}{}^T Q_{m+1}{}^{-1} H_{m+1}$ separately, the recursive least squares estimation for the load inertia can be expressed in (9) when the number m+1 set of test data is observed.

$$\begin{aligned} \hat{J}_{loadLSW}(m+1) &= \hat{J}_{loadLSW}(m) + K_{m+1}\left[M_{load(m+1)} - h_{m+1}\hat{J}_{LSW}(m)\right] \\ K_{m+1} &= P_{m+1}{}^{-1} h_{m+1}{}^T q_{m+1}{}^{-1} \\ P_{m+1}{}^{-1} &= P_m{}^{-1} - P_m{}^{-1} h_{m+1}{}^T \left(q_{m+1} + h_{m+1} P_m{}^{-1} h_{m+1}{}^T\right)^{-1} h_{m+1} P_m{}^{-1} \end{aligned} \quad (9)$$

The recursive least squares estimation for the load inertia is finished until the number of estimation is equal to that of observational data. The recursive least squares estimation algorithm for the load inertia can be shown in Figure 4.

4. Real-Time Estimation of the Load Parameters Based on the Recursive Least-Square Algorithm

4.1. Estimation of the Load Parameters. The torque-displacement characteristics of the new clutch can be received which the input torque of the clutch and the output torque of the transmission can be tested using the mechanical efficiency instrument. The experiment methods are described as follows. Firstly, the critical point of the clutch empty stroke is tested on the condition of no load. Secondly, the different half engagement point and the transmitting torque of the clutch can be tested and recorded under the first gear position with the help of the mechanical efficiency instrument and the loading device of which the different resistance torque can be simulated by changing the hydraulic pressure of the hydraulic station. Then, the clutch torque-displacement characteristics in (1) can be expressed with the test data using the method of least squares fit.

On the basis of above methods, the new clutch torque-displacement characteristics in this paper can be expressed approximately in (10) if the friction coefficient μ' is 0.25.

$$T_c(x_c) = 17.2431(x_c - x_{ce}) + 21.0131(x_c - x_{ce})^2 + 0.2839(x_c - x_{ce})^3 \quad (10)$$

The starting process under the first gear is shown in Figure 5 on the condition of unknown load inertia and unknown load resistance torque. Based on (2) and (10), the clutch transmitting torque is 100 N·m. The mechanical efficiency under the first gear is 90 percent by using the efficiency instrument. For this reason, the load resistance torque is 1285.20 N·m.

The zero time of the parameter estimation is determined when the half engagement point is tested. The eighteen sets of observation data with the clutch transmitting torque and the output shaft speed are used to estimate the load inertia. The Gaussian random white noise N(0,1) is added on the basis of

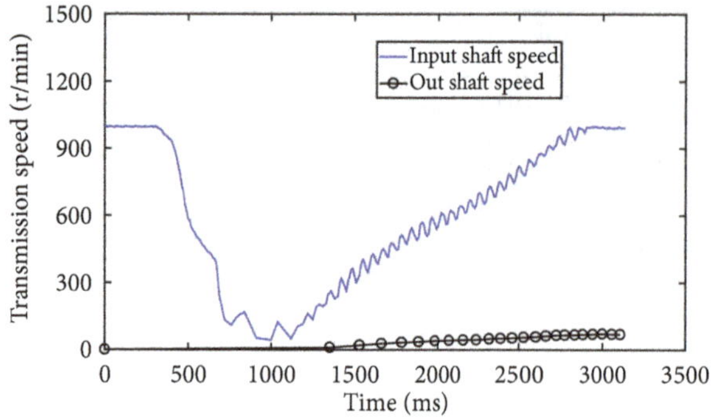

FIGURE 5: The transmission speeds during AMT startup under the first gear.

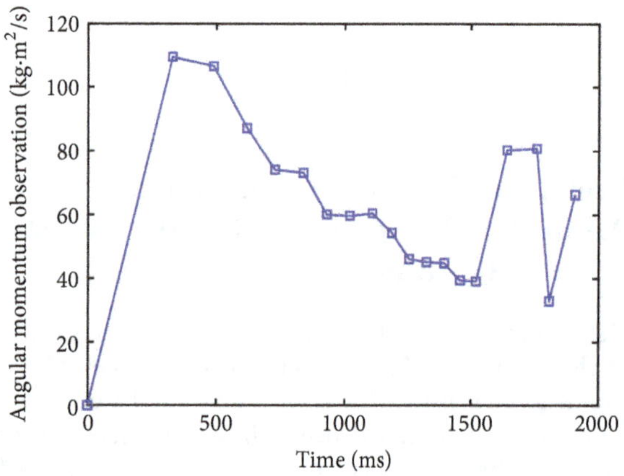

FIGURE 6: Load momentum observation during AMT startup under the first gear.

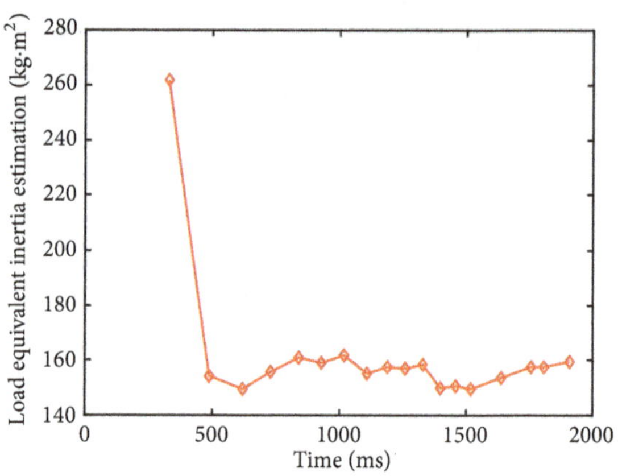

FIGURE 7: RLSE of the load equivalent inertia during AMT startup under the first gear.

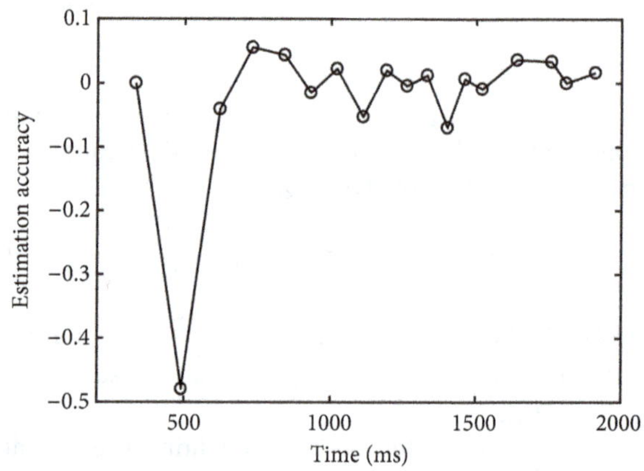

FIGURE 8: Estimation accuracy during AMT startup under the first gear.

the observation data. The recursive least squares method is applied in (5) of which the initial value of the load equivalent recursive least squares estimation (RLSE) derived from the top four sets of test data is 261.88 kg·m^2. The estimation accuracy is defined as the former estimation divided by the difference value of the current estimation value and the former estimation value. The estimation results containing the load momentum observation on the transmission output shaft, the load equivalent inertia estimation (the recursive least squares estimation of the load equivalent inertia on the transmission output shaft), and the estimation accuracy are shown in Figures 6, 7, and 8 respectively.

On the same working condition above, the starting process under the third gear is shown in Figure 9. Forty-one sets of observation data with the clutch transmitting torque and the output shaft speed during the starting process under the third gear are used to estimate the load inertia using the recursive least squares method. The zero time of the parameter estimation is still determined when the half engagement point is tested. The Gaussian random white noise N(0,1) is added on the basis of the observation data. The recursive least squares method is applied in (5) of which the starting value of RLSE derived from the top four sets of test data is 69.76 kg·m^2. The estimation results of the load momentum observation, the recursive least squares

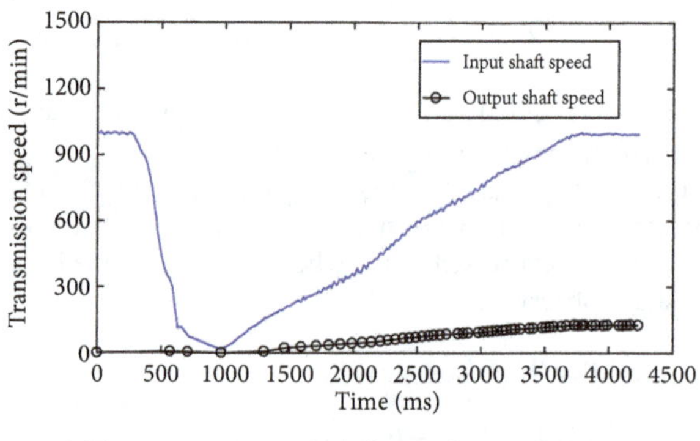

FIGURE 9: The transmission speeds during the starting process under the third gear.

estimation of the load equivalent inertia on the transmission output shaft, and the estimation accuracy are betrayed in Figures 10, 11, and 12 separately.

4.2. Analyzation of Estimation Results. The value of the actual test bed inertia plate is 130 kg·m^2. As a result, the actual load equivalent inertia on the transmission output shaft under the

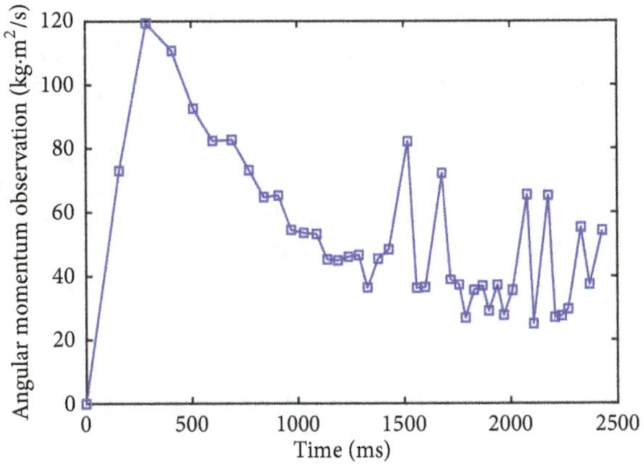

FIGURE 10: Load momentum observation during AMT startup under the third gear.

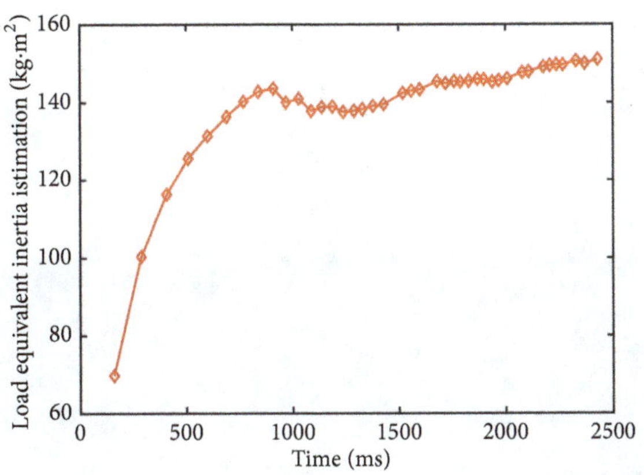

FIGURE 11: RLSE of the load equivalent inertia during AMT startup under the third gear.

first gear is 167.53 kg·m^2 and that under the third gear is 148.36 kg·m^2.

With the increase of the observation data, the recursive result is more and more stable under the first gear which is betrayed in Figure 7. The final RLSE of the load equivalent inertia on the transmission output shaft under the first gear is 159.49 kg·m^2. Obviously, there is a difference with 8.04 kg·m^2 between the final RLSE of the load equivalent inertia and the actual load equivalent inertia on the transmission output shaft under the first gear and the percentage of the estimation error is 4.80%. As time goes on, the estimation difference of the load equivalent inertia is lower, so the value of the estimation accuracy becomes lower that it is only 0.0161 or 1.61% at the end of the estimation in Figure 8. That is to say, the estimation accuracy is more and more stable as time goes on.

As is shown in Figure 9, the acceleration process is longer than that under the first gear. The final RLSE of load equivalent inertia on the transmission output shaft under the third gear is 151.10 kg·m^2 in Figure 11. As a consequence, the

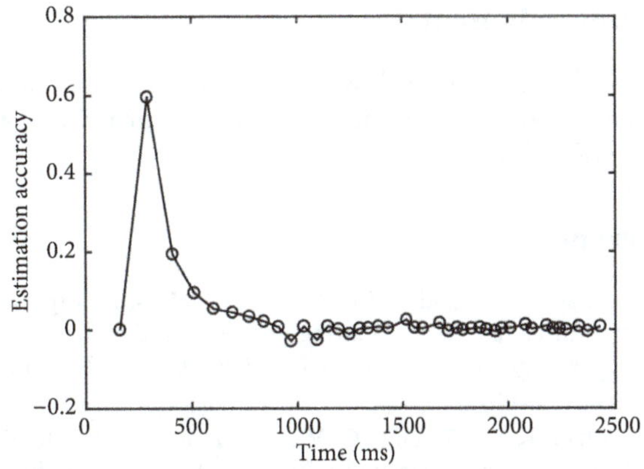

FIGURE 12: Estimation accuracy during AMT startup under the third gear.

estimation error between the final RLSE and the actual load equivalent inertia is 2.74 kg·m^2 which is lower than that under the first gear. The percentage of the estimation error is 1.85%. The estimation accuracy is only 0.0074 or 0.74% at the end of the estimation in Figure 12.

As can be seen, the more the data is, the better the estimation result is. Consequently, the load estimation method is effective. The reasons of the estimation errors can be found from both sides. On the one hand, the pressure fluctuation of the loading device can result in the fluctuation of the load resistance torque. On the other hand, the clutch wear leads to the fitting error of the clutch torque-displacement characteristics.

5. Conclusion

The load parameters estimation methods are studied during AMT startup as the soft starter for belt conveyor in the paper. The main conclusions are as follows. Firstly, the mathematical models of the load parameters estimation are established according to the mechanical characteristics during AMT startup. Secondly, the recursion least squares method of the load inertia is analyzed. Thirdly, based on the clutch torque-displacement characteristics and the test data during AMT startup, the load resistance torque and the load equivalent inertia are identified. The on-line estimation results show that the load parameters can be effectively recognized by the use of the above identification methods of the load parameters which provides the theoretical and experimental evidences to further study the clutch control strategy and the shift schedule for heavy-duty AMT.

Conflicts of Interest

The authors declare that they have no conflicts of interest.

Acknowledgments

This work was supported by the Key Research and Development Program of Shandong Province (China, Grant no. 2015GGX105005).

References

[1] Z. Zhao, J. Gu, and L. He, "Estimation of torques transited by twin-clutch during shifting process for dry dual clutch transmission," *Journal of Mechanical Engineering*, vol. 53, no. 14, pp. 77–87, 2017.

[2] Z. Zhao, L. He, Y. Yang, C. Wu, X. Li, and J. Karl Hedrick, "Estimation of torque transmitted by clutch during shifting process for dry dual clutch transmission," *Mechanical Systems and Signal Processing*, vol. 75, pp. 413–433, 2016.

[3] Z. Zhao, C. Wu, Y. Yang, and H. Chen, "Optimal robust control of shifting process for hybrid electric car with dry dual clutch transmission," *Journal of Mechanical Engineering*, vol. 52, no. 18, pp. 105–117, 2016.

[4] M. Pettersson and L. Nielsen, "Gear shifting by engine control," *IEEE Transactions on Control Systems Technology*, vol. 8, no. 3, pp. 495–507, 2000.

[5] M. Pettersson and L. Nielsen, "Diesel engine speed control with handling of driveline resonances," *Control Engineering Practice*, vol. 11, no. 3, pp. 319–328, 2003.

[6] J. Baumann, D. D. Torkzadeh, A. Ramstein, U. Kiencke, and T. Schlegl, "Model-based predictive anti-jerk control," *Control Engineering Practice*, vol. 14, no. 3, pp. 259–266, 2006.

[7] K. Yi, B. Shin, and K. Lee, "Estimation of turbine torque of automatic transmissions using nonlinear observers," *Journal of Dynamic Systems, Measurement, and Control*, vol. 122, no. 2, pp. 276–283, 2000.

[8] D.-Q. Wang, H.-B. Liu, and F. Ding, "Highly efficient identification methods for dual-rate hammerstein systems," *IEEE Transactions on Control Systems Technology*, vol. 23, no. 5, pp. 1952–1960, 2015.

[9] S. Zhang, D. Wang, and F. Liu, "Separate block-based parameter estimation method for Hammerstein systems," *Royal Society Open Science*, vol. 5, no. 6, Article ID 172194, 2018.

[10] M. Qing-Rui and H. You-Fu, "Effect of oil film squeezing on hydro-viscous drive speed regulating start," *Tribology International*, vol. 43, no. 11, pp. 2134–2138, 2010.

[11] M. H. Ghayesh, "Nonlinear transversal vibration and stability of an axially moving viscoelastic string supported by a partial viscoelastic guide," *Journal of Sound and Vibration*, vol. 314, no. 3-5, pp. 757–774, 2008.

[12] G. Lodewijks, "Two decades dynamics of belt conveyor systems," *Bulk Solids Handling*, vol. 22, no. 2, pp. 124–173, 2002.

[13] L. K. Nordell, "Zisco installs world's longest troughed belt 15.6 km horizontally curved overland conveyor," in *Proceedings of the International Materials Handling Conference, Beltcon*, pp. 1–11, 1997.

[14] L. K. Nordell and Z. P. Ciozda, "Transient belt stresses during starting and stopping: elastic response simulated by finite element methods," *Bulk Solids Handling*, vol. 4, no. 1, pp. 93–98, 1984.

[15] M. P. Singh, "Role of drive system technology in maximizing the performance and economics of long belt conveyors," *Bulk Solids Handling*, vol. 14, no. 4, pp. 695–702, 1994.

[16] Y. Li, Z. Wang, and X. Cong, "Constant acceleration control for AMT's application in belt conveyor's soft starting," *International Journal of Control and Automation*, vol. 6, no. 6, pp. 315–326, 2013.

[17] Y. Li, Z. Wang, W. Peng, and Z. Zheng, "Experimental study on synchronous shifting for AMT without synchronizer based on three-phase induction motor," *International Journal of Smart Home*, vol. 10, no. 2, pp. 197–212, 2016.

8

An Equivalent Circuit Model for Lithium Battery of Electric Vehicle considering Self-Healing Characteristic

Zhaodi Pei,[1] **Xiaoxu Zhao**,[1] **Huimei Yuan**,[1] **Zhen Peng**,[2] **and Lifeng Wu**[1,3,4]

[1]*College of Information Engineering, Capital Normal University, Beijing 100048, China*
[2]*Information Management Department, Beijing Institute of Petrochemical Technology, Beijing 102617, China*
[3]*Beijing Engineering Research Center of High Reliable Embedded System, Capital Normal University, Beijing 100048, China*
[4]*Beijing Advanced Innovation Center for Imaging Technology, Capital Normal University, Beijing 100048, China*

Correspondence should be addressed to Lifeng Wu; wulifeng@cnu.edu.cn

Academic Editor: Darong Huang

Considering the self-healing phenomenon of lithium batteries during intermittent discharge, a self-healing characteristic-based equivalent circuit model of lithium batteries is proposed. The mathematical description of the lithium battery in the self-healing process is obtained through the analysis of the equivalent circuit model. Based on experimental platform, an experiment considering self-healing characteristic was performed. Result shows that the self-healing characteristic-based lithium battery equivalent circuit model can describe the voltage of the lithium battery accurately during the self-healing process.

1. Introduction

Energy is the foundation of the society progress and development and it is vital to develop a new energy storage that is efficient and environmentally friendly and more necessarily put it to effect [1]. Electric vehicles have received extensive attention due to their features such as cleanliness and high efficiency [2, 3]. The lithium-ion battery is one of the best choices for electric vehicle battery because of its small volume, light weight, and low self-discharge rate [4]. However, power batteries have become an important factor that influences the performance and stability of electric vehicles [5].Therefore, the research on the safety and reliability of lithium-ion batteries has become the focus of attention [6–8]. Huang et al. [9] proposed a new method to estimate the state of charge (SOC) and state of health (SOH) of a battery by using the amount of voltage change per unit time during the discharge process of battery. Based on the battery cycle-life test data, it is found that the state of health has a linear relationship with the reciprocal of the unit time voltage drop, which is a function of the state of charge. This method realizes the online estimation of SOC and SOH and the model is proved to have robustness. Zou et al. [10] proposed a combined SOC and SOH estimation method through the lifetime of a lithium-ion battery. Two Extended Kalman Filters of two time scales are used to estimate SOC in real time and update SOH offline, which provides accurate SOC and SOH estimations. Xiong et al. [11] used adaptive extended Kalman filter (AEKF) to estimate the state of charge of lithium batteries. The AEKF algorithm improves the traditional KF and solves the defect of inaccuracy on nonlinear problems. Besides, the AEKF algorithm can improve the prediction accuracy by adaptively updating the process and measurement noise covariance. However, the above researches focus on the ideal working condition and do not take the intermittent discharge of the lithium battery in the actual operation into account. In fact, self-healing phenomenon could be found during the intermittent discharge of lithium batteries [12]. In other words, the capacity of the battery will rebound when standing. There is no doubt that the self-healing phenomenon could be a benefit to prolong the useful life of lithium batteries [6]. Therefore, further research is needed under this condition.

Currently, the models of batteries can be divided into two categories: electrochemical models and equivalent circuit models [13]. Based on the electrochemical theory,

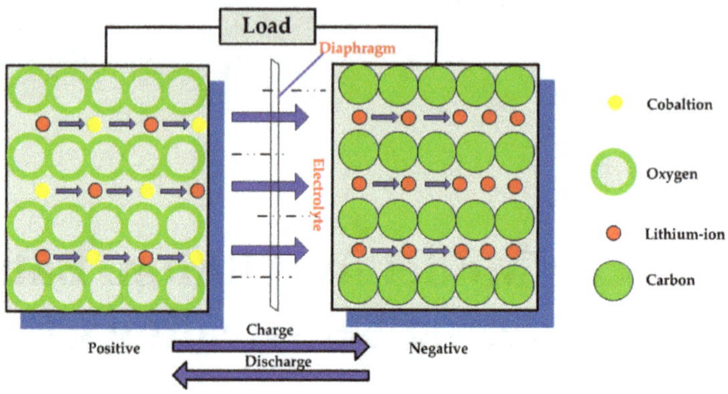

FIGURE 1: Lithium battery charge and discharge microscopic diagram.

the electrochemical model uses mathematical methods to describe the reaction process inside the battery. By using this model, the relationship between the cell performance and microscopic quantities can be established. However, this method is computationally complex and difficult to determine the value of parameters [14, 15]. The equivalent circuit model based on the external characteristics of the battery can avoid analyzing the internal reactions and complex parameter calculations. Comprehensively considering the accuracy and practicality of the battery model, equivalent circuit model is widely used in battery performance estimation of battery management systems [16, 17]. He et al. [18] proposed a model-based online estimation method for $LiFePO_4$ battery, which uses the Thevenin equivalent circuit model to simulate the $LiFePO_4$ battery and gives the corresponding mathematical expression equation. Feng et al. [19] proposed a new lithium battery equivalent circuit model by adding a moving average noise to the first-order resistor-capacitor circuit model. This new model simplifies the computational complexity of model parameters in a much more accurate way. Sun et al. [20] proposed a model-based dynamic multiparameter method for peak power estimation of lithium-ion batteries. This method refines the ohmic resistance of the Thevenin model for the lithium-ion battery, which improves the accuracy of the model. It is indicated that the model-based dynamic multiparameter method can be used to calculate the available power more accurately, whereas the above models cannot describe the change in self-healing phenomenon of lithium batteries. Therefore, in order to better describe the state of lithium battery in self-healing process, an equivalent circuit model of lithium battery is proposed based on self-healing characteristics.

The remainder of the paper is organized as follows. Section 2 introduces the self-healing phenomenon, which is very common in the actual use of electric vehicles. In Section 3, the equivalent circuit model based on self-healing characteristic is proposed and the analysis of the model is given. Section 4 describes the experiment platform and the experiment schedule based on LabView. In Section 5, the parameters of the model are identified and the simulation results are compared with the experiment. In addition, the error between the simulation and experiment is analyzed. In the final section, some conclusions are given.

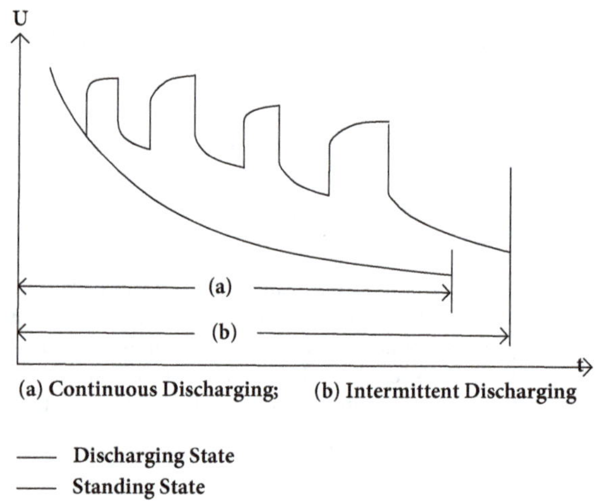

FIGURE 2: Comparison of lithium battery continuous discharge and intermittent discharge.

2. Self-Healing Process

In practice, three states are often used in the lithium battery. They are charging state, discharging state and standing state. As is shown in Figure 1, when the battery is charged, the positive electrode of the battery will generate lithium ions and at the same time lithium ions will pass through the electrolyte. Finally the lithium ions, generated at the positive electrode of the battery, arrive at the negative cathode of the battery and combine with graphite. And the capacity of the battery is related to the amount of lithium ions arriving at the cathode of the battery. When the battery is discharged, lithium ions in the negative electrode of the battery are separated from the graphite and returned to the positive electrode of the battery. The capacity of the battery is decreased.

In actual use, after each charge is completed, the lithium battery is repeatedly switched between discharging state and standing state, that is, intermittent discharging, as shown in Figure 2. When standing, the battery will self-heal. This

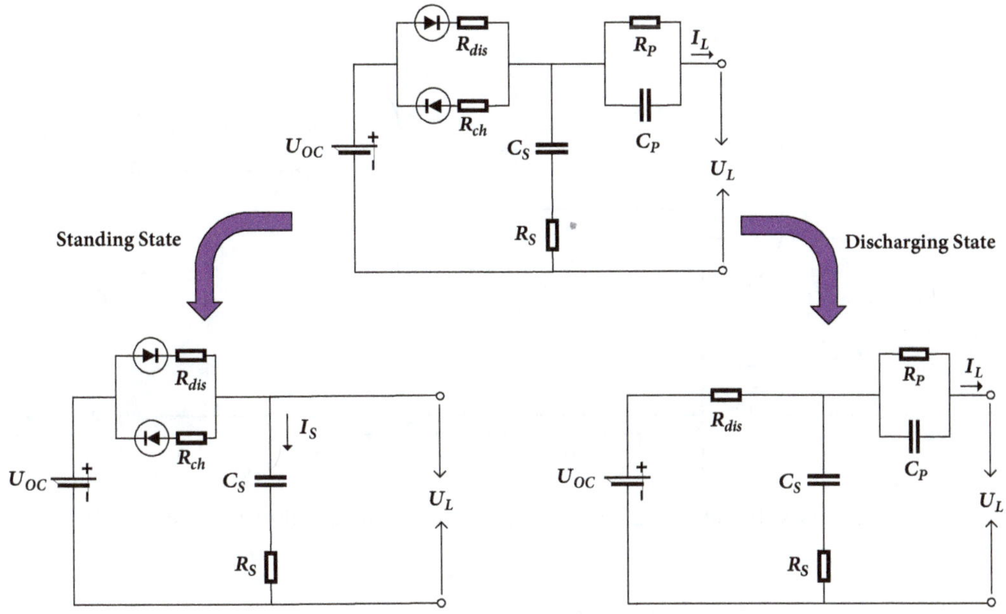

FIGURE 3: Lithium battery equivalent circuit model circuit diagram.

phenomenon is the reverse process of battery discharge, which is similar to the charging process of the battery, and the battery's state of charge will increase [21]. And, in the long term, this phenomenon will finally influence the state of health of lithium battery [22]. Mostly the electric vehicle is in a running and stagnant state, and the lithium battery is in a discharging and standing state so self-healing phenomenon is very common. The strength of self-healing is related to the time the battery is in standing state and it will change as the battery cycle times increase. Besides, the self-healing phenomenon will affect the remaining useful life of the battery. Therefore, it is essential to establish a battery model which could be used to describe the self-healing phenomenon of lithium-ion batteries. This paper will introduce a lithium battery equivalent circuit model based on self-healing characteristic.

3. Battery Model

3.1. Proposal of the Model. In order to better describe the voltage change of the lithium battery in the standing state, a lithium battery equivalent model based on self-healing characteristics is proposed. As shown in Figure 3, the self-healing characteristic-based model includes the open-circuit voltage U_{OC}, the internal resistance R_i of the battery, a capacitor C_S, a resistor R_S, and a RC network. The internal resistance R_i, respectively, includes a resistor R_{dis} in discharging state and a resistor R_{ch} in charging state. And I_S is the current pass through the capacitance C_S, I_L is the load current (positive for discharging state and negative for charging state), and U_L is the terminal voltage. Among them, the open-circuit voltage U_{OC} describes the voltage characteristics of the battery, and the capacitance C_S describes the capacity characteristics of the battery. The model can be divided into two parts. One part is mainly to describe the voltage of the lithium battery in discharging state and the other part is mainly to describe the voltage of self-healing phenomenon of the lithium battery in standing state.

3.2. Description of the Model. When the battery reaches a stable state, the battery voltage no longer changes. At this time, the terminal voltage is equal to the open-circuit voltage U_{OC} of the battery, which is also equal to the voltage of the capacitor C_S, as shown in

$$U_L = U_{OC} = U_{Cs} \quad (1)$$

3.2.1. Description of the Model in Discharging State. The Hybrid Pulse Power Characteristic test is conducted to describe the discharging state of the lithium battery equivalent circuit model. When the battery passes the pulse current, the voltage of the battery undergoes two stages, faster voltage conversion and slower voltage conversion. Figure 4 shows the battery voltage changes when the HPPC test is carried out.

In the faster voltage conversion, the rapid drop in voltage is mainly affected by the internal resistance R_i of the battery. The value of internal resistance R_i is related to the change in voltage, as show in (2). The slower voltage conversion in voltage is assigned to the part of the RC network including diffusion resistance R_P. And (3) describe the voltage change of the slower conversion.

$$\Delta U_{L1} = U_{OC} - U_{L\text{-}F} = U_{OC} * \frac{R_i}{R_L//R_S + R_i} \quad (2)$$

$$\begin{aligned}\Delta U_{L2} &= U_{L\text{-}F} - U_{L\text{-}S} \\ &= U_{OC} * \left(\frac{R_L//R_S}{R_L//R_S + R_i} - \frac{R_L}{R_L + R_P + R_i} \right)\end{aligned} \quad (3)$$

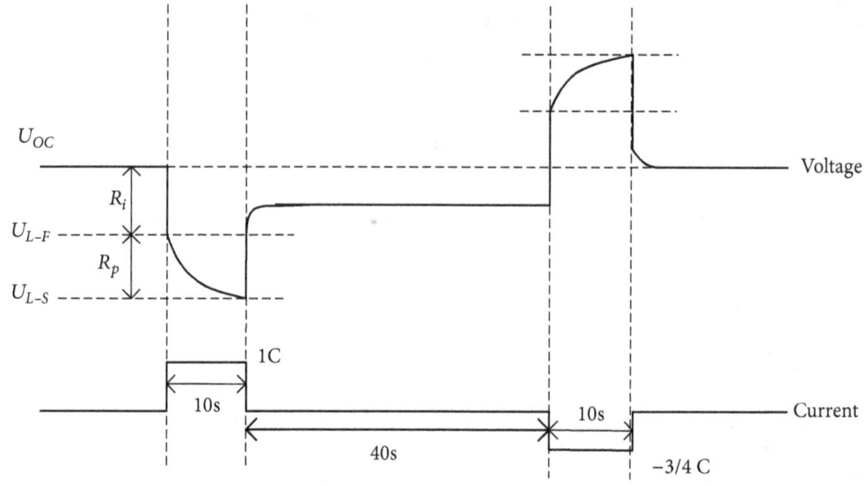

FIGURE 4: The lithium battery voltage and current changes in the HPPC test.

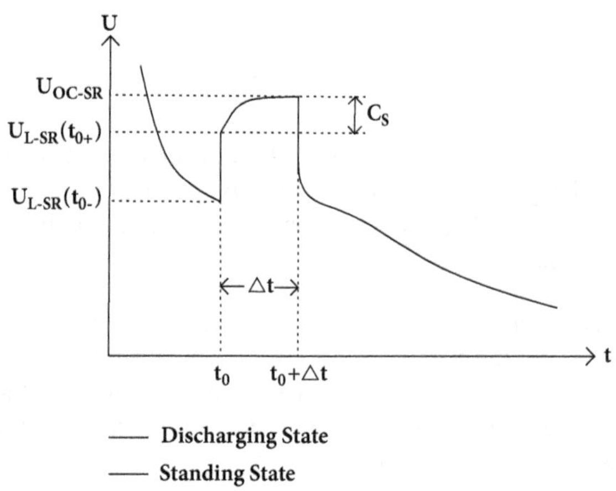

FIGURE 5: Self-healing test profile.

3.2.2. *Description of the Model in Standing State.* In order to describe the self-healing phenomenon of the battery, it is necessary to perform the self-healing test of the battery. And the self-healing test profile is shown in Figure 5. One self-healing test is performed within a single charge and discharge cycle of a lithium battery. When the charging process is completed, the lithium battery will start to discharge. At the point of t_0 when the voltage reaches the level of $U_{L\text{-}SR}$, the battery will be switched into the standing state. And the self-healing process with time interval Δt is performed. When the battery is in the standing state, no current flows through the *RC* network (resistance R_P and capacitance C_P), so the model can be simplified to the form in Figure 3.

At $t_0(\text{-})$, the lithium battery is still in the state of discharging, the voltage over the capacitor C_S can be expressed by (4); at $t_0(+)$, the battery has been switched to the standing state, and the voltage of the capacitor C_S can be expressed by (5). Because the voltage of the capacitor cannot be abruptly changed, the voltage of capacitor C_S at time $t_0(\text{-})$ is equal to the voltage at time $t_0(+)$, as shown in

$$U_{CS}(t_{0-}) = U_{L\text{-}SR}(t_{0-}) + I_L * R_P = U_{OC\text{-}SR} - U_i(t_{0-}) \quad (4)$$

$$U_{Cs}(t_{0+}) = U_{L\text{-}SR}(t_{0+}) - U_{Rs}(t_{0+}) \quad (5)$$

$$U_{Cs}(t_0) = U_{Cs}(t_{0-}) = U_{Cs}(t_{0+}) \quad (6)$$

From (4), (5), and (6), how the voltage changes when the battery is switched from the discharging state to the standing one can be deduced.

$$\Delta U_{L\text{-}SR} = U_{Rs}(t_{0+}) + I_L * R_P \quad (7)$$

When the battery is still for an enough long period of time, the battery is in a stable state. It can be obtained by (1) that the voltage of the capacitor C_S will be equal to the open-circuit voltage of the battery $U_{OC\text{-}SR}$, while the voltage of capacitor C_S at t_0 is less than $U_{OC\text{-}SR}$. Comparing (4) and (8), it can be concluded that the lithium battery is charged by the ideal voltage source U_{OC} during the interval time Δt when the battery is in the standing state. Where the current flowing through capacitor C_S is marked as I_S, the time constant is τ and the quantity of the battery is Q_{CS}:

$$U_{Cs}(\Delta t \longrightarrow \infty) = U_{OC\text{-}SR} \quad (8)$$

$$I_S = C_S * \frac{dU_{Cs}}{dt} \quad (9)$$

$$\tau = C_S * (R_i + R_S) \quad (10)$$

$$Q_{Cs} = U_{Cs} * C_S \quad (11)$$

When the lithium battery is switched from the discharging state to the standing state, the voltage of capacitor Cs satisfies

$$U_{Cs} = U_{OC\text{-}SR} - [U_{OC\text{-}SR} - (U_{L\text{-}SR}(t_{0-}) + I_L * R_P)] * \exp\left[-\frac{t/C_S}{(R_S + R_i)}\right] \quad (12)$$

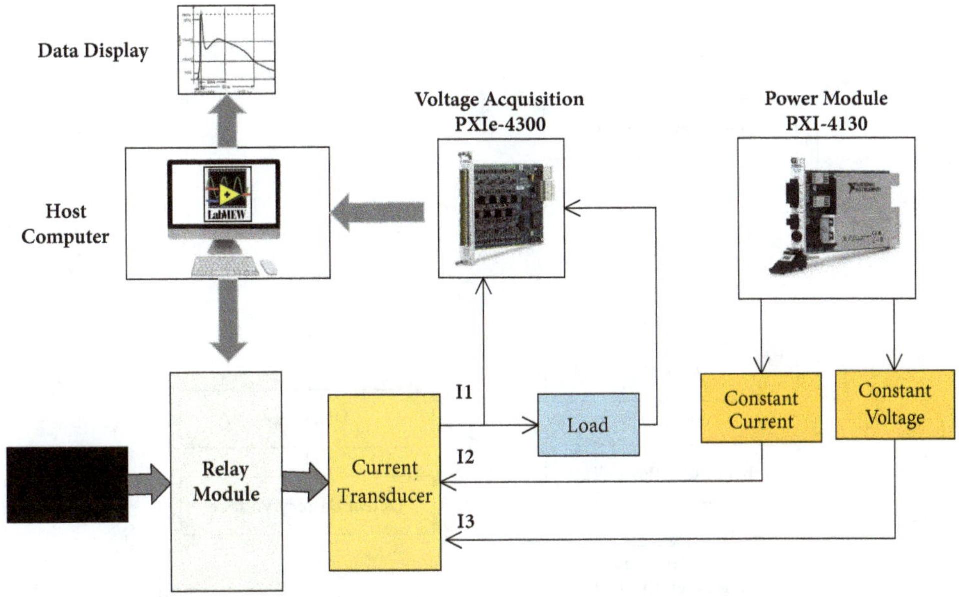

FIGURE 6: Lithium battery monitoring system based on *LabView*.

Meanwhile, the relationship between the terminal voltage $U_{L\text{-}SR}$ and the voltage of capacitor U_{Cs} accords with (5). So from (5), (7), (9), and (12), the function of the terminal voltage $U_{L\text{-}SR}$ can be derived as follows:

$$U_{L\text{-}SR} = U_{Cs} + \frac{R_S * C_S}{Cs * (R_i + R_S)} * \left[U_{OC\text{-}SR} - U_{L\text{-}SR}(t_{0-}) - I_L * R_P\right] * \exp\left(-\frac{t}{\tau}\right) \quad (13)$$

4. Experiments

4.1. Experiment Platform. In order to monitor the discharging state and the standing state of the battery, a lithium battery charge/discharge monitoring system based on *LabView2012* is established, as shown in Figure 6. The system uses *LabView 2012* as a software development platform, including upper computer, *NI* data acquisition cards, power supply modules, and relay modules. The system realizes the programming control of the discharge process of the lithium battery, simulates the transition from the discharging state to the standing state during the actual use of the lithium battery, and monitors the self-healing phenomenon of the lithium battery. The upper computer can accomplish the real-time monitoring of the lithium battery and is responsible for switching the state of the circuit. Voltage acquisition card *PXIe-4300* contains 8 analog input channels; each channel is equipped with an independent analog-to-digital converter of 16-bit resolution and can be simultaneously sampled to ensure the efficiency of data acquisition. The current sensor has a measurement range of 50 mA to 10A and is responsible for transmitting the current signal to a channel of the voltage acquisition card. The voltage, current, and other information of the battery are transmitted to the upper computer through the *PXI* bus, ensuring the accuracy and high efficiency of data transmission. *PXI-4130* is used to charge lithium batteries. It is a programmable power module that can control output constant voltage and constant current.

4.2. Experiment Schedule. In order to analyze the self-healing phenomenon during the intermittent discharge process, a set of experiments are carried out on the 18650 lithium battery and the experiment schedule is shown in Figure 7. For the purpose of obtaining the initial capacity of the lithium battery, the SCT test was performed and repeated three times. After that, the battery is discharged and the HPPC test is performed. At the same time, the test data is recorded. The key to the experiment schedule is to conduct self-healing characteristic test on the lithium battery. Among them, U^* represents the state transition voltage. When the voltage drops to U^* for the first time, the working state of the battery changes, switching from the discharging state to the standing state. The value of the state transition voltage represents the depth of discharge, which could affect the strength of the self-healing phenomenon. So, for an evident observation of the self-healing phenomenon, the state transition voltage is selected to be 3 V in this experiment. That is, when the battery voltage drops to 3 V for the first time, the battery enters the standing state and performs the self-healing test. The value of the SOC at this time can be obtained by the method of coulomb counting by the data logger of the experimental system. Experiments of three times scales are performed, respectively, 15minutes, 60 minutes, and 90 minutes. In other words, the period in the standing state is 15 min, 60 min, and 90 min. And each self-healing test is repeated three times.

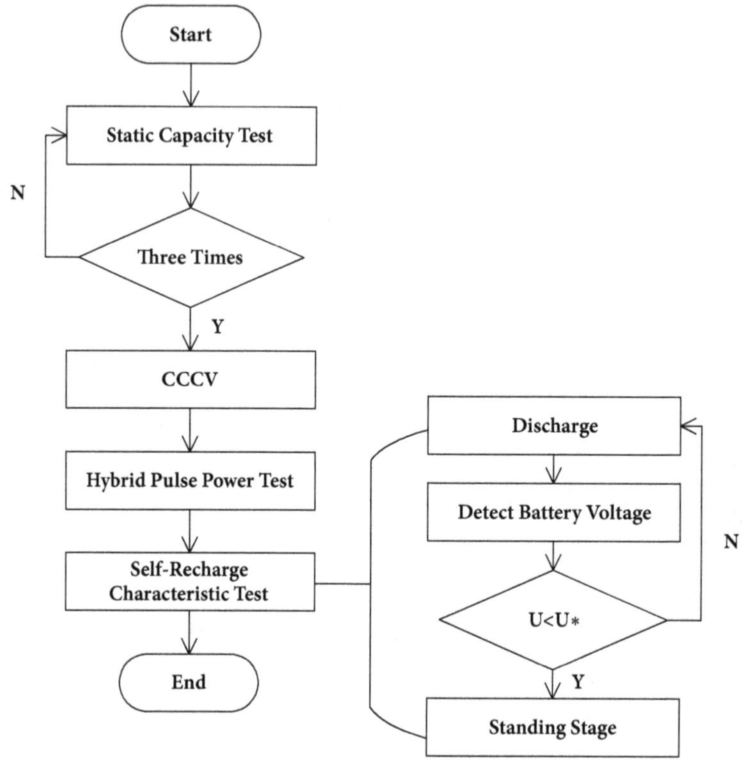

FIGURE 7: Experiment schedule.

TABLE 1: Parameter Identification Results at $U^* = 3$ V.

SOC	*Rdis*	*Rch*	*Rp*	*Cs*	*Rs*
12.23%	82.3 mΩ	81.7 mΩ	16.1 mΩ	400.851 F	0.506 Ω

5. Results and Discussion

In order to make clear the relationship between self-healing characteristics and self-healing time, this paper selects the first test data with self-healing time of 60 minutes to identify the model parameters and compares the simulated value of the battery model with measured value during the standing state. In the meantime, the parameters obtained from this experiment are applied to the other experiments with the same self-healing time and also to multiple sets of experiments with different self-healing times. The results are as follows.

5.1. Parameter Identification. Using the test data to identify the unknown parameters of the battery model is the basis for analysis and discussion of the results. MATLAB 2018 provides a variety of methods to achieve the fitting functions, one of which is Cftool. Cftool has a visual interactive interface and provides numbers of fitting functions, so this paper selects Cftool as the fitting tool. The parameter identification results are shown in Table 1.

5.2. Evaluation on the Battery Model. After the model parameters are obtained from the first experiment test data with self-healing time of 60 minutes, the simulated value of the model is compared with the measurements. Figures 8(a), 8(b), and 8(c) represent the results of the comparison between the simulations and the experiment measurements of three times, of which the self-healing time is 60 min. After the lithium battery is switched from the discharging state to the standing state, the model is able to describe the rising status of the lithium battery voltage. And when the standing time is long enough, the simulation value is still very close to the terminal voltage of the lithium battery. Repeat the experiment with a standing time of 60 min and compare the test data with the simulated values of the model. Result shows that the model can simulate the process of the other experiments with the same self-healing time, as shown in Figures 8(b) and 8(c).

Figure 9 shows the error characteristic curve between model simulation values and experiment measurements when the self-healing time is 60 min. When the battery starts to switch from the discharging state to the standing state, the battery voltage suddenly changes. In this short period, the model has a larger error. But, after a brief change in voltage, the model can better describe the status of the lithium battery. And as the battery voltage tends to be flat, the model error decreases and approaches the level of zero.

In order to prove that this model can describe the state of lithium batteries with different self-healing time, experiments with self-healing time of 15 min and 90 min are designed. And the parameters obtained from the experiment with a self-healing time of 60 min are applied to experiments with self-healing time of 15 min and 90 min. As is shown in Figures 10 and 11, the model can also describe the status of the battery.

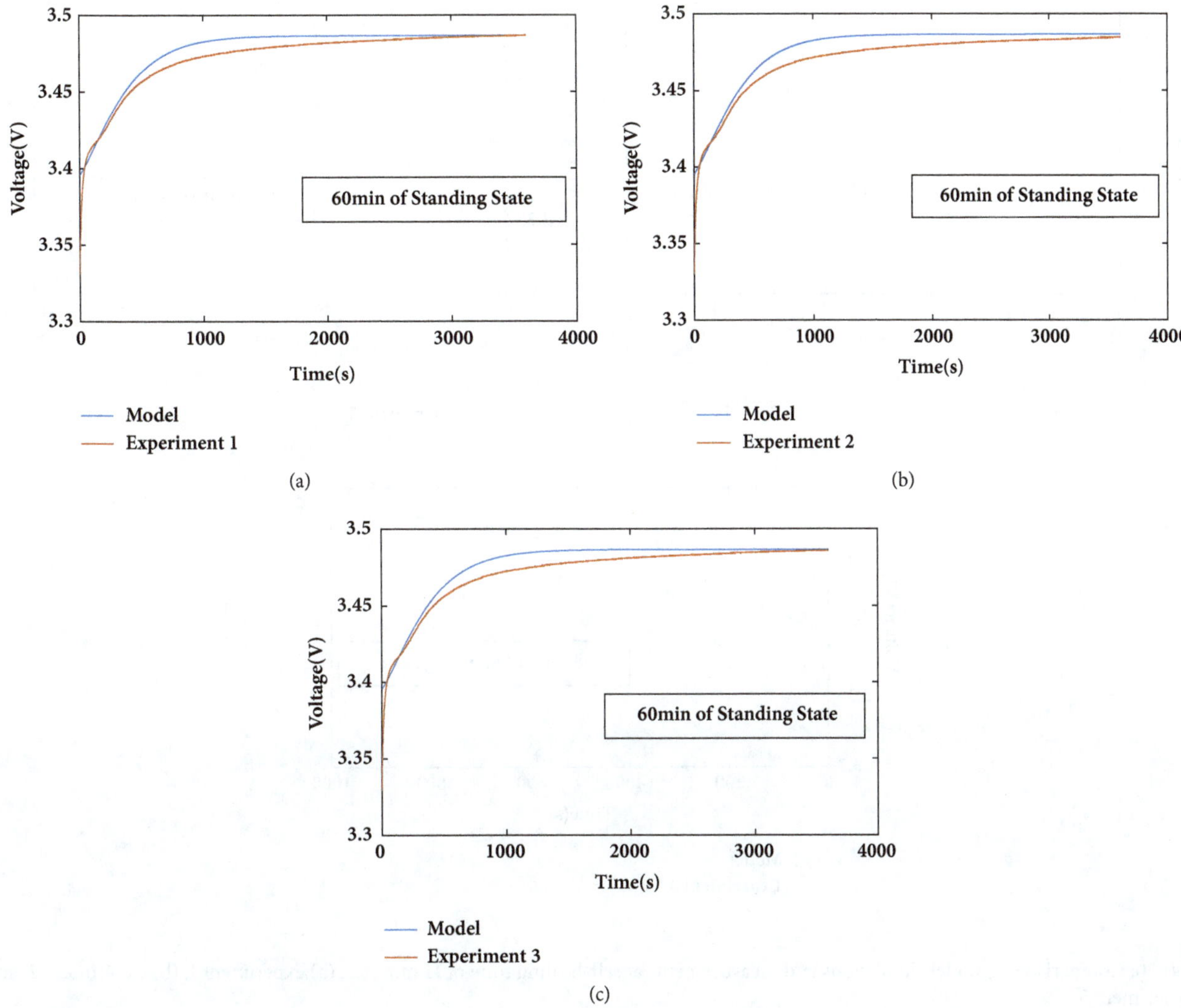

FIGURE 8: Comparison of model simulations and measurements at self-healing time of 60 minutes: (a) experiment 1, (b) experiment 2, and (c) experiment 3.

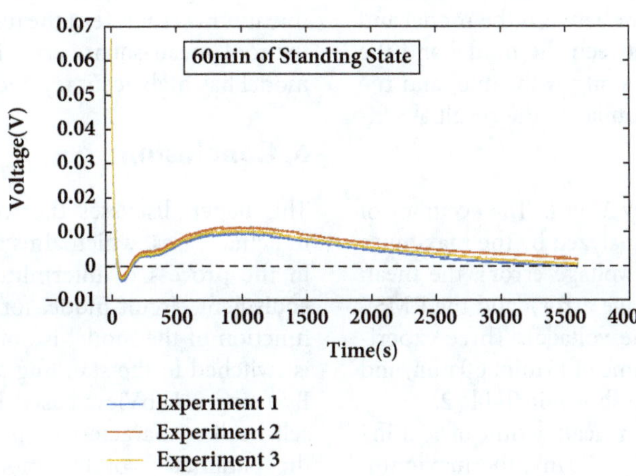

FIGURE 9: Comparison of the error between the model simulations and experiments measurements at self-healing time of 60 minutes.

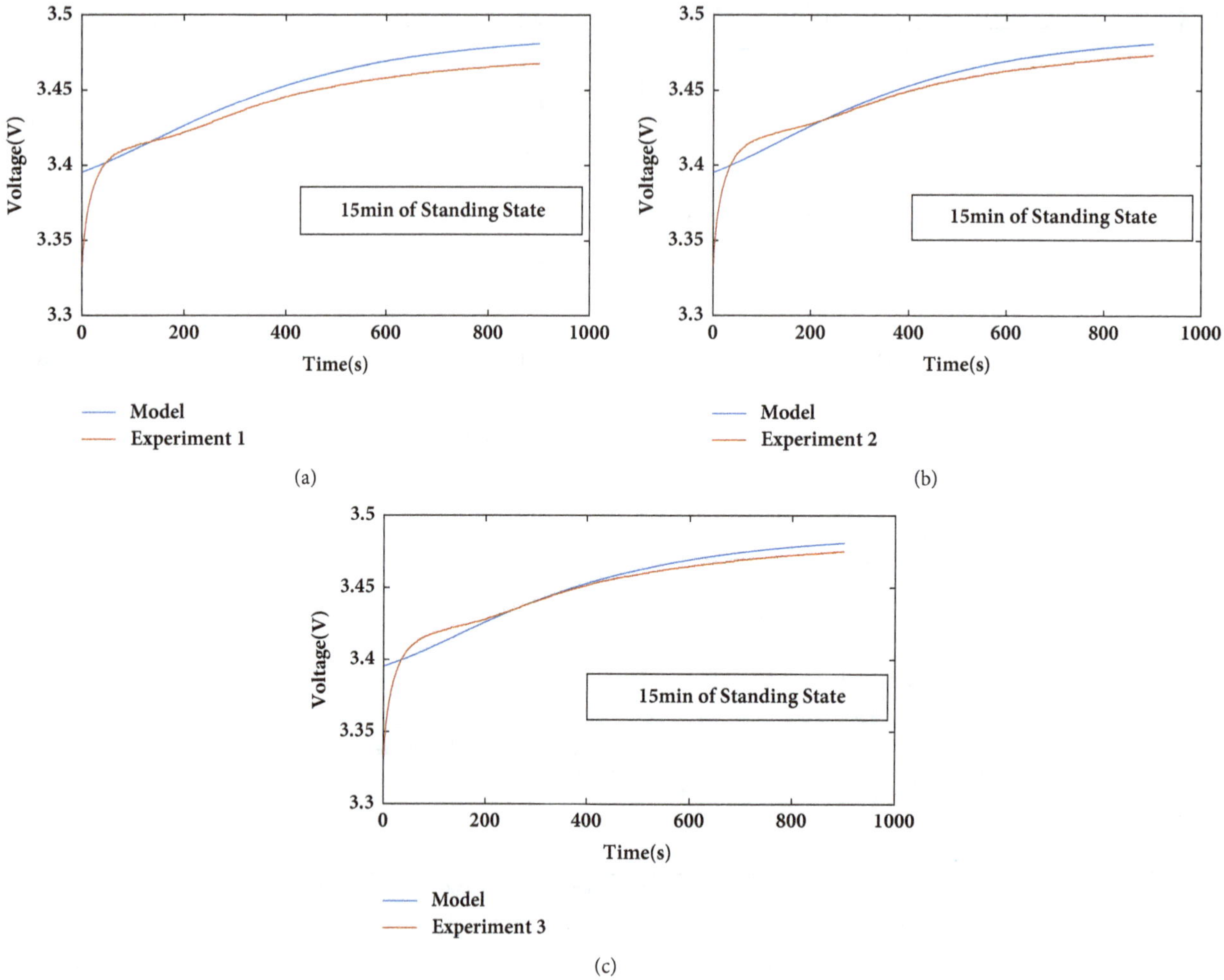

FIGURE 10: Comparison of model simulations and measurements at self-healing time of 15 minutes, (a) experiment 1, (b) experiment 2, and (c) experiment 3.

Figures 12 and 13 are the error curves between the model simulations and the experiment measurements when the self-healing time is 15 min and 90 min, respectively. In the state transition stage, there is a large error between the model and the measurement but the error between the model and the measurement gradually decreases along with time, and the error tends to be stable, which is similar to the result at self-healing time of 60 min.

5.3. Accuracy Analysis of the Battery Model. The accuracy of the lithium battery model will be analyzed by the maximum error (the maximum value of the voltage error), the mean error (the average value of the voltage error), and the RMSE (the root-mean-square error of the voltage). Three experiment test datasets for self-healing time of 15 min, 60 min, and 90 min are analyzed. The results are shown in Table 2.

In the three experiments with self-healing time of 60 minutes, the maximum error is less than 70 mV, the maximum error rate is also less than 2%, and the root-mean-square error is less than 8 mV. When using this model for accuracy analysis of test data with self-healing time of 15 min and 90 min, the maximum error does not exceed 70 mV. Except that the maximum error rate of the third group of experiments with self-healing time of 90 minutes is slightly more than 2%, the maximum error rate of the rest of the tests is less than 2%. And the root-mean-square error is less than 13 mV. Therefore, the model has high accuracy and efficiency.

6. Conclusion

This paper discusses the working state of lithium battery in actual cases, which aims at the self-healing phenomenon in the process of intermittent discharge and proposes an equivalent circuit model for self-healing phenomenon. The function of the model is obtained when the lithium battery is switched to the standing state from the discharging state. By using a LabView-based lithium battery test system, the self-healing characteristic-based experiment is designed, and the parameters of the model are identified through test data. Experiment results prove that the model can accurately describe the voltage change of the battery in the standing state with different self-healing time. The error between

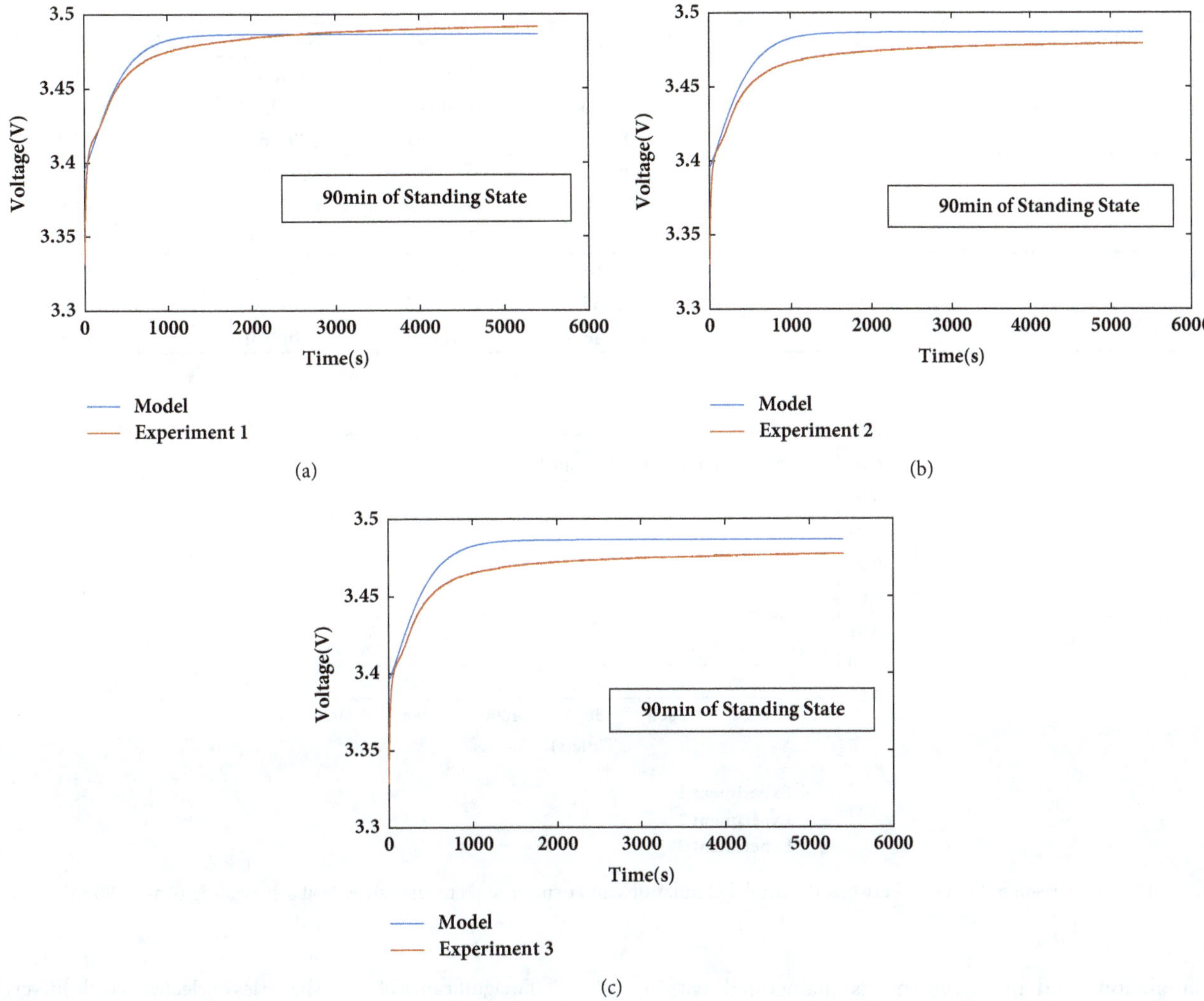

FIGURE 11: Comparison of model simulations and measurements at self-healing time of 90 minutes, (a) experiment 1, (b) experiment 2, and (c) experiment 3.

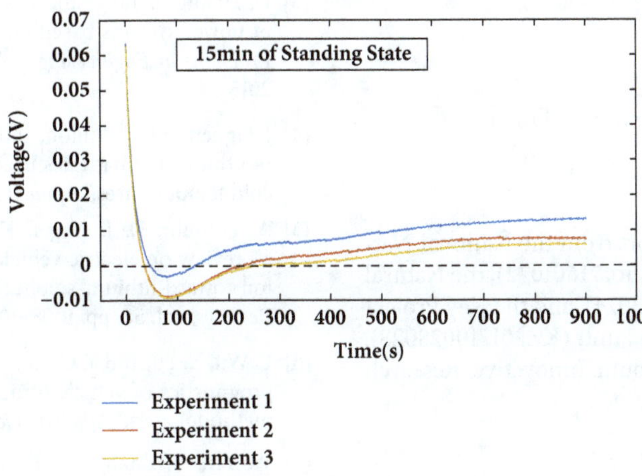

FIGURE 12: Comparison of the error between the model simulations and experiments measurements at self-healing time of 15 minutes.

TABLE 2: Model error analysis.

Time of Standing State	Number of Experiments	Maximum/V	Mean/V	RMSE/V	Max.Error Rate(%)
60 min	1	0.06464	0.004535	0.006093	1.94
	2	0.06556	0.006485	0.007588	1.97
	3	0.06655	0.005302	0.006689	1.99
15 min	1	0.06342	0.008516	0.010537	1.91
	2	0.06277	0.003718	0.007682	1.88
	3	0.06241	0.002196	0.006796	1.87
90 min	1	0.06466	0.000354	0.004623	1.94
	2	0.06555	0.010587	0.011145	1.97
	3	0.06746	0.012269	0.012731	2.03

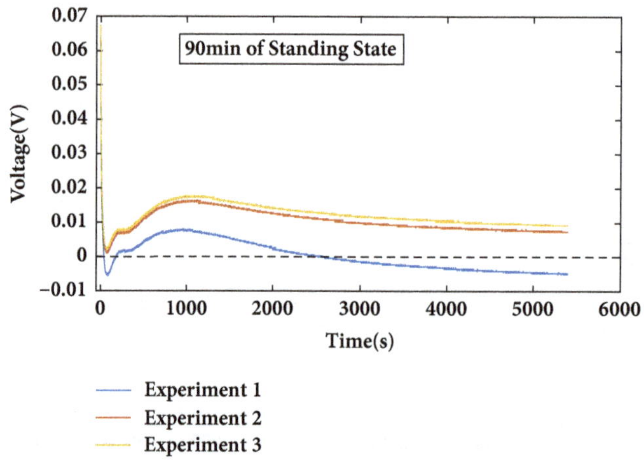

FIGURE 13: Comparison of the error between the model simulations and experiments measurements at self-healing time of 90 minutes.

the simulations and measurements is maintained within a stable range and the model is highly accurate. In spite of the advantages of the equivalent circuit model for self-healing characteristic, the description for the working state of lithium batteries is complicated to some degree and the identification of battery parameters needs more experiments. Future works aim at the optimization of the model and parameter identifications.

Conflicts of Interest

All authors declare they have no conflicts of interest.

Acknowledgments

This work received financial support from the National Natural Science Foundation of China (no. 71601022), the Natural Science Foundation of Beijing (4173074), and the Key Project B Class of Beijing Natural Science Fund (KZ201710028028). And the work is supported by Youth Innovative Research Team of Capital Normal University.

References

[1] J. B. Dunn, L. Gaines, J. C. Kelly, C. James, and K. G. Gallagher, "The significance of Li-ion batteries in electric vehicle life-cycle energy and emissions and recycling's role in its reduction," *Energy & Environmental Science*, vol. 8, no. 1, pp. 158–168, 2015.

[2] J. Hu, H. Morais, T. Sousa, and M. Lind, "Electric vehicle fleet management in smart grids: a review of services, optimization and control aspects," *Renewable & Sustainable Energy Reviews*, vol. 56, pp. 1207–1226, 2016.

[3] D. Huang, J. Tang, and L. A. Zhao, "A fault diagnosis method of power systems based on gray system theory," *Mathematical Problems in Engineering*, vol. 2015, Article ID 971257, 11 pages, 2015.

[4] J. Jaguemont, L. Boulon, and Y. Dubé, "A comprehensive review of lithium-ion batteries used in hybrid and electric vehicles at cold temperatures," *Applied Energy*, vol. 164, pp. 99–114, 2016.

[5] A. Fotouhi, D. J. Auger, K. Propp, S. Longo, and M. Wild, "A review on electric vehicle battery modelling: from lithium-ion toward lithium-sulphur," *Renewable & Sustainable Energy Reviews*, vol. 56, pp. 1008–1021, 2016.

[6] L. Wu, X. Fu, and Y. Guan, "Review of the remaining useful life prognostics of vehicle lithium-ion batteries using data-driven methodologies," *Applied Sciences*, vol. 6, no. 6, p. 166, 2016.

[7] W. Yan, B. Zhang, W. Dou, D. Liu, and Y. Peng, "Low-cost adaptive lebesgue sampling particle filtering approach for real-time li-ion battery diagnosis and prognosis," *IEEE Transactions*

on Automation Science and Engineering, vol. 14, no. 4, pp. 1601–1611, 2017.

[8] M. E. Orchard, P. Hevia-Koch, B. Zhang, and L. Tang, "Risk measures for particle-filtering-based state-of-charge prognosis in lithium-ion batteries," *IEEE Transactions on Industrial Electronics*, vol. 60, no. 11, pp. 5260–5269, 2013.

[9] S.-C. Huang, K.-H. Tseng, J.-W. Liang, C.-L. Chang, and M. G. Pecht, "An online SOC and SOH estimation model for lithium-ion batteries," *Energies*, vol. 10, no. 4, 2017.

[10] Y. Zou, X. Hu, H. Ma, and S. E. Li, "Combined State of Charge and State of Health estimation over lithium-ion battery cell cycle lifespan for electric vehicles," *Journal of Power Sources*, vol. 273, pp. 793–803, 2015.

[11] R. Xiong, F. Sun, X. Gong, and H. He, "Adaptive state of charge estimator for lithium-ion cells series battery pack in electric vehicles," *Journal of Power Sources*, vol. 242, pp. 699–713, 2013.

[12] A. Eddahech, O. Briat, and J.-M. Vinassa, "Lithium-ion battery performance improvement based on capacity recovery exploitation," *Electrochimica Acta*, vol. 114, pp. 750–757, 2013.

[13] X. S. Hu, S. B. Li, and H. Peng, "A comparative study of equivalent circuit models for Li-ion batteries," *Journal of Power Sources*, vol. 198, pp. 359–367, 2012.

[14] X. Lai, Y. Zheng, and T. Sun, "A comparative study of different equivalent circuit models for estimating state-of-charge of lithium-ion batteries," *Electrochimica Acta*, vol. 259, pp. 566–577, 2018.

[15] H. He, R. Xiong, and J. Fan, "Evaluation of lithium-ion battery equivalent circuit models for state of charge estimation by an experimental approach," *Energies*, vol. 4, no. 4, pp. 582–598, 2011.

[16] X. Chen, W. Shen, Z. Cao, A. Kapoor, and I. Hijazin, "Adaptive gain sliding mode observer for state of charge estimation based on combined battery equivalent circuit model in electric vehicles," in *Proceedings of the IEEE 8th Conference on Industrial Electronics and Applications (ICIEA '13)*, pp. 114–123, IEEE, June 2013.

[17] H. W. He, R. Xiong, H. Q. Guo, and S. C. Li, "Comparison study on the battery models used for the energy management of batteries in electric vehicles," *Energy Conversion and Management*, vol. 64, no. 4, pp. 113–121, 2012.

[18] H. He, R. Xiong, and H. Guo, "Online estimation of model parameters and state-of-charge of LiFePO4 batteries in electric vehicles," *Applied Energy*, vol. 89, no. 1, pp. 413–420, 2012.

[19] T. Feng, L. Yang, X. Zhao, H. Zhang, and J. Qiang, "Online identification of lithium-ion battery parameters based on an improved equivalent-circuit model and its implementation on battery state-of-power prediction," *Journal of Power Sources*, vol. 281, pp. 192–203, 2015.

[20] F. Sun, R. Xiong, H. He, W. Li, and J. E. E. Aussems, "Model-based dynamic multi-parameter method for peak power estimation of lithium-ion batteries," *Applied Energy*, vol. 96, no. 3, pp. 378–386, 2012.

[21] T. Qin, S. Zeng, J. Guo, and Z. Skaf, "A rest time-based prognostic framework for state of health estimation of lithium-ion batteries with regeneration phenomena," *Energies*, vol. 9, no. 11, p. 896, 2016.

[22] T. Qin, S. Zeng, J. Guo, and Z. Skaf, "State of health estimation of li-ion batteries with regeneration phenomena: a similar rest time-based prognostic framework," *Symmetry*, vol. 9, no. 1, p. 4, 2016.

9

The Bioinspired Model-Based Hybrid Sliding-Mode Formation Control for Underactuated Unmanned Surface Vehicles

Mingyu Fu and Duansong Wang

Harbin Engineering University, Harbin, Heilongjiang, China

Correspondence should be addressed to Duansong Wang; wangduansong@hrbeu.edu.cn

Academic Editor: Sing Kiong Nguang

In this paper, a novel hybrid strategy is proposed for unmanned surface vehicle (USV) formation control. The strategy is divided into two subsystems: a virtual velocity controller based on the bioinspired model and a dynamic controller based on the sliding-mode model. The proposed control scheme solves the problem of a speed jump that occurs in the traditional backstepping method when the margin of error increases suddenly, and it also satisfies the actuator control constraint. Additionally, a dynamic controller is designed, combining the sliding mode with the proposed virtual controller, to avoid the traditional chattering problem. System stability is proven by the Lyapunov theory. Simulation results verify the effectiveness of the proposed controller.

1. Introduction

The formation control of USVs is receiving increasing interest in control science and engineering [1–3]. Multiple USVs can accomplish challenging and dangerous tasks, such as mine clearance, patrol, investigation, and the transportation of strategic materials, which can not only decrease personal injury but also achieve tasks that a single USV cannot complete [4]. To achieve a desirable formation pattern, several methods including leader-follower approach [5], virtual structure strategy [6], behavioral-based approach [7], and artificial potential function [8] have been proposed. Among these methods, the leader-follower strategy is commonly used because of simplicity and scalability. To address the formation problem, many control schemes have been implemented. An adaptive backstepping method for a group of underactuated autonomous vehicles was studied in [9, 10], and the course angle error between leader and follower was considered to guarantee the follower's course angle stability. A Linear Quadratic Regulator Proportional Integral (LQR PI) controller was implemented in [11] for centralized heterogeneous leader-follower architecture. With this control scheme, formation geometry can be switched to any shape while flying, and obstacle avoidance can be realized. In [12], a sliding-mode control law for controlling multiple USVs in arbitrary formations was proposed, mesh stability and parameter uncertainty in the dynamic model and wave disturbance were considered in designing the controllers, and the effectiveness of the controller was verified by the computer simulation. Though it has an outstanding characteristic insensitivity to parameter variations, one major drawback of sliding-mode control is the inherent problem of chattering. In addition, many intelligent methods were used to achieve formation control. A bounded neural network control law was constructed with the aid of a saturated function for the USV while over a closed curve in [13]. An adaptive neural network formation controller combined with a dynamic surface technique for USVs was studied in [14], which solved the problem of "explosion of complexity" by introducing the first-order filter. In [15], a robust adaptive formation control law is proposed for multiple autonomous surface vehicles moving in a leader-follower formation. However, while all works mentioned above have common styles, the procedure for designing a controller is complex, and too many parameters needed to be adjusted. Based on graph-theoretic concepts and locally distributed information, a neural fuzzy formation controller was designed with the capability of online learning in [16]. Mu and Wang [17] proposed fuzzy-based path following control for USV to solve the problem of unknown control gain coefficient. The fuzzy rules-based formation control approaches can solve the problem of large initial robot velocities, but formulating the fuzzy rules is not easy because they are usually obtained by trial and error based human knowledge [18]. A neural network needs either online learning or offline training procedures, either of which could be computational complicated.

The main contribution of this work is to solve the problem of the speed jump and actuator control constraints of USV formation. The leader-follower strategy is used since it is easy to implement. When the controller is designed, a virtual USV is introduced to produce a predefined path, and the other USV keeps the desired distance and angle with it to achieve the desired formation pattern. A bioinspired model-based hybrid sliding-mode controller is designed, inspired by [19], and on the basis of the previous literature [4]. It can not only avoid the traditional chattering problem and smooth the input signal but also simplify the control law.

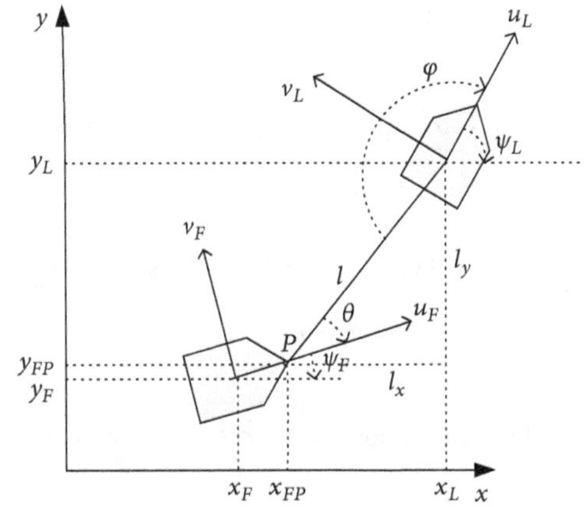

Figure 1: Leader-follower formation configuration with the control point P.

2. Formation Model of USV

2.1. Single USV Model. Consider a class of networked multi-agent systems consisting of n-unmanned surface underactuated vehicles. The dynamical model of the i-*th* USV is given by [20]

$$
\begin{aligned}
\dot{x}_i &= u_i \cos\psi_i - v_i \sin\psi_i \\
\dot{y}_i &= u_i \sin\psi_i + v_i \cos\psi_i \\
\dot{\psi}_i &= r_i \\
\dot{u}_i &= \frac{m_{22i}}{m_{11i}} v_i r_i - \frac{d_{11i}}{m_{11i}} u_i + \frac{1}{m_{11i}} \tau_{ui} + \frac{d_u}{m_{11i}} \\
\dot{v}_i &= -\frac{m_{11i}}{m_{22i}} u_i r_i - \frac{d_{22i}}{m_{22i}} v_i + \frac{d_v}{m_{22i}} \\
\dot{r}_i &= \frac{m_{11i} - m_{22i}}{m_{33i}} u_i v_i - \frac{d_{33i}}{m_{33i}} r_i + \frac{1}{m_{33i}} \tau_{ri} + \frac{d_r}{m_{33i}};
\end{aligned} \tag{1}
$$

The signals x_i, y_i, and ψ_i denote position and orientation (yaw angle), respectively, in the earth fixed frame. u_i, v_i, and r_i represent the surge, sway, and yaw velocities with respect to body frame, respectively. m_{ii} denotes the masses including added masses in the surge, sway and yaw axes. The damping term in the body coordinate is described by d_{ii}. The signals $\tau = (\tau_{ui}, 0, \tau_{ri})^T$ are the surge force and yaw torque inputs provided by thrusters. $W = [d_u, d_v, d_r]^T$ denotes the loads induced by waves, wind, and ocean currents along the surge, sway, and yaw axes, respectively [21].

2.2. USV Formation Model. In this work, the formation control problem for USV is addressed by a distributed strategy based on virtual leader strategy. For convenience, only 2 USVs are considered as a group. The mathematical model is formed as in Figure 1: where l is the distance between leader USV and follower USV, φ is described as the relative angle between follower and leader, l_d, φ_d are the desired distance and angle, and l_x, l_y represent the components of l in the earth fixed frame on x- and y-axes. p is the bow position of USV, x_{Fp}, y_{Fp}, ψ_{Fp} are the position and orientation of p, and d is the distance between p and the mass center of the follower. We can construe from Figure 1 that

$$
\begin{aligned}
x_{Fp} &= x_F + d\cos\psi_F, \\
y_{Fp} &= y_F + d\sin\psi_F
\end{aligned} \tag{2}
$$

To achieve the desired formation pattern, the following inequality should be guaranteed:

$$
\begin{aligned}
\lim_{t\to\infty} l - l_d &\le \zeta_1, \\
\lim_{t\to\infty} \varphi - \varphi_d &\le \zeta_2
\end{aligned} \tag{3}
$$

ζ_1, ζ_2 are positive constants that can be arbitrarily small. l_d, φ_d are the desired distance and angle yet bounded and there exist a positive constant, that is, ρ_1, ρ_2, such that $\|\dot{l}_d\| \le \rho_1$, $\|\dot{\varphi}_d\| \le \rho_2$.

If the value of l, φ can be confirmed, then the value of l_x, l_y can be confirmed, and then the value of l, φ can be transformed to control l_x, l_y. As seen from the picture,

$$
\begin{aligned}
l_x &= l\cos\theta = x_L - x_F - d\cos\psi_F, \\
l_y &= l\sin\theta = y_L - y_F - d\sin\psi_F
\end{aligned} \tag{4}
$$

Then,

$$
\begin{aligned}
\dot{l}_x &= u_L\cos\psi_L - v_L\sin\psi_L - (u_F\cos\psi_F - v_F\sin\psi_F) \\
&\quad + d r_F \sin\psi_F \\
\dot{l}_y &= u_L\sin\psi_L + v_L\cos\psi_L - (u_F\sin\psi_F + v_F\cos\psi_F) \\
&\quad - d r_F \cos\psi_F
\end{aligned} \tag{5}
$$

The desired distance between 2 USVs is considered as l_d, and the projection weight in the earth fixed frame is l_x^d, l_y^d, so

$$
\begin{aligned}
l_x^d &= -l_d\cos(\varphi_d + \psi_L), \\
l_y^d &= -l_d\sin(\varphi_d + \psi_L)
\end{aligned} \tag{6}
$$

Differentiating (6) yields

$$\begin{aligned}\dot{l}_x^d &= -\dot{l}_d \cos(\varphi_d + \psi_L) + l_d\dot{\varphi}_d \sin(\varphi_d + \psi_L) \\ &\quad + l_d r_L \sin(\varphi_d + \psi_L) \\ \dot{l}_y^d &= -\dot{l}_d \sin(\varphi_d + \psi_L) + l_d\dot{\varphi}_d \cos(\varphi_d + \psi_L) \\ &\quad + l_d r_L \cos(\varphi_d + \psi_L)\end{aligned} \tag{7}$$

The errors of the formation model can be defined as

$$\begin{bmatrix} e_x \\ e_y \\ e_\psi \end{bmatrix} = \begin{bmatrix} \cos\psi_F & \sin\psi_F & 0 \\ -\sin\psi_F & \cos\psi_F & 0 \\ 0 & 0 & 1 \end{bmatrix} \begin{bmatrix} l_x - l_x^d \\ l_y - l_y^d \\ \psi_L - \psi_F \end{bmatrix} \tag{8}$$

From what has been discussed above, the formation mathematics model can be written as follows:

$$\begin{aligned}\dot{e}_x &= e_y r_F - u_F + u_L \cos e_\psi - v_L \sin e_\psi \\ &\quad - l_d r_L \sin(\varphi_d + e_\psi) + \mu_1 \\ \dot{e}_y &= -e_x r_F - v_F + u_L \sin e_\psi - v_L \cos e_\psi - d r_F \\ &\quad + l_d r_L \cos(\varphi_d + e_\psi) + \mu_2 \\ \dot{e}_\psi &= r_L - r_F \\ \dot{u}_F &= \frac{m_{22}}{m_{11}} v_F r_F - \frac{d_{11}}{m_{11}} u_F + \frac{1}{m_{11}} \tau_{uF} + \frac{d_u}{m_{11}} \\ \dot{v}_F &= -\frac{m_{11}}{m_{22}} u_F r_F - \frac{d_{22}}{m_{22}} v_F + \frac{d_v}{m_{22}} \\ \dot{r}_F &= \frac{m_{11} - m_{22}}{m_{33}} u_F v_F - \frac{d_{33}}{m_{33}} r_F + \frac{1}{m_{33}} \tau_{rF} + \frac{d_r}{m_{33}}\end{aligned} \tag{9}$$

where

$$\begin{aligned}\mu_1 &= \dot{l}_d \cos(\varphi_d + e_\psi) - l_d \dot{\varphi}_d \sin(\varphi_d + e_\psi), \\ \mu_2 &= \dot{l}_d \sin(\varphi_d + e_\psi) + l_d \dot{\varphi}_d \cos(\varphi_d + e_\psi),\end{aligned} \tag{10}$$

and we can get that μ_1, μ_2 are bounded since $|\sin(\varphi_d + e_\psi)| \le 1$, $|\cos(\varphi_d + e_\psi)| \le 1$, l_d, φ_d are bounded, and $\|\dot{l}_d\| \le \rho_1$, $\|\dot{\varphi}_d\| \le \rho_2$.

Based on [4], we have the following assumptions.

Assumption 1. The surge velocity $u_i \ge 0$.

Assumption 2. Trajectory (x_l, y_l) produced by the virtual leader satisfies that its derivatives with respect to t exit up to second order.

Assumption 3. The terms d_u, d_v, d_r that satisfy $d_u \le |D_u|, d_v \le |D_v|, d_r \le |D_r|$, D_u, D_v, D_r are constants.

Assumption 4. The follower can receive the position and velocity information from the leader by sensors in time.

In summary, the underactuated USV formation control problem can be divided into two parts: kinematics control and dynamics control.

For kinematics control, virtual surge, and sway velocities are designed to make

$$\begin{aligned}\lim_{t\to\infty} l_x &\longrightarrow l_x^d, \\ \lim_{t\to\infty} l_y &\longrightarrow l_y^d\end{aligned} \tag{11}$$

For dynamics control, surge force and yaw torque are designed to make actual velocities approximate to virtual velocities.

3. Hybrid Control Strategy for Unmanned Surface Vehicles

3.1. Virtual Velocity Controller. The virtual velocity controller based on the backstepping approach can be defined as

$$\begin{bmatrix} u_d \\ v_d \\ r_d \end{bmatrix} = \begin{bmatrix} k_1 e_x + u_L \cos e_\psi - v_L \sin e_\psi + e_y r_F - l_d r_L \sin(\varphi_d + e_\psi) + \mu_1 \\ k_2 e_y - e_x r_F + u_L \sin e_\psi - v_L \cos e_\psi - d r_F + l_d r_L \cos(\varphi_d + e_\psi) + \mu_2 \\ r_L + k_3 e_\psi \end{bmatrix} \tag{12}$$

Here, u_d, v_d, r_d are the virtual surge, sway, and yaw motion speed of the following USVs, u_L, v_L, r_L are the desired velocity in the body-fixed frame, and k_1, k_2, k_3 are a positive constant, respectively.

3.2. Bioinspired Velocity Controller. With the analysis of (12), the virtual speed is directly related to the state error. The traditional backstepping method will generate a sharp speed jump when a sudden tracking error occurs. This means that large acceleration and forces/moments are required that make exceeding the control constraint practically impossible. To solve the speed jump and control constraint problems, a bioinspired model is introduced to design the virtual speed controller.

The bioinspired neural dynamics model was first put forward by Grossberg [21]. It can describe the online adaptive behavior of individuals. It was originally derived based on the membrane model proposed by Hodgkin and Huxley [22] for a patch of membrane using electrical elements. The dynamics of voltage across the membrane can be described in the membrane model, using state equation technique as

$$C_m \frac{dV_m}{dt} = -\left(E_p + V_m\right) g_p + \left(E_{Na} - V_m\right) g_{Na} - \left(E_k + V_m\right) g_k \tag{13}$$

where V_m is the neural activity of the j-th neuron in the neural network, the parameters $E_p + V_m$, $E_{Na} - V_m$, $E_k + V_m$ are nonnegative constants representing the passive decay rate, the upper and lower bounds of the neural activity, respectively, and the variables the excitatory and inhibitory inputs to the neuron.

The bioinspired model can be defined as following form:

$$\dot{V}_i = -AV_i + \left(B - V_i\right) f\left(e_i\right) - \left(D + V_i\right) g\left(e_i\right) \tag{14}$$

where

$$f\left(e_i\right) = \begin{cases} e_i, & e_i \geq 0 \\ 0, & e_i < 0, \end{cases}$$
$$g\left(e_i\right) = \begin{cases} -e_i, & e_i \leq 0 \\ 0, & e_i > 0. \end{cases} \tag{15}$$

In this work, the tracking errors e_x, e_y, e_ψ are chosen as the input of the neural dynamic model, and the outputs V_i will substitute the error of e_x, e_y, e_ψ.

$$\begin{aligned} \dot{V}_x &= -A_1 V_x + \left(B_1 - V_x\right) f\left(e_x\right) - \left(D_1 + V_x\right) g\left(e_x\right) \\ \dot{V}_y &= -A_2 V_y + \left(B_2 - V_y\right) f\left(e_y\right) - \left(D_2 + V_y\right) g\left(e_y\right) \\ \dot{V}_\psi &= -A_3 V_\psi + \left(B_3 - V_\psi\right) f\left(e_\psi\right) - \left(D_3 + V_\psi\right) g\left(e_\psi\right) \end{aligned} \tag{16}$$

A_i $(i = 1, 2, 3)$ are nonnegative constants on behalf of the neurons of attenuation rate. B_i, D_i $(i = 1, 2, 3)$ are considered nonnegative constants that denote the upper and lower bounds of neurons dynamic, which can restrict the outputs to $[-D_i, B_i]$.

Therefore, the proposed virtual speed controller is as follows:

$$\begin{aligned} \Gamma_u &= kV_x + u_L \cos V_\psi - \upsilon_L \sin V_\psi + V_y r_F - l_d r_L \sin\left(\varphi_d + e_\psi\right) + \mu_1 \\ \Gamma_\upsilon &= kV_y - V_x r_F + u_L \sin V_\psi - \upsilon_L \cos V_\psi - d r_F + l_d r_L \cos\left(\varphi_d + V_\psi\right) + \mu_2 \\ \Gamma_r &= r_L + k_3 V_\psi \end{aligned} \tag{17}$$

k, k_3 are the same parameters as (12). Based on the description above, V_x, V_y, V_ψ are all bounded and smooth without any sharp jumps when the inputs suddenly change.

3.3. Sliding-Mode Controller. After the virtual speed controller has been designed, a sliding-mode controller is introduced to produce the control force to make the USV arrive at the virtual speed.

Generally, the process of designing the sliding-mode controller is divided into two parts: defining a sliding manifold and designing a control law to move toward the sliding manifold.

The sliding manifold is selected as

$$s_1 = \Gamma_{eu} + \lambda_1 \int_0^t \Gamma_{eu}(\tau)\, d\tau \tag{18}$$

where $\Gamma_{eu} = u_F - \Gamma_u$ is the error between the reality surge velocity and the virtual velocity of the leader. From the derivation of (18), we can get

$$\dot{s}_1 = \lambda_1 \Gamma_{eu} + \dot{\Gamma}_{eu} = \lambda_1 \left(u_F - \Gamma_u\right) + \dot{u}_F - \dot{\Gamma}_u \tag{19}$$

let $\dot{s}_1 = 0$; we can get the equivalent control law

$$\tau_{uFeq} = m_{11}\dot{\Gamma}_u - \lambda_1 m_{11}\left(u_F - \Gamma_u\right) - m_{22}\upsilon_F r_F + d_{11}u_F - d_u \tag{20}$$

Then, the switching control law can be selected as

$$\tau_{uFsw} = -\eta_1 \operatorname{sgn}\left(s_1\right) \tag{21}$$

We can obtain the surge force control law

$$\tau_{uF} = \tau_{uFeq} + \tau_{uFsw} \tag{22}$$

Next, the yaw control torque will be designed.

The yaw control torque is based on a two-order sliding manifold, and the sliding manifold is defined according to the sway speed error of the underactuated unmanned surface vehicles:

$$s_2 = \dot{\Gamma}_{ev} + 2\lambda_2 \Gamma_{ev} + \lambda_2^2 \int_0^t \Gamma_{ev}(\tau)\, d\tau \tag{23}$$

where $\Gamma_{ev} = \upsilon_F - \Gamma_\upsilon$ and λ_2 is a positive constant.

By derivation (23), we can get

$$\dot{s}_2 = \ddot{\Gamma}_{ev} + 2\lambda_2 \dot{\Gamma}_{ev} + \lambda_2^2 \Gamma_{ev} \tag{24}$$

$$\dot{\Gamma}_{ev} = \dot{\upsilon}_F - \dot{\Gamma}_\upsilon \tag{25}$$

$$\begin{aligned} \ddot{\Gamma}_{ev} &= \ddot{\upsilon}_F - \ddot{\Gamma}_\upsilon \\ &= -\frac{m_{11}}{m_{22}}\left(\dot{u}_F r_F + u_F \dot{r}_F\right) - \frac{d_{22}}{m_{22}}\dot{\upsilon}_F + \frac{\dot{d}_v}{m_{22}} - \ddot{\Gamma}_v \end{aligned} \tag{26}$$

Considering the difficulty of computing $\ddot{\Gamma}_{ev}$, a feedback control input of acceleration error is introduced:

$$\ddot{\Gamma}_v = -k_v \dot{\Gamma}_v \tag{27}$$

let $\dot{s}_2 = 0$; then the equivalent control law can be designed as

$$\tau_{rFeq} = -\frac{m_{22}m_{33}}{m_{11}u_F}\left(\frac{d_{22}}{m_{22}}\dot{\upsilon}_F - \frac{\dot{d}_v}{m_{22}} + k_v \dot{\Gamma}_v - 2\lambda_2 \dot{\Gamma}_{ev} - \lambda_2^2 \Gamma_{ev}\right) - \frac{\dot{u}_F r_F m_{33}}{u_F} - \left(m_{11} - m_{22}\right) u_F \upsilon_F + d_{33} r_F - d_r \tag{28}$$

The switching control law is

$$\tau_{rFsw} = -\frac{m_{22}m_{33}}{m_{11}u_F}\left(\eta_2 \operatorname{sgn}(s_2)\right) \quad (29)$$

so, the yaw control torque is

$$\tau_{rF} = \tau_{rFeq} + \tau_{rFsw} \quad (30)$$

The controller based on the backstepping method can be described as

$$\tau_{uj} = m_{1j}\left(-k_6 u_e - \frac{m_{2j}}{m_{1j}}v_j r_j + \frac{d_{1j}}{m_{1j}}u_j + \dot{u}_j^{\alpha}\right) \quad (31)$$

$$\tau_{rj} = m_{3j}\left(-k_7 r_e - \frac{m_{1j}-m_{2j}}{m_{3j}}u_j v_j + \frac{d_{3j}}{m_{3j}}r_j + \dot{r}_j^{\alpha}\right) \quad (32)$$

where u_j^{α}, r_j^{α} are virtual velocity controller.

4. Stability Analysis

We choose a Lyapunov function as

$$V_1 = \frac{1}{2}\left(e_x^2 + e_y^2 + e_\psi^2\right) + \frac{k}{2B}\left(V_x^2 + V_y^2\right) + \frac{k_3}{2B}V_\psi^2 \geq 0 \quad (33)$$

k, B, *and* k_3 are positive constants.

Let $V_{10} = (1/2)(e_x^2 + e_y^2) + (k/2B)(V_x^2 + V_y^2) \geq 0$, $V_{11} = (1/2)e_\psi^2 + (k_3/2B)V_\psi^2$.

Then,

$$\dot{V}_{10} = e_x\dot{e}_x + e_y\dot{e}_y + \frac{k}{B}\left(V_x\dot{V}_x + V_y\dot{V}_y\right) \quad (34)$$

Substituting (9) and (12) into (32), we can get

$$\begin{aligned}
\dot{V}_{10} &= -ke_xV_x - ke_yV_y + \frac{k}{B}\left[V_x(-AV_x\right.\\
&\quad + (B - V_x) f(e_x) - (D + V_x) g(e_x)) + V_y\left(-AV_y\right.\\
&\quad \left.\left. + (B - V_y) f(e_y) - (D + V_y) g(e_y)\right)\right] = -ke_xV_x\\
&\quad - ke_yV_y + \frac{k}{B}\left[-AV_x^2 - f(e_x)V_x^2 - g(e_x)V_x^2\right.\\
&\quad + Bf(e_x)V_x - Dg(e_x)V_x - AV_y^2 - f(e_y)V_y^2\\
&\quad \left.- g(e_y)V_y^2 + Bf(e_y)V_y - Dg(e_y)V_y\right]\\
&= -ke_xV_x - ke_yV_y + \frac{k}{B}\left[(-A - f(e_x) - g(e_x))\right.\\
&\quad \cdot V_x^2 + (Bf(e_x) - Dg(e_x))V_x + \left(-A - f(e_y)\right.\\
&\quad \left.\left.- g(e_y)\right)V_y^2 + \left(Bf(e_y) - Dg(e_y)\right)V_y\right]
\end{aligned} \quad (35)$$

and if $B = D$,

$$\begin{aligned}
\dot{V}_{10} &= -ke_xV_x - ke_yV_y\\
&\quad + \frac{k}{B}\left[(-A - f(e_x) - g(e_x))V_x^2\right]\\
&\quad + k\left[f(e_x) - g(e_x)\right]V_x\\
&\quad + \frac{k}{B}\left[-A - f(e_y) - g(e_y)\right]V_y^2\\
&\quad + k\left[f(e_y) - g(e_y)\right]V_y\\
&= \frac{k}{B}\left[(-A - f(e_x) - g(e_x))V_x^2\right]\\
&\quad + k\left[f(e_x) - g(e_x) - e_x\right]V_x\\
&\quad + \frac{k}{B}\left[-A - f(e_y) - g(e_y)\right]V_y^2\\
&\quad + k\left[f(e_y) - g(e_y) - e_y\right]V_y
\end{aligned} \quad (36)$$

In the same way,

$$\begin{aligned}
\dot{V}_{11} &= e_\psi\dot{e}_\psi + \frac{k_3}{B}V_\psi\dot{V}_\psi\\
&= -k_3e_\psi V_\psi + \frac{k_\psi}{B}\left[-A - f(e_\psi) - g(e_\psi)\right]V_\psi^2;
\end{aligned} \quad (37)$$

as defined in (14), if $e_i \geq 0$, then $f(e_i) = e_i$, $g(e_i) = 0$, $A + f(e_i) + g(e_i) = A + e_i > 0$, $f(e_i) - g(e_i) - e_i = e_i - e_i = 0$.

If $e_i < 0$, then $f(e_i) = 0$, $g(e_i) = -e_i$,

$$\begin{aligned}
&A + f(e_i) + g(e_i) = A - e_i > 0,\\
&f(e_i) - g(e_i) - e_i = e_i - e_i = 0.
\end{aligned} \quad (38)$$

We can determine that $A + f(e_i) + g(e_i)$ is a positive constant and $f(e_i) - g(e_i) - e_i$ equal to 0. So, $\dot{V}_{10} < 0$, $\dot{V}_{11} < 0$; therefore $\dot{V}_1 = \dot{V}_{10} + \dot{V}_{11} < 0$, so V_1 is monotonically decreasing and $\lim_{t\to\infty}V_1 = 0$; we can obtain $\lim_{t\to\infty}e_x = 0$, $\lim_{t\to\infty}e_y = 0$, $\lim_{t\to\infty}e_\psi = 0$. Based on Lyapunov stability, the outer loop is stable.

To guarantee the surge speed $u_F \longrightarrow \Gamma_u$, a candidate Lyapunov function is chosen as follows:

$$V_2 = \frac{1}{2}m_{11}s_1^2 \quad (39)$$

and, then, derivation of (36),

$$\dot{V}_2 = m_{11}s_1\dot{s}_1 = -\eta_1\left|s_1\right| < 0 \quad (40)$$

where η_1 is a positive constant. This means that s_1 has asymptotic stability, so the trajectory can reach the origin, $\Gamma_{eu} \longrightarrow 0$, $u_F \longrightarrow \Gamma_u$.

Choose another Lyapunov function as

$$V_3 = \frac{1}{2} m_{22} m_{33} s_2^2 \quad (41)$$

With Assumption 1, we can get

$$\dot{V}_3 = m_{22} m_{33} s_2 \dot{s}_2 = -\frac{m_{22} m_{33}}{m_{11} u_F} \eta_2 |s_2| < 0 \quad (42)$$

Therefore, s_2 is asymptotic stability, $\Gamma_{ev} \longrightarrow 0$, $v_F \longrightarrow \Gamma_v$. The surge speed and sway speed of unmanned surface vehicles are bounded, which has been proven above, and the yaw angle speed is also bounded, which will be verified below.

Consider the augmented Lyapunov function:

$$V_4 = \frac{1}{2} m_{33} r_F^2; \quad (43)$$

then,

$$\begin{aligned} \dot{V}_4 &= m_{33} r_F \dot{r}_F \\ &= r_F \left((m_{11} - m_{22}) u_F v_F - d_{33} r_F + \tau_{rF} + d_r \right) \end{aligned} \quad (44)$$

If $(m_{11} - m_{22}) u_F v_F + \tau_{rF} + d_r < d_{33} r_F$ is satisfied, then $\dot{V}_4 < 0$, V_4 is a decreasing function. Since $V_4 = (1/2) m_{33} r_F^2 \geq 0$, r_F is also decreasing and the largest value must exist. u_F, v_F, τ_{rF} is bounded, which has been proven above, and d_r is also bounded, as clarified in Assumption 3, so r_F must be bounded.

5. Simulation Results and Analysis

5.1. Straight Line Formation Control. To verify the effectiveness of the proposed strategy, the mathematical value of 3 USVs is used for simulation from [23].

$m_{11} = 120 \times 10^3$ kg, $m_{22} = 217.9 \times 10^3$ kg, $m_{33} = 636 \times 10^5$ kg,

$d_{11} = 215 \times 10^2$ kg • s^{-1}, d22 = 117×10^3 kg • s^{-1}, d33 = 802×10^4 kg • s^{-1}.

A simple case of straight line formation control is considered first. In the simulation, USV 2 and USV 3 are followers; USV 1 is a virtual leader. The system initial state is as follows: $\boldsymbol{\eta}_2 = [0.1, 2, \pi/2]^T$, $\boldsymbol{\eta}_3 = [2, -5, \pi/2]^T$; the desired position and orientation with respect to USV 1 are as follows: ld_{12}=5m, ld_{13}=5m, $\varphi_{d12} = \pi/2$, $\varphi_{d13} = -\pi/2$. We select k_1=k_2=k=3, k_3=3, A_1=A_2=A_3=4, B_1=D_1=6, B_2=D_2=5, B_3=D_3=5, $\lambda_1 = 1$, $\eta_1 = 150$, $\lambda_2 = 1.3$, $\eta_2 = 300$. The disturbance is defined as time-varying disturbance $\mathbf{W} = 4 \times 10^3 [\sin(0.2t), 1 + 2\sin(0.3t), 2]^T$.

The motion curve of three USVs under the proposed scheme is shown in Figure 2. As seen from Figures 2–4, at first, the leaders adjust their control input, making them approach the desired distance and orientation. After a few seconds, the leader and the follower have reached the desired index. During this procedure, we can see from Figure 3 that the yaw angle changes slowly; this is attributed to the bioinspired model constraining the control input when the initial error was large and included the coordinate transformation of the control point. Figure 4 shows that the control input of the USV is smooth and different from the traditional method in [24]. As the controller based on the backstepping approach, this controller causes the sharp speed jumps when tracking errors change suddenly at initial time. For example, the surge speed of the backstepping method jumps to more than 10 m/s, but the biological inspired method is just about 5 m/s in Figures 5 and 6. Therefore, the proposed bioinspired method is more practical and effective.

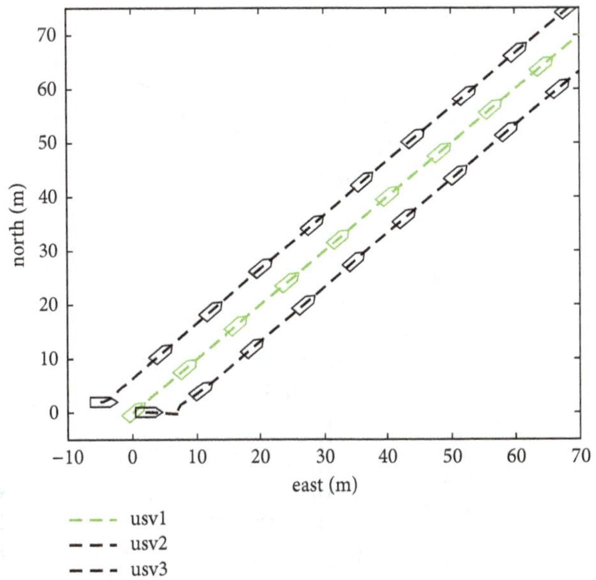

FIGURE 2: The USV motion curve of straight line formation.

5.2. Circular Formation Control. A typical circular path was studied in this section. The initial state $\eta_1 = [0.1, 0.1, \pi/4]^T$, $\eta_2 = [6, 0.1, 0.1]^T$, $\eta_3 = [0.1, 6, \pi/2]^T$. The desired distance and orientation are ld_{12}=8m, ld_{13}=8m, $\varphi_{d12} = \pi/2$, $\varphi_{d13} = -\pi/2$. We select $k_1 = k_2 = k_3 = k = 1$, $A_1 = A_2 = A_3 = 8$, $B_1 = D_1 = 10$, $B_2 = D_2 = 6$, $B_3 = D_3 = 6$. $\lambda_1 = 1.6$, $\eta_1 = 200$, $\lambda_2 = 1$, $\eta_2 = 200$. The disturbance is defined as time-varying disturbance $\mathbf{W} = 4 \times 10^3 [\sin(0.2t), 1 + 2\sin(0.3t), 2]^T$. The simulation results are as shown in Figures 5–7. We can also see that, after the first 20 seconds, the followers USV 2 and USV 3 produce a control force and torque based on the bioinspired hybrid control method, making them move in the error-decreasing direction. Though the distance error and orientation error are large at first, the control input changes slowly in the first few seconds, which leads to the error tending toward zero at a reasonable speed in Figures 8 and 9.

6. Conclusion and Future Work

The development of a bioinspired model-based hybrid sliding-mode formation controller for underactuated unmanned surface vehicles has been presented. The kinematics and dynamic equation of USV formation are first established; then a bioinspired-based, sliding-mode hybrid control strategy, including a virtual speed controller

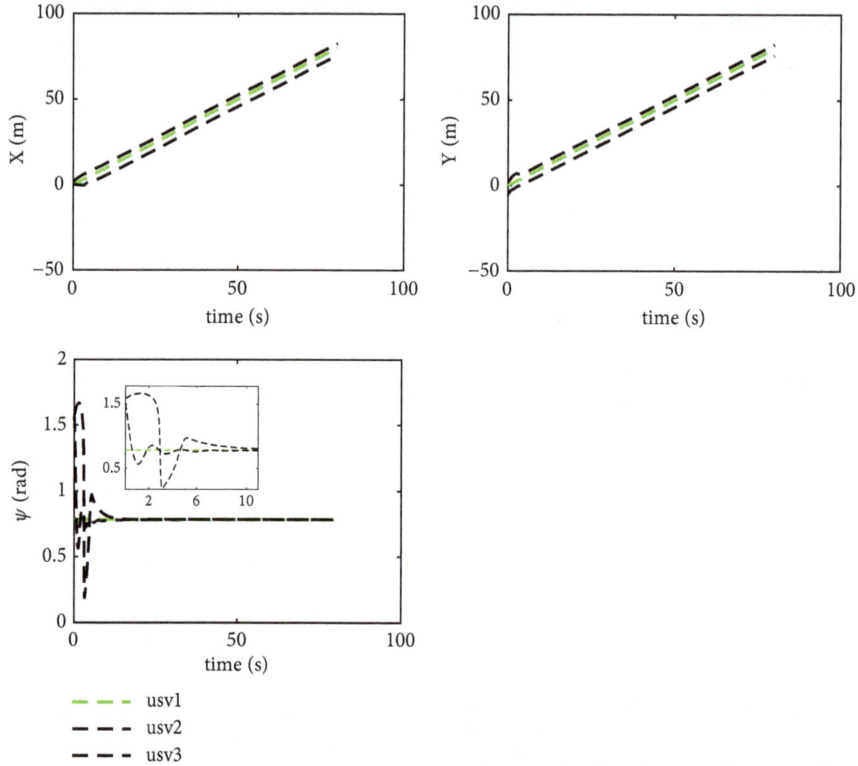

Figure 3: The position and attitude of the three USV.

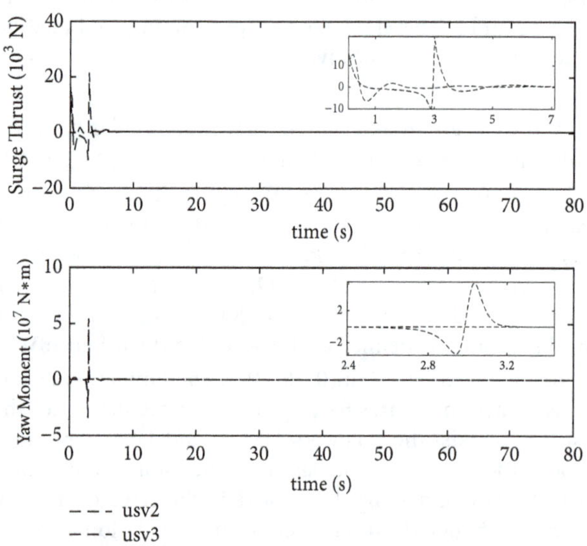

Figure 4: The force and torque of USV 2 and USV 3.

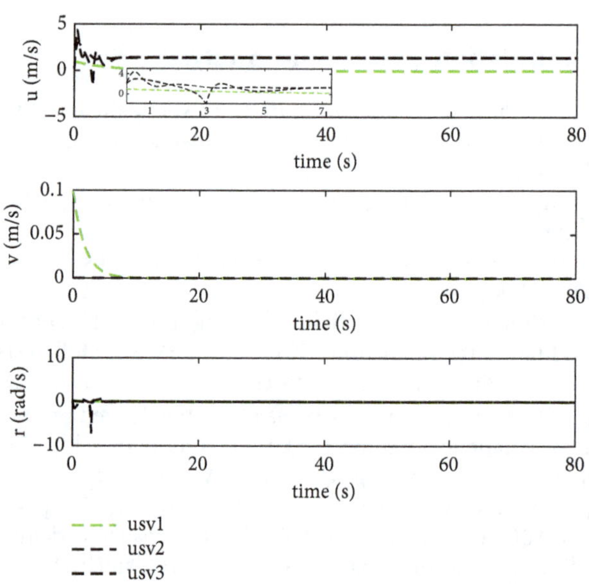

Figure 5: The velocity of USV with the proposed method.

and a sliding-mode controller, is proposed. The stability and effectiveness of the controller have been proven by the Lyapunov theory and verified through MATLAB simulation. The controller addresses the problem of speed jump and controller saturation caused by the large initial error and guarantees yaw angle stability through coordinate transformation. However, collision avoidance and communication between the follower and the leader should be considered in future work. In addition, distributed formation control and constrained control of underactuated marine vehicles are also highly desirable; related research is most concerned with the full actuated surface vehicles so far [25, 26].

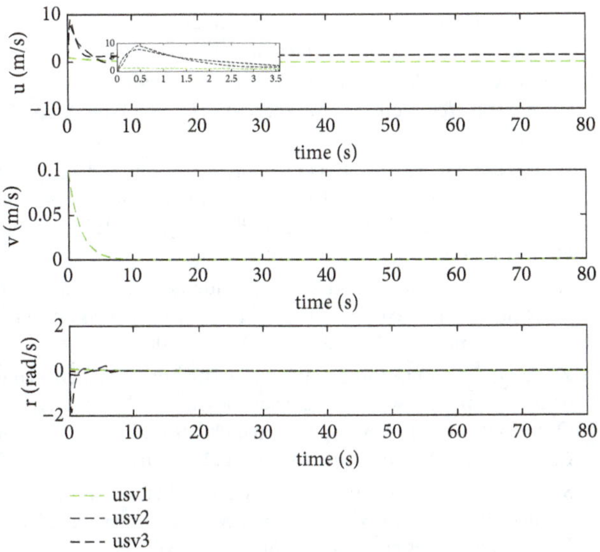

FIGURE 6: The velocity of USV with backstepping method.

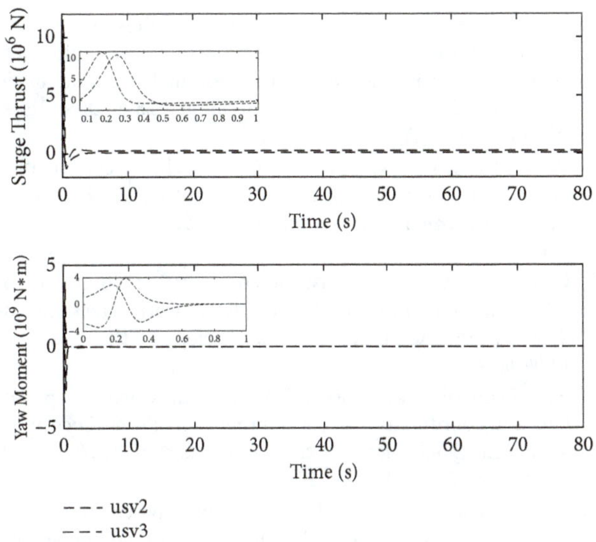

FIGURE 8: The force and torque of USV 2 and USV 3.

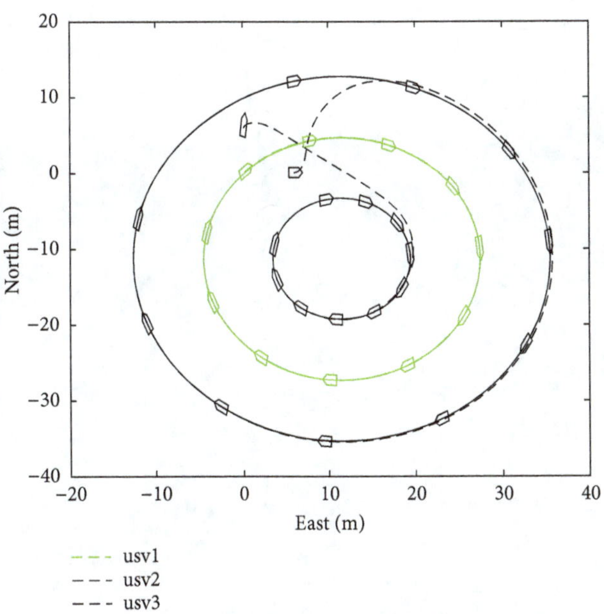

FIGURE 7: The USV motion curve of straight line formation.

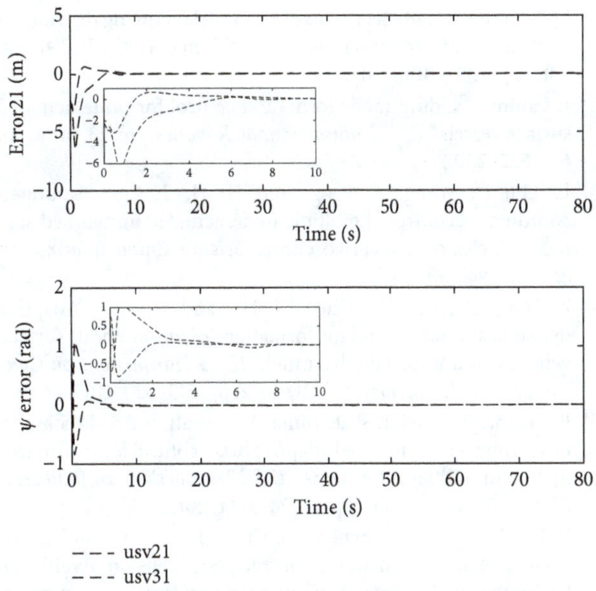

FIGURE 9: The position and attitude of the three USV.

Conflicts of Interest

The authors declare that they have no conflicts of interest.

Acknowledgments

This work was partially supported by the National Nature Science Foundation of China (Grant no. 51309062).

References

[1] K. D. Do, "Synchronization Motion Tracking Control of Multiple Underactuated Ships with Collision Avoidance," *IEEE Transactions on Industrial Electronics*, vol. 63, no. 5, pp. 2976–2989, 2016.

[2] X. Wang and S. Li, "Distributed finite-time consensus algorithms for leader-follower second-order multi-agent systems with mismatched disturbances," in *Proceedings of the American Control Conference*, pp. 2814–2819, IEEE, 2016.

[3] Z.-H. Peng, D. Wang, W.-Y. Lan, and G. Sun, "Robust leader-follower formation tracking control of multiple underactuated surface vessels," *China Ocean Engineering*, vol. 26, no. 3, pp. 521–534, 2012.

[4] M.-Y. Fu, D.-S. Wang, and C.-L. Wang, "Formation Control for Water-Jet USV Based on Bio-Inspired Method," *China Ocean Engineering*, vol. 32, no. 1, pp. 117–122, 2018.

[5] X. Wang, S. Li, and X. Yu, "Distributed finite-time consensus algorithms for leader-follower second-order multi-agent systems with mismatched disturbances," in *Proceedings of the American Control Conference (ACC)*, pp. 2814–2819, IEEE, Boston, MA, USA, 2016.

[6] R. W. Beard, J. Lawton, and F. Y. Hadaegh, "A coordination architecture for spacecraft formation control," *IEEE Transactions on Control Systems Technology*, vol. 9, no. 6, pp. 777–790, 2001.

[7] C. Ma and Q. Zeng, "Distributed formation control of 6-DOF autonomous underwater vehicles networked by sampled-data information under directed topology," *Elsevier Science Publishers*, 2015.

[8] N. E. Leonard and E. Fiorelli, "Virtual leaders, artificial potentials and coordinated control of groups," in *Proceedings of the 40th IEEE Conference on Decision and Control (CDC)*, vol. 3, pp. 2968–2973, IEEE, USA, 2001.

[9] J. Ghommam and M. Saad, "Backstepping-based cooperative and adaptive tracking control design for a group of underactuated AUVs in horizontal plan," *International Journal of Control*, vol. 87, no. 5, pp. 1076–1093, 2014.

[10] L. Ding and G. Guo, "A backstepping method for ship formation control," *Control and Decision*, vol. 27, no. 2, pp. 299–303, 2012.

[11] Q. Ali and S. Montenegro, "Explicit model following distributed control scheme for formation flying of mini UAVs," *IEEE Access*, vol. 4, pp. 397–406, 2016.

[12] F. Fahimi, "Sliding-mode formation control for underactuated surface vessels," *IEEE Transactions on Robotics*, vol. 23, no. 3, pp. 617–622, 2007.

[13] L. Liu, D. Wang, Z. Peng, and H. H. T. Liu, "Saturated coordinated control of multiple underactuated unmanned surface vehicles over a closed curve," *Science China Information Sciences*, vol. 60, no. 7, 2017.

[14] Z. Peng, D. Wang, Z. Chen, X. Hu, and W. Lan, "Adaptive dynamic surface control for formations of autonomous surface vehicles with uncertain dynamics," *IEEE Transactions on Control Systems Technology*, vol. 21, no. 2, pp. 513–520, 2013.

[15] B. Ranjbar-Sahraei, F. Shabaninia, A. Nemati, and S.-D. Stan, "A novel robust decentralized adaptive fuzzy control for swarm formation of multiagent systems," *IEEE Transactions on Industrial Electronics*, vol. 59, no. 8, pp. 3124–3134, 2012.

[16] Y. H. Chang, C. L. Chen, W. S. Chan, H. W. Lin, and C. W. Chang, "Fuzzy Formation Control and Collision Avoidance for Multiagent Systems," *Mathematical Problems in Engineering*, vol. 2013, no. 8, Article ID 908180, pp. 61-61, 2013.

[17] D.-D. Mu, G.-F. Wang, Y.-S. Fan, Y.-M. Bai, and Y.-S. Zhao, "Fuzzy-Based Optimal Adaptive Line-of-Sight Path Following for Underactuated Unmanned Surface Vehicle with Uncertainties and Time-Varying Disturbances," *Mathematical Problems in Engineering*, pp. 1–12, 2018.

[18] B. Sun, D. Zhu, and S. X. Yang, "A bio-inspired cascaded approach for three-dimensional tracking control of unmanned underwater vehicles1," *International Journal of Robotics and Automation*, vol. 29, no. 4, pp. 349–358, 2014.

[19] Z. Peng, D. Wang, and X. Hu, "Robust adaptive formation control of underactuated autonomous surface vehicles with uncertain dynamics," *Iet Control Theory & Applications*, vol. 5, no. 12, pp. 1378–1387, 2011.

[20] R. Yang and D. Zhu, "Plane trajectory tracking control of autonomous underwater vehicles based on a biologically inspired shunting model," *Journal of Shanghai Maritime University*, 2011.

[21] T. I. Fossen, *Marine Control Systems: Guidance, Navigation, and Control of Ships, Rigs and Underwater Vehicles*, 2002.

[22] S. Grossberg, "Nonlinear neural networks: principles, mechanisms, and architectures," *Neural Networks*, vol. 1, no. 1, pp. 17–61, 1988.

[23] A. L. Hodgkin and A. F. Huxley, "A quantitative description of membrane current and its application to conduction and excitation in nerve," *The Journal of Physiology*, vol. 117, no. 4, pp. 500–544, 1952.

[24] K. D. Do, Z. P. Jiang, and J. Pan, "Underactuated ship global tracking under relaxed conditions," *IEEE Transactions on Automatic Control*, vol. 47, no. 9, pp. 1529–1536, 2002.

[25] Z. Peng, J. Wang, and D. Wang, "Distributed Maneuvering of Autonomous Surface Vehicles Based on Neurodynamic Optimization and Fuzzy Approximation," *IEEE Transactions on Control Systems Technology*, vol. 99, pp. 1–8, 2017.

[26] S.-L. Dai, S. He, H. Lin, and C. Wang, "Platoon Formation Control With Prescribed Performance Guarantees for USVs," *IEEE Transactions on Industrial Electronics*, vol. 99, pp. 1-1, 2017.

10

Nonlinear Uncertainties Canceling in Multi-Agent Systems Enabled by Cooperative Adaptation

Cyuansi Shih and Chengyu Cao

Department of Mechanical Engineering, The University of Connecticut, Storrs, CT 06269, USA

Correspondence should be addressed to Cyuansi Shih; cyuan-si.shih@uconn.edu

Academic Editor: Radek Matušů

This paper deals with uncertainties problem in multi-agent systems with novel cooperative adaptation approach. Since uncertainties in multi-agent systems are interconnected, local agent often faces uncertainties not only from itself but also from neighbors. The proposed approach is that a local agent estimates uncertainties from itself and neighboring agents and then changes control strategy. The uncertainties or the equivalences of neighbors can be estimated based on their available outputs; thus, the local agent can adapt to them to cancel out these effects. Stability analysis is also derived that characterizes the transient and steady state performance of multi-agent system. The simulation presents the details of the proposed cooperative adaptation mechanism by compared typical cooperative control.

1. Introduction

The cooperative control in multi-agent system has attracted compelling attention in recent years from various applications including scheduling satellite formation, cooperative unmanned air vehicles, connected autonomous vehicles, and multiple underwater vehicle coordination. In these applications, the agents are required to collaborate with each other in order to finish cooperative mission. Consensus in multi-agent systems is a typical example that agents have common interest and then make negotiations to reach consensus by appropriate algorithmic methods. Consensus problems with various dynamic systems have already been successfully addressed in literature. The consensus in multi-agent systems with single-integrator and double-integrator kinematics is studied in [1–3]. The general linear dynamics of consensus in multi-agent systems is also discussed in [4–10]. In [11–16], the authors further extended to nonlinear dynamics for multi-agent systems.

In practical applications, the models of agents are difficult to obtain because of sensing or measurement limitation. Also, disturbances often exist in interconnected network systems, such as channel noise and communication noise. Thus, it is necessary to consider uncertainty effects in multi-agent systems. The typical direct and indirect adaptive control [17] are two nonlinear schemes explicitly designed for signal system uncertainties. In [18], the model reference adaptive control approach is used to ensure asymptotic tracking of a defined reference model for a class of multi-agent systems with constant unknown parameters. Utilizing the universal approximation capability of artificial neural network [19], adaptive control could deal with nonlinear functions instead of linear parameters in multi-agent systems, and this leads to the neural network controller design. Consensus problem of second-order systems with Lipschitz nonlinearity is addressed [16]. The uncertainties of nonlinear multi-agent systems under jointly connected directed switching networks are also addressed in [20, 21]. The results in [9, 18] deal with high-order system with Lipschitz nonlinearity. In these literatures, the results are restricted to local stability and certain type of network structure.

This paper proposed a novel cooperative adaptation approach to deal with uncertainties issue of network consensus. The cooperative law drives agents to collaborate and then cooperative adaptation law eliminates uncertainties from network. The idea is that the local agent adjusts his own control strategies based on the uncertainty estimation of itself and neighbors. The uncertainties from other agents can be treated as external disturbances and are not available to the local agent. Thus, other agents uncertainties can be estimated based on their available output and then local agent cancels out these uncertainties by adapting to them. This paper also provided transient and steady state performance analysis in the presence of uncertainties for multi-agent systems.

The article is organized as follows. The problem formulation is given in Section 2. The adaptive cooperation control scheme is introduced in Section 3. In Section 4, the stability analysis and performance bounds for multi-agent systems are presented in Section 4. The simulation results are presented in Section 5, while Section 6 concludes this paper.

2. Problem Formulation

There are N agents moving in m dimensional Euclidean space, where the dynamics of each agent is described by

$$\dot{x}_i(t) = Ax_i(t) + B\left(u_i(t) + \sigma_i(x_i, t)\right) \quad (1)$$

where $x_i \in \mathbb{R}^m$ is the state vector, such as position and velocity vector of agent i, $u_i \in \mathbb{R}^m$ are the control input, and $\sigma_i : \mathbb{R}^m \times \mathbb{R}^m \times \mathbb{R} \longmapsto \mathbb{R}^m$ represents the unknown, nonlinear, and time-varying uncertainties.

The objective is to design decentralized control inputs that ensure multi-agent systems with uncertainties achieve consensus. Consensus in this paper means all agents converge to common attitude or speed. Each agent is subjected to individual uncertainties, such as dynamics uncertainty, affected from neighborhood agents. The decentralized control input plays the role not only cooperating but also canceling uncertainty among agents. The network connectivity has a fairly general assumption that creates a spanning tree in the network connection. The uncertainties are restricted to being Lipschitz and allow less restriction on structure of nonlinearity.

The connection among agents are specified by a graph $\mathcal{G} = (\mathcal{V}, \mathcal{E})$ which consists of a set of vertices denoted by $\mathcal{V} = \{1, 2, \ldots, N\}$ and a set of edges denoted by $\mathcal{E} \subseteq \{(i, j) : i, j \in \mathcal{V}, i \neq j\}$. A vertex represents an agent, and each edge represents a connection. The adjacency matrix $\mathcal{A}$ with elements a_{ij} from the graph denotes the connections, where $a_{ij} = 1$ means a path from agent j to agent i; otherwise $a_{ij} = 0$. The Laplacian matrix $\mathcal{L}$ is defined as $l_{ii} = \sum_{j=1}^{N} a_{ij}$ and $l_{ij} = -a_{ij}$ when $i \neq j$. M is defined as the normalized Laplacian matrix, and thus $M = I_N - \mathbf{1}r^T$. The rank of M is $N-1$, and M has $N-1$ eigenvalues at 1 and one eigenvalue at 0. The connection graph is undirected if $a_{ij} = a_{ji}$ and thus the adjacency matrix is symmetric; otherwise the graph is directed. We consider the undirected graph among agents communication in this paper.

Assumption 1. The uncertainties depended on agent's state and its changes. For local agent $x_i > 0$, there exist $L > 0$, $\beta_0 > 0$, and $\beta > 0$ such that

$$\left\|\sigma_i\left(\xi_i(t_1)\right) - \sigma_i\left(\xi_i(t_2)\right)\right\| < L\left\|\xi_i(t_1) - \xi_i(t_2)\right\|_\infty + \beta_0 \quad (2)$$

where $\xi_i = x_i - \sum_{j\in N_i} r_j x_j$ and r_j is normalized edge of graph. For overall network, we have

$$\left\|\sigma\left(\xi(t_1)\right) - \sigma\left(\xi(t_2)\right)\right\| < LM\left\|\xi(t_1) - \xi(t_2)\right\|_\infty + \beta \quad (3)$$

Assumption 2. The nonlinear uncertainties $\sigma_i(\xi_i, t)$ are bounded with respect to state ξ_i. This means the uncertainties do not become unbounded even system state keep changing. The partial derivatives of $\sigma_i(\xi_i, t)$ with respect to ξ_i and t are bounded, i.e., $\|\partial\sigma_i(\xi_i, t)/\partial\xi_i\| \le d_{\sigma_{i_{\xi_i}}}$ and $\|\partial\sigma_i(\xi_i, t)/\partial t\| \le d_{\sigma_{i_t}}$. We also further have $\|\partial\sigma(\xi, t)/\partial\xi\| \le d_{\sigma_\xi}$ and $\|\partial\sigma(\xi, t)/\partial t\| \le d_{\sigma_t}$.

Assumption 3. The undirected graph $\mathcal{G}$ is connected and thus creates a spanning tree in the graph. The Laplacian matrix $\mathcal{L}$ has a single eigenvalue at 0 and all other eigenvalues are with positive real part.

3. Uncertainties Canceling Algorithm

In this section, we introduce a novel cooperative adaptation law to drive multi-agent systems as (1) achieve global objective, x_i toward neighbor agents x_j. The algorithm first deals with the agent's collaboration by the cooperative law that is based on local neighbors' relative information. The algorithm further handles uncertainties issue among network through cooperative adaptation law. The overall control law is represented as below:

$$u_i = u_{i_c} + u_{i_a} \quad (4)$$

where u_{i_c} is cooperative control law and u_{i_a} is cooperative adaptation law.

First, we introduced the state transformation to describe the relation between individual agent and network as follows:

$$\xi_i = x_i - \sum_{j\in N} r_j x_j \quad (5)$$

The network dynamics between local agent and neighbors is described as

$$\begin{aligned}\dot{\xi}_i = A\xi_i &+ B\left(u_{i_c} - \sum_{j\in N} r_j u_{j_c}\right) + B\left(u_{i_a} - \sum_{j\in N} r_j u_{j_a}\right) \\ &+ B\left(\sigma_i - \sum_{j\in N} r_j \sigma_j\right)\end{aligned} \quad (6)$$

The overall network transformation follows

$$\xi = x - \left(\mathbf{1}r^T \otimes I_n\right)x = M \otimes BI_n x \quad (7)$$

Cooperative Law. In order to achieve the network consensus, the cooperative control law design is moving to the average state of neighbors. In other words, it uses neighbors' relative information as feedback and is presented as

$$u_{i_c} = -k\left(x_i - \sum_{j\in N} r_j x_j\right) \quad (8)$$

where the selection of k will make sure to stabilize the overall network system that is described as

$$\dot{\xi} = (I_N \otimes A - \mathscr{L} \otimes Bk)\xi + (M \otimes B)(u_a + \sigma) \quad (9)$$

where $\xi = [\xi_1 \dots \xi_n]$, $u_a = [u_1 \dots u_n]$, $\sigma = [\sigma_1 \dots \sigma_n]$, $A_m = (I_N \otimes A - \mathscr{L} \otimes Bk)$, and $B_m = M \otimes B\mathbf{1}_n$. Thus, A_m is Hurwitz matrix by k selection.

Cooperative Adaptation Law. We further design the decentralized cooperative adaptation law to deal with uncertainty issue among network. This disturbance rejection includes state predictor, adaptive law, and control as below. Since uncertainty for dynamical systems is made in real time, all uncertainties from network can be estimated based on their available output. Hence, a local agent can adjust its own control law based on the estimation of network uncertainty. Thus, the local agent estimates these uncertainties through transferred dynamics. With network dynamics from (9), we consider the following state predictor:

$$\dot{\hat{\xi}} = A_m\hat{\xi} + B_m(u_a + \hat{\sigma}) \quad (10)$$

where $\hat{\xi} \in \mathbb{R}^m$ is the estimated state vector of ξ in (9) and $\hat{\sigma}$ is the estimated uncertainty of σ. These parameters vectors are updated by adaptive law below.

Fast adaptation mechanism is applied in here as a high-gain feedback for state predictor. It is noted that both state predictor and adaptation work together and there is no time-delay or at most one integration time-step in the loop. Here, the predicted state, $\hat{\xi}_i$, is used to generate the adaptive parameter, $\hat{\sigma}_i$, based on the network error dynamics. The piece-wise constant adaptive law is used in here. In order to drive $\hat{\xi}$ to track ξ. The update law is given by

$$\hat{\sigma}(t) = -\Phi^{-1}(T)e^{A_m T}\tilde{\xi}(kT), \quad t \in [kT, (k+1)T] \quad (11)$$

where $\Phi(T) = \int_0^T e^{A_m(T-\tau)}d\tau$, $T > 0$ is the adaptation step size, and $\tilde{\xi} = \hat{\xi} - \xi$. The adaptive law drives $\tilde{\xi}$ to be small and as a result, $\hat{\sigma}$ will approach σ. The information of $\hat{\sigma}$ will be used in the control law design. The error dynamics for network system is described as

$$\dot{\tilde{\xi}} = A_m\tilde{\xi} + B_m(\hat{\sigma} - \sigma) \quad (12)$$

To prevent aggressive signals in real application, a low-pass filtering mechanism is added in control law, which filters away aggressive signals and recovers robustness. The adaptive control law design is

$$u_a(t) = -C(s)\hat{\sigma}(t) \quad (13)$$

where $C(s)$ is low-pass filter to filter the control signal. The introduction of low-pass filter at this point in the control loop allows us to decouple control and estimation by canceling the overall network uncertainty. Furthermore, the bandwidth of filter can be chosen to attenuate frequencies based on the variation of uncertainties. The closed-loop network system from (1) and (13) becomes

$$\dot{\xi} = A_m\xi - C(s)B_m\hat{\sigma} + B_m\sigma \quad (14)$$

Remark 4. The objective of cooperative adaptation law is to make the adaptive parameter $\hat{\sigma}(t)$ approach the network disturbance $\sigma(t)$. The effects of the control law $u_a(t)$ cancel the effects of $\sigma(t)$ and as a result $\tilde{\xi}(t)$ in (12) will become very small. Thus, the agents can achieve consensus without uncertainties.

4. Network Stability Analysis

The analysis first verifies the network stability by introducing a reference network system. The reference system will asymptotically track desired system when the bandwidth of low-pass filter is large enough. The network convergence analysis also proved the stability conditions. Further, the real network system is proven to track the reference system and its stability condition is provided. Finally, the estimation errors between real network and predict network system are shown to be bounded and the uncertainties canceling is achieved by decreasing the size of time-step.

4.1. Nonadaptive Reference System. We consider a reference network dynamics for (9) as

$$\dot{\xi}_r = A_m\xi_r + B_m\left(u_{r_a} + \sigma_r\right) \quad (15)$$

where $\xi_r = [\xi_{r_1}, \dots, \xi_{r_n}]^T$, $\xi_{r_i} = x_{r_i} - \sum_{j\in N_i} r_j x_{r_j}$, $\sigma_r = [\sigma_{r_1}, \dots, \sigma_{r_n}]^T$ and the reference control law

$$u_{r_a} = -C(s)\sigma_r \quad (16)$$

The closed-loop reference network system from (15) and (16) becomes

$$\dot{\xi}_r = A_m\xi_r + (1 - C(s))B_m\sigma_r \quad (17)$$

Lemma 5. *For the closed-loop network reference system in (15) and (16) subject to the $\mathscr{L}_1$ gain upper bound*

$$\frac{\Theta_{r_1}\beta + \left\|\xi_{r_0}\right\|_\infty}{1 - \Theta_{r_1}L\left(\rho_r + \gamma_\xi\right)} < \rho_r \quad (18)$$

$$\Theta_{r_1}L\left(\rho_r + \gamma_\xi\right) < 1 \quad (19)$$

where $\Theta_{r_1} = \|(sI_N - A_m)^{-1}(I_N - C(s)I_N)B_m\|_{\mathscr{L}_1}$.

$$\left\|\xi_{r_0}\right\| < \rho_r \quad (20)$$

Then

$$\left\|\xi_r\right\|_\infty < \rho_r \quad (21)$$

$$\left\|u_{r_a}\right\|_\infty < \rho_{u_r} \quad (22)$$

where ρ_r *is introduced in (18) and (19)*

$$\rho_{u_r} = \|C(s)M\|\left(L\left(\rho_r\right)\rho_r + \beta\right) \quad (23)$$

Proof. Because $\|\xi_{r_0}\|_\infty < \rho_r$ and $\xi_r(t)$ is continuous, to prove $\|\xi_r\|_\infty < \rho_r$, assuming the opposite is true, it implies there exists a time t' such that

$$\xi_r(t') = \rho_r, \qquad \left\|\xi_{r_{t'}}\right\|_\infty \le \rho_r \tag{24}$$

Take Laplace transform for (17); we have

$$\xi_r(s) = (1 - C(s))(sI_N - A_m)^{-1} B_m \sigma_r(s) + \xi_{r_0} \tag{25}$$

and its upper bound has

$$\left\|\xi_{r_{t'}}\right\|_\infty \le \left\|(1 - C(s))(sI_N - A_m)^{-1} B_m\right\|_{\mathscr{L}_1} \|\sigma_r(s)\|_\infty + \left\|\xi_{r_0}\right\|_\infty \tag{26}$$

Using Assumption 1 and upper bound in (24), we arrived at

$$\left\|\sigma_{r_{t'}}\right\|_\infty \le L(\rho_r)\left\|\xi_{r_{t'}}\right\|_\infty + \beta \tag{27}$$

Substituting (27) into (26), we obtain

$$\left\|\xi_{r_{t'}}\right\|_\infty \le \frac{\Theta_{r_1}\beta + \left\|\xi_{r_0}\right\|_\infty}{1 - \Theta_{r_1} L(\rho_r)} \tag{28}$$

where $\Theta_{r_1} = \|(sI_N - A_m)^{-1}(I_N - C(s)I_N)B_m\|_{\mathscr{L}_1}$. The stability condition in (18), together with (28), implies that $\|\xi_{r_{t'}}\|_\infty < \rho_r$, which clearly contradicts assumption in (24) since there is no t' to make assumption true. This proves (21). From (16), the network control law is

$$u_{r_a t'} = -C(s) M \sigma_r \tag{29}$$

It follows from (29), (21), and (27) that we have

$$\left\|u_{r_a t'}\right\|_\infty < \|C(s) M\|_{\mathscr{L}_1} (L(\rho_r)\rho_r + \beta) < \rho_{u_r} \tag{30}$$

which proves (22) and concludes the proof. □

Lemma 6. *Consider the reference system in (17). If the bandwidth of low-pass filter $C(s)$ is large enough, the control law in (16) can make ξ_r asymptotically track the desired network system described by*

$$\dot{\xi}_d = A_m \xi_d \tag{31}$$

where $\xi_d = [\xi_{d_1}, \ldots, \xi_{d_n}]$ *and* $\xi_{d_i} = x_{d_i} - \sum_{j \in N_i} r_j x_{d_j}$ *are desired network state,* $A_m = I_N \otimes A - \mathscr{L} \otimes Bk$.

Proof. For a given network, if the bandwidth of low-pass filter $C(s) = a/(s + a)$ can be chosen large enough, the disturbance effect in (17) will be canceled and (17) become as

$$\dot{\xi}_r = A_m \xi_r \tag{32}$$

This further implies

$$\xi_r(s) = (sI_N - A_m)^{-1} \xi_{r_0} \tag{33}$$

We further obtain that

$$\lim_{s \to 0} \xi_r(s) = \lim_{s \to 0} \xi_d(s) \tag{34}$$

Thus, the reference system will track the desired network system, which concludes the proof. □

4.2. Network Convergence Analysis. Since network $\mathscr{G}$ is connected, it follows from Assumption 3 that 0 is a simple eigenvalue of $\mathscr{L}$ and the other eigenvalues have positive real part. Thus, there exists a nonsingular matrix T such that $T^{-1}\mathscr{L}T$ is in the Jordan canonical form [9]. Because the left and right eigenvectors of $\mathscr{L}$ corresponding to zero eigenvalue are r and $\mathbf{1}$, respectively, the following can be chosen:

$$T = [\mathbf{1} \quad U], \qquad T^{-1} = \begin{bmatrix} r^T \\ V \end{bmatrix} \tag{35}$$

where $U \in \mathbf{C}^{N\times(N-1)}$ and $V \in \mathbf{C}^{N\times(N-1)}$ such that

$$T^{-1}\mathscr{L}T = J = \begin{bmatrix} 0 & 0 \\ 0 & \Delta \end{bmatrix} \tag{36}$$

where $\Delta \in \mathbf{C}^{N\times(N-1)}$ is an upper-triangular matrix, where the diagonal entries are the nonzero eigenvalues of $\mathscr{L}$. Let $\zeta_d = (T^{-1} \otimes I_n)\xi_d$. Then, (31) can be presented as

$$\dot{\zeta}_d = (I_N \otimes A - J \otimes Bk)\zeta_d \tag{37}$$

Lemma 7. *Suppose Assumption 3 holds. The distributed control law in (16) with agents in (15) solves consensus problem if and only if all the matrices $A - \lambda_i B$, $i = 2, \ldots, N$, are Hurwitz, where λ_i are the nonzero eigenvalues of Laplacian matrix $\mathscr{L}$.*

Proof. The elements of the state matrix in (37) are either block diagonal or block upper-triangular. Thus, ξ_i, $i = 2, \ldots, N$, converge asymptotically to zero if and only if the $N - 1$ subsystems along the diagonal as

$$\dot{\zeta}_{d_i} = (A - \lambda_i B)\zeta_{d_i}, \quad i = 2, \ldots, N \tag{38}$$

are asymptotically stable. The stability of closed-loop system in (32) and (31) is equal to the system in (37) that converges asymptotically to zero. The system consensus is achieved. This concludes the proof. □

Lemma 8. *Suppose Assumption 3 holds and the cooperative control law (8) satisfies Lemma 7; then*

$$\lim_{t \to \infty} x_{d_i} = \left(r^T \otimes e^{At}\right) r \begin{bmatrix} x_{d_1}(0) \\ \vdots \\ x_{d_N}(0) \end{bmatrix} \tag{39}$$

where $r \in \mathbf{R}^N$ *such that* $r^T\mathscr{L} = 0$ *and* $r^T\mathbf{1} = 1$.

Proof. The desired network dynamics (31) can transform as

$$\dot{x}_d = (I_N \otimes A - \mathscr{L} \otimes Bk)x_d \tag{40}$$

where $\xi_d = \mathscr{L}x_d$. The solution of (40) with Lemma 7 can be obtained as

$$\begin{aligned} x_d(t) &= e^{(I_N \otimes A - \mathscr{L} \otimes Bk)t} x_d(0) \\ &= (T \otimes I_n) e^{I_N \otimes A - J \otimes Bk} (T^{-1} \otimes I_n) x_d(0) \\ &= (T \otimes I_n) \begin{bmatrix} e^{At} & 0 \\ 0 & e^{(I_{N-1} \otimes A - \Delta \otimes Bk)t} \end{bmatrix} (T^{-1} \otimes I_n) x_d(0) \end{aligned} \quad (41)$$

From Lemma 7, $I_N \otimes A - \Delta \otimes Bk$ is Hurwitz. Thus,

$$\begin{aligned} \lim_{t \to \infty} e^{(I_N \otimes A - \mathscr{L} \otimes Bk)t} &= (\mathbf{1} \otimes I_n) e^{At} (r^T \otimes I_n) \\ &= (\mathbf{1} r^T) \otimes e^{At} \end{aligned} \quad (42)$$

Thus, the final consensus value is

$$\lim_{t \to \infty} x_d(t) = (\mathbf{1} r^T \otimes e^{At}) x_d(0) \quad (43)$$

This concludes the proof. □

4.3. Network Transient Performance. We discussed the relation between reference network system and desired network system and then perform the network convergence analysis in last section. We further prove the real network system can track reference network system in this section. Before this proof, we will establish the error bounds between the network state predictor and real network system and between the adaptive law and the signal it estimates.

Lemma 9. *Given the network system in (9) with $\mathscr{L}_1$ adaptive control law in (13) and the reference system in (15) with (16) subject to stability conditions in (18) and (19), if $\|\xi_{r_0}\| < \rho_r$ and we choose T in (11) and low-pass filter C(s) in (13) to ensure*

$$\gamma_0(T) < \gamma_{\tilde{\xi}} \quad (44)$$

$$\gamma_{\xi_e}(T) < \gamma_\xi \quad (45)$$

where $\gamma_{\tilde{\xi}}$ is arbitrary positive constant and γ_ξ is introduced in (19); then we have

$$\|\tilde{\xi}\|_\infty < \gamma_{\tilde{\xi}} \quad (46)$$

$$\|\xi - \xi_r\|_\infty < \gamma_\xi \quad (47)$$

$$\|u - u_{r_a}\|_\infty < \gamma_u \quad (48)$$

Suppose there exist upper bounds for the truncated $\mathscr{L}_\infty$ norms of $\xi(t)$ and $\sigma(t) - \hat{\sigma}(t)$ at some time, t_1

$$\|\xi_{t1}\|_\infty < \rho \quad (49)$$

Define

$$\beta_1 = \sqrt{\lambda_{max}(A_m^T A_m)} \quad (50)$$

$$\beta_2 = \int_0^T \sqrt{\lambda_{max} \Phi(T-\tau)^T \Phi(T-\tau) B_m} d\tau \cdot \left\|\Phi^{-1}(T) e^{AT}\right\| \quad (51)$$

$$\beta_3 = \int_0^T \sqrt{\lambda_{max} \Phi(T-\tau)^T \Phi(T-\tau) B_m} d\tau \quad (52)$$

Proof. Because $\xi(t), \tilde{\xi}(t)$ is continuous, to prove $\|\tilde{\xi}\|_\infty < \gamma_{\tilde{\xi}}$, assume that the opposite is true assuming $\|\tilde{\xi}\|_\infty \geq \gamma_{\tilde{\xi}}$. This implies that there exists a time t' such that

$$\|\tilde{\xi}(t')\|_\infty = \gamma_{\tilde{\xi}} \quad (53)$$

$$\|\xi(t') - \xi_r(t')\|_\infty = \gamma_\xi, \quad or \quad (54)$$

$$\|u_a(t') - u_{r_a}(t')\| = \gamma_u, \quad or \quad (55)$$

The solution of network error dynamics (12) over the interval $[jT, (j+1)T]$ is described as

$$\begin{aligned} \tilde{\xi}(t) &= \Phi(T)\tilde{\xi}(jT) + \int_{jT}^t \Phi(t-\tau) B_m \hat{\sigma}(t) d\tau \\ &\quad - \int_{jT}^t \Phi(t-\tau) B_m \sigma(t) d\tau, \quad j = 0, 1, 2, \cdots \end{aligned} \quad (56)$$

where $\Phi(T) = e^{A_m T}$, $(j+1)T < t'$ and τ is a dummy variable. At $t = (j+1)T$, it becomes

$$\begin{aligned} \tilde{\xi}((j+1)T) &= \Phi(T)\tilde{\xi}(jT) \\ &\quad + \int_0^T \Phi(T-\tau) B_m \hat{\sigma}(jT) d\tau \\ &\quad - \int_0^T \Phi(T-\tau) B_m \sigma(jT+\tau) d\tau \end{aligned} \quad (57)$$

We can choose $\hat{\sigma}$ such that its effects on the system derive $\tilde{x}$ to zero at the next time-step. Then, we have

$$\tilde{\xi}((j+1)T) = -\int_0^T \Phi(T-\tau) B_m \sigma(jT+\tau) d\tau \quad (58)$$

It follows from Assumption 1, Lemma 5, and (54) that

$$\|\sigma(t)\|_\infty \leq L(\rho_r + \gamma_\xi) + \beta \leq b_\sigma \quad (59)$$

Moreover, following Lemma 5 and (55), we have

$$\|u_a(t)\|_\infty \leq \rho_{ur} + \gamma_u \leq b_u \quad (60)$$

The upper bound in (59) and (60) allows for the following upper bound:

$$\|\tilde{\xi}(j+1)T\|_\infty \leq \beta_3 b_\sigma \leq \kappa(T) \quad (61)$$

for all j such that $(j+1)T < t'$. From (56), we arrived at the following upper bound:

$$\left\|\tilde{\xi}(jT+t)\right\|_{\infty} \le \beta_1\kappa + \beta_2\kappa + \beta_3 b_\sigma \le \gamma_0(T) \quad (62)$$

where β_1, β_2, and β_3 are defined in (50), (51), and (52), respectively. Then, for all $t \in [0,t']$, following from (44), we have

$$\left\|\tilde{\xi}(t)\right\| \le \gamma_0(T) < \gamma_{\tilde{\xi}} \quad (63)$$

The results clearly contradict assumption in (53) since there is no t' to make assumption true. This proves (46).

Next, we will drive the bounds for the estimation error of adaptive law, $\tilde{\sigma} = \hat{\sigma} - \sigma$, which are used to further obtain the bounds of $\|\xi - \xi_r\|$. We derive the bound for $\dot{\sigma}(x,t)$ first and then get

$$\dot{\sigma}(t) = \frac{\partial\sigma}{\partial\xi}^T \dot{\xi}(t) \quad (64)$$

From (14), we have

$$\dot{\xi} = A_m\xi - C(s)B_m\hat{\sigma} + B_m\sigma \quad (65)$$

To establish the bound for this signal, we need to derive a bound for $\xi(s)$. Using (11) and (53) for all $t < t_1$, then we have

$$\left\|\hat{\sigma}_{t1}\right\|_{\infty} \le \left\|\Phi^{-1}(T)e^{A_mT}\right\|\gamma_{\tilde{\xi}} \le b_{\hat{\sigma}} \quad (66)$$

Using (64), we can now write the

$$\begin{aligned}\|\dot{\sigma}(t)\|_{\infty} \le L\Big\{&\|A_m\|_{\mathscr{L}_1}\left(\rho_r+\gamma_\xi\right) + \|C(s)B_m\|_{\mathscr{L}_1} b_{\hat{\sigma}} \\ &+ \|B_m\|_{\mathscr{L}_1}\left[L\left(\rho_r+\gamma_\xi\right)+\beta\right]\Big\} \le b_{d\sigma}\end{aligned} \quad (67)$$

Then, control signal upper bound

$$\|\dot{u}\|_{\infty} \le \|sC(s)M\otimes\mathbf{1}_n\| b_{\hat{\sigma}} \le b_{du} \quad (68)$$

Note that the adaptive law of (11) is designed such that

$$\Phi(T)\tilde{\xi}(jT) + \int_0^T \Phi(T-\tau)d\tau B_m\hat{\sigma}(jT) = 0 \quad (69)$$

For the time $t \in [(j-1)T, jT]$, we also have

$$\tilde{\xi}(jT) = -\int_0^T \Phi(T-\tau)B_m\sigma((j-1)T+\tau)d\tau \quad (70)$$

Following from the definition of adaptive law, we obtain

$$\begin{aligned}\tilde{\xi}(jT) = &-\int_0^T \Phi(T-\tau)A_m\tilde{\xi}(jT)d\tau \\ &-\int_0^T \Phi(T-\tau)B_m\hat{\sigma}(jT)d\tau\end{aligned} \quad (71)$$

Thus, (70) and (71) give us

$$\begin{aligned}&\int_0^T \Phi(T-\tau)B_m\sigma((j-1)T+\tau)d\tau \\ &= \int_0^T \Phi(T-\tau)\left(A_m\tilde{\xi}(jT) + B_m\hat{\sigma}(jT)\right)d\tau\end{aligned} \quad (72)$$

which imply there exist a time, $t_p \in [(j-1)T, jT]$, such that

$$B_m\hat{\sigma}(jT) + A_m\tilde{\xi}(jT) = B_m\sigma(t_p) \quad (73)$$

This imply that for any $t \in [jT, (j+1)T]$

$$\begin{aligned}&\|B_m(\hat{\sigma}(t)-\sigma(t))\| \\ &\le \|B_m\|\left(\left\|\hat{\sigma}(jT)-\sigma(t_p)\right\| + \left\|\sigma(t)-\sigma(t_p)\right\|\right) \\ &\le \beta_1\left\|\tilde{\xi}(jT)\right\| + \int_{t_p}^t \|\dot{\sigma}(\tau)\|d\tau \le \beta_1\gamma_0 + 2\left(b_{d\sigma}\right)T \\ &\le b_{\tilde{\sigma}}\end{aligned} \quad (74)$$

Let $\tilde{\sigma} = \hat{\sigma} - \sigma$; then we have

$$\|B_m\tilde{\sigma}\|_{\infty} \le b_{\tilde{\sigma}}\left(T, \rho_b, \rho_\xi\right) \quad (75)$$

Next, we derive the bounds of $\|u_a - u_{r_a}\|$. From network control law in (8) and (13), we have

$$u_a(s) = -C(s)\hat{\sigma} \quad (76)$$

From (29), we have

$$u_{r_a}(s) = -C(s)\sigma_r \quad (77)$$

Thus, apply (76) and (77), we arrived

$$u_a(s) - u_{r_a}(s) = -C(s)(\sigma-\sigma_r) - C(s)\tilde{\sigma} \quad (78)$$

where $\hat{\sigma} = \sigma + \tilde{\sigma}$. Let $\xi_e(t) = \xi(t) - \xi_r(t)$; from Assumption 1 and (45), we can arrive at the following upper bound:

$$\|(\sigma-\sigma_r)\|_{\infty} \le L\|\xi_e\|_{\infty} + \beta \quad (79)$$

Now, we derive the bounds of $\|\xi - \xi_r\|$ and then we have

$$\dot{\xi}_e(t) = A_m\xi_e + B_m((1-C(s))(\sigma-\sigma_r) - C(s)\tilde{\sigma}) \quad (80)$$

The solution of (80) is

$$\begin{aligned}&\xi_e(s) \\ &= (sI_N - A_m)^{-1}B_m((1-C(s))(\sigma-\sigma_r) - C(s)\tilde{\sigma})\end{aligned} \quad (81)$$

Substituting (79) and (55) to (81), we further get

$$\gamma_{\xi_e} = \frac{\Theta_{e_1}\beta + \Theta_{e_2}b_{\tilde{\sigma}}}{1-\Theta_{e_1}L\left(\rho_r+\gamma_\xi\right)} \quad (82)$$

and

$$\|\xi_e(t)\|_{\infty} \le \Theta_{e_1}\left(L\|\xi_e\|_{\infty} + \beta\right) + \Theta_{e_2}b_{\tilde{\sigma}} \le \gamma_{\xi_e} \quad (83)$$

where $\Theta_{e_1} = \|(sI_N - A_m)^{-1}B_m(1-C(s))\|_{\mathscr{L}_1}$ $\Theta_{e_2} = (sI_N - A_m)^{-1}C(s)$.

$$\|\xi_e(t)\|_{\infty} \le \gamma_{\xi_e} \le \gamma_\xi \quad (84)$$

Substituting (75) and (79) to (78), the upper bound is

$$\left\|u_a(s) - u_{r_a}(s)\right\| \le \|C(s)\| \left(L\gamma_{\xi_e} + \beta + b_{\bar{\sigma}}\right) < \gamma_u \tag{85}$$

The stability condition in (18), together with (84), implies that $\|\xi - \xi_r\|_\infty < \gamma_\xi$, which clearly contradicts assumption in (55) since there is no t' to make assumption true. This proves (47). If we choose $T = 0$ and the bandwidth of low-pass filters $C(s)$ large enough, (84) with the condition become $\|\xi_{e_t}\|_\infty < \gamma_\xi$. Thus, it always exists γ_ξ to satisfy (84). This completes the proof of Lemma 9. □

Lemma 10. *The error bounds from Lemma 9 depend on the time-step T. When T approaches 0, the error approaches 0 as well.*

$$\lim_{T\longrightarrow 0} \gamma_{\bar{\xi}}(T, \rho_b) \longrightarrow 0 \tag{86}$$

$$\lim_{T\longrightarrow 0} \gamma_{\xi}(T, \rho_b) \longrightarrow 0 \tag{87}$$

$$\lim_{T\longrightarrow 0} \gamma_{u}(T, \rho_b) \longrightarrow 0 \tag{88}$$

Proof. The inside of integration in (61) is a bounded function; so when $T \longrightarrow 0$, then $\gamma_{\bar{\xi}} \longrightarrow 0$. Since (61) holds, the first term in (74) goes to 0 as $T \longrightarrow 0$ and the second term clearly goes to 0 as well. Thus, $\gamma_\xi \longrightarrow 0$ when $T \longrightarrow 0$. The last statement is γ_u. Because (74) holds, we can clearly see from (48) that when $T \longrightarrow 0$, $\gamma_u \longrightarrow 0$. This completes the proof. □

5. Simulation

The simulation will show the details of proposed adaptive cooperation algorithm by compared with typical cooperative control. The network systems are included uncertainties and disturbances in these examples. The agent dynamics under consideration is a vertical motion of a quadrotor. There are 10 quadrotors that try to maintain in the specified altitude and velocity for cooperation observation in simulation scenarios. The dynamics described by a second-order state space as

$$\dot{x}_i = \begin{bmatrix} 0 & 1 \\ 0 & 0 \end{bmatrix} x_i + \begin{bmatrix} 0 \\ \dfrac{1}{m} \end{bmatrix} (u_i + \sigma_i) \tag{89}$$

where $x_i = [z_i \ w_i]^T$ are vertical position and velocity, $m = 1kg$ is the mass of quadrotor, u_i is the thrust generated by rotors, and $\sigma_i(x_i, t)$ is the force produced by wind on quadrotor.

The first example shows the cooperative control algorithm from (8). The integration step in simulation is set as $T = 10^{-2}$ seconds; the initial altitude and velocity $x(0)_i$ are chosen randomly within the range $0 \sim 10m$ and $-1 \sim 1m/s$ individually. The control gain is chosen as $k = [1 \ 1]$. The communication links are considered as undirected ring network topology as Figure 1. The Laplacian matrix is given by

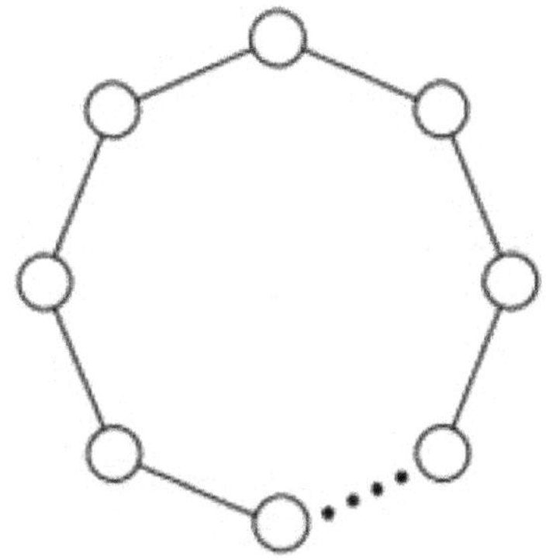

FIGURE 1: Undirected ring network.

$$L = \begin{bmatrix} 2 & -1 & 0 & & 0 & -1 \\ -1 & & & \ddots & 0 & 0 \\ 0 & & \ddots & \ddots & \ddots & \\ & \ddots & \ddots & \ddots & & 0 \\ 0 & 0 & \ddots & & 2 & -1 \\ -1 & 0 & & 0 & -1 & 2 \end{bmatrix}. \tag{90}$$

The eigenvalues of L only have one zero eigenvalues and other eigenvalues are positive that means Assumption 3 is satisfied. The uncertainties considered in agent dynamics are globally Lipschitz as

$$\sigma_i = c * e^{-x_i} (\sin(x_i) + \cos(x_i)) \tag{91}$$

Although the cooperative control tries to drive quadrotors to common position, the uncertainties affect the effectiveness of algorithm (8). The cooperative control signals from Figure 3 show the fluctuation and diverge gradually. Thus, the consensus from Figure 2 cannot be achieved by cooperative control (8) for the group of quadrotors.

The second example shows the proposed adaptive cooperation algorithm from (8) and (13). The integration step in simulation is set as $T = 10^{-2}$ seconds. The initial altitude and velocity $x(0)_i$ are chosen randomly within the range $0 \sim 10m$ and $-1 \sim 1m/s$ individually. The filter of adaptive cooperation controller for agents is chosen as $C(s) = 10/(s + 10)$. The communication links are also considered as undirected ring network topology as Figure 1.

Figure 4 shows the consensus achieved with the same position and velocity for the group of quadrotors, and the uncertainties in network are also canceled in Figure 6. Figure 5 shows the adaptive cooperation control signal of quadrotors. Although network systems create uncertainties, the transient performance is still maintained well by (8) and (13). The simulation results with Figures 4 and 5 demonstrate Lemmas 6–8. Figure 6 shows the mismatched dynamics that means the state predictor tracks the real state well and adapts to uncertainty quickly. The results in Figure 6 thus demonstrate Lemma 10. Further, the results in Figures 4–6

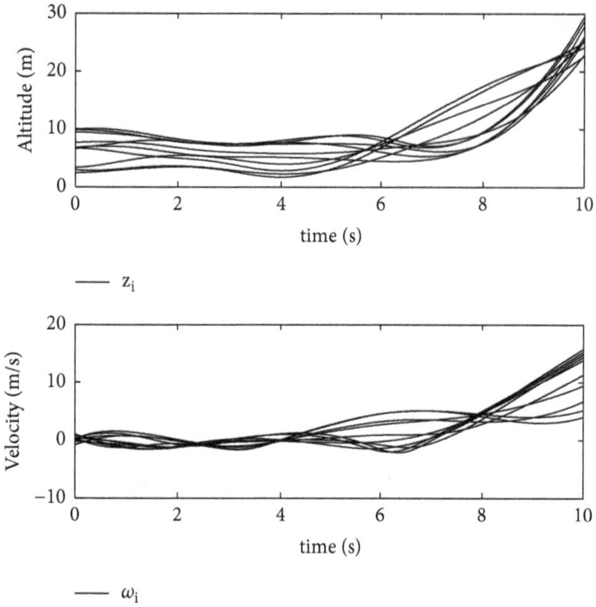

FIGURE 2: The trajectory of quadrotors.

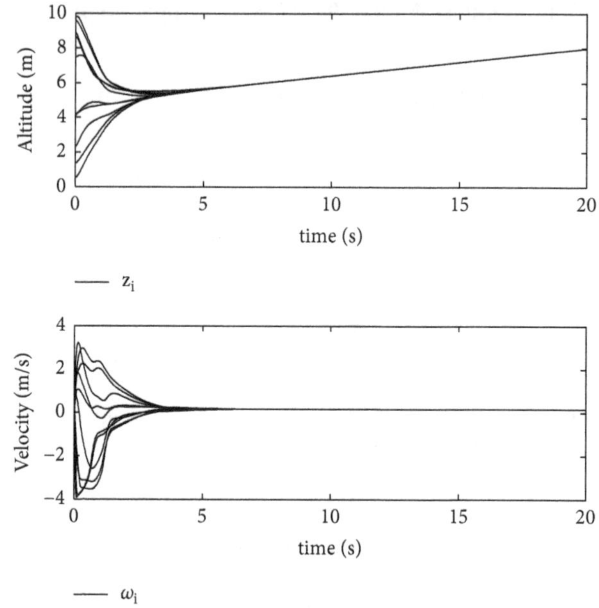

FIGURE 4: The trajectory of quadrotors.

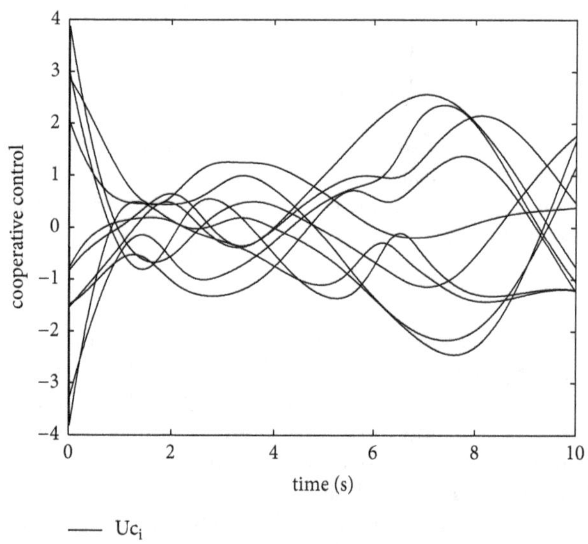

FIGURE 3: Cooperative control input.

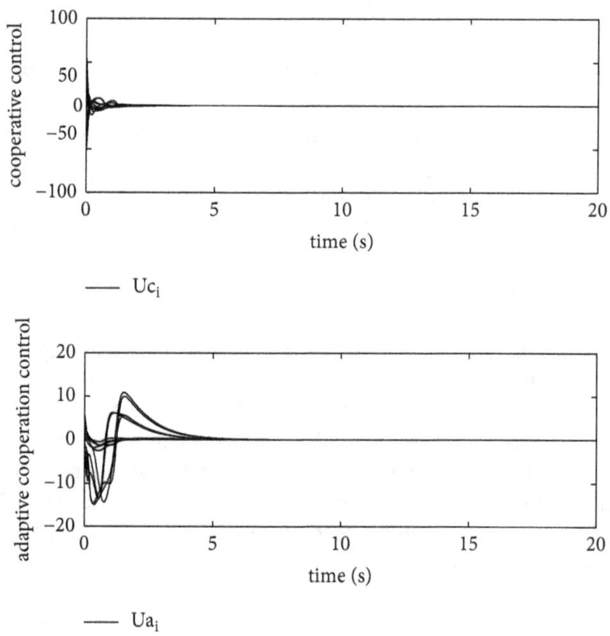

FIGURE 5: Adaptive cooperation control input.

also show the robustness of proposed adaptive cooperation control design.

6. Conclusion

This paper presents a novel cooperative adaptation approach to deal with uncertainty of multi-agent systems. The typical cooperative control was used to achieve collaboration in multi-agent systems but it also causes interconnected uncertainties among network. Then, the cooperative adaptation approach was proposed to cancel uncertainties in multi-agent systems. The cooperative adaptation algorithm features as fast adaptation scheme and can be treated as high-gain approach. To prevent aggressive changing in real control signal, a low-pass filtering mechanism is added in control law, which filters away aggressive signals and recovers robustness. The stability condition and tracking performance for closed-loop network system are presented. The estimation errors between predicted states and real state are proved to be arbitrarily small. Then, the network stability and performance bounds between real system and reference system can be made arbitrarily small by decreasing the integration time-step. The simulation results show the details of proposed scheme. This approach can be further extended to deal with uncertainties issue in multi-unmanned vehicle systems.

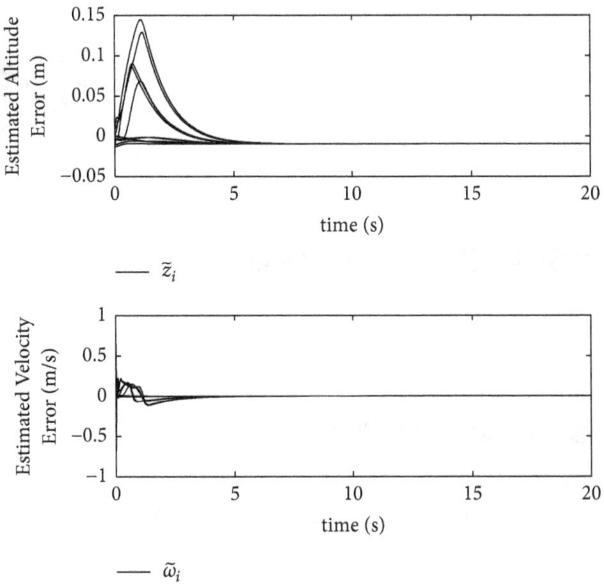

FIGURE 6: Estimation error of states.

Conflicts of Interest

The authors declare that they have no conflicts of interest.

Acknowledgments

This work was supported by NSF Grant IIS-1208499.

References

[1] R. Olfati-Saber and R. M. Murray, "Consensus problems in networks of agents with switching topology and time-delays," *IEEE Transactions on Automatic Control*, vol. 49, no. 9, pp. 1520–1533, 2004.

[2] R. Olfati-Saber, J. A. Fax, and R. M. Murray, "Consensus and cooperation in networked multi-agent systems," *Proceedings of the IEEE*, vol. 95, no. 1, pp. 215–233, 2007.

[3] W. Ren and R. W. Beard, "Consensus seeking in multiagent systems under dynamically changing interaction topologies," *IEEE Transactions on Automatic Control*, vol. 50, no. 5, pp. 655–661, 2005.

[4] C.-Q. Ma and J.-F. Zhang, "Necessary and sufficient conditions for consensusability of linear multi-agent systems," *IEEE Transactions on Automatic Control*, vol. 55, no. 5, pp. 1263–1268, 2010.

[5] H. Kim, H. Shim, and J. H. Seo, "Output consensus of heterogeneous uncertain linear multi-agent systems," *IEEE Transactions on Automatic Control*, vol. 56, no. 1, pp. 200–206, 2011.

[6] S. E. Tuna, "Conditions for synchronizability in arrays of coupled linear systems," *IEEE Transactions on Automatic Control*, vol. 54, no. 10, pp. 2416–2420, 2009.

[7] L. Scardovi and R. Sepulchre, "Synchronization in networks of identical linear systems," *Automatica*, vol. 45, no. 11, pp. 2557–2562, 2009.

[8] J. H. Seo, H. Shim, and J. Back, "Consensus of high-order linear systems using dynamic output feedback compensator: low gain approach," *Automatica*, vol. 45, no. 11, pp. 2659–2664, 2009.

[9] Z. Li, Z. Duan, G. Chen, and L. Huang, "Consensus of multiagent systems and synchronization of complex networks: a unified viewpoint," *IEEE Transactions on Circuits and Systems I: Regular Papers*, vol. 57, no. 1, pp. 213–224, 2010.

[10] T. Yang, S. Roy, Y. Wan, and A. Saberi, "Constructing consensus controllers for networks with identical general linear agents," *International Journal of Robust and Nonlinear Control*, vol. 21, no. 11, pp. 1237–1256, 2011.

[11] K. Liu, G. Xie, W. Ren, and L. Wang, "Consensus for multiagent systems with inherent nonlinear dynamics under directed topologies," *Systems & Control Letters*, vol. 62, no. 2, pp. 152–162, 2013.

[12] G. Wen, Z. Duan, G. Chen, and W. Yu, "Consensus tracking of multi-agent systems with Lipschitz-type node dynamics and switching topologies," *IEEE Transactions on Circuits and Systems I: Regular Papers*, vol. 61, no. 2, pp. 499–511, 2014.

[13] E. Nuno, R. Ortega, L. Basanez, and D. Hill, "Synchronization of networks of nonidentical Euler-Lagrange systems with uncertain parameters and communication delays," *Institute of Electrical and Electronics Engineers Transactions on Automatic Control*, vol. 56, no. 4, pp. 935–941, 2011.

[14] A. Das and F. L. Lewis, "Distributed adaptive control for synchronization of unknown nonlinear networked systems," *Automatica*, vol. 46, no. 12, pp. 2014–2021, 2010.

[15] H. Su, G. Chen, X. Wang, and Z. Lin, "Adaptive second-order consensus of networked mobile agents with nonlinear dynamics," *Automatica*, vol. 47, no. 2, pp. 368–375, 2011.

[16] W. Yu, G. Chen, and M. Cao, "Consensus in directed networks of agents with nonlinear dynamics," *IEEE Transactions on Automatic Control*, vol. 56, no. 6, pp. 1436–1441, 2011.

[17] I. D. Landau, R. Lozano, M. M'Saad, and A. Karimi, *Adaptive Control*, vol. 51, Springer, New York, NY, USA, 1998.

[18] Y. Liu and Y. Jia, "Adaptive leader-following consensus control of multi-agent systems using model reference adaptive control approach," *IET Control Theory & Applications*, vol. 6, no. 13, pp. 2002–2008, 2012.

[19] G.-X. Wen, C. L. Chen, Y.-J. Liu, and Z. Liu, "Neural-network-based adaptive leader-following consensus control for second-order non-linear multi-agent systems," *IET Control Theory & Applications*, vol. 9, no. 13, pp. 1927–1934, 2015.

[20] W. Liu and J. Huang, "Adaptive leader-following consensus for a class of higher-order nonlinear multi-agent systems with directed switching networks," *Automatica*, vol. 79, pp. 84–92, 2017.

[21] W. Liu and J. Huang, "Cooperative global robust output regulation for nonlinear output feedback multiagent systems under directed switching networks," *Institute of Electrical and Electronics Engineers Transactions on Automatic Control*, vol. 62, no. 12, pp. 6339–6352, 2017.

11

Smart Shift Decision Method Based on Stacked Autoencoders

Zengcai Wang,[1,2] Yazhou Qi,[1,2] Guoxin Zhang,[1,2] and Lei Zhao[1,2]

[1]*School of Mechanical Engineering, Shandong University, Jinan, China*
[2]*Key Laboratory of High Efficiency and Clean Mechanical Manufacture (Shandong University), Ministry of Education, Jinan, China*

Correspondence should be addressed to Zengcai Wang; wangzc@sdu.edu.cn

Academic Editor: Enrique Onieva

Manual calibration and testing on real vehicles are common methods of generating shifting schedules for newly developed vehicles. However, these methods are time-consuming. Shifting gear timing is an important operating parameter that affects shifting time, power loss, fuel efficiency, and driver comfort. The stacked autoencoder (SAE) algorithm, a type of artificial neural network, is used in this study to predict shifting gear timing on the basis of throttle percentage, vehicle velocity, and acceleration. Experiments are conducted to obtain training and testing data. Different neural networks are trained with experimental data on a real vehicle under different road conditions collected using the CANcaseXL device and control AMESim simulation model, which was constructed based on real vehicle parameters. The input number of SAE is determined through a comparison between two and three parameters. The output type of SAE is determined through a comparative experiment on pattern recognition and multifitting. Meanwhile, the network structure of SAE is determined through a comparative experiment on simple and deep-learning neural networks. Experimental results demonstrate that using the SAE intelligent shift control strategy to determine shift timing not only is feasible and accurate but also saves time and development costs.

1. Introduction

Shift control is a key technology in automotive automatic transmission control because it directly affects the fuel economy and dynamic performance of vehicles [1]. Current methods of gear decision-making consist of two categories. One is the traditional method, in which an optimal method for solving the shift schedule is established after the shift control parameter is selected according to a certain performance index. The other method involves using intelligent shift, driving experience, and expert knowledge. Shift control in the traditional basic shift schedule is based on different parameters (e.g., single-, two-, or three-parameter shift schedules). However, regardless of the number of parameters, a shift schedule can only reflect the running state of a vehicle and does not fully consider the intention of the driver and the automotive environment. Manual calibration and testing on real vehicles are common means of generating shifting schedules for newly developed vehicles; however, these methods are time-consuming [2]. Current research on intelligent shift decision-making mainly includes two methods, namely, those based on fuzzy logic and those based on neural networks [3–6]. Alternate method based on artificial intelligence techniques implemented with fuzzy controllers is proposed [7]. Elnashar used the fuzzy theory and neural network to establish the automatic variable speed controller of the automobile and carried out the computer simulation [8]. Gear shift strategy is based on stochastic dynamic programming [9]. Mechatronic gear shift system has been developed and tested in a passenger car reducing fuel consumption and improving the vehicle drivability [10]. Casavola et al. have fully analyzed and here described genetic and fuzzy algorithm gear shifting strategy [11]. The artificial neural network (ANN) has strong nonlinear mapping and generalization capabilities and can even be trained to adapt to the dynamic conditions of vehicles, road conditions, and changes in the characteristics of drivers [12]. A feed-forward neural network that calculates optimal dynamic and economical shift schedules using a three-parameter shift schedule has been presented [13]. The fuzzy neural network has been adopted for the construction of vehicles with multiparameter transmission [14]. A neural network has also been developed to evaluate downhill engine brake necessity and make correct shifting decisions similar to those made by experienced drivers [15]. In [16], a multilayer perceptron algorithm was applied to predict shifting gear timing. However, this algorithm can only be implemented on the AMESim transmission model and not on vehicles with transmission. Moreover, it does not consider the deviation of the optimized values.

In this study, actual vehicle data obtained from an experimental vehicle under different road conditions were analyzed using the CANcaseXL device. The network implemented on an AMESim model of the experimental vehicle with automated transmission was sufficiently trained. A stacked autoencoder (SAE) method was developed for the gear recognition rate and compared with the traditional ANN.

2. Experimental Vehicle Model

Figure 1 shows a diagram of neural network shift control. The vehicle model was established based on an actual test vehicle using the AMESim software. The driver and neural network controller models were established using the Simulink software. The CANcaseXL device was used to analyze the actual vehicle data of the experimental vehicle. The model described was modified based on the test data. The engine output and torque characteristics of the engine mode were changed based on the experimental data. The basic parameters of the vehicle, such as rated power, maximum engine torque, engine speed range, total mass loaded, wheel rolling radius, transmission mechanical efficiency, air resistance coefficient, and rolling resistance coefficient, were set. The transmission model's shifting logic based on a shift table is shown in Table 1, together with the setting of the solenoid valve and shifting clutch. Figure 2 presents the physical model constructed using AMESim to model the proposed vehicle. Figure 3 presents the driver and neural network controller models constructed using MATLAB to model the behavior of the driving and transmission control unit. The driver model controls the accelerator and brake pedals and outputs the signal of throttle and brake percentages with time. The neural network controller model is a sufficient training network that uses a neural network with a three-parameter shift. The details of the neural network controller model are described in the subsequent sections.

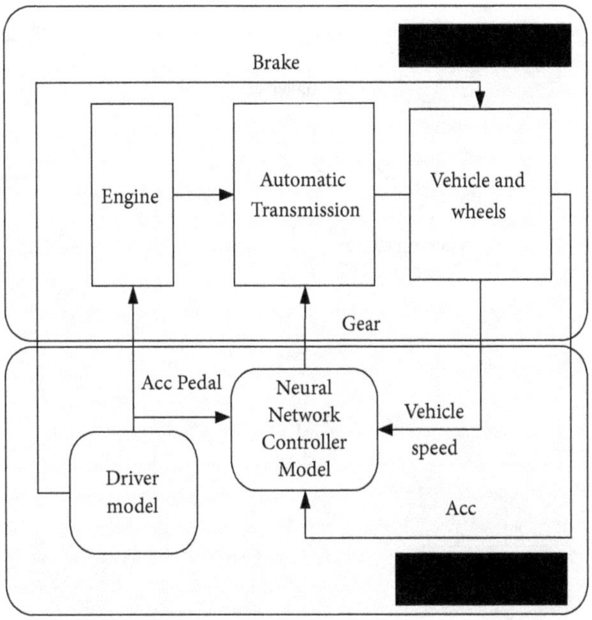

FIGURE 1: Diagram of neural network shift control.

TABLE 1: Shift table.

Gear	C4-5-6	C3-5-R	C2-6	CL-R	C1-2-3-4
1st				Clutch	Clutch
2nd			Clutch		Clutch
3rd		Clutch			Clutch
4th	Clutch				Clutch
5th	Clutch	Clutch			
6th	Clutch		Clutch		
Gear	C4-5-6	C3-5-R	C2-6	CL-R	C1-2-3-4

3. ANN and SAE

3.1. ANN. Artificial intelligence consists of two major branches. Expert systems and the study of ANN do not require detailed system information and work similar to a black box model. However, the relationships between the input parameters and controlled and uncontrolled variables are learned by studying previously tested data, similar to nonlinear regression. Another advantage of using ANN is its ability to handle large and complex systems with many interrelated parameters [17]. A suitable shifting strategy based on the BP neural network algorithm has been established [18]. Reference [19] presented a neural-networked, three-parameter gear-shift control. In [18, 19], the back-propagation learning algorithm was used in a feed-forward process, but these studies did not use curve fitting or pattern recognition of neural networks, which were used in [20, 21]. In this study, the use of curve fitting and pattern recognition was discussed. This study selected a multilayer ANN structure, as shown in Figure 4.

The input data were a combination of throttle percentage, vehicle velocity, and acceleration. Gear was the desired output. p was set as the sample number, and n was set as the number of samples. With a certain input x_i, the network output is y_i, and the node output is o_{pj}. The node input is

$$I_{jp} = \sum_{i=1}^{n} w_{ij} o_{pi} - \theta_{pi}, \quad (1)$$

where w_{ij} are the weights of the connection of the intermediate layer j node to the node on level i and θ_{pi} is the node threshold. The output of each node is a nonlinear function of the input desirable sigmoid function.

$$f(x) = \frac{1}{1+e^{-x}}. \quad (2)$$

For the set of input samples $X_{p1}, \ldots, X_{pn}$, network output set $O_{p1}, O_{p2}, \ldots, O_{pn}$, and target output set $Y_{p1}, Y_{p2}, \ldots, Y_{pn}$,

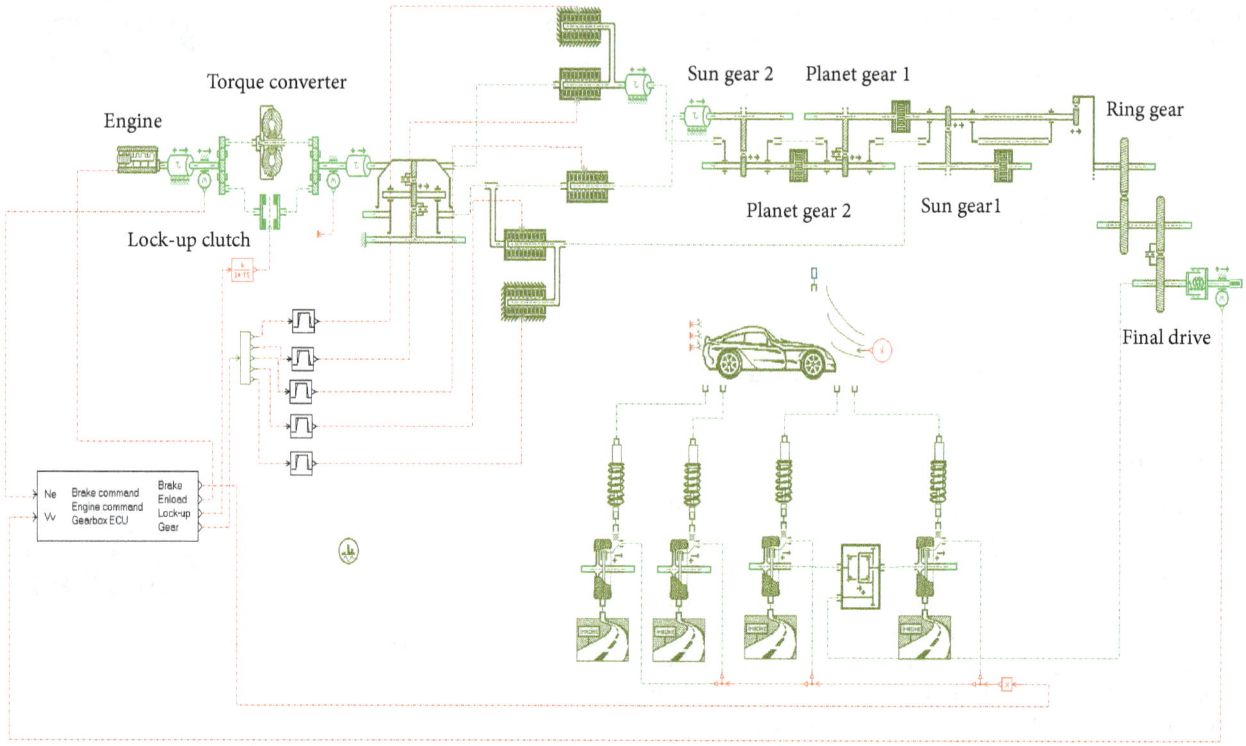

FIGURE 2: AMESim model.

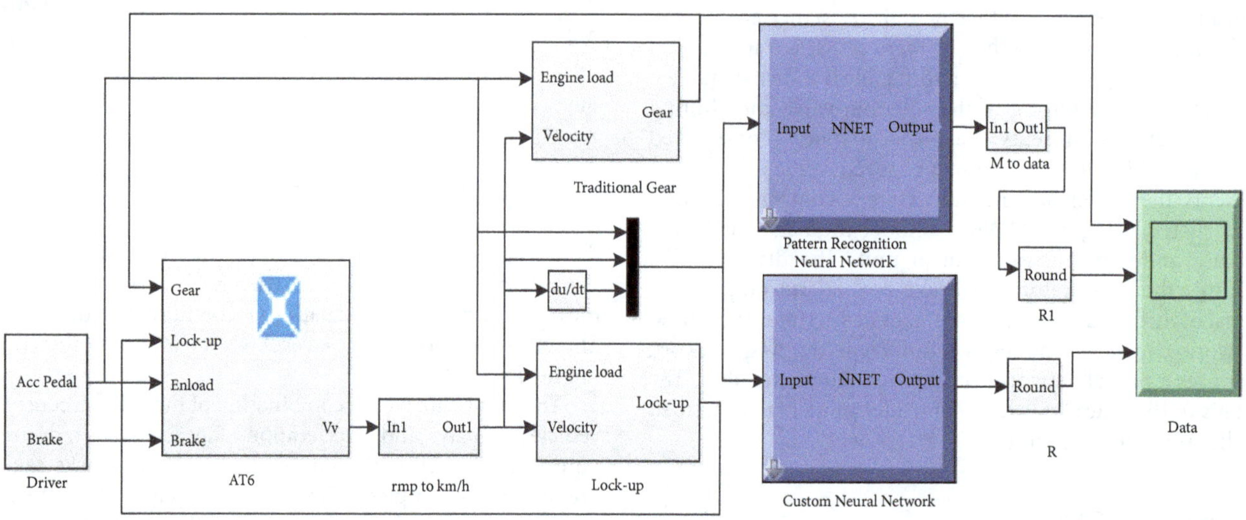

FIGURE 3: Simulink model.

the error function of the sample training network can be defined as follows:

$$E_p = \frac{1}{2}\sum \frac{\left(Y_{pi} - O_{pi}\right)^2}{2}, \tag{3}$$

where Y_{pi} is the ideal output value of the ith output node of sample p and O_{pi} is the output value of node i which stands at the output layer with the influence of sample input p. With the most rapid descent method, the algorithm used to change w_{ij} and θ_{pi} when E_p is zero is derived as

$$\begin{aligned} \Delta_p w_{ij} &= -\mu \frac{\partial E_P}{\partial W_{ij}} = -\mu \delta_{pi} O_{pj}, \\ \Delta_p \theta_{ij} &= -\mu \frac{\partial E_P}{\partial \theta_{ij}} = -\mu \delta_{pi}, \end{aligned} \tag{4}$$

where μ is the learning ratio or epoch, δ_{pi} is the error signal of the ith node of sample p, and Δ_p is incremental weight of w_{ij} in sample p.

If i is the output node, then

$$\delta_{pi} = \left(Y_{pi} - O_{pi}\right) f'\left(I_{pi}\right). \tag{5}$$

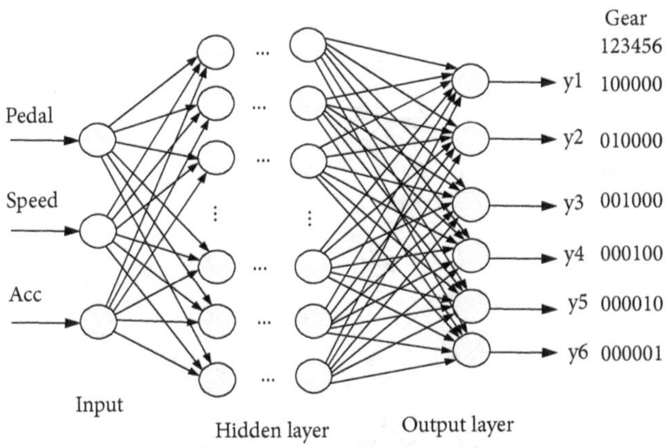

FIGURE 4: Three-parameter neural network structure.

If i is not the output node, then

$$\delta_{pi} = \left(\sum_j \delta_{pi} w_{ij} \right) f' \left(I_{pi} \right). \tag{6}$$

In the first training stage, the inputs and desired outputs are provided to the neural network (NN). The weights are modified to minimize the error between the NN predictions and the expected outputs. Training aims to minimize errors and consequently optimize the NN solution. Each iterative step where the weights are recalculated is called an epoch. When the minimum is achieved, the weights are fixed, and the training process ends. Once an NN has been trained to a satisfactory level, it is used as an NN controller for the AMESim model.

In this work, the back-propagation learning algorithm was used in a feed-forward, and a function fitting neural network was developed and compared with a pattern recognition neural network. An ANN that is trained and simulated with validated vehicle data was proposed and compared. The result showed that the function fitting neural network predicted gear timing better than the pattern recognition neural network. Accurate data are discussed in the next section. The three-parameter shift schedule based on neural network model was compared with the two-parameter shift schedule. The result of the comparison showed that the proposed three-parameter gear shift conformed to driver experience or practice better and was smoother for gear-shift surface changes. Accurate data are discussed in the next sections.

3.2. SAE. At present, manufacturers should improve their services, products, and technological base to achieve a sustainable value proposition, become highly efficient and effective in the market, and satisfy the needs of users. Significant emerging technologies being discussed in various research studies include wearable technologies, networked and smart environments connected by the Internet of Things, evolving tools, tangible interfaces, human-robot collaborations, processes and interactions, virtual reality, ubiquitous use of machine learning, and deep-learning algorithms [22]. In the last few years, applications based on deep neural networks have performed various tasks, such as speech recognition or image classification, to levels that even exceed human abilities [23–25]. SAE is a deep-learning algorithm. It is better than a traditional neural network. According to [26, 27], SAE-based deep learning uses limited shallow neural network learning capability; thus, the depth of the network should be known when SAE is utilized. SAE consists of multiple layers of sparse autoencoders, in which the outputs of each layer are the inputs of the successive layer. Formally, we consider an SAE with n layers. $W^{(k,1)}$, $W^{(k,2)}$, $b^{(k,1)}$, and $b^{(k,2)}$ denote the parameters of $W^{(1)}$, $W^{(2)}$, $b^{(1)}$, and $b^{(2)}$ for the kth autoencoder. The encoding for the SAE is obtained by implementing the encoding step of each layer in a forward order as follows:

$$\begin{aligned} a^{(l)} &= f\left(z^{(l)}\right), \\ z^{(l+1)} &= W^{(l,1)} a^{(l)} + b^{(l,1)}. \end{aligned} \tag{7}$$

The decoding step is obtained by running the decoding stack of each autoencoder in a reverse order as follows:

$$\begin{aligned} a^{(n+l)} &= f\left(z^{(n+l)}\right), \\ z^{(n+l+1)} &= W^{(n-l,2)} a^{(n+l)} + b^{(n-l,2)}. \end{aligned} \tag{8}$$

The information of interest is contained in $a^{(n)}$, which is the activation of the deepest layer of hidden units. This vector represents the input in terms of higher-order features.

An autoencoder neural network is an unsupervised learning algorithm that applies back-propagation, and it sets the target values to be equal to the inputs. The autoencoder attempts to learn an approximation to the identity function; thus, the output is similar to the input.

First, a sparse autoencoder is trained based on the raw inputs of throttle percentage, vehicle velocity, and acceleration to learn the primary features $h_n^{(1)}$ on the raw input (Figure 5). Second, the raw input is fed to the trained sparse autoencoder, thereby obtaining the primary feature activations $h_n^{(1)}$ for each of the inputs. Third, these primary features are used as the "raw input" to another sparse autoencoder to

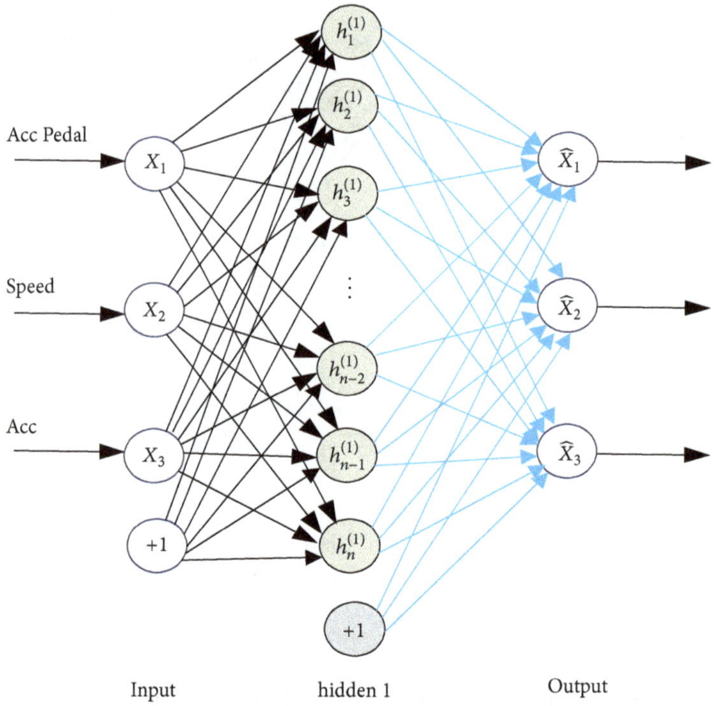

FIGURE 5: Training of hidden layer 1.

learn the secondary features $h_n^{(2)}$ on these primary features (Figure 6). Following this method, the primary features are fed to the second sparse autoencoder to obtain the secondary feature activations $h_n^{(2)}$ for each of the primary features $h_n^{(1)}$. These secondary features are treated as a "raw input" to the softmax classifier, which trains it to map secondary features to digit labels (Figure 7). Finally, all three layers are combined to form an SAE with three parameter inputs (Figure 8).

4. Results and Discussion

4.1. Data Extraction. A vehicle can only shift gears according to a fixed shift control strategy under different driving environments because the basic shift rule is preset in the controller. Thus, the given gear of the vehicle may not be the optimum. The basic shift strategy presents a poor control performance when the vehicle operates on a ramp. The vehicle encounters a cycle shift when climbing, and its speed increases with a high gear ratio. Hence, the engine brakes are not fully utilized. In this work, the CANcaseXL device was used to analyze the actual data of the experimental vehicle when shifting gears manually under different road conditions. We collected information on the driver's shift under seven different road conditions. The information was then used to train the neural network. Useful vehicle data are shown in Figures 9–15. When the neural network with three-parameter shift was trained, throttle percentage, vehicle velocity, and acceleration were treated as input data, and the gear was treated as target data. When the neural network with two-parameter shift was trained, throttle percentage and vehicle velocity were treated as input data, and the gear was treated as target data.

4.2. SAE with Data Fitting. The function fitting neural network was compared with the pattern recognition neural network to predict gear timing. The function fitting neural network is an example of a fitting problem in which inputs are matched to associated target outputs to create a neural network that not only estimates the known targets of given inputs but also accurately generalizes the estimated outputs. The regression across a sample of the road test is plotted in Figure 16. The regression plot shows the actual network outputs plotted in terms of the associated target values. If the network has learned to fit the data well, then the linear fit to this output-target relationship should closely intersect in the bottom-left and top-right corners of the plot. If the case is the opposite, then further training or training a network with more hidden neurons would be advisable. The value of R is an indication of the relationship between the outputs and targets. If R is close to one, then the outputs and targets have an exact linear relationship. If R is close to zero, then the outputs and targets have no linear relationship.

Figure 17 shows the regression of the shift schedule with three parameters and the shift schedule with two parameters. Figure 17 shows that R of the three-parameter shift schedule is relatively closer to 1 than that of the two-parameter shift schedule. The comparison of the two- and three-parameter neural networks showed that three-parameter gear shift conforms to driver's experience or practice better. Moreover, as the data points increase, the fitting effect decreases. This

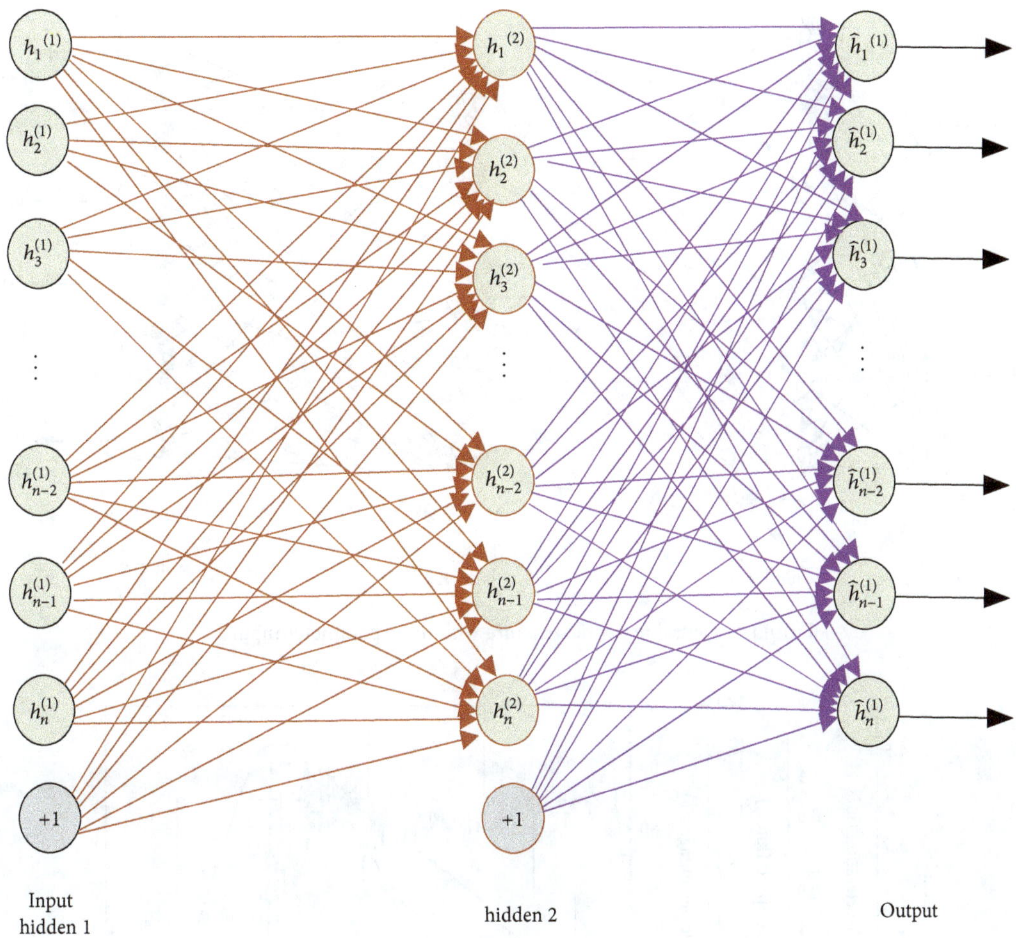

FIGURE 6: Training of hidden layer 2.

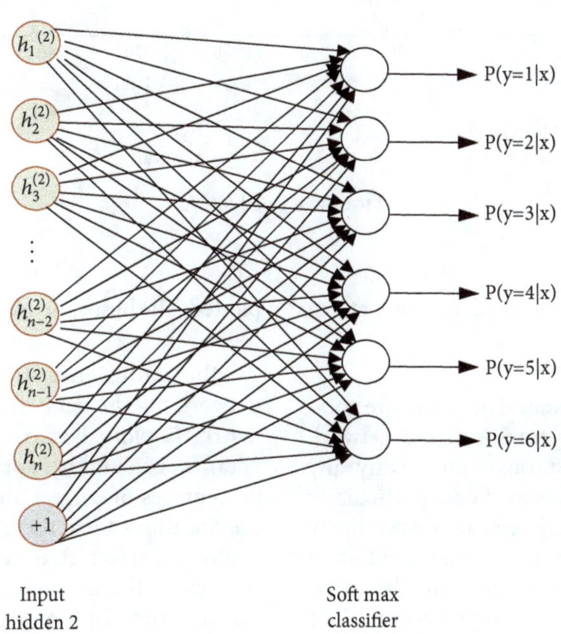

FIGURE 7: Training of the softmax classifier.

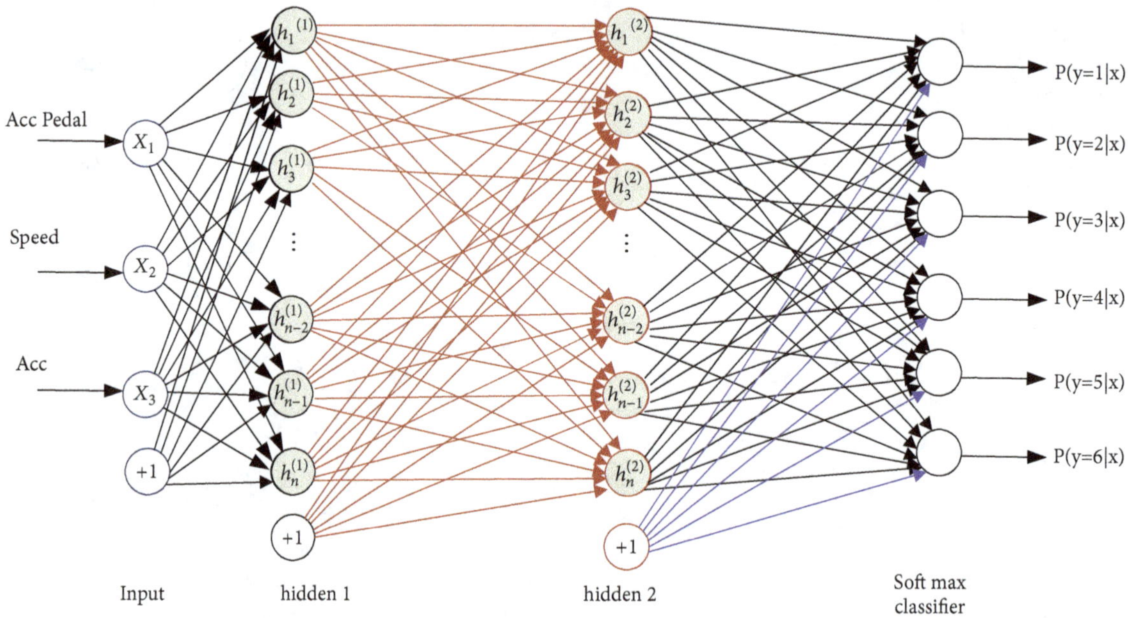

FIGURE 8: Neural network structure with three parameter inputs.

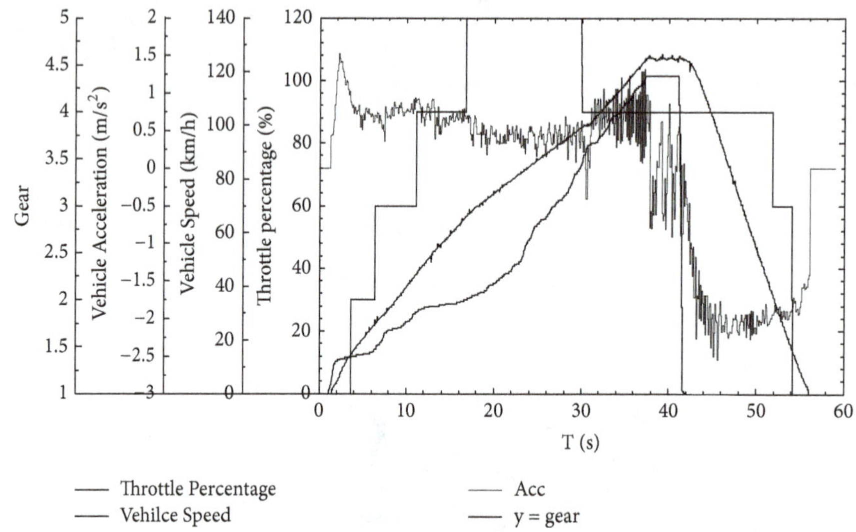

FIGURE 9: Road test 1.

finding made us think of a new method of the pattern recognition neural network.

4.3. SAE with Pattern Recognition. Neural networks are good at solving pattern recognition problems. A neural network with sufficient elements (called neurons) can classify any data with arbitrary accuracy. These networks are particularly well suited for complex gear shifting decisions over many variables. Therefore, neural networks are good candidates for solving gear classification problems. Then, the data were used to train the pattern recognition neural network. The performance curve is given in Figure 18. The plot shows the value of the performance function versus the iteration number (epochs). Training, validation, and test performances are plotted. The best validation performance is 0.023685 at epoch 1000.

The confusion matrix explains how the trained neural network fits the data, as shown in Figure 19. The confusion matrix is plotted across all the training data of the two-parameter neural network. The confusion matrix shows the percentages of correct and incorrect classifications. Correct classifications form the green squares on the matrices diagonally. Incorrect classifications are represented by the red squares. If the network is trained well and has learned to classify properly, then the percentages in the green squares should be large, indicating good classifications. If the case is the opposite, then further training or training a network with more hidden neurons would be advisable. The number

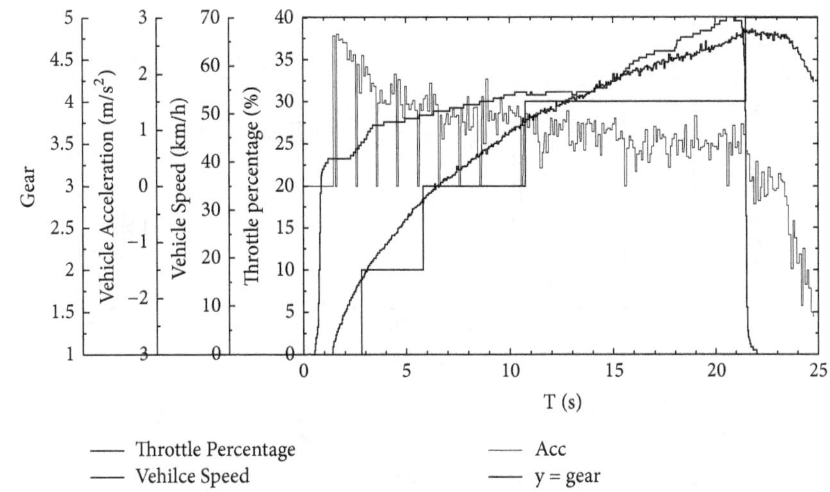

FIGURE 10: Road test 2.

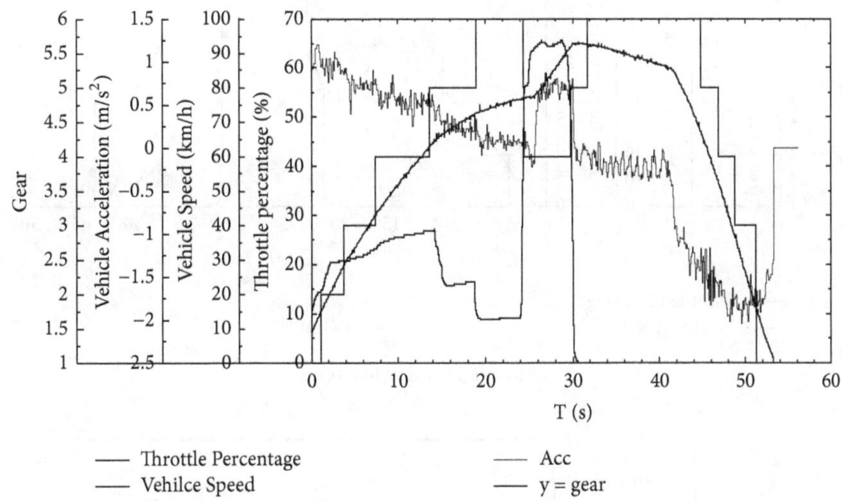

FIGURE 11: Road test 3.

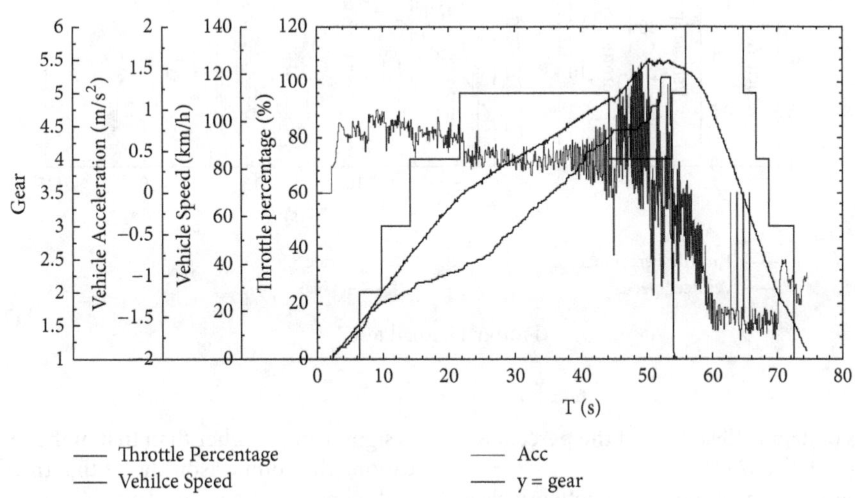

FIGURE 12: Road test 4.

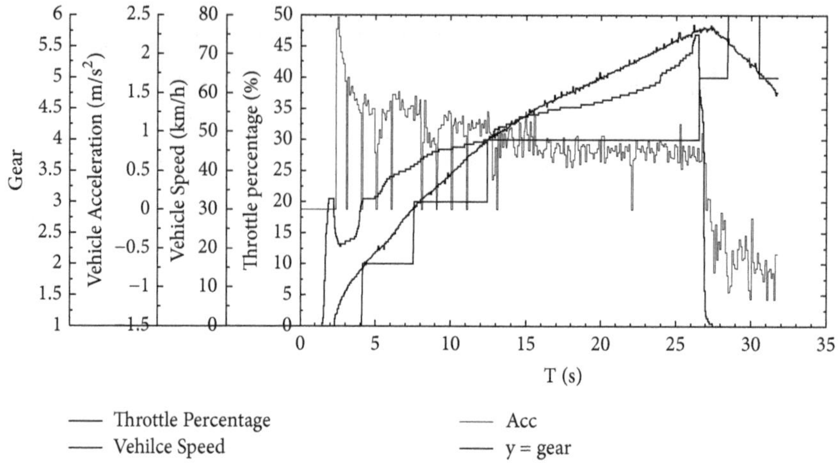

FIGURE 13: Road test 5.

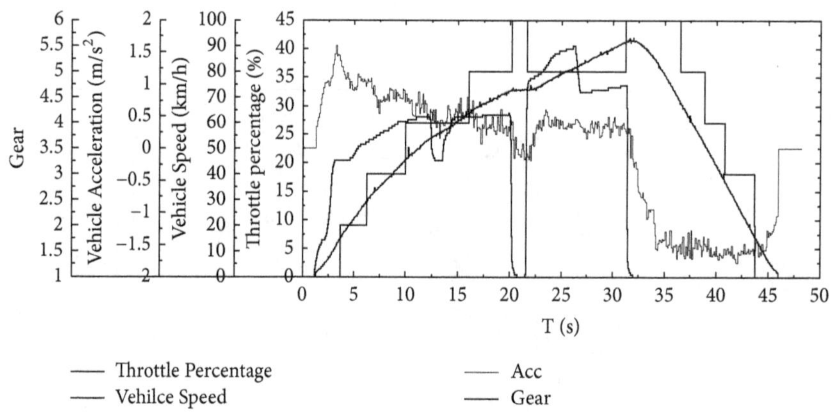

FIGURE 14: Road test 6.

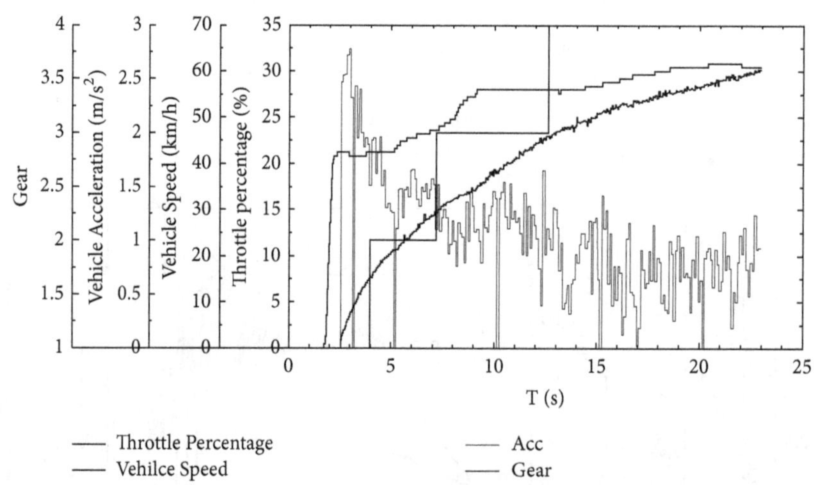

FIGURE 15: Road test 7.

of experimental pieces of data is 189,808, and the percentage of correct classification is 94.383759%.

Figure 20 shows the accuracy of the neural network classification with three and two parameters. The accuracy of the three-parameter classification is relatively close to 100% and is significantly higher than that of the two-parameter classification. The comparison shows that the three-parameter gear shift conforms better to driver's experience or practice than the two-parameter gear shift. Moreover, as the data points increase, the percentages of correct classification decrease.

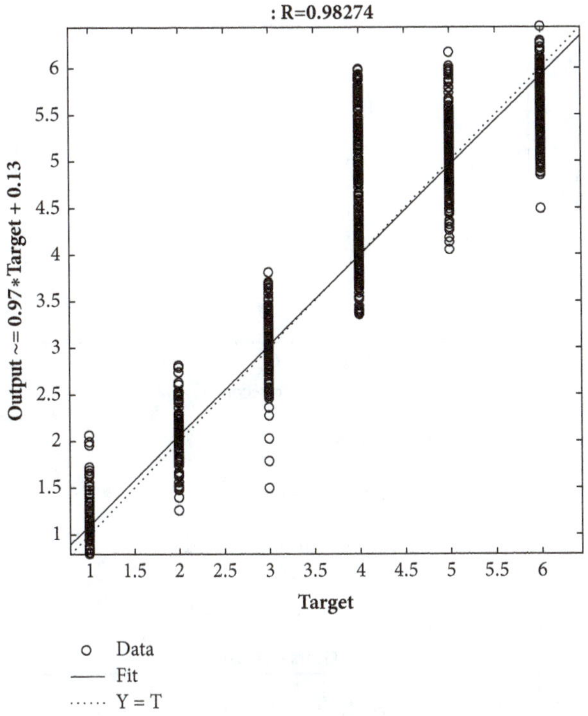

FIGURE 16: Regression *R*.

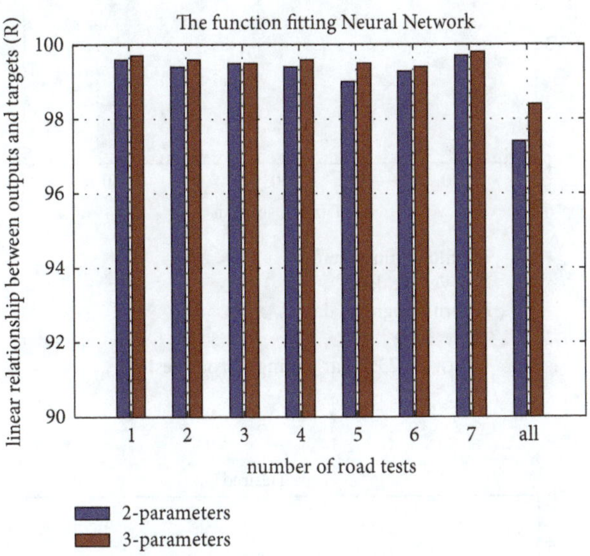

FIGURE 17: Regression of the test road.

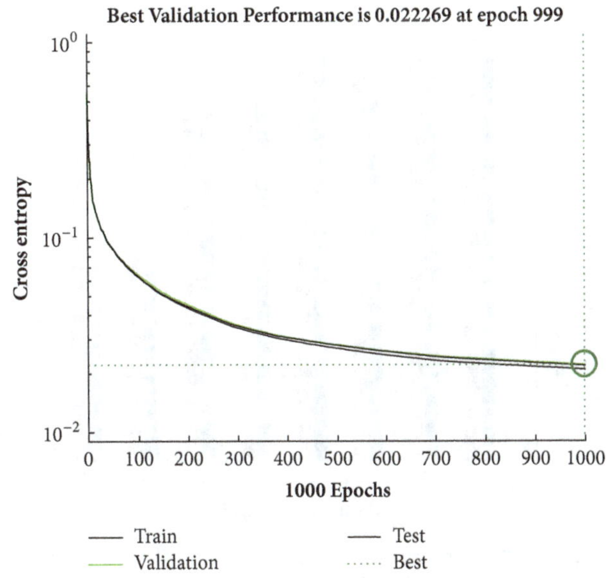

FIGURE 18: Training error curve of the neural network.

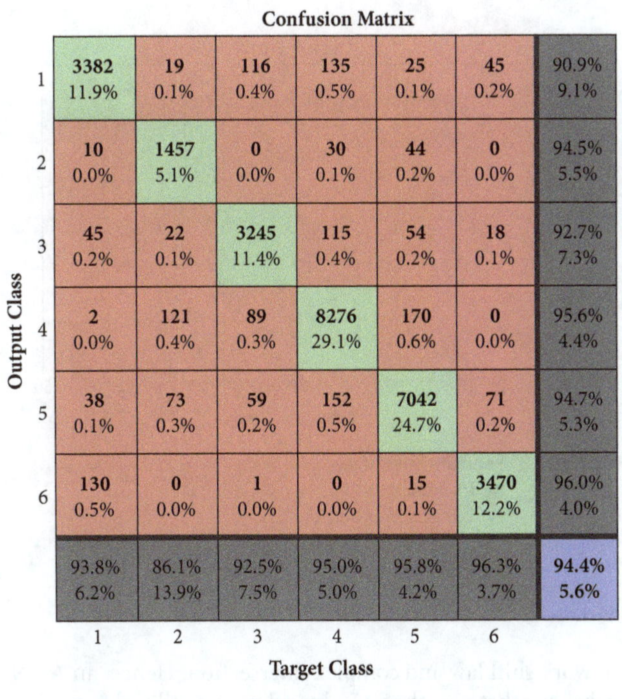

FIGURE 19: Accuracy of neural network classification.

4.4. Comparison of Pattern Recognition and Data Fitting under the Same Network Structure. Two types of neural networks were trained and simulated with the validated vehicle data to determine whether a neural network with fitting predicts gear timing better than pattern recognition. Figures 21–27 show the output of a neural network with the input data of seven types of roads. Three curves are drawn in each figure to represent the actual vehicle gear timing, the pattern recognition network output gear timing, and the function fitting output gear timing. The comparison shows that the function fitting neural network predicts gear timing better than the pattern recognition neural network.

4.5. Comparison between SAE and a Shallow Neural Network. Figure 17 shows that the fitting effect decreases as the data points increase. Figure 20 shows that the percentages of correct classification decrease as the data points increase. As the data points increase, the learning capability of the shallow neural network becomes limited. A large amount of data for network training is required to apply the neural

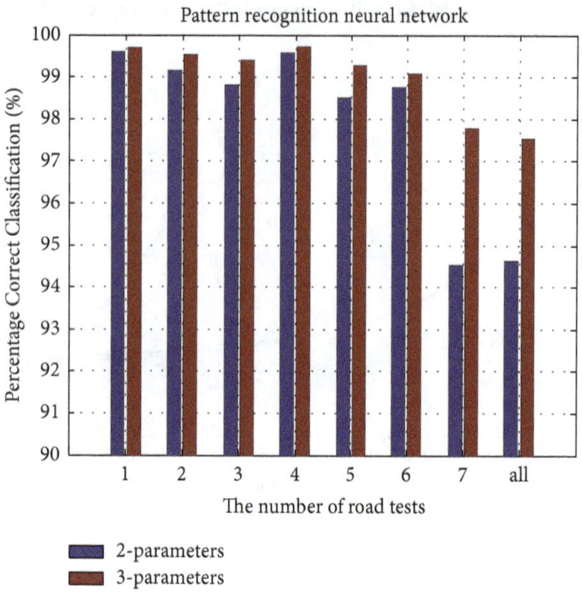

FIGURE 20: Accuracy of neural network classification.

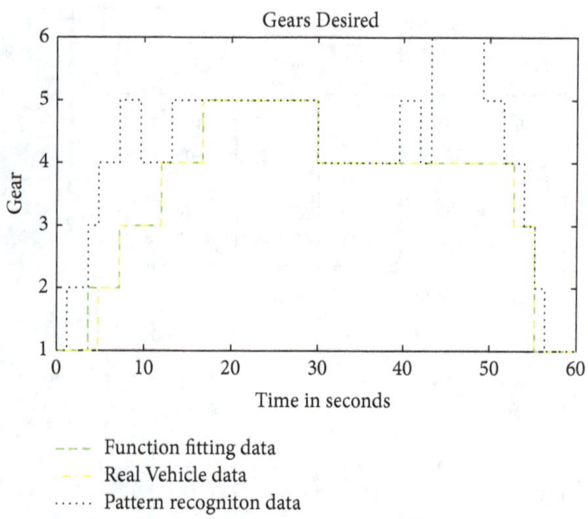

FIGURE 21: Gear timing of road test 1.

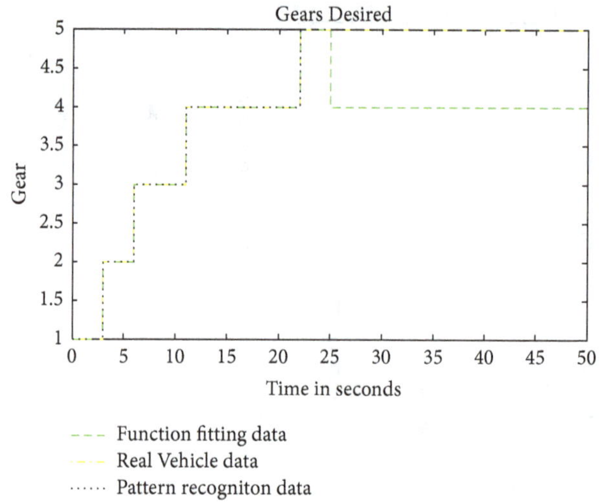

FIGURE 22: Gear timing of road test 2.

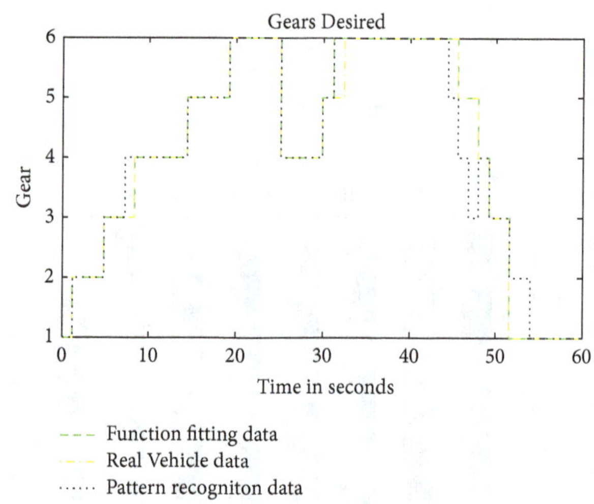

FIGURE 23: Gear timing of road test 3.

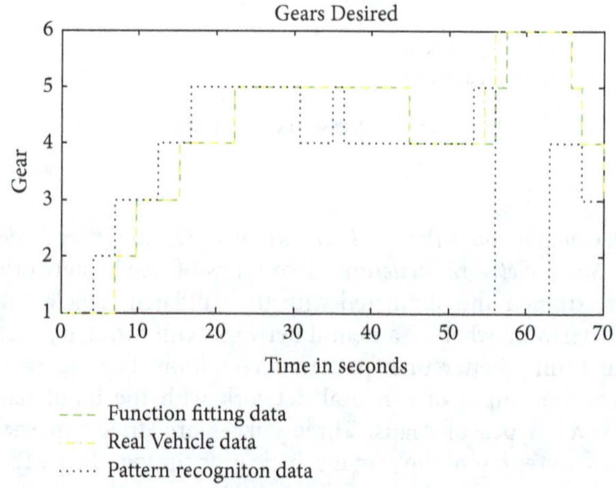

FIGURE 24: Gear timing of road test 4.

network shift law and complex intersection. Hence, an ANN technique that uses the SAE algorithm was utilized to predict shifting gear timing based on throttle percentage, vehicle velocity, and acceleration. Table 2 shows that neural network with pretraining performs better than the algorithm without pretraining.

5. Conclusion

Manual calibration and testing on real vehicles are traditional ways of generating shifting schedules for newly developed vehicles. However, these methods are time-consuming. In this study, the back-propagation learning algorithm was used in feed-forward progress, and a polynomial fitting neural network was developed and compared with the pattern

TABLE 2: Comparison of algorithms with and without pretraining.

Experiment	Road test 1	Road test 2	Road test 3	Road test 4	Road test 5	Road test 6	Road test 7	Road test 8
Number of data points	40713	17127	38719	51430	8565	33254	15824	205632
BP	0.987	0.993	0.990	0.980	0.997	0.982	0.995	0.974
BP + pretraining	0.997	0.996	0.995	0.994	0.997	0.993	0.997	0.990

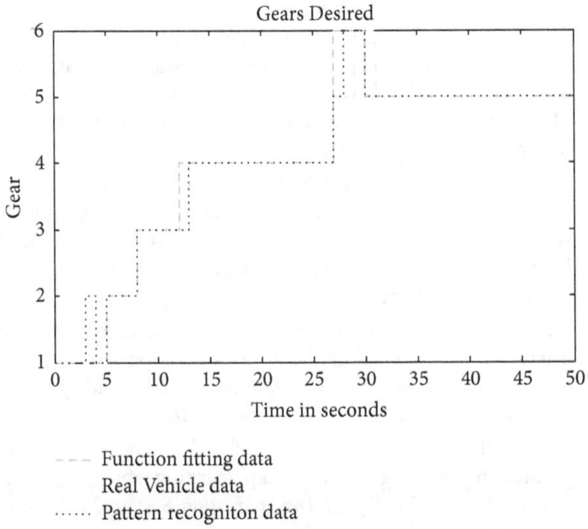

FIGURE 25: Gear timing of road test 5.

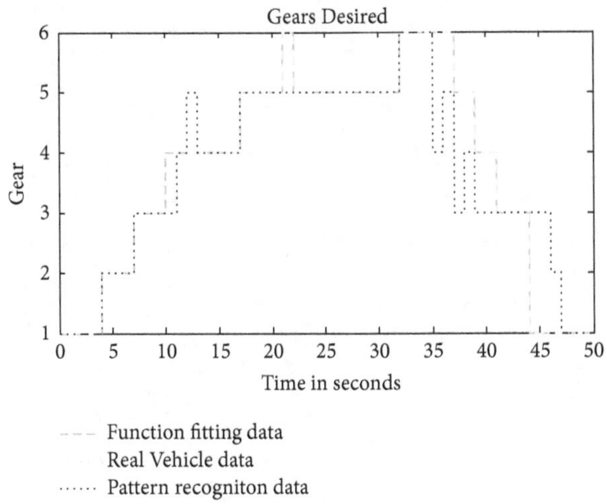

FIGURE 26: Gear timing of road test 6.

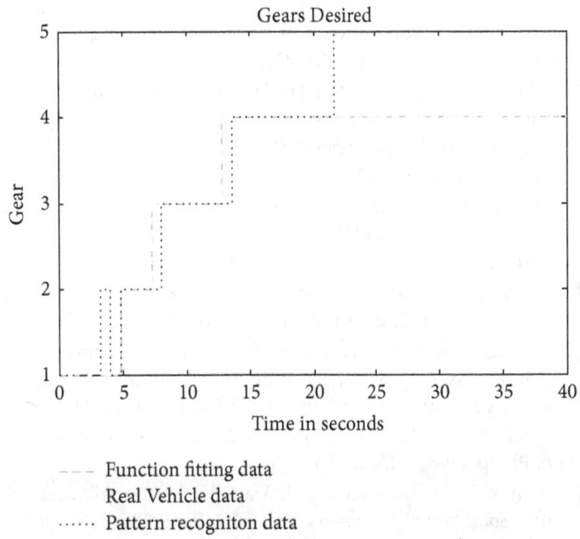

FIGURE 27: Gear timing of road test 7.

recognition neural network. The polynomial fitting neural network predicts gear timing better than the pattern recognition neural network. The polynomial fitting neural network and pattern recognition neural network with three parameters perform better than those with two parameters.

The accuracy of the BP neural network decreases with the increase in the experimental sample points in terms of the fitting effect and correct rate after network training, and meeting the requirements of intelligent shift control to cope with complicated conditions is difficult. However, its depth learning stack self-coding is good. At the experimental sample point of 205,632, the accuracy of SAE is 1.6% higher than that of the BP neural network. Therefore, the SAE neural network with the three-parameter intelligent shift control strategy not only correctly predicts shift timing and saves time but also adapts to the different driving habits of drivers.

Conflicts of Interest

The authors declare that they have no conflicts of interest.

Acknowledgments

This work was supported by the Open Foundation of State Key Laboratory of Automotive Simulation and Control (China, Grant no. 20161105) and the Natural Foundation of Shandong Province (China, Grant no. ZR2018MEE015).

References

[1] Z. Sun and K. Hebbale, "Challenges and opportunities in automotive transmission control," in *Proceedings of the 2005 American Control Conference, ACC*, pp. 3284–3289, usa, June 2005.

[2] J. Wang, Y. Lei, A. Ge, and X. Lu, "Shift strategy research on off-road vehicle," *SAE Technical Papers*, 2009.

[3] A. Ghazi Zadeh, A. Fahim, and M. El-Gindy, "Neural network and fuzzy logic applications to vehicle systems: literature survey," *International Journal of Vehicle Design*, vol. 18, no. 2, pp. 132–193, 1997.

[4] B. Bosacchi and I. Masaki, "Fuzzy logic technology & the Intelligent Highway System (IHS)," in *Proceedings of the Second IEEE International Conference on Fuzzy Systems*, pp. 65–70, April 1993.

[5] K. Kondo and H. Goka, "Adaptive shift scheduling strategy introducing neural network in automatic transmission," *JSAE Review*, vol. 16, no. 4, pp. 411–414, 1995.

[6] Y. H. Yamaguchi, H. Narita, H. Takahashi, and Y. Katou, "Automatic transmission shift schedule control using fuzzy logic," SAE Technical Papers 930674, 1993.

[7] F. Alonge, F. Raimondi, and L. Ciancimino, "A new approach to the intelligent gear shift using artificial intelligence and fuzzy logic," in *Proceedings of the 6Th Wseas Intern. Conf. on Mathem. Methods Comput. Techn. in Electrical Engineer*, 2004.

[8] G. Elnashar, *Intelligent controllers based on fuzzy systems and neural networks*, The Catholic University of America, 2000.

[9] V. D. Ngo, J. A. Colin Navarrete, T. Hofman, M. Steinbuch, and A. Serrarens, "Optimal gear shift strategies for fuel economy and driveability," *Proceedings of the Institution of Mechanical Engineers, Part D: Journal of Automobile Engineering*, vol. 227, no. 10, pp. 1398–1413, 2013.

[10] J. Jantos, K. Praznowski, and J. Mamala, "Automatic gear shift unit," *Solid State Phenomena*, vol. 147-149, pp. 185–190, 2009.

[11] A. Casavola, G. Prodi, and G. Rocca, "Efficient gear shifting strategies for green driving policies," in *Proceedings of the 2010 American Control Conference, ACC 2010*, pp. 4331–4336, July 2010.

[12] Q. H. Chen and D. T. Qin, "Neural-networked three-parameter gear-shift schedule for automated-manual-transmission car," *Control Theory & Applications*, vol. 11, p. 023, 2010.

[13] J. Tan, X. Yin, Y. Lei, and A. Ge, "Research on a neural network model based automatic shift schedule with dynamic 3-parameters," SAE Technical Paper 2005-01-1597, 2005.

[14] S. Wu and H. Lu, "Fuzzy neural network control in automatic transmission of construction vehicle," in *Proceedings of the 6th World Congress on Intelligent Control and Automation, WCICA 2006*, pp. 2512–2516, chn, June 2006.

[15] Y. Li, F. Yu, and C. Wu, "Application of neural network to downhill shift strategy for automatic transmission," *Journal of Shanghai Jiaotong University (Science)*, vol. 15, no. 4, pp. 498–503, 2010.

[16] M. S. Barakat and M. Abdelaziz, "Neural Network Transmission Control," *SAE Technical Papers*, 2016.

[17] V. Shree, V. Ganesan, P. Tamil Porai, and N. V. Mahalakshmi, "Optimization of fuel injection timing of a gasoline engine using artificial neural network," *SAE Technical Papers*, vol. 12, 2013.

[18] Q. Dai, "Training and simulation on gear position decision for vehicle based on optimal algorithm of network," *Chinese Journal of Mechanical Engineering*, vol. 38, no. 11, pp. 124–127, 2002.

[19] C. Qing-hong and Q. Da-tong, "Neural-networked three-parameter gear-shift schedule for automated-manual-transmission car," *Control Theory & Applications*, vol. 11, p. 023, 2010.

[20] M. Peña-Cabrera, V. Lomas-Barrie, I. López-Juárez, and R. Osorio-Comparán, "Contour object generation method for object recognition using FPGA," *International Journal of Automation Technology*, vol. 7, no. 2, pp. 182–189, 2013.

[21] M. Kato and T. Yamamoto, "Development of Neural-Net-Based Decision Support System for Mattress Patterns Using Particle Swarm Optimization," *International Journal of Automation Technology*, vol. 4, no. 2, pp. 178–183, 2010.

[22] S. Wiesner, E. Marilungo, and K.-D. Thoben, "Cyber-physical product-service systems - challenges for requirements engineering," *International Journal of Automation Technology*, vol. 11, no. 1, pp. 17–28, 2017.

[23] N. Yalta, K. Nakadai, and T. Ogata, "Sound source localization using deep learning models," *Journal of Robotics and Mechatronics*, vol. 29, no. 1, pp. 37–48, 2017.

[24] J. A. Garcia Sanchez, K. Ohnishi, A. Shibata, F. Dong, and K. Hirota, "Deep level emotion understanding using customized knowledge for human-robot communication," *Journal of Advanced Computational Intelligence and Intelligent Informatics*, vol. 19, no. 1, pp. 91–99, 2015.

[25] L.-F. Chen, Z.-T. Liu, M. Wu, F. Dong, and K. Hirota, "Multi-robot behavior adaptation to humans' intention in human-robot interaction using information-driven fuzzy friend-Q learning," *Journal of Advanced Computational Intelligence and Intelligent Informatics*, vol. 19, no. 2, pp. 173–184, 2015.

[26] W. Li, H. Fu, L. Yu et al., "Stacked Autoencoder-based deep learning for remote-sensing image classification: a case study of African land-cover mapping," *International Journal of Remote Sensing*, vol. 37, no. 23, pp. 5632–5646, 2016.

[27] Y. Lv, Y. Duan, W. Kang, Z. Li, and F.-Y. Wang, "Traffic flow prediction with big data: a deep learning approach," *IEEE Transactions on Intelligent Transportation Systems*, vol. 16, no. 2, pp. 865–873, 2015.

12

Deep Sparse Autoencoder for Feature Extraction and Diagnosis of Locomotive Adhesion Status

Changfan Zhang,[1] Xiang Cheng,[1,2] Jianhua Liu,[1] Jing He,[1] and Guangwei Liu[1]

[1]*College of Electrical and Information Engineering, Hunan University of Technology, Zhuzhou 412000, China*
[2]*CHNV Technology Co., Ltd., Dongguan 523000, China*

Correspondence should be addressed to Jing He; hejing@263.net

Academic Editor: Qiang Liu

The model is difficult to establish because the principle of the locomotive adhesion process is complex. This paper presents a data-driven adhesion status fault diagnosis method based on deep learning theory. The adhesion coefficient and creep speed of a locomotive constitute the characteristic vector. The sparse autoencoder unsupervised learning network studies the input vector, and the single-layer network is superimposed to form a deep neural network. Finally, a small amount of labeled data is used to fine-tune training the entire deep neural network, and the locomotive adhesion state fault diagnosis model is established. Experimental results show that the proposed method can achieve a 99.3% locomotive adhesion state diagnosis accuracy and satisfy actual engineering monitoring requirements.

1. Introduction

Precise diagnosis of the wheel-rail adhesion state is an important prerequisite for adhesion control. Currently, the wheel-rail adhesion state of a locomotive is mostly diagnosed based on the detection and analysis of relevant parameters to determine the type of adhesion state and the degree of adhesion [1]. For the diagnosis of adhesion states, a sampling eigenvector should be generated based on the creeping speed of the driving wheel and the wheel-rail cohesion coefficient, a sample feature should be extracted, and the feature should be coded; then, various intelligent algorithms should be used to classify the eigenvector [2]. Many studies on the use of neural networks in the adhesion field have been reported. For example, Castillo et al. [3] used a neural network to estimate the adhesion state in an ABS system. Castillo [4] trained an artificial neural network to calculate the best creep operating point for each road on the basis of traffic information collected by a vehicle sensor. Li Ningzhou [5] studied the adhesion feature of the air brake of a locomotive and used the optimized recursive neural network to optimize the parameters of the adhesive controller and improve the utilization rate of locomotive adhesion, thereby obtaining a good experimental result.

These methods are more convenient and intelligent than the general mechanism analysis method. However, these methods still belong to the supervised learning area [6]. Thus, they require sufficient data for feature extraction. Meanwhile, extracting the right features is often relatively complex and difficult. To obtain labels, experiments and rich professional knowledge are required. With the artificial participation factors, the uncertainty of feature extraction and optimization is greatly increased, thereby making the diagnosis of the right adhesion state difficult. Furthermore, a traditional neural network essentially uses hidden-layer neurons for nonlinear transformation [7]. It can learn potential features from a given sample and fit out an approximation function [8]. Taking the classical BP neural network as an example, obtaining high-precision features becomes more difficult while the layers are few. If the number of layers is excessive, then the gradient may disappear, and the local optimal solution is another defect that is difficult to overcome [9].

Sample feature extraction is a key step in determining the accuracy of fault diagnosis [10]. The change in adhesion

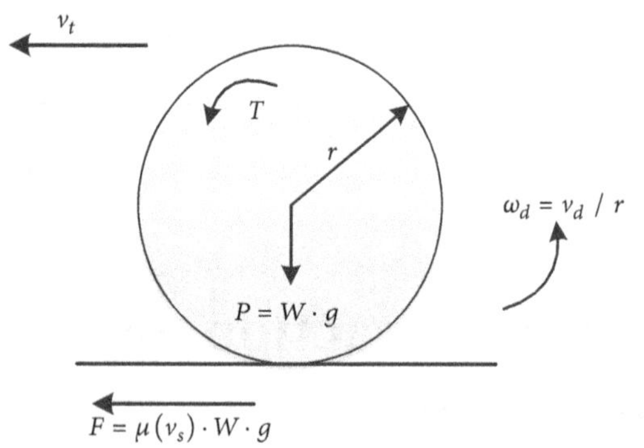

FIGURE 1: Schematic of adhesion.

state is a complex process that is affected by multiple factors, producing a complex nonlinear relation between factors and outcomes. Fault prediction and analysis are particularly challenging. The introduction of deep learning [11] has made a breakthrough in research on high-precision feature extraction. As an unsupervised learning algorithm, deep neural network not only has an excellent feature extraction ability but can also overcome the common problem of obtaining sample labels [12]. Thus, deep neural network has become a popular research area in the field of fault diagnosis [13–15]. This paper proposes a sparse autoencoder deep neural network with dropout to diagnose the wheel-rail adhesion state of a locomotive. This deep neural network can significantly reduce the adverse effect of overfitting, making the learned features more conducive to classification and identification.

The rest of this paper is organized as follows: Section 2 describes the adhesion principle and characteristics. Section 3 describes the principle and process of the deep neural network algorithm. Section 4 discusses the comparative experimental research and result analysis. Section 5 presents the conclusions.

2. Description of Adhesion Status

Adhesion is the ultimate manifestation of locomotive driving force in the wheel-rail relationship and the fundamental motive force for locomotives [16]. The wheel pair rolls forward when subjected to a tangential traction, and the rolling pressure causes deformation between the wheel and the rail. Simultaneously, the gravity of the car body imposed on the rail keeps the contact surface between the wheel and the rail relatively stable. This phenomenon is called adhesion. As shown in Figure 1, the contact point between the wheel and the rail is elastically deformed under the action of the wheel load (P). The wheel rolls forward under the action of the driving torque (T), the original contact surface deformation develops into a new elliptical deformation, and the tractive effort at the wheel rim (F) is generated.

Adhesion coefficient μ is typically defined as the ratio of traction to axle weight (see (1)), where W is the axle weight (kg) and g is the gravitational acceleration (m/s^2).

$$\mu = \frac{F}{W \cdot g} \tag{1}$$

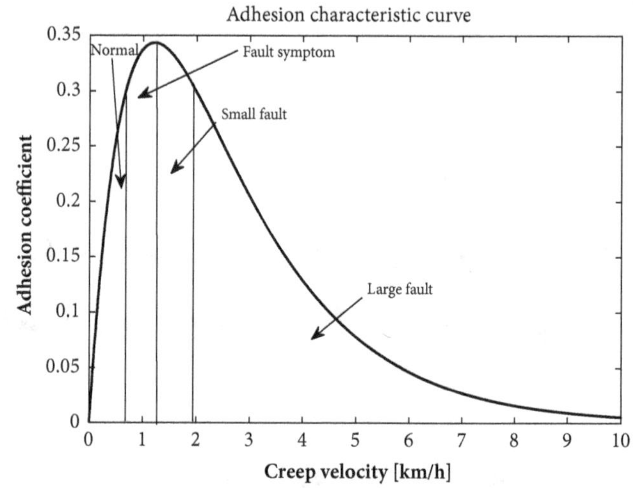

FIGURE 2: Refined adhesion characteristic curve.

In the process of normal movement, the train body speed (v_t) is always less than the wheel speed (v_d) due to the wheel-rail microsliding generated by the deformation. This phenomenon is called creep, and the speed difference between them is defined as creep speed v_s.

$$v_s = v_d - v_t \tag{2}$$

Creep is a slight wheel-spin phenomenon produced by the locomotive drive system. The adhesion coefficient of the rail contact surface rises constantly with the creep speed within a certain range [17], and the locomotive has a great available traction. Once the range is exceeded, the wheel-rail adhesion coefficient drops sharply with the increase in the creep speed.

Figure 2 shows the adhesion characteristic curve of the locomotive. The adhesion peak point is taken as the boundary in which the left side is called the creep region and the right side is called the slid region [4]. However, when the adhesion state is divided into two categories, abnormal adhesion can be identified, but the predicted foundation of potential creep failure cannot be provided. To this end, this paper further refines the adhesion features: normal (N0), fault symptom (N1), small fault (F1), and large fault (F2).

The adhesion state is divided into four categories. When a minor fault is encountered, fault tolerant control methods [18] can be adopted to prevent serious system performance deterioration [1].

3. Deep Neural Network

Unsupervised learning can be used to automatically learn potential features from the samples without labels [19, 20]. This method has a significant advantage when addressing complex problems, such as adhesion state recognition. The sparse autoencoder is an unsupervised algorithm, and this deep neural network can effectively extract the characteristics that reflect the adhesion state [21, 22].

3.1. Sparse Autoencoder. From the structural point of view, the autoencoder is an axisymmetric single hidden-layer neural network [23]. The autoencoder encodes the input sensor data by using the hidden layer, approximates the minimum error, and obtains the best-feature hidden-layer expression [24]. The concept of the autoencoder comes from the unsupervised computational simulation of human perceptual learning [25], which itself has some functional flaws. For example, the autoencoder does not learn any practical feature through copying and inputting memory into implicit layers, although it can reconstruct input data with high precision. The sparse autoencoder inherits the idea of the autoencoder and introduces the sparse penalty term, adding constraints to feature learning for a concise expression of the input data [26, 27].

For the adhesion state identification of locomotive, k sets of monitoring data $\{x_1, x_2, x_3, \ldots, x_n\}$ exist, which are reconstructed into a N × M data set $\{x(1), x(2), x(3), \ldots, x(N)\}$, $x(i) \in R^M$. These data are used as input matrix X. The input data encoded by the automatic encoder are used to construct a mapping relationship. In this paper, the activation function of the autoencoder is sigmoid, which is designed to obtain a better representation of input data: $h(X, W, b) = \sigma(WX + b)$. A sparse penalty term is added to the sparse autoencoder cost function to limit the average activation value of the hidden-layer neuron. Normally, when the output value of a neuron is 1, it is active, and the neuron is inactive when its output value is 0. The purpose of enforcing sparsity is to limit the undesired activation. $a_j(x)$ is set as the *jth* activation value. In the process of feature learning, the activation value of the hidden-layer neuron is usually expressed as $a = sigmoid(WX + b)$, where W is the weight matrix and b is the deviation matrix. The mean activation value of the *jth* neuron in the hidden layer is defined as

$$\rho_j = \frac{1}{n}\sum_{i=1}^{n}\left[a_j\left(x\left(i\right)\right)\right]. \tag{3}$$

The hidden layer is kept at a lower value to ensure that the average activation value of the sparse parameter is defined as ρ, and the penalty term is used to prevent ρ_j from deviating from parameter ρ. The Kullback–Leibler (KL) divergence [28] is used in this study as the basis of punishment. The mathematical expression of KL divergence is as follows:

$$KL\left(\rho \parallel \rho_j\right) = \rho \ln \frac{\rho}{\rho_j} + (1-\rho)\ln\frac{1-\rho}{1-\rho_j}. \tag{4}$$

When ρ_j does not deviate from parameter ρ, the KL divergence value is 0; otherwise, the KL divergence value will gradually increase with the deviation. The cost function of the neural network is set as C (W, b). Then, the cost function of adding the sparse penalty term is

$$C_{sparse} = C\left(W, b\right) + \beta\sum_{j=1}^{S_2} KL\left(\rho \parallel \rho_j\right) \tag{5}$$

where S_2 is the number of neurons in the implicit layer and β is the weight of the sparse penalty term. The training essence of a neural network is to find the appropriate weight and threshold parameter (W, b). After the sparse penalty term is defined, the sparse expression can be obtained by minimizing the sparse cost function.

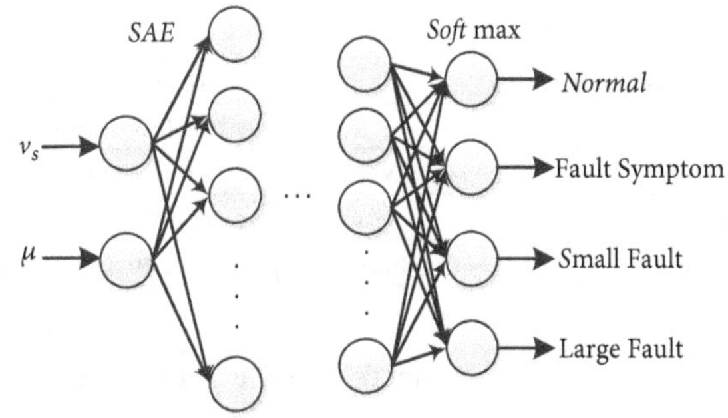

FIGURE 3: Schematic of the network structure.

3.2. Softmax Regression. The sparse autoencoder can form the deep network structure through the multilayer stack, which can be used for feature learning and clustering of the adhesion data collected by the sensor. However, this autoencoder has no ability to classify. Therefore, this paper presents a deep neural network architecture that combines the stacked sparse self-encoder and softmax regression. The schematic of the network structure is shown in Figure 3.

Softmax regression is an extension of the logistic regression model on multiple classifications [29]. The category tag of the logistic regression can only take two values, whereas the softmax tag can take on multiple values [30]. Let us suppose m training samples of adhesion state $\{(x^{(1)}, y^{(1)}), (x^{(2)}, y^{(2)}), \cdots, (x^{(m)}, y^{(m)})\}$, $y^{(i)} \in \{1, 2, \ldots, n\}$. The hypothesis function is used to estimate the probability value $p(y = j \mid x)$ for each category j. The softmax output is defined as follows:

$$h_\theta\left(x^{(i)}\right) = \begin{bmatrix} p\left(y^{(i)} = 1 \mid x^{(i)}; \theta\right) \\ p\left(y^{(i)} = 2 \mid x^{(i)}; \theta\right) \\ \vdots \\ p\left(y^{(i)} = k \mid x^{(i)}; \theta\right) \end{bmatrix} = \frac{1}{\sum_{j=1}^{k} e^{\theta_j^T x^{(i)}}} \begin{bmatrix} e^{\theta_1^T x^{(i)}} \\ e^{\theta_2^T x^{(i)}} \\ \vdots \\ e^{\theta_k^T x^{(i)}} \end{bmatrix} \tag{6}$$

where $\theta_1 \theta_2, \ldots \theta_k \in \Re^{n+1}$ is the model parameter. $1/\sum_{j=1}^{k} e^{\theta_j^T x^{(i)}}$ normalizes the probability distribution such that the sum of all probabilities is 1.

3.3. Overfitting and L2 Regularization. L2 regularization is a way of effectively reducing the neural network overfitting [31]. In this study, this method is used to avoid overlearning on features caused by synergies. The basic principles of L2 regularization are as follows:

$$C'_{sparse} = C_{sparse} + \alpha \left\|w\right\|^2 \tag{7}$$

Table 1: Description of adhesion state of locomotive wheel rail.

Different types of States	State representation
Normal	Running with a low adhesion
Fault symptom	Near the peak adhesion coefficient
Small fault	Across the peak adhesion point
Large fault	Serious slip running unsafe already

Table 2: State type.

Different types of states	State encoding
Normal (N0)	000
Fault symptom (N1)	001
Small fault (F1)	010
Large fault (F2)	100

where C_{sparse} is the cost function of the neural network, α is coefficient, and $\|w\|^2$ is the penalty term. The greater the coefficient, the deeper the weight attenuation.

3.4. Framework of the Algorithm. The feature learning ability of the single sparse autoencoder is limited. To construct a model with improved feature extraction capacity, we stacked the sparse autoencoders into a deep structure (SAE). In this process, the output of the upper layer of the encoder is taken as the input of the next layer to achieve a multilearning sample feature. The flowchart of the deep neural network algorithm is shown in Figure 4 and subsequently described.

(1) The Creep Speed (v_s) and Adhesion Coefficient μ Monitored by the Sensor Are Used to Train the Sparse Autoencoder

(1) The parameters, such as the network learning rate and the dropout parameter, are set; weights W and thresholds b are initialized.

(2) The number of iterations is set, and the mean activation value ρ_j and the sparse cost function are calculated according to (3)–(5); the network parameters are updated based on the backpropagation (BP) algorithm.

(2) The Deep Neural Network Is Fine-Tuned with a Small Number of Labeled Samples

(1) With the abovementioned step, the parameters of threshold b and weight W learned from the network are saved.

(2) The L2 regularization and learning rates are set, and the mean square error is calculated.

(3) The BP algorithm is used to update the weights of the network and fine-tune the entire network.

(3) The Performance of the Identification Model Is Tested

(1) The sample size is usually 30% of the total number of samples.

(2) According to the L2 coefficient, the weight of neural network is attenuated while performing the front-propagation algorithm.

(3) The output of DNN is compared with the sample labels, and comparative statistics are made.

4. Experimental Research and Analysis

For performance comparison with this dropout-based deep neural network, a BP neural network and an optimized BP neural network based on a genetic algorithm (GA-BP) are used as state recognition models. Taking the creep speed (v_s) and the adhesion coefficient (μ) of the locomotive as characteristic signals, the adhesion features include normal operation zone, wheel-spin warning zone, slight wheel-spin zone, and serious wheel-spin. Detailed information is shown in Table 1. The test sample added a Gaussian noise with a mean of 0 and a variance of 0.02 for an improved fit construction.

The changes in the state of adhesion directly affect the running safety of a locomotive. These changes are reflected in the sensor monitoring data. In this section, the adhesion state of the locomotive is identified and diagnosed according to the inherent characteristics of the sample data of deep neural network based on the sparse autoencoder. In general, the original data samples of sensors are divided into training and test sets according to a 7:3 ratio. A total of 700 training samples and 300 test samples are used in this experiment.

4.1. Simulation of the BP Neural Network. We need to set parameters before the experiment. The number of hidden-layer nodes is set according to the empirical formula $l = 2n + 1$, where n is the input dimension and l is the number of hidden-layer nodes. In this study, the creep speed (v_s) and the adhesion coefficient (μ) are used as inputs, so that l should be 5, and the adhesion state is expressed in binary form, as shown in Table 2. This BP neural network should be a $2 \times 5 \times 3$ structure (Figure 5). W is the weight and β is the bias of the BP neural network. The mean square error curve is shown in Figure 6.

To demonstrate the advantages of the proposed algorithm, a genetic algorithm- (GA-) optimized BP neural network is used as a contrast experiment. The crossover operator uses a single point crossover, the crossover probability is 0.7, and the mutation probability is 0.01. The evolutionary process of the GA is shown in Figure 7, and the error descent curve of the BP neural network that is optimized by the GA is shown in Figure 8.

4.2. Simulation of Sparse Autoencoder Deep Neural Network. The visualization of the target classification is shown in Figure 9 to provide a clear analysis of the classification ability of the proposed algorithm. The plane between the yellow and blue modules is the desired classification plane. In this experiment, 1,000 sets of monitoring data are selected as experimental samples, and the 7:3 ratio is used to divide the samples into training and test samples. The actual results are shown in Figure 10, which shows that the actual classification plane is basically consistent with the expected one.

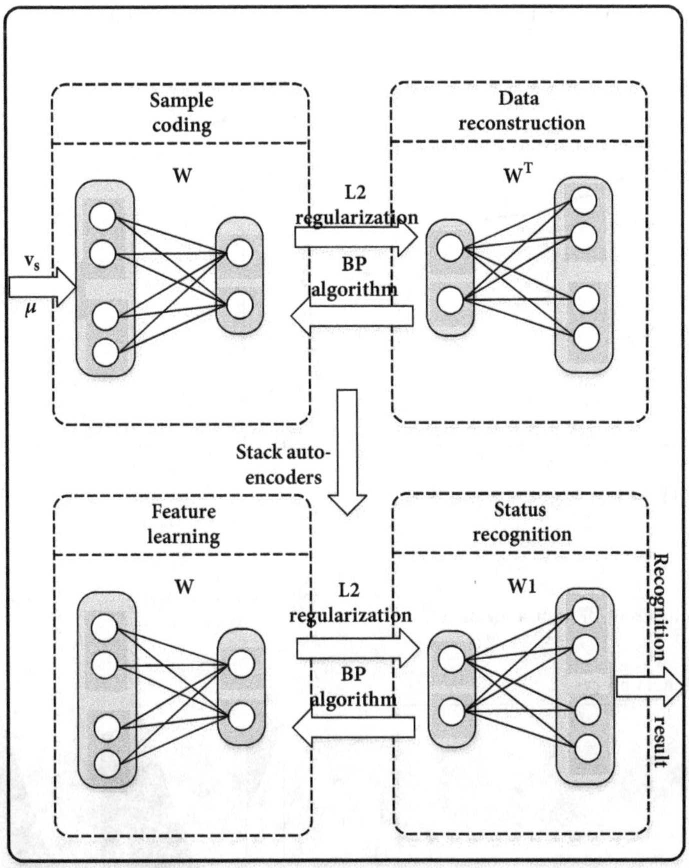

FIGURE 4: Flowchart of deep neural network algorithm.

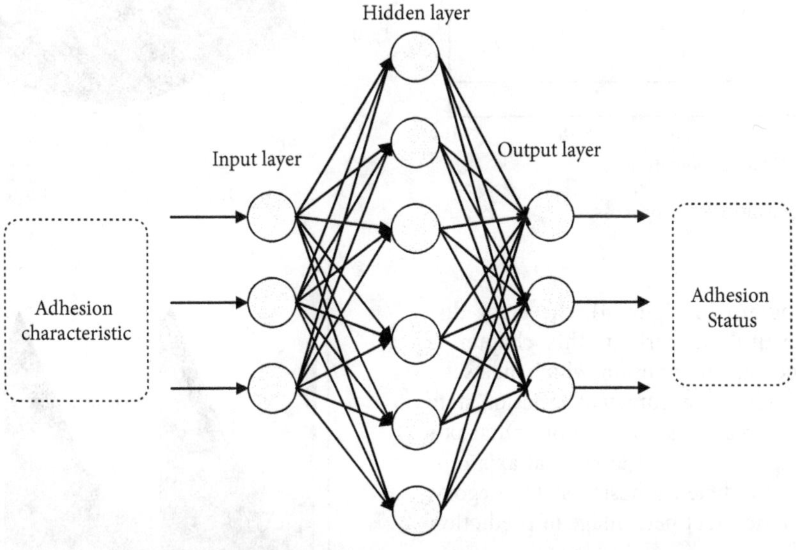

FIGURE 5: Schematic of the structure of the BP neural network.

Figure 9 shows the visualization of the classification target of adhesion status. In general, it is necessary to divide the adhesion status into four different statuses. Three classification planes are needed to achieve this (yellow and blue junction in the figure). The requirement of training neural network to accurately classify the adhesion state of locomotives is to make the test dataset also present a clear classification plane. The actual state division result is shown in Figure 10. The classifying plane of the adhesion state is clear.

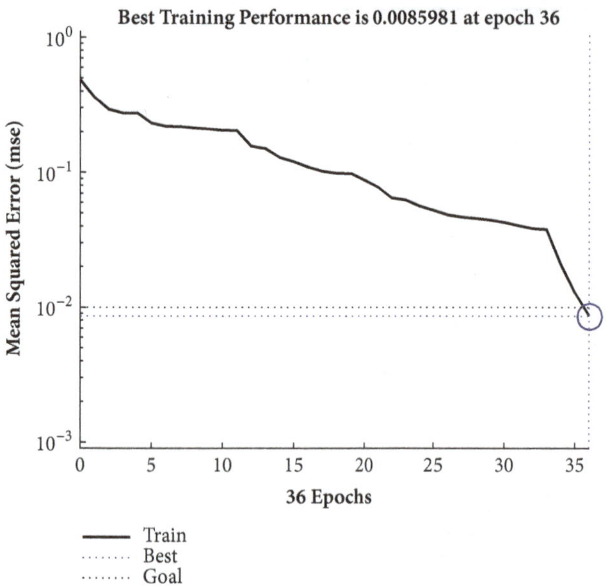

FIGURE 6: Simulation error curves of BP neural network.

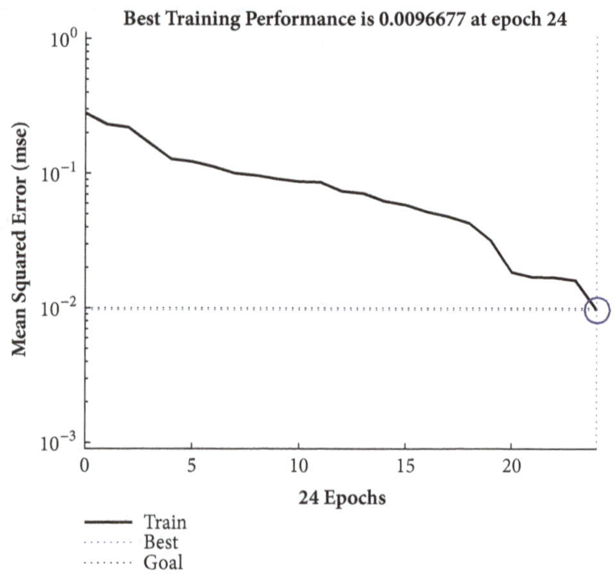

FIGURE 8: Simulation error curves of the GA-BP neural network.

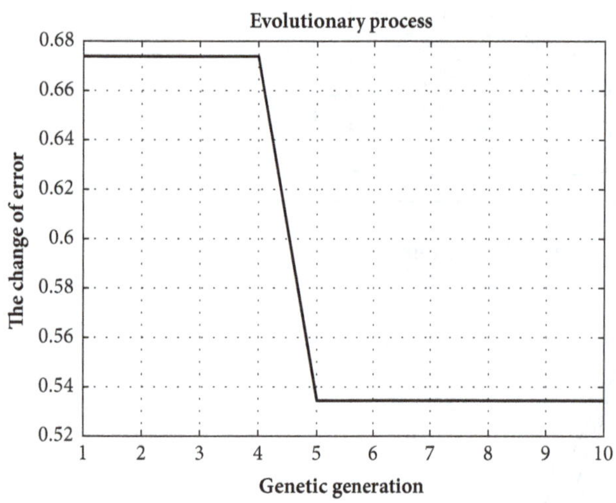

FIGURE 7: Evolutionary process.

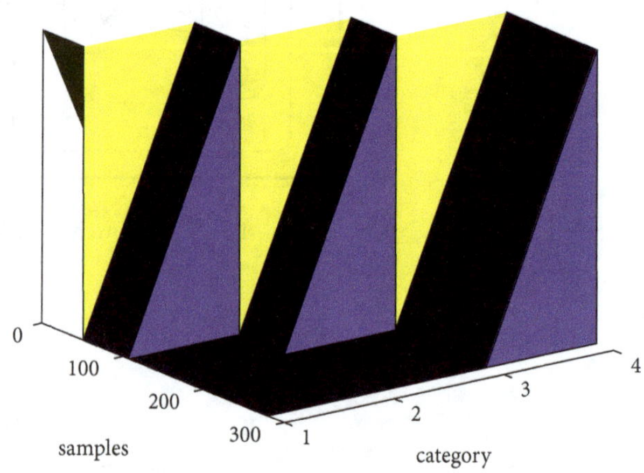

FIGURE 9: Classification target visualization.

From the error histogram in Figure 11, the error distribution of the deep neural network in this chapter is basically in line with the normal distribution, which meets the needs of practical application. The accuracy of adhesion state recognition is shown in Figures 12 and 13. The horizontal axis represents the desired target category, the vertical axis represents the experimentally predicted adhesion state category, and the gray block shows the exact percentage of prediction and expectation.

Figure 12 shows that the accuracy of the deep neural network adhesion state recognition is 96.1%. The overfitting of the neural network generally appears as the trained neural network does not accurately identify the test samples. The adhesion state is a continuously changing process; in order to ensure the safety of driving, the identification of the adhesion state must be as accurate as possible. Since the recognition accuracy rate does not reach the ideal state and there are

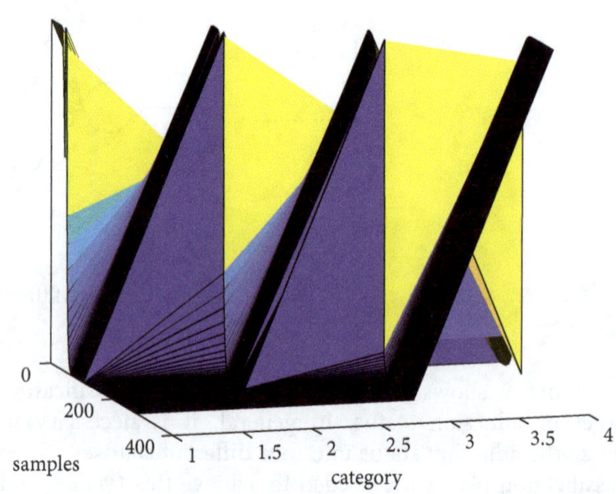

FIGURE 10: Contrast experiment with sparse coefficient 0.08.

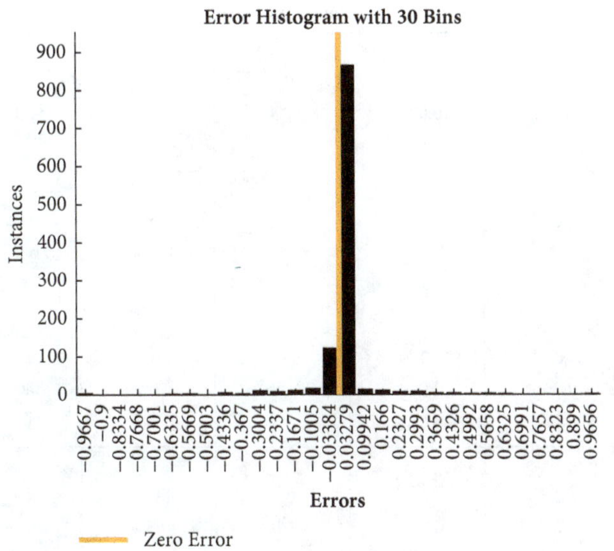

FIGURE 11: Error histogram.

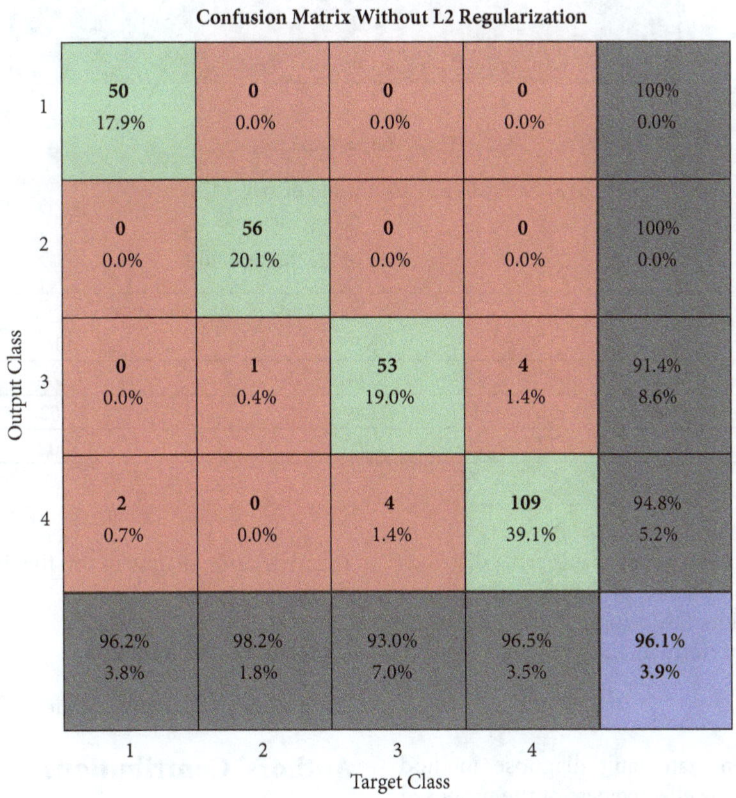

FIGURE 12: Confusion matrix without L2 regularization.

significantly more training samples than test samples, there is ample reason to speculate that the deep neural network used in this section has been overfitted. To improve this phenomenon, the L2 regularization method was used to attenuate the weights of the deep neural network. Figure 13 shows the results of the deep neural network adhesion state recognition after L2 regularization. The accuracy rate reaches 99.6%. The accuracy rate of the neural network for the adhesion state test set is improved, and the proposed L2 regularization can improve the overfitting phenomenon that may occur in the adhesion state recognition based on deep neural network.

The experimental results show that the SAE-based locomotive adhesion diagnosis can meet the requirements of

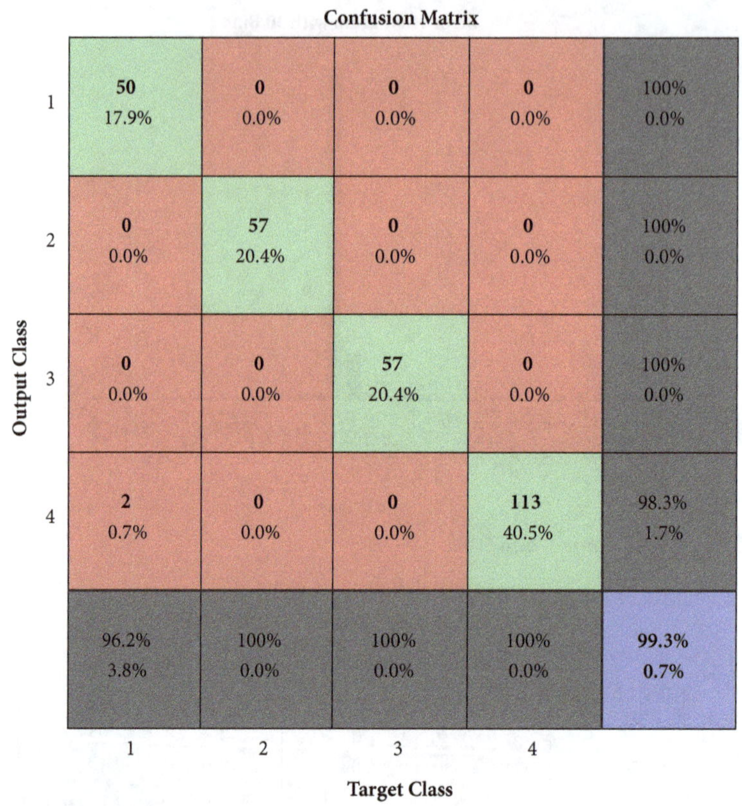

FIGURE 13: Confusion matrix.

TABLE 3: Test accuracy of the four methods.

Method	Type of adhesion state			
	N0	N1	F1	F2
BP	83.33	57.14	77.78	95.45
GA-BP	84.23	55.49	88.62	97.66
DNN	96.2	100	100	99.3

high-accuracy recognition under a reasonable error distribution. Table 3 presents the comparison of the three algorithms in this paper with the traditional BP neural network and GA-optimized neural network performance.

5. Conclusions

In this paper, an adhesion state fault diagnose method based on SAE is proposed. The effectiveness of the proposed method is validated by computer simulation. The conclusions are elaborated as follows:

(1) The adhesion state is divided into four categories, which provide a strong basis for wheel skid warning.

(2) The sparse automatic encoder can extract data features effectively, make classification easier, and extract more robust data features.

(3) Compared with the traditional BP neural network, the deep neural network of the sparse autoencoders can ensure effective fault diagnosis of the locomotive adhesion condition.

Conflicts of Interest

The authors declare no conflicts of interest.

Authors' Contributions

Changfan Zhang and Xiang Cheng conceived and designed the research; Changfan Zhang and Xiang Cheng performed the research; Changfan Zhang, Xiang Cheng, Jing He, Jianhua Liu, and Guangwei Liu wrote the paper.

Acknowledgments

This work was supported by the National Natural Science Foundation of China (nos. 61773159, 61473117), Hunan

Provincial Natural Science Foundation of China (nos. 2018JJ2093, 2017JJ4031), Key Laboratory for Electric Drive Control and Intelligent Equipment of Hunan Province (no. 2016TP1018), and Science and Technology Innovative Research Team in Higher Educational Institutions of Hunan Province.

References

[1] C. Zhang, J. Sun, J. He, and L. Liu, "Online Estimation of the Adhesion Coefficient and Its Derivative Based on the Cascading SMC Observer," *Journal of Sensors*, vol. 2017, Article ID 8419295, 11 pages, 2017.

[2] C. Zhang, X. Cheng, J. He, and G. Liu, "Automatic recognition of adhesion states using an extreme learning machine," *International Journal of Robotics and Automation*, vol. 32, no. 2, pp. 194–200, 2017.

[3] J. J. Castillo, J. A. Cabrera, A. J. Guerra, and A. Simón, "A Novel Electrohydraulic Brake System with Tire-Road Friction Estimation and Continuous Brake Pressure Control," *IEEE Transactions on Industrial Electronics*, vol. 63, no. 3, pp. 1863–1875, 2016.

[4] J. J. C. Aguilar, J. A. C. Carrillo, A. J. G. Fernández, and E. C. Acosta, "Robust road condition detection system using in-vehicle standard sensors," *Sensors*, vol. 15, no. 12, pp. 32056–32078, 2015.

[5] N. Li, X. Feng, and X. Wei, "Optimized adhesion control of locomotive airbrake based on GSA-RNN," in *Proceedings of the 7th International Conference on Intelligent Human-Machine Systems and Cybernetics, IHMSC 2015*, pp. 157–161, China, August 2015.

[6] A. Stanescu and D. Caragea, "An empirical study of ensemble-based semi-supervised learning approaches for imbalanced splice site datasets," *BMC Systems Biology*, vol. 9, no. 5, article no. S1, 2015.

[7] M. Madhiarasan and S. N. Deepa, "Comparative analysis on hidden neurons estimation in multi layer perceptron neural networks for wind speed forecasting," *Artificial Intelligence Review*, vol. 48, no. 4, pp. 449–471, 2017.

[8] N. J. Guliyev and V. E. Ismailov, "A single hidden layer feedforward network with only one neuron in the hidden layer can approximate any univariate function," *Neural Computation*, vol. 28, no. 7, pp. 1289–1304, 2016.

[9] M. Cai and J. Liu, "Maxout neurons for deep convolutional and LSTM neural networks in speech recognition," *Speech Communication*, vol. 77, pp. 53–64, 2016.

[10] Y. Liu, Y. Li, X. Ma, and R. Song, "Facial expression recognition with fusion features extracted from salient facial areas," *Sensors*, vol. 17, no. 4, 2017.

[11] G. E. Hinton, S. Osindero, and Y. Teh, "A fast learning algorithm for deep belief nets," *Neural Computation*, vol. 18, no. 7, pp. 1527–1554, 2006.

[12] X. Ma, Z. Dai, Z. He, J. Ma, Y. Wang, and Y. Wang, "Learning traffic as images: a deep convolutional neural network for large-scale transportation network speed prediction," *Sensors*, vol. 17, no. 4, 2017.

[13] Xiao He, Zidong Wang, Gang Li, Zhijie Zhou, and Youqing Wang, "Fault Diagnosis and Application to Modern Systems," *Journal of Control Science and Engineering*, vol. 2017, pp. 1–3, 2017.

[14] C. Li, R.-V. Sánchez, G. Zurita, M. Cerrada, and D. Cabrera, "Fault diagnosis for rotating machinery using vibration measurement deep statistical feature learning," *Sensors*, vol. 16, no. 6, article no. 895, 2016.

[15] W. Zhang, G. Peng, C. Li, Y. Chen, and Z. Zhang, "A new deep learning model for fault diagnosis with good anti-noise and domain adaptation ability on raw vibration signals," *Sensors*, vol. 17, no. 2, 2017.

[16] T. Ishrat, G. Ledwich, M. Vilathgamuwa, and P. Borghesani, "Wheel slip control based on traction force estimaton of electric locomotives," in *Proceedings of the Australasian Universities Power Engineering Conference, (AUPEC '16)*, Australia, 2016.

[17] M. Spiryagin, P. Wolfs, C. Cole et al., "Theoretical investigation of the effect of rail cleaning by wheels on locomotive tractive effort," in *Core*, p. V001T10A003, 2016.

[18] Xiao He, Yamei Ju, Yang Liu, and Bangcheng Zhang, "Cloud-Based Fault Tolerant Control for a DC Motor System," *Journal of Control Science and Engineering*, vol. 2017, pp. 1–10, 2017.

[19] H. Elghazel and A. Aussem, "Unsupervised feature selection with ensemble learning," *Machine Learning*, vol. 98, no. 1-2, pp. 157–180, 2013.

[20] F. Anselmi, J. Z. Leibo, L. Rosasco, J. Mutch, A. Tacchetti, and T. Poggio, "Unsupervised learning of invariant representations," *Theoretical Computer Science*, vol. 633, pp. 112–121, 2016.

[21] W. Zhou, Z. Shao, C. Diao, and Q. Cheng, "High-resolution remote-sensing imagery retrieval using sparse features by auto-encoder," *Remote Sensing Letters*, vol. 6, no. 10, pp. 775–783, 2015.

[22] O. Tsinalis, P. M. Matthews, and Y. Guo, "Automatic Sleep Stage Scoring Using Time-Frequency Analysis and Stacked Sparse Autoencoders," *Annals of Biomedical Engineering*, vol. 44, no. 5, pp. 1587–1597, 2016.

[23] M. Kang, K. Ji, X. Leng, X. Xing, and H. Zou, "Synthetic aperture radar target recognition with feature fusion based on a stacked autoencoder," *Sensors*, vol. 17, no. 1, 2017.

[24] J. Leng and P. Jiang, "A deep learning approach for relationship extraction from interaction context in social manufacturing paradigm," *Knowledge-Based Systems*, vol. 100, no. C, pp. 188–199, 2016.

[25] B. A. Olshausen and D. J. Field, "Emergence of simple-cell receptive field properties by learning a sparse code for natural images," *Nature*, vol. 381, no. 6583, pp. 607–609, 1996.

[26] D. Yang, J. Lai, and L. Mei, "Deep representations based on sparse auto-encoder networks for face spoofing detection," *Springer International Publishing*, 2016.

[27] J. Xu, L. Xiang, Q. Liu et al., "Stacked sparse autoencoder (SSAE) for nuclei detection on breast cancer histopathology images," *IEEE Transactions on Medical Imaging*, vol. 35, no. 119, 2016.

[28] A. Youssef, C. Delpha, and D. Diallo, "An optimal fault detection threshold for early detection using Kullback-Leibler Divergence for unknown distribution data," *Signal Processing*, vol. 120, pp. 266–279, 2016.

[29] J. Yang, Y. Bai, G. Li, M. Liu, and X. Liu, "A novel method of diagnosing premature ventricular contraction based on sparse auto-encoder and softmax regression," *Bio-Medical Materials and Engineering*, vol. 26, pp. S1549–S1558, 2015.

[30] M. M. A. Rahhal, Y. Bazi, H. Alhichri, N. Alajlan, F. Melgani, and R. R. Yager, "Deep learning approach for active classification of electrocardiogram signals," *Information Sciences*, vol. 345, pp. 340–354, 2016.

[31] E. Phaisangittisagul, "An Analysis of the Regularization Between L2 and Dropout in Single Hidden Layer Neural Network," in *Proceedings of the 7th International Conference on Intelligent Systems, Modelling and Simulation, ISMS 2016*, pp. 174–179, Thailand, January 2016.

13

A New Synergistic Forecasting Method for Short-Term Traffic Flow with Event-Triggered Strong Fluctuation

Darong Huang, Zhenping Deng, and Bo Mi

Institute of Information Science and Engineering, Chongqing Jiaotong University, Chongqing 400074, China

Correspondence should be addressed to Darong Huang; drhuang@cqjtu.edu.cn

Academic Editor: Zhijie Zhou

Directing against the shortcoming of low accuracy in short-term traffic flow prediction caused by strong traffic flow fluctuation, a novel method for short-term traffic forecasting based on the combination of improved grey Verhulst prediction algorithm and first-order difference exponential smoothing is proposed. Firstly, we constructed an improved grey Verhulst prediction model by introducing the Markov chain to its traditional version. Then, based on an introduced dynamic weighting factor, the improved grey Verhulst prediction method, and the first-order difference exponential smoothing technique, the new method for short-term traffic forecasting is completed in an efficient way. Finally, experiment and analysis are carried out in the light of actual data gathered from strong fluctuation environment to verify the effectiveness and rationality of our proposed scheme.

1. Introduction

In recent years, the popularization of seamless links among heterogeneous traffic equipment brought about higher requirements on the real-time and reliability of short-term traffic flow prediction. With continuous improvement of traffic information processing, how to predict the short-term traffic flow accurately and effectively has aroused wide attention of scholars domestically and abroad [1–3], whereupon numerous of prominent research results have emerged. So far, relevant academic circles mainly focus on the construction and optimization of prediction models in terms of time series, linear regression, historical average model, Kalman filtering, grey theory, chaos theory, nonparametric regression, neural network, support vector machine, dynamic traffic assignment model, and so forth [4–9]. These algorithms and models mentioned above are relatively mature, and their prediction effects are acceptable under the environment of favorable traffic flow stability. However, once the traffic data are seriously fluctuated, one single model can neither guarantee the prediction accuracy nor break through certain limitations of operations to predict short-term traffic flow under the environments of heterogeneous traffic equipment transmission.

In order to solve the aforementioned problems, domestic and foreign scholars have contrived an improved model to realize fusion prediction with the advantages of different models integrated of short-term traffic flow. For example, Xie et al. [10] improved the search efficiency of K-nearest neighbor algorithm by cooperating the multivariate statistical regression model and the pattern distance search method based on the analysis of original K-nearest neighbor algorithm for short-term traffic flow prediction. Their experimental results indicated that the prediction effect is better when the K values are reasonable. Fan et al. [4] exploited the characteristics of time-varying and nonlinearity of traffic flow and proposed a new hybrid forecasting model based on the nonparametric regression model and the BP neural network model; they also employed fuzzy control to determine the weight of each single model. However, the basic support platform of the hybrid model is the traffic flow database, so the demand of data volume is large. Once the data volume is insufficient, the prediction accuracy will be depressed. Xiao et al. [11] proposed an improved binding cycle truncation accumulated generating operation seasonal grey rolling forecasting model based on the properties of similar seasonality within intraday and weekly trends. The model weakens the random disturbance and highlights the intrinsic grey exponent rule after accumulating the sequence, so that the model has better performance under different traffic flow conditions. Lin-chao et al. [12] proposed a short-term traffic flow prediction model based on support vector regression which is suitable for real-time monitoring and then analyzed the model parameters by grid search under the premise of considering the influence of space-time factors. The model can achieve ideal prediction effect even if the real-time traffic data are insufficient. Nevertheless, the accuracy of the model will decrease when the traffic status changes. In order to solve such problem, Ma et al. [13] proposed a two-dimensional prediction method by using the Kalman filtering theory based on historical data. The advantage of this method is that the two predicted values are fused by using an equation with weight coefficients where the weight coefficients can be generated in real time in the process of prediction. Experimental results show that the model has an admirable predictive effect. Chan et al. [14] used exponential smoothing to preprocess traffic data that is taken as the input of network and then used Levenberg Marquardt (LM) variant algorithm to train the network weights, making the generalization ability of network enhanced.

However, thanks to the diversified developments of traffic information processing and data transmission techniques within heterogeneous traffic network, as well as the impact caused by dynamic changes of road topology, traffic accidents, severe weather, driving styles, and so forth, short-term traffic data are instantaneous and the irregular volatility is always changing [15, 16]. In this case, the above combinational algorithm can just solve the short-term traffic prediction problem in certain circumstance, and the prediction effect can hardly be achieved in the light of the real-time requirement. Hence, it is urgent to construct new models and algorithms to deal with the short-term traffic forecasting problem aiming at the strong fluctuation triggered by events. Thus, in this paper, we intend to propose a novel method for short-term traffic prediction based on the energetically grey Verhulst prediction algorithm and the first-order difference exponential smoothing technique to solve the problem of low prediction accuracy caused by traffic event-triggering strong fluctuation.

The layout of this paper is arranged as follows. Firstly, based on the introduction of the traditional grey Verhulst model, an improved grey Markov forecasting model is devised by introducing Markov chain. Secondly, combining the advantages of forecasting by utilizing first-order difference exponential smoothing algorithm and introducing a dynamic weighting factor, a new method for short-term traffic forecasting is concretely constructed according to the afore contrived models. Finally, comparative analysis of the examples illustrated that our proposed model and algorithm are more effective.

2. Short-Term Traffic Flow Prediction Model Based on Grey Difference Exponential Smoothing

2.1. Short-Term Traffic Flow Prediction Model Based on Grey Markov Theory

2.1.1. Introduction of Grey Verhulst Model. The grey system theory was first put forward by Professor Deng Julong, a Chinese scholar, in 1980s [17]. Its quantitative model is mainly based on the structure of number generating, which makes the prediction effect no longer affected by the empirical statistical law gained from the analysis of mass data. At the same time, it surmounted the limitation of white system and the black system relying solely on the probability and statistics method. Therefore, the grey system theory has been widely used in the fields of agricultural production, industrial control, traffic management, and so forth [18].

Classical grey system theory mainly includes GM (1, 1) model, GM (2, 1) model, grey Verhulst model, and so forth [19]. Considering the strong randomness and nonlinearity of traffic flow with strong fluctuation in heterogeneous information model, the data sequence significantly gives expression to the characteristics of nonmonotonic oscillatory development or the S-shaped property of saturation trend, which makes the grey GM (1, 1) prediction model unsuitable. Therefore, we choose the grey Verhulst model in this paper to forecast short-term traffic. The basic modeling process is described as follows.

The nonnegative data sequence is defined as

$$x^{(0)} = \left(x_1{}^{(0)}, x_2{}^{(0)}, \ldots, x_n{}^{(0)}\right). \tag{1}$$

while $x^{(1)}$ stands for the one-time accumulation sequence (1-Ago):

$$x^{(1)} = \left(x_1{}^{(1)}, x_2{}^{(1)}, \ldots, x_n{}^{(1)}\right),$$
$$x_k{}^{(1)} = \sum_{i=1}^{k} x_i{}^{(0)} \quad (k = 1, 2, \ldots, n). \tag{2}$$

Then, assuming that $z^{(1)}$ represents the consecutive neighbor sequence of $x^{(1)}$,

$$z^{(1)} = \left(z_1{}^{(1)}, z_2{}^{(1)}, \ldots, z_n{}^{(1)}\right),$$
$$z_k{}^{(1)} = 0.5\left(x_k{}^{(1)} + x_{k-1}{}^{(1)}\right) \quad (k = 2, 3, \ldots, n). \tag{3}$$

The grey Verhulst model and its whitening equation can be defined as follows.

Definition 1 (see [20, 21]). The grey Verhulst model is

$$x^{(0)} + az^{(1)} = b\left(z^{(1)}\right)^2, \tag{4}$$

where a and b are the parameters of the equation.

Definition 2 (see [20, 21]). The whitening equation of the grey Verhulst model is

$$\frac{dx^{(1)}}{dt} + ax^{(1)} = b\left(x^{(1)}\right)^2, \tag{5}$$

where a, b, and t are the parameters of the whitening equation.

As deduced in [20, 21], the following conclusions are drawn.

Theorem 3. *If the grey Verhulst model is defined as formula (4) and $\widehat{a} = (a, b)^T$ represents the parameter column, let*

$$B = \begin{bmatrix} -z_2^{(1)} & \left(z_2^{(1)}\right)^2 \\ -z_3^{(1)} & \left(z_3^{(1)}\right)^2 \\ \vdots & \vdots \\ -z_n^{(1)} & \left(z_n^{(1)}\right)^2 \end{bmatrix}, \quad Y = \begin{bmatrix} x_2^{(0)} \\ x_3^{(0)} \\ \vdots \\ x_n^{(0)} \end{bmatrix}. \tag{6}$$

The least squares estimator of the parameter column satisfies

$$\widehat{\alpha} = \left(B^T B\right)^{-1} B^T Y. \tag{7}$$

Theorem 4. *If the grey Verhulst model is defined as definition (2), the solution of the whitening equation can be deduced as*

$$x^{(1)}(t) = \frac{a x_0^{(1)}}{b x_0^{(1)} + \left(a - b x_0^{(1)}\right) e^{at}}. \tag{8}$$

According to Theorems 3 and 4, the following conclusions are drawn.

Inference 5. The time response sequence of grey Verhulst model can be defined as

$$\widehat{x}_{k+1}^{(1)} = \frac{a x_0^{(1)}}{b x_0^{(1)} + \left(a - b_0^{(1)}\right) e^{at}}, \quad (k = 1, 2, \ldots, n). \tag{9}$$

Let $x_0^{(1)} = x_1^{(0)}$; we iterate formula (9) to be

$$\widehat{x}_{k+1}^{(1)} = \frac{a x_1^{(0)}}{b x_1^{(0)} + \left(a - b x_1^{(0)}\right) e^{at}}, \quad (k = 1, 2, \ldots, n). \tag{10}$$

The reduction formula can then be defined as

$$\widehat{x}_{k+1}^{(0)} = \widehat{x}_{k+1}^{(1)} - \widehat{x}_k^{(1)}, \quad (k = 1, 2, \ldots, n). \tag{11}$$

Pointing at the traffic flow with the trend of increasing saturation, numerous traffic prediction algorithms have been proposed taking advantage of the grey Verhulst model to achieve preferable prediction effects. However, as a complicated nonlinear system involving multitudinous uncertainties, the probability of fortuitous events on urban roads is highly fluctuant, which leads to the deviation of forecasting results. Therefore, it is necessary to take the nonlinearity and time-varying characteristics of overall interactions into account in line with various influential factors. So, we improve the algorithm aiming at the accuracy of prediction model next.

2.1.2. Short-Term Traffic Flow Prediction Model Based on Grey Markov Theory. During the process of traffic data aggregation, which is coordinated by complex human-vehicle-environment interaction, the current traffic flow is often affected by previous moments. Therefore, in order to improve the prediction accuracy of grey Verhulst model, mathematical description of traffic flow aggregation is given in advance by constructing the Markov state transition probability matrix [22].

It is obvious that the state transition probability matrix should be updated over time when Markov is used to optimize the grey model. That is to say, at time point $t+1$, the state transition probability of Markov chain should be recalculated according to real-time traffic data along with the state transition probability matrix updated synchronously. Therefore, at time point $t+1$, the probability of traffic flow aggregation from state i to j can be described as

$$P_{ij} = p\left(X_{t+1} = j \mid X_t = i\right). \tag{12}$$

And the updated Markov transition probability matrix is formulated to be

$$P = \begin{bmatrix} p_{11} & p_{12} & \cdots & p_{1n} \\ p_{21} & p_{22} & \cdots & p_{2n} \\ \vdots & \vdots & & \vdots \\ p_{n1} & p_{n2} & \cdots & p_{nn} \end{bmatrix}, \tag{13}$$

where each element p_{ij} is constrained as $p_{ij} \geq 0$, $i, j \in I$, and $\sum_{j \in I} p_{ij} = 1$, $i \in I$.

The grey Markov forecasting model is shown in Figure 1. According to the flow chart of this model, the value of grey Verhulst model to be predicted is $\widehat{x}$ at time point $t + 1$. In order to make the predicted value as close to the true value as possible, the true value at the time point t is utilized and represented as x_{ij}, and the transition probability from the true value of time point t to the predictive value is denoted by p_{ij}. Thus, the cumulative sum of the product of the traffic flow average value and its corresponding probability from the current state to other states can be taken as the predicted value of the grey Markov model:

$$G_{t+1} = \begin{cases} \widehat{x} + \phi & \left(p_{ij} = 0\right) \\ \widehat{x} \cdot p_{ij} + \phi + \sum\limits_{k=1, k \neq j}^{n} p_{ik} \cdot \overline{x}_{ik} & \left(p_{ij} \neq 0\right), \end{cases} \tag{14}$$

where ϕ is the adjustment factor and $\phi = \widehat{x}_t - \widehat{x}_{t-1}$.

The adjustment factor ϕ is used to solve the data fluctuation problem of collaborative traffic flow prediction in terms of human, vehicle, and environment interaction. However, the smoothness of short-term traffic flow forecasting remains to be improved once events trigger strong fluctuations caused by traffic flow aggregation. Therefore, in order to continuously improve the prediction smoothness together with accuracy, a grey difference exponential smoothing short-term traffic prediction method is further constructed by introducing the first-order difference exponential smoothing model in the next subsection.

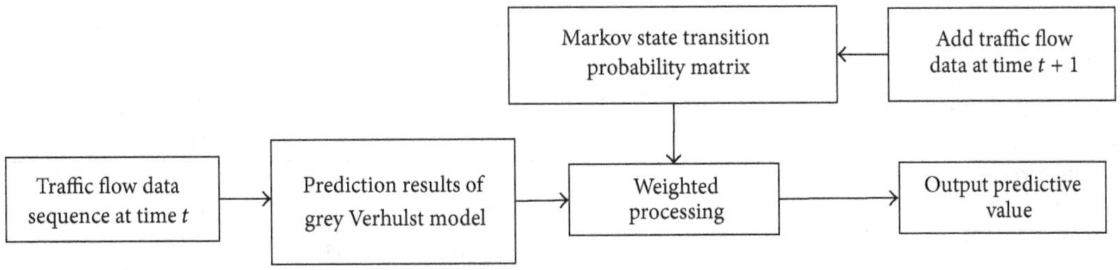

FIGURE 1: Grey forecasting model based on Markov chain.

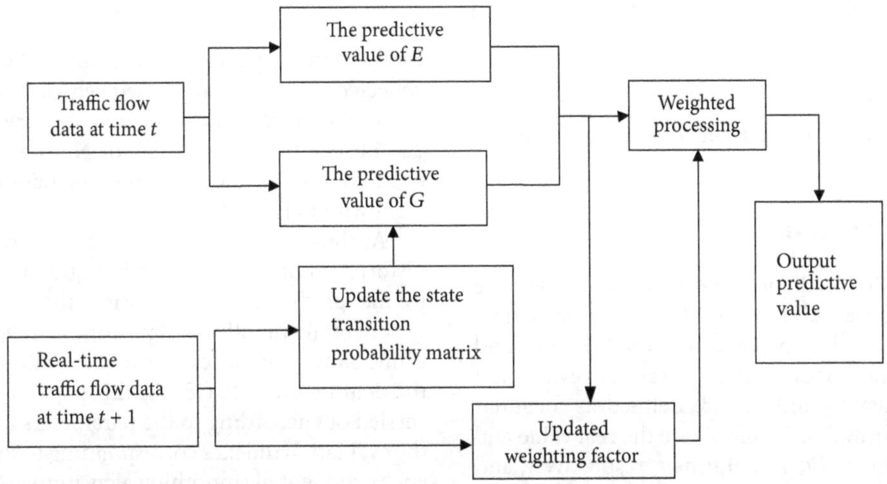

FIGURE 2: Grey difference exponential smoothing forecasting model.

2.2. Short-Term Flow Traffic Prediction Model Based on Grey Difference Exponential Smoothing. Exponential smoothing [23] is a widely used sequential prediction algorithm, whose predictive values are obtained by assigning different weights to historical records. Similar to the exponential smoothing algorithm, the first-order difference exponential smoothing algorithm achieves the predicted value by giving different weights to different data sequences. The basic modeling process is shown as follows.

For data sequence (1), the difference sequence is represented as

$$\nabla \widehat{E}_t = x_{t+1}^{(0)} - x_t^{(0)}, \quad (t = 1, 2, \ldots, n-1). \quad (15)$$

And the first-order difference exponential smoothing equation is

$$E_{t+1} = \nabla E_t + x_t^{(0)}, \quad (t = 1, 2, \ldots, n), \quad (16)$$

where $\nabla E_t = \alpha \nabla \widehat{E}_{t-1} + (1-\alpha)\nabla E_{t-1}$. Therefore, the forecast result should be

$$E_{t+1} = \alpha \sum_{j=1}^{t-1} (1-\alpha)^{j-1} \nabla \widehat{E}_{t-j} + (1-\alpha)^{t-1} \nabla E_1 + x_t^{(0)}. \quad (17)$$

Herein, $0 < \alpha < 1$ and $(1-\alpha)^{t-1}\nabla E_1 \to 0$ when $t \to \infty$. Thus, formula (17) can be rewritten as

$$E_{t+1} = \alpha \sum_{j=1}^{t-1} (1-\alpha)^{j-1} \nabla \widehat{E}_{t-j} + x_t^{(0)}. \quad (18)$$

We use dynamic weighting factors to weight the models of grey Markov model and first-order difference exponential smoothing model. The predictive value of the grey exponential smoothing model is thus achieved:

$$Y_{t+1} = \varphi G_{t+1} + \gamma E_{t+1}, \quad (\varphi + \gamma = 1). \quad (19)$$

In formula (19), G_{t+1} and E_{t+1} are the predictive value of grey Markov and first-order difference exponential smoothing, respectively, while φ and γ are their corresponding weight factors. The weight factors are calculated as follows:

$$\varphi = \frac{\left|x_{t+1}^{(0)} - E_{t+1}\right| + \zeta/2}{\left|x_{t+1}^{(0)} - G_{t+1}\right| + \left|x_{t+1}^{(0)} - E_{t+1}\right| + \zeta},$$
$$\gamma = \frac{\left|x_{t+1}^{(0)} - G_{t+1}\right| + \zeta/2}{\left|x_{t+1}^{(0)} - G_{t+1}\right| + \left|x_{t+1}^{(0)} - E_{t+1}\right| + \zeta}. \quad (20)$$

In formula (20), the weight of φ and γ depends on the absolute value of the difference between the predicted results of different models and the short-term traffic flow. In the meantime, in order to avoid the failure of formula (20), when the two prediction errors are 0, ζ tends to zero. The pseudocode of grey difference exponential smoothing prediction model is shown in Algorithm 1.

According to the synthetic forecasting algorithm, the prediction process of grey difference exponential smoothing prediction model is shown in Figure 2.

Input: Historical traffic flow data set $x^{(0)}$
Output: Short-term traffic flow forecasting result set Y
Initial state probability matrix $P, \alpha, \varphi, \gamma, \xi$
for $t \leftarrow 1$ to n **do**
  Calculate forecasting result E_{t+1} **and** G_{t+1}
  $Y_{t+1} = \varphi G_{t+1} + \gamma E_{t+1}, (\varphi + \gamma = 1)$
  Calculate the error ΔE_{t+1} **and** ΔG_{t+1}
  $\Delta E_{t+1} = \left|x^{(0)}{}_{t+1} - E_{t+1}\right|, \Delta G_{t+1} = \left|x^{(0)}{}_{t+1} - G_{t+1}\right|$
  Update the weights of φ **and** γ
  $\varphi = \dfrac{\Delta E_{t+1} + \xi/2}{\Delta G_{t+1} + \Delta E_{t+1} + \xi}, \gamma = \dfrac{\Delta G_{t+1} + \xi/2}{\Delta G_{t+1} + \Delta E_{t+1} + \xi}$
  Update the state transition matrix P **by** X_{t+1}
end
return Y

ALGORITHM 1: Pseudocode of grey differential exponent smoothing.

3. Example and Analysis

In order to verify the prediction effect of our model, the benchmarks of mean absolute error (MAE), mean square error (MSE), mean absolute percentage error (MAPE), and mean square percentage error (MSPE) are taken as evaluation indexes. Those evaluation indexes are defined as formulas (21)–(24). In those formulas, X_t and $\widehat{Y}_t$ are the real value and predicted value of the traffic flow at time t, respectively, and n stands for the number of samples in the short-term traffic flow.

(1) Mean absolute error is

$$\text{MAE} = \frac{1}{n}\sum_{t=1}^{n}\left|X_t - \widehat{Y}_t\right|. \tag{21}$$

(2) Mean square error is

$$\text{MSE} = \frac{1}{n}\sqrt{\sum_{t=1}^{n}\left(X_t - \widehat{Y}_t\right)^2}. \tag{22}$$

(3) Mean absolute percentage error is

$$\text{MAPE} = \frac{1}{n}\sum_{t=1}^{n}\left|\frac{X_t - \widehat{Y}_t}{X_t}\right|. \tag{23}$$

(4) Mean square percentage error is

$$\text{MSPE} = \frac{1}{n}\sqrt{\sum_{t=1}^{n}\left(\frac{X_t - \widehat{Y}_t}{X_t}\right)^2}. \tag{24}$$

The urban expressway in Nan'an District of Chongqing Municipality in China is selected in observation phase to verify the actual prediction effect of the grey difference exponential smoothing model, which is composed of 6 bidirectional lanes. The schematic diagram of the section observation is shown in Figure 3. The data detected include the traffic volume, speed, and occupancy of each lane, which are collected via microwave-based vehicle sensors.

TABLE 1: Evaluation index of first-order difference exponential smoothing algorithm.

α	Index			
	MAE	MSE	MAPE	MSPE
$\alpha = 0.05$	**26.820**	**2.563**	**0.389**	**0.046**
$\alpha = 0.08$	27.191	2.601	0.392	0.046
$\alpha = 0.10$	27.451	2.627	0.394	0.046
$\alpha = 0.20$	28.828	2.760	0.408	0.047
$\alpha = 0.50$	33.704	3.220	0.470	0.052

The road traffic flow is counted every 5 minutes, and the 226 observation data points of No. 1 Lane are used as the basic data for verification. The fluctuation trend of the data is shown in Figure 4.

As illustrated in Figure 4, there is a strong fluctuation in historical traffic flow, which brought about great difficulties to the prediction. When using the first-order difference exponential smoothing algorithm to predict the short time traffic flow, it is necessary to determine the value of α in the definite formula (18). In this paper, five experiments are carried out according to the parameters listed in Table 1, and the evaluation indexes corresponding to the first-order difference exponential smoothing algorithm are obtained in terms of different benchmarks.

According to the results in Table 1, when $\alpha = 0.05$, the first-order difference exponential smoothing algorithm produces the best effect, and the fitting patterns between the predicted value and actual value are shown in Figure 5.

The initial state probability transfer matrix is calculated from the historical traffic sequence when the grey Markov model is used to predict the short-term traffic flow. The state is divided according to the same interval that every 30 vehicles are used as a state. In the following prediction process, we added real-time traffic data to update Markov state probability transfer matrix in real time. The prediction result of grey Markov model is determined by formula (14), and the fitting effect of predicted value and actual value is shown in Figure 6.

Comparing Figure 6 with Figure 5, the grey Markov model can commendably reflect the trend of traffic flow. However, urban road traffic as a complex nonlinear system shows high uncertainty at different times. The grey Markov chain model cannot accurately predict once sharp fluctuations occur. Meanwhile, compared with the grey Markov chain prediction model, the first-order difference exponential smoothing model is more accordant with the trend of original data, but the prediction results of the algorithm depend on the value of α and, more importantly, are affected by time delays.

Therefore, combining the advantages of grey Markov model and first-order difference exponential smoothing, a new predictive value is obtained by using dynamic weighting factors according to formula (19) and formula (20). The prediction results are shown in Figure 7.

Based on the short-term traffic data of the observed road sections, the test data are predicted, respectively, by formulas

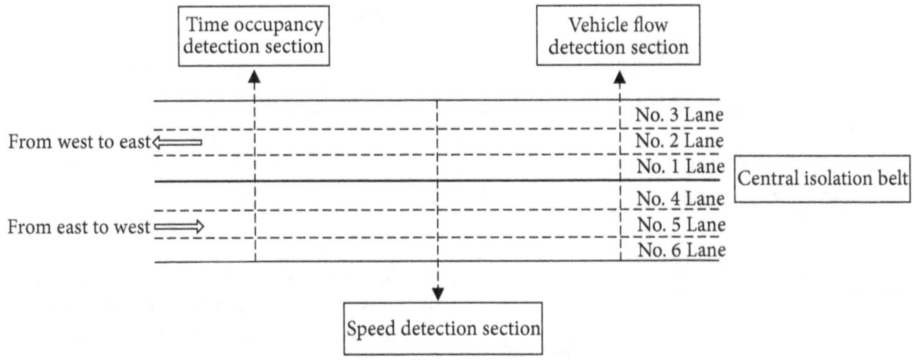

FIGURE 3: Road observation diagram.

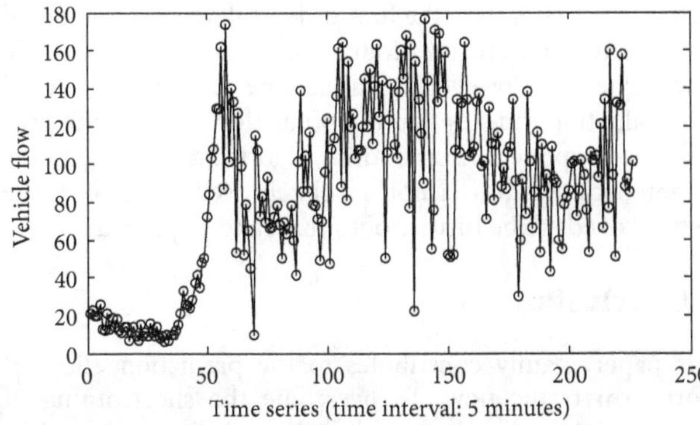

FIGURE 4: Historical traffic flow data.

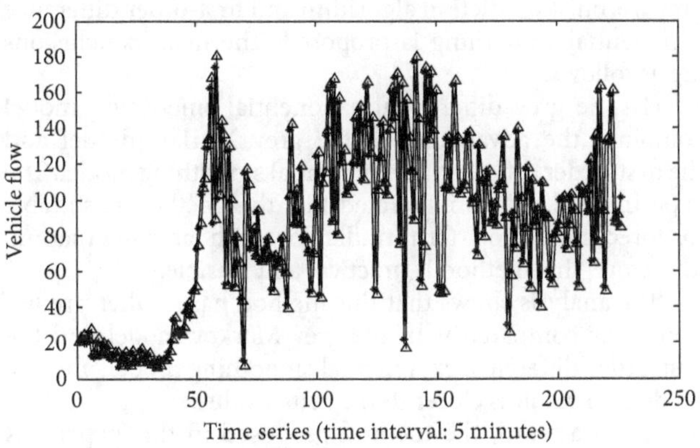

FIGURE 5: Predictive results of first-order difference exponential smoothing model.

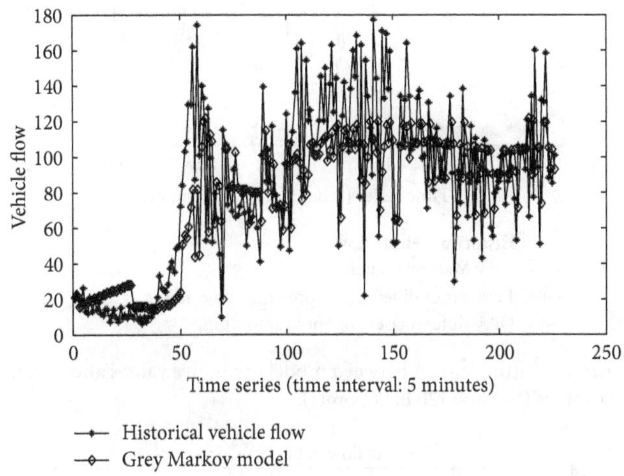

FIGURE 6: Predictive results of grey Markov model.

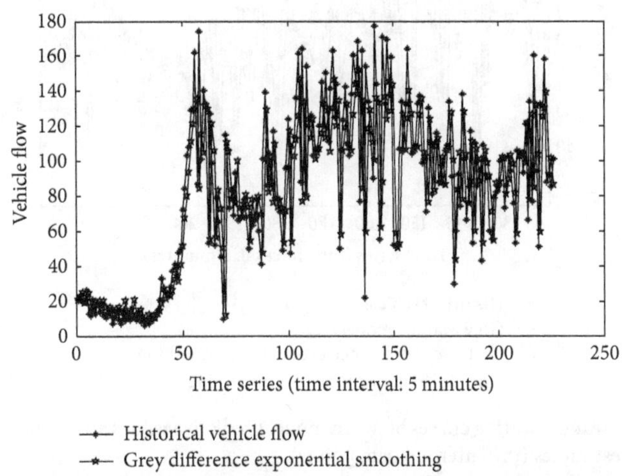

FIGURE 7: Predictive results of grey difference exponential smoothing model.

(14), (18), and (19). The fitting results of the first 120 data points and the last 106 data points are separately illustrated in Figures 8 and 9.

The evaluation indexes of the above prediction models are shown in Table 2 for the first-order difference exponential smoothing parameter $\alpha = 0.05$. Simulation results showed that the grey difference exponential smoothing algorithm can fit short-term traffic flow well and can be applied to urban traffic signal control for short-term traffic flow data with strong fluctuation.

Compared with the above models, the first-order difference exponential smoothing algorithm can closely reflect the

TABLE 2: Evaluation index of forecasting model.

Model	Index			
	MAE	MSE	MAPE	MSPE
Grey Markov model	25.520	2.220	0.445	0.049
First-order difference exponential smoothing model	26.820	2.563	0.389	0.046
Grey difference exponential smoothing model	**23.943**	**2.237**	**0.368**	**0.043**

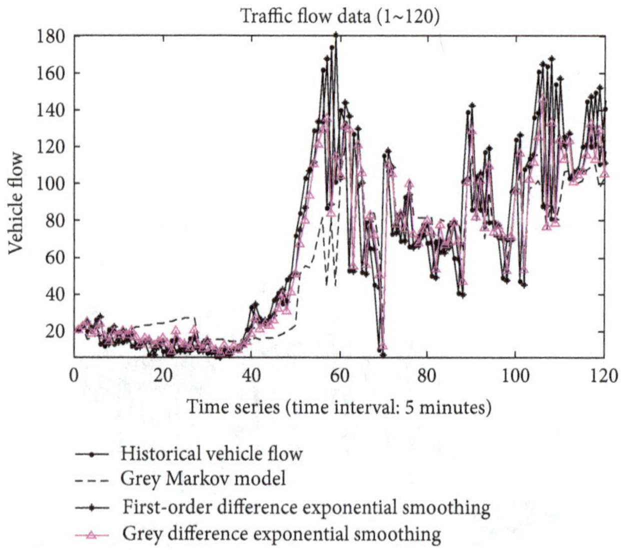

FIGURE 8: Fitting curves between model predictive values and actual test values (the first 120 data points).

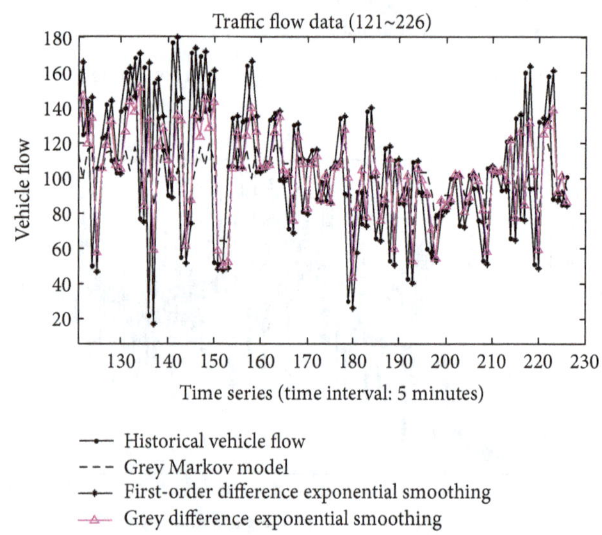

FIGURE 9: Fitting curves between model predictive values and actual test values (the latter 106 data points).

trend of the original data sequence in the prediction trend of the data, but its predicted value depends on the value of α and the trend of the data sequence. For the abrupt data points, the accuracy of its prediction will often be affected to distinct degrees. Compared with the first-order difference exponential smoothing model, the prediction accuracy of grey Markov model considers the development trend of the whole data and the state transition probability of historical traffic flow data. Therefore, the model can reflect the overall equilibrium level for data series with large fluctuation. The grey difference exponential smoothing model combines the advantages of the two models and uses the dynamic weight to obtain the predictive value. The performance evaluation indexes MAE, MSE, MAPE, and MSPE of the improved method are better than the first-order difference exponential smoothing model and grey Markov model for the same traffic flow sequence. However, it can also be seen from Figure 6 and evaluation indexes that the prediction effect of the grey difference exponential smoothing algorithm using dynamic weighted calculation is not absolutely well. Therefore, the method needs to be further optimized and improved.

4. Conclusion

This paper mainly contributes to the prediction effect of short-term traffic flow. To overcome the shortcoming of low accuracy in short-term traffic flow prediction caused by strong traffic flow fluctuation, a novel method for short-term traffic forecasting based on the combination of improved grey Verhulst prediction algorithm and first-order difference exponential smoothing is proposed. The main conclusions are as follows.

(1) The grey differential exponential smoothing model combines the advantages of the grey Markov model and the first-order differential exponential smoothing model. The experimental results illustrated that this method is suitable for forecasting short-term traffic flow with large fluctuation. Therefore, this method is practical and feasible.

The analysis shows that this method has smaller prediction error compared with the grey Markov model and the first-order difference exponential smoothing model, and the prediction result is closer to the actual value.

(2) In a word, the method proposed in this paper has obvious advantages. However, due to high uncertainty and nonlinearity of the short-term traffic flow, how to improve the prediction accuracy with fluctuation remains an open problem. So this will be our continuous research orientation.

Conflicts of Interest

The authors declare that they have no conflicts of interest.

Acknowledgments

This work is supported by the National Science Foundation of China (NSFC) under Grants 61573076 and 61703063; the Scientific Research Foundation for the Returned Overseas Chinese Scholars under Grant 2015-49; the Program for Excellent

Talents of Chongqing Higher School under Grant 2014-18; Science and Technology Research Project of Chongqing Municipal Education Commission of China under Grants KJ1705121 and KJ1705139; the Chongqing Natural Science Foundation of China under Grant CSTC2017jcyjA1665.

References

[1] Z. Hong-bin, S. Xiao-duan, and H. Yu-long, "Analysis and prediction of comples dynamical characteristics of short—term traffic flow," *Acta Physica Sinica*, vol. 63, no. 4, Article ID 080502, pp. 51–58, 2014 (Chinese).

[2] D.-R. Huang, J. Song, S.-Q. Li, and H.-Y. Xiang, "Control technology review of traffic congestion in urban road network under networked dynamic scheduling and control," *Journal of Traffic and Transportation Engineering*, vol. 13, no. 5, pp. 105–114, 2013 (Chinese).

[3] S. Oh, Y.-J. Byon, K. Jang, and H. Yeo, "Short-term travel-time prediction on highway: a review on model-based approach," *KSCE Journal of Civil Engineering*, vol. 3, pp. 1–13, 2017.

[4] N. Fan, X.-M. Zhao, M. Dai, and Y.-S. An, "Short-term traffic flow prediction model," *Journal of Traffic and Transportation Engineering*, vol. 12, no. 4, pp. 114–119, 2012 (Chinese).

[5] Y.-A. Daraghmi, C.-W. Yi, and T.-C. Chiang, "Negative binomial additive models for short-term traffic flow forecasting in Urban areas," *IEEE Transactions on Intelligent Transportation Systems*, vol. 15, no. 2, pp. 784–793, 2014.

[6] Y.-S. Jeong, Y.-J. Byon, M. M. Castro-Neto, and S. M. Easa, "Supervised weighting-online learning algorithm for short-term traffic flow prediction," *IEEE Transactions on Intelligent Transportation Systems*, vol. 14, no. 4, pp. 1700–1707, 2013.

[7] K. Y. Chan, T. Dillon, E. Chang, and J. Singh, "Prediction of short-term traffic variables using intelligent swarm-based neural networks," *IEEE Transactions on Control Systems Technology*, vol. 21, no. 1, pp. 263–274, 2013.

[8] M. Lippi, M. Bertini, and P. Frasconi, "Short-term traffic flow forecasting: an experimental comparison of time-series analysis and supervised learning," *IEEE Transactions on Intelligent Transportation Systems*, vol. 14, no. 2, pp. 871–882, 2013.

[9] N. G. Polson and V. O. Sokolov, "Deep learning for short-term traffic flow prediction," *Transportation Research Part C: Emerging Technologies*, vol. 79, pp. 1–17, 2017.

[10] H.-H. Xie, X.-H. Dai, and Y. Qi, "Improved K-nearest neighbor algorithm for short-term traffic flow forecasting," *Journal of Traffic and Transportation Engineering*, vol. 14, no. 3, pp. 87–94, 2014 (Chinese).

[11] X. Xiao, J. Yang, S. Mao, and J. Wen, "An improved seasonal rolling grey forecasting model using a cycle truncation accumulated generating operation for traffic flow," *Applied Mathematical Modelling*, vol. 51, pp. 386–404, 2017.

[12] L. Lin-chao, H. Shang-lu, and Z. Jian, "Online short—term traffic flow prediction considering the impact of temporal—spatial features," *Journal of Transportation Systems Engineering and Information Technology*, vol. 16, no. 5, pp. 165–171, 2016 (Chinese).

[13] M. Ma, S. Liang, H. Guo, and J. Yang, "Short-term traffic flow prediction using a self-adaptive two-dimensional forecasting method," *Advances in Mechanical Engineering*, vol. 9, no. 8, pp. 1–12, 2017.

[14] K. Y. Chan, T. S. Dillon, J. Singh, and E. Chang, "Neural-network-based models for short-term traffic flow forecasting using a hybrid exponential smoothing and levenberg-marquardt algorithm," *IEEE Transactions on Intelligent Transportation Systems*, vol. 13, no. 2, pp. 644–654, 2012.

[15] X. Wang, A. J. Khattak, J. Liu, G. Masghati-Amoli, and S. Son, "What is the level of volatility in instantaneous driving decisions?" *Transportation Research Part C: Emerging Technologies*, vol. 58, pp. 413–427, 2015.

[16] W. Qing-bo, H. Zhao-cheng, and Z. Xi-shuang, "Prediction of Urban Traffic Performance Index Considering Multiple Factors," *Journal of Transportation Systems Engineering and Information Technology*, vol. 17, no. 1, pp. 74–81, 2017 (Chinese).

[17] S. Liu, Y. Yang, N. Xie et al., "New progress of grey system theory in the new millennium," *The Journal of Grey System*, vol. 6, no. 1, pp. 2–31, 2016.

[18] R. Rajesh, V. Ravi, and R. Venkata Rao, "Selection of risk mitigation strategy in electronic supply chains using grey theory and digraph-matrix approaches," *International Journal of Production Research*, vol. 53, no. 1, pp. 238–257, 2015.

[19] H. Da-rong and H. Li-fen, "Present situation and development of grey system theory in fault forecast application," *Journal of Gun Launch & Control*, vol. 3, pp. 88–92, 2009 (Chinese).

[20] D. Song, D. Yao-guo, X. Ning et al., "Optimized background value in grey Verhulst model and its application," *Control and Decision*, vol. 30, no. 10, pp. 1835–1840, 2015 (Chinese).

[21] Y. Zhang, "Improved background value of grey Verhulst model and its application," *System Engineering-Theory and Practice*, vol. 33, no. 12, pp. 3168–3171, 2013.

[22] Z. Yi-fan and C. Wen-ying, "Markov chain analysis of metro network in responding to emergency," *China Safety Science Journal*, vol. 25, no. 2, pp. 165–170, 2015 (Chinese).

[23] Q. Chi and H. Zhong-sheng, "Application of adaptive sing—exponent smoothing for short—term traffic flow prediction," *Control Theory & Applications*, vol. 29, no. 4, pp. 465–469, 2012 (Chinese).

14

Neural Back-Stepping Control of Hypersonic Flight Vehicle with Actuator Fault

Qi Wu [1] and Yuyan Guo [2]

[1] *Department of Computer Science, Guangdong Police College, Guangzhou 510230, China*
[2] *School of Automation, Northwestern Polytechnical University, Xi'an 710072, China*

Correspondence should be addressed to Qi Wu; wuqi888@126.com

Academic Editor: Youqing Wang

This paper addresses the fault-tolerant control of hypersonic flight vehicle. To estimate the unknown function in flight dynamics, neural networks are employed in controller design. Moreover, in order to compensate the actuator fault, an adaptive signal is introduced in the controller design to estimate the unknown fault parameters. Simulation results demonstrate that the proposed approach could obtain satisfying performance.

1. Introduction

Fault diagnosis and fault-tolerant control have been a hot issue since its significance on actual systems with actuator faults. By identifying or estimating unknown faults, the influence of faults could be considered and eliminated in controller design. Representative study on this issue could be found in [1, 2], while in this paper, fault diagnosis have been studied and applied on hypersonic flight vehicle (HFV) whose control methods have attracted a lot of attention. Hypersonic flight vehicle has a much higher speed than traditional aircraft, however due to its special flight environment as well as structure, the controller design of HFV faces more challenges [3].

Currently, most studies on HFV control did not take the actuator fault into consideration. However, unknown fault such as actuator dead-zone is quite common in nonlinear systems and may cause serious consequences [4]. In order to remove the influence caused by actuator fault, several approaches have been studied such as fuzzy control [5], robust control [6, 7], and adaptive compensation [8, 9]. The model of the HFV also contains uncertain nonlinearity. As a result, studies on HFV have concentrated on the estimation of the uncertainty. Among all the methods, neural network has been widely used due to its good performance in approximating unknown nonlinear functions [10].

The model of HFV is highly nonlinear; meanwhile the altitude subsystem could be considered as a strict-feedback system [11]. Therefore, back-stepping method [12–14] in HFV control has been widely studied. To eliminate continuous derivatives of each virtual control in back-stepping design, multiple methods such as dynamic surface control [15] and differentiator [14] are combined with back-stepping which makes the approach practicable. It is noticed that in [16] the multiple actuator fault is considered and the robust design is presented to guarantee the system stability. Moreover, in [17], the input dead-zone is considered where the Nussbaum design is included to make the adaptive design available. In this paper, we try to estimate the parameters of system fault in the hypersonic flight dynamics and then the "dynamic inversion" of the actuator fault can be included in the control signal. The whole design is using a back-stepping control law with neural networks and adaptive method.

The paper is organized in 6 parts. In Section 2, the control-oriented model (COM) of HFV considered in this paper is simply introduced. In Sections 3 and 4, the back-stepping controller based on neural networks and adaptive fault estimation is designed and system stability is analysed.

The simulation results of the designed controller are shown in Section 5. Finally the summary is given in Section 6.

2. Longitudinal Dynamics of HFV with Actuator Fault

The COM in [18] is employed for study:

$$\dot{V} = \frac{T\cos\alpha - D}{m} - g\sin\gamma$$
$$\dot{h} = V\sin\gamma$$
$$\dot{\gamma} = \frac{L + T\sin\alpha}{mV} - \frac{g\cos\gamma}{V} \tag{1}$$
$$\dot{\alpha} = q - \dot{\gamma}$$
$$\dot{q} = \frac{M_{yy}}{I_{yy}}.$$

The detail of the dynamics can be found in [18].

Consider the following actuator fault model:

$$\delta_e = \begin{cases} u - b & \text{if } u \geq b \\ 0 & \text{if } -b < u < b \\ u + b & \text{if } u \leq -b, \end{cases} \tag{2}$$

where u is the designed control input and $b < 0$ is an unknown fault parameter to be estimated.

Remark 1. In HFV systems, the fault will cause error between designed and actual control input and it will result in tracking error or even flight instability.

3. Adaptive Back-Stepping Controller

The strict-feedback form altitude subsystem based on the hypersonic flight dynamics is considered

$$\dot{x}_1 = f_1 + g_1 x_2$$
$$\dot{x}_2 = x_3 \tag{3}$$
$$\dot{x}_3 = f_3 + g_3\delta_e,$$

where $x_1 = \gamma$, $x_2 = \gamma + \alpha$, and $x_3 = q$. f_i, g_i are nonlinear functions of the HFV model.

Assumption 2. In altitude subsystem (3), g_3 is a bounded function.

Flight path angle (FPA) tracking error $\tilde{x}_1$ is defined as

$$\tilde{x}_1 = x_1 - x_{1d}, \tag{4}$$

where x_{1d} is the command signal of FPA which is designed via altitude reference signal.

Step 1. Choose virtual control of x_2 as

$$x_{2d} = \frac{1}{g_1}\left(-f_1 - \varsigma_1\tilde{x}_1 + \dot{x}_{1d}\right), \tag{5}$$

where ς_1 is the control gain.

To avoid the continuous derivative of virtual control, the following first-order differentiator is designed:

$$\dot{\varpi}_{2c} = -l_{20}\sqrt{\left|\varpi_{2c} - x_{2d}\right|}\,\mathrm{sgn}\left(\varpi_{2c} - x_{2d}\right) + \varpi_{2d}$$
$$\dot{\varpi}_{2d} = -l_{21}\,\mathrm{sgn}\left(\varpi_{2d} - \dot{\varpi}_{2c}\right), \tag{6}$$

where l_{20} and l_{21} are positive designed parameters.

Step 2. Define the tracking error $\tilde{x}_2$:

$$\tilde{x}_2 = x_2 - x_{2d}. \tag{7}$$

Choose virtual control of x_3 as

$$x_{3d} = -\varsigma_2\tilde{x}_2 + \dot{\varpi}_{2c} - g_1\tilde{x}_1. \tag{8}$$

where ς_2 is the control gain.

The following first-order differentiator is designed:

$$\dot{\varpi}_{3c} = -l_{30}\sqrt{\left|\varpi_{3c} - x_{3d}\right|}\,\mathrm{sgn}\left(\varpi_{3c} - x_{3d}\right) + \varpi_{3d}$$
$$\dot{\varpi}_{3d} = -l_{31}\,\mathrm{sgn}\left(\varpi_{3d} - \dot{\varpi}_{3c}\right), \tag{9}$$

where l_{30} and l_{31} are positive designed parameters.

Step 3. Define the tracking error $\tilde{x}_3$:

$$\tilde{x}_3 = x_3 - x_{3d}. \tag{10}$$

Due to the uncertainty caused by imprecise model, the nonlinear function f_3 may be unknown. Therefore its NN-based estimation value is employed in the control law:

$$u_0 = \frac{1}{g_3}\left(-\hat{\omega}_3^T\Theta_3 - \varsigma_3\tilde{x}_3 - \tilde{x}_2 + \dot{\varpi}_{3c}\right), \tag{11}$$

where ς_3 is the control gain and $\hat{\omega}_3$ is the estimation of the optimal NN weights ω_3^*, which is obtained via adaptive law:

$$\dot{\hat{\omega}}_3 = \Gamma_3\tilde{x}_3\Theta_3\left(\overline{x}_3\right) - \Gamma_3\delta_3\hat{\omega}_3, \tag{12}$$

where Γ_3 and δ_3 are positive parameters. Θ_3 is obtained via radial basis function. Define $\tilde{\omega}_3 = \omega_3^* - \hat{\omega}_3$.

In order to eliminate the influence of unknown constant b, signal $\hat{b}$ is introduced in controller design to compensate the actuator:

$$u = u_0 + \hat{b}\,\mathrm{sgn}\left(u_0\right). \tag{13}$$

The adaptive law is designed as

$$\dot{\hat{b}} = -\Gamma_b g_3\tilde{x}_3\mathrm{sgn}\left(u_0\right) - \Gamma_b\sigma_b\hat{b}, \tag{14}$$

where Γ_b and σ_b are positive designed parameters. Define the estimation error $\tilde{b} = b - \hat{b}$.

Remark 3. In previous work on HFV dead-zone fault control [17], the paper regarded the dead-zone as a part of the compound disturbance, where robust technique is employed. In our paper, we proposed an adaptive law to estimate the unknown fault parameter and added a compensating signal in the controller so that the influence of dead-zone fault could be directly eliminated, which is shown by Figures 3 and 5 in our manuscript.

Define the velocity tracking error as

$$\widetilde{V} = V - V_r, \tag{15}$$

where V_r is the reference signal. Then the following PID controller is designed:

$$\Phi = k_{pv}\widetilde{V} + k_{iv}\int \widetilde{V}(t)\,dt + k_{dv}\dot{\widetilde{V}}. \tag{16}$$

4. Stability Analysis

Theorem 4. *Consider the HFV altitude system (3) with control laws (5), (8), (11), and (13) and adaptive laws (12) and (14); all of the error signals are uniformly ultimately bounded.*

Proof. The Lyapunov function candidate is chosen as

$$V_L = \sum_{i=1}^{3} V_i, \tag{17}$$

where

$$\begin{aligned} V_1 &= \frac{1}{2}\widetilde{x}_1^2 \\ V_2 &= \frac{1}{2}\widetilde{x}_2^2 \\ V_3 &= \frac{1}{2}\widetilde{x}_3^2 + \frac{1}{2}\widetilde{\omega}_3^T\Gamma_3^{-1}\widetilde{\omega}_3 + \frac{1}{2}\Gamma_b^{-1}\widetilde{b}^2. \end{aligned} \tag{18}$$

The derivative of V_L is obtained as

$$\dot{V}_L = \sum_{i=1}^{3}\widetilde{x}_i\left(\dot{x}_i - \dot{x}_{id}\right) - \widetilde{\omega}_3^T\Gamma_3^{-1}\dot{\hat{\omega}}_3 - \Gamma_b^{-1}\widetilde{b}\dot{\hat{b}}. \tag{19}$$

According to the conclusion in [19], $\hat{\omega}_{2c}$, $\hat{\omega}_{3c}$ in differentiators (6), (9) could estimate $\dot{x}_{2d}$ and $\dot{x}_{3d}$ to arbitrary accuracy. Therefore there exists

$$\dot{x}_{id} = \hat{\omega}_{ic} + \chi_i, \quad i = 2, 3, \tag{20}$$

where $\chi_i \le |\chi_{im}|$, χ_{im} is a positive constant. Substitute the control laws and adaptive laws into (19); there exists

$$\begin{aligned} \dot{V}_L = &-\sum_{i=1}^{3}\varsigma_i\widetilde{x}_i^2 + \delta_3\left(\widetilde{\omega}_3^T\omega_3^* - \widetilde{\omega}_3^T\widetilde{\omega}_3\right) + \varepsilon_3\widetilde{x}_3 \\ &+ b\widetilde{x}_3 g_3\left[\operatorname{sgn}(u_0) - \operatorname{sgn}(u)\right] + \sigma_b\left(b\widetilde{b} - \widetilde{b}^2\right) \\ &-\sum_{i=2}^{3}\widetilde{x}_i\chi_i, \end{aligned} \tag{21}$$

where ε_3 is the bounded inevitable NN reconstruction error satisfying $|\varepsilon_3| \le \varepsilon_{3m}$, where ε_{3m} is a positive constant. Define $\eta = bg_3[\operatorname{sgn}(u_0) - \operatorname{sgn}(u)]$; consider Assumption 1; η is a bounded signal satisfying $|\eta| \le \eta_m$, where η_m is a positive constant.

The following inequalities hold:

$$\begin{aligned} \delta_3\left(\widetilde{\omega}_3^T\omega_3^* - \widetilde{\omega}_3^T\widetilde{\omega}_3\right) &\le \frac{1}{2}\delta_3\left(\|\omega_3^*\|^2 - \widetilde{\omega}_3^T\widetilde{\omega}_3\right) \\ \sigma_b\left(b\widetilde{b} - \widetilde{b}^2\right) &\le \frac{1}{2}\sigma_b\left(b^2 - \widetilde{b}^2\right) \\ -\sum_{i=2}^{3}\widetilde{x}_i\chi_i &\le \sum_{i=2}^{3}\left(\frac{1}{2}\widetilde{x}_i^2 + \frac{1}{2}\chi_{im}^2\right) \\ \varepsilon_3\widetilde{x}_3 &\le \frac{1}{2}\varepsilon_{3m}^2 + \frac{1}{2}\widetilde{x}_3^2 \\ \eta\widetilde{x}_3 &\le \frac{1}{2}\eta_m^2 + \frac{1}{2}\widetilde{x}_3^2. \end{aligned} \tag{22}$$

The derivative of V_L satisfies

$$\begin{aligned} \dot{V}_L \le &-\varsigma_1\widetilde{x}_1^2 - \left(\varsigma_2 - \frac{1}{2}\right)\widetilde{x}_2^2 - \left(\varsigma_3 - \frac{3}{2}\right)\widetilde{x}_3^2 - \frac{1}{2}\delta_3\widetilde{\omega}_3^T\widetilde{\omega}_3 \\ &- \frac{1}{2}\sigma_b\widetilde{b}^2 + \frac{1}{2}\delta_3\|\omega_3^*\|^2 + \frac{1}{2}\varepsilon_{3m}^2 + \frac{1}{2}\eta_m^2 \\ &+ \frac{1}{2}\sigma_b b^2 + \sum_{i=2}^{3}\frac{1}{2}\chi_{im}^2 \\ \le &-\sigma V_L + P, \end{aligned} \tag{23}$$

where $\sigma = \min\{2\varsigma_1, (2\varsigma_2 - 1), (2\varsigma_2 - 3), \delta_3\Gamma_3, \sigma_b\Gamma_b\}$ and $P = (1/2)\delta_3\|\omega_3^*\|^2 + (1/2)\varepsilon_{3m}^2 + (1/2)\eta_m^2 + (1/2)\sigma_b b^2 + \sum_{i=2}^{3}(1/2)\chi_{im}^2$. Then all of the signals in V_L are uniformly ultimately bounded. This completes the proof. □

5. Simulation

Let altitude increase 152.4 m from the initial value; notice that the signal will pass the following filter to make sure the reference signal is smooth enough for tracking:

$$\frac{\Delta_r}{\Delta} = \frac{\omega_1\omega_2^2}{(s+\omega_1)\left(s^2 + 2\xi\omega_2 s + \omega_2^2\right)}, \tag{24}$$

where $\omega_1 = 0.8$, $\omega_2 = 0.5$, and $\xi = 0.1$. The FPA command signal is obtained as

$$x_{1d} = \arcsin\left(\frac{-k_h\widetilde{h} - k_i\int\widetilde{h}\,dt + \dot{h}_r}{V}\right), \tag{25}$$

where $k_h = 0.5$, $k_i = 0.05$, $\widetilde{h} = h - h_r$ is the altitude tracking error, and h_r is the altitude reference signal.

The parameters in the controller are set as $\varsigma_1 = 2$, $\varsigma_2 = 2$, $\varsigma_3 = 3$, $k_{pv} = 5$, $k_{dv} = 0.01$, and $k_{iv} = 0.01$. The parameters of first-order differentiator and adaptive laws are chosen as $l_{20} = 120$, $l_{30} = 100$, $l_{21} = 0.9$, $l_{31} = 0.05$, $\Gamma_3 = 0.5$, $\delta_3 = 0.002$,

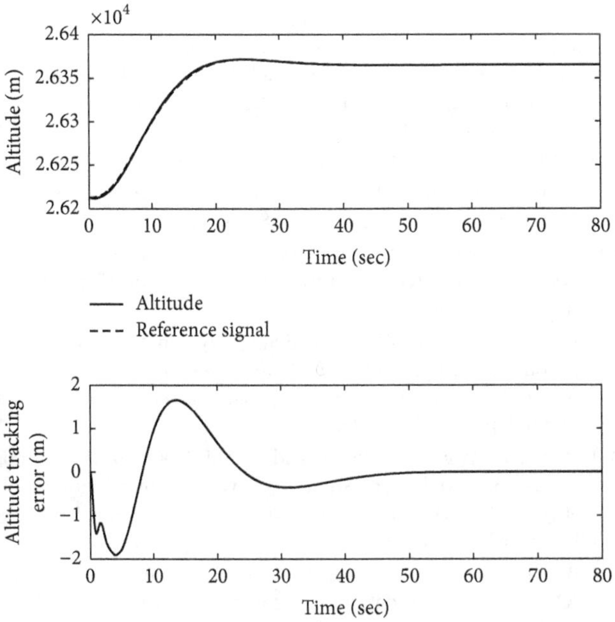

FIGURE 1: Altitude.

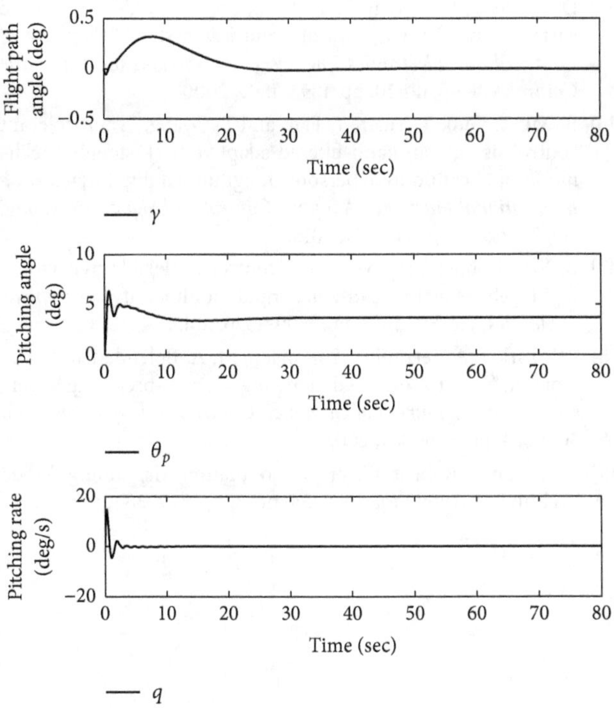

FIGURE 2: FPA, pitching angle, and pitching rate.

$\Gamma_b = 0.113$, and $\sigma_b = 0.005$. Take the initial states as $h_0 = 26212$ m, $V_0 = 2743$ m/s, $\gamma_0 = 0$ deg, $\alpha_0 = 0$ deg, and $q_0 = 0$ deg/s. The fault parameter is set as $b = 0.05$. The results are as follows.

Figure 1 indicates that the designed controller could obtain good performance on trajectory tracking. Figure 2 shows that the all state variables are bounded. The response of designed adaptive signal $\hat{b}$ and NN weights norm are shown

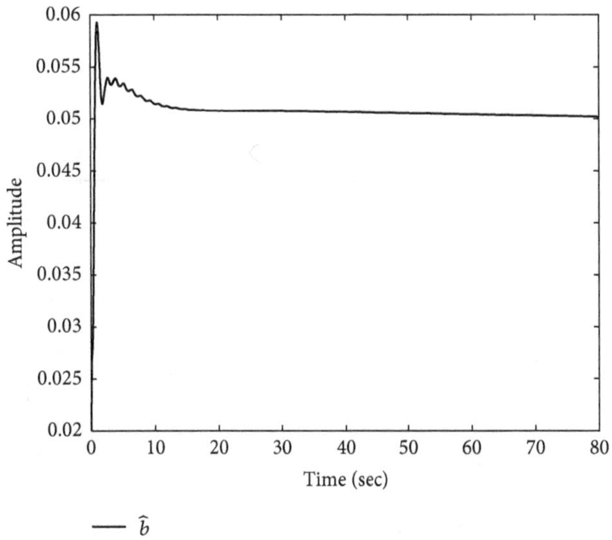

FIGURE 3: Response of $\hat{b}$.

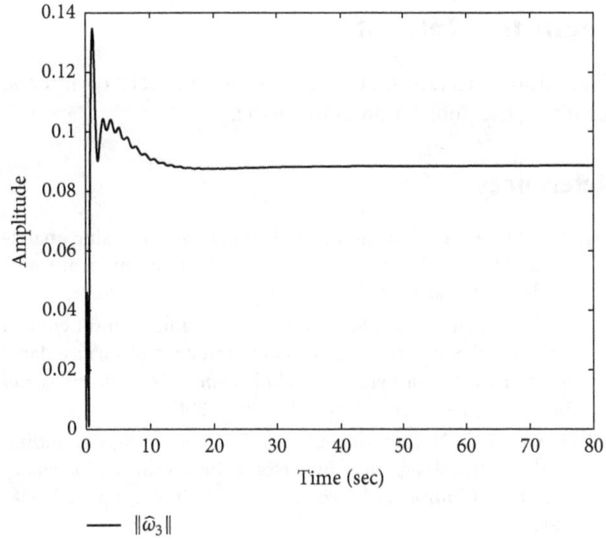

FIGURE 4: NN weights norm.

in Figures 3 and 4. The designed control input and actual control input are shown in Figure 5, respectively. The results verify that the adaptive compensation design could eliminate the influence of actuator fault.

6. Conclusion

This paper proposes an adaptive back-stepping control law with NN learning for HFV control. The influence of actuator fault is eliminated by constructing an adaptive compensation signal. Meanwhile, the unknown nonlinearity is estimated by neural networks. The simulation results clearly present the consequence of the above design and verify that the approach could reach the desired tracking performance when actuator fault and model uncertainty exist.

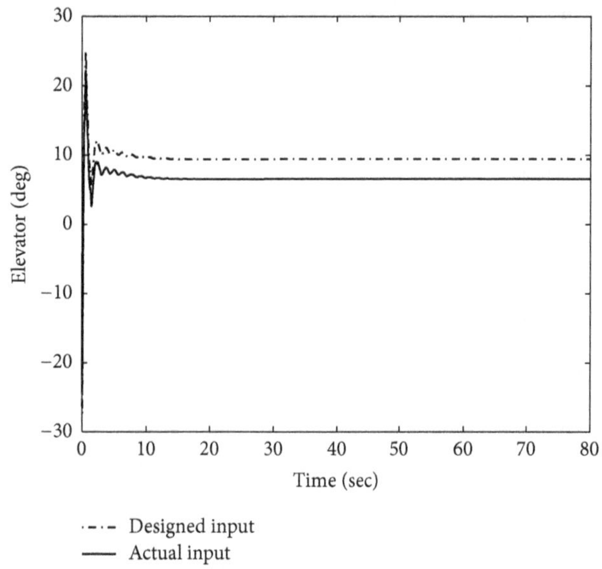

Figure 5: Control input.

Conflicts of Interest

The authors declare that there are no conflicts of interest regarding the publication of this paper.

References

[1] P. M. Frank, "Fault diagnosis in dynamic systems using analytical and knowledge-based redundancy: a survey and some new results," *Automatica*, vol. 26, no. 3, pp. 459–474, 1990.

[2] D. Xu, B. Jiang, and P. Shi, "Robust NSV fault-tolerant control system design against actuator faults and control surface damage under actuator dynamics," *IEEE Transactions on Industrial Electronics*, vol. 62, no. 9, pp. 5919–5928, 2015.

[3] H. Xu, M. D. Mirmirani, and P. A. Ioannou, "Adaptive sliding mode control design for a hypersonic flight vehicle," *Journal of Guidance, Control, and Dynamics*, vol. 27, no. 5, pp. 829–838, 2004.

[4] S. Ibrir, W. F. Xie, and C.-Y. Su, "Adaptive tracking of nonlinear systems with non-symmetric dead-zone input," *Automatica*, vol. 43, no. 3, pp. 522–530, 2007.

[5] Q. Shen, B. Jiang, and V. Cocquempot, "Fault diagnosis and estimation for near-space hypersonic vehicle with sensor faults," *Proceedings of the Institution of Mechanical Engineers, Part I: Journal of Systems and Control Engineering*, vol. 226, no. 3, pp. 302–313, 2012.

[6] Y. Li, S. Tong, Y. Liu, and T. Li, "Adaptive fuzzy robust output feedback control of nonlinear systems with unknown dead zones based on a small-gain approach," *IEEE Transactions on Fuzzy Systems*, vol. 22, no. 1, pp. 164–176, 2014.

[7] D. Zhao, Y. Wang, Y. Li, and S. X. Ding, "H∞ fault estimation for 2-D linear discrete time-varying systems based on krein space method," *IEEE Transactions on Systems, Man, and Cybernetics Systems*, vol. PP, no. 99, Article ID 2723623, pp. 1–10, 2017.

[8] D. Zhao, D. Shen, and Y. Wang, "Fault diagnosis and compensation for two-dimensional discrete time systems with sensor faults and time-varying delays," *International Journal of Robust and Nonlinear Control*, vol. 27, no. 16, pp. 3296–3320, 2017.

[9] Y. Wang, D. Zhao, Y. Li, and S. . Ding, "Unbiased minimum variance fault and state estimation for linear discrete time-varying two-dimensional systems," *Institute of Electrical and Electronics Engineers Transactions on Automatic Control*, vol. 62, no. 10, pp. 5463–5469, 2017.

[10] S. Zhang, C. Li, and J. Zhu, "Composite dynamic surface control of hypersonic flight dynamics using neural networks," *Science China Information Sciences*, vol. 58, no. 7, 2015.

[11] J.-L. Wu, "Stabilizing controllers design for switched nonlinear systems in strict-feedback form," *Automatica*, vol. 45, no. 4, pp. 1092–1096, 2009.

[12] B. Xu, X. Huang, D. Wang, and F. Sun, "Dynamic surface control of constrained hypersonic flight models with parameter estimation and actuator compensation," *Asian Journal of Control*, vol. 16, no. 1, pp. 162–174, 2014.

[13] B. Xu, D. Wang, Y. Zhang, and Z. Shi, "DOB based neural control of flexible hypersonic flight vehicle considering wind effects," *IEEE Transactions on Industrial Electronics*, vol. 64, no. 11, pp. 8676–8685, 2017.

[14] B. Xu, D. Yang, Z. Shi, Y. Pan, B. Chen, and F. Sun, "Online recorded data-based composite neural control of strict-feedback systems with application to hypersonic flight dynamics," *IEEE Transactions on Neural Networks and Learning Systems*, vol. PP, no. 99, Article ID 2743784, pp. 1–11, 2017.

[15] D. Swaroop, J. K. Hedrick, P. P. Yip, and J. . Gerdes, "Dynamic surface control for a class of nonlinear systems," *Institute of Electrical and Electronics Engineers Transactions on Automatic Control*, vol. 45, no. 10, pp. 1893–1899, 2000.

[16] B. Xu, Y. Guo, Y. Yuan, Y. Fan, and D. Wang, "Fault-tolerant control using command-filtered adaptive back-stepping technique: application to hypersonic longitudinal flight dynamics," *International Journal of Adaptive Control and Signal Processing*, vol. 30, no. 4, pp. 553–577, 2016.

[17] B. Xu, "Robust adaptive neural control of flexible hypersonic flight vehicle with dead-zone input nonlinearity," *Nonlinear Dynamics*, vol. 80, no. 3, pp. 1509–1520, 2015.

[18] J. T. Parker, A. Serrani, S. Yurkovich, M. A. Bolender, and D. B. Doman, "Control-oriented modeling of an air-breathing hypersonic vehicle," *Journal of Guidance, Control, and Dynamics*, vol. 30, no. 3, pp. 856–869, 2007.

[19] A. Levant, "Robust Exact Differentiation via Sliding Mode Technique," *Automatica*, vol. 34, no. 3, pp. 379–384, 1998.

15

PID Control for Electric Vehicles Subject to Control and Speed Signal Constraints

Amanda Danielle O. da S. Dantas,[1]
André Felipe O. de A. Dantas,[2] **João Tiago L. S. Campos**,[2]
Domingos L. de Almeida Neto,[2] **and Carlos Eduardo T. Dórea**[1]

[1]*Department of Automation and Computing, CT, Federal University of Rio Grande do Norte, 59078-970 Natal, RN, Brazil*
[2]*Master of Engineering in Oil and Gas, Potiguar University, 59054-180 Natal, Brazil*

Correspondence should be addressed to André Felipe O. de A. Dantas; andre.dantas@unp.br

Academic Editor: Darong Huang

A PID control for electric vehicles subject to input armature voltage and angular velocity signal constraints is proposed. A PID controller for a vehicle DC motor with a separately excited field winding considering the field current constant was tuned using controlled invariant set and multiparametric programming concepts to consider the physical motor constraints as angular velocity and input armature voltage. Additionally, the integral of the error, derivative of the error constraints, and λ were considered in the proposed algorithm as tuning parameters to analyze the DC motor dynamic behaviors. The results showed that the proposed algorithm can be used to generate control actions taking into account the armature voltage and angular velocity limits. Also, results demonstrate that a controller subject to constraints can improve the electric vehicle DC motor dynamic; and at the same time it protects the motor from overvoltage.

1. Introduction

Some researchers state that electric vehicles can be one of the renewable solutions to energy and environmental problems caused by oil based vehicles due to the various advantages associated with the use of electric energy, such as low cost [1–5]. In this scenario, direct current (DC) motors are one of the most used actuators in the construction of electric vehicles [6]. This type of actuator has numerous advantages, such as low cost, high reliability, easy maintenance, and simple control for both speed and position variables with PID being one of the main used controllers [7, 8].

The Proportional Integral Derivative Controller (PID) has been widely used for most industrial process, due to its simplicity and effectiveness in control [9, 10]. This type of controller is commonly used in level, flow, temperature, and vehicular systems, as well as electric motors [10–12]. In addition, the design of the PID controller is considered easy to implement, since it is only necessary to tune three parameters K_p, K_i, and K_d and tuning methods can be performed automatically [13]. Some of most used PID tuning methods in control engineering literature are Ziegler and Nichols, Cohen and Conn, Relay method, and Relatus Apparatus. These methods are effective and achieve excellent results when controlling unconstrained monovariable systems although some of these ones are also applicable for multivariate systems [9].

Despite all advantages of PID controllers, most of tuning methods do not consider the process constraints. Thus, many researches tried to consider these conditions in the control loop using antireset windup, control signal saturation, and integrator constraints. These techniques aim to limit the control action to suit the controller to constrained processes [14, 15]. However, these methods still do not take into account the constraints while tuning the controller and, therefore, such methods are not totally appropriate; i.e., they do not lead to an optimal control signal for the constrained system.

In order to solve the optimal constrained problem many controllers are being proposed. One solution consists of maintaining the system trajectory within λ-contractive controlled invariant polyhedron set defined in the state space. This set contains all states for which there is a state feedback control law that maintains the trajectory of the dynamic system within Ω [16, 17]. The state feedback control law can be calculated online, from the solution of a linear programming (LP) problem, or offline, by solving a multiparametric linear programming problem (mp-LP) [17]. This optimal solution represents an explicit PWA (PieceWise Affine) state feedback control law defined under a set of polyhedral regions in state space [14]. In a complementary way to feedback control recent research has shown that there is a state space form that allows the tuning of PID controllers using the Linear Quadratic Regulator (LQR) [13, 18], and this may allow us to combine both strategies making a new tuning method that considers constraints.

Within this context, in this paper a design of a new type of gain-scheduling PID controller to control angular velocity of electric vehicle DC motors subject to constraints in angular velocity and input voltage and PID states is proposed. To this end, the formulations in the state space of the PID controller are used, as well as the concept of controlled invariant sets together with the solution of a multiparametric programming problem [6, 9, 19, 20]. In this case, we use the same techniques applied to obtain explicit controllers (which take into account system constraints) to tune similar PID controllers (mp_PID) to constrained systems.

This work is organized as follows: At first, we will approach the concept of the λ-contractive controlled invariant set. In sequence, the problem of linear multiparametric programming will be described. After that, we will introduce how to tune PID controllers from multiparametric linear programming technique. An overview of electrical vehicle DC motors will be discussed later. Finally, a set of simulations will be carried out with the objective of proving the functionality of the proposed algorithm and the concept of mp_PID in the control of electrical vehicle DC motors, i.e., specified to work with electric cars.

2. Controlled Invariant Sets

The concept of controlled invariant sets has become important in the design of controllers for linear discrete-time systems subject to constraints since it represents a fundamental condition to maintain system stability ensuring that the constraints are not violated [21].

Consider the linear time-invariant discrete-time system described by

$$x(k+1) = Ax(k) + Bu(k), \tag{1}$$

where $k \in \mathbb{N}$ is the sample time, $x \in \mathbb{R}^n$ is the state of the system with $x \in \Omega = \{x : Gx \leq \rho\}$ (where $G \in \mathbb{R}^{g \times n}$ and $\rho \in \mathbb{R}^g$), and $u(k) \in \mathbb{R}^m$ is the control input subject to the constraints $u(k) \in \mathfrak{U} = \{u : Vu \leq \varphi\}$.

A nonempty closed set $\Omega \subset \mathbb{R}^n$ is controlled invariant with respect to the system described in (1), if there exists a control signal u such that $x(k+1)$ remains inside it for every $x(k)$ belonging to the closed set. Moreover, if a given contraction rate $0 < \lambda < 1$ is considered, a set $\Omega \subset \mathbb{R}^n$ is said to be λ-contractive controlled invariant set with respect to system (1) if there exists a control signal u such that $x(k+1)$ belongs to the set $\lambda\Omega$, for every $x(k)$ belonging to the closed set [16, 21]. In general, the set of constraints Ω defined in state space is not a controlled invariant set; i.e., there is not necessarily a control law ($u(k) \in \mathfrak{U}$) which maintains the trajectory of the state vector completely contained in the set of constraints. However, it is possible to compute a controlled invariant set Ω_c, to be as large as possible, contained within the set of constraints Ω [22]. Therefore, before starting the controller synthesis process, it is necessary to define a controlled invariant set and then to compute a suitable control law that is able to restrict the state vector to a controlled invariant set $\forall x \in \Omega_c$.

By defining the maximal contractive controlled invariant set ($\Omega_c = \{G_c x \leq \rho_c\}$) [16, 22], a state feedback control law ($u(k) \in \mathfrak{U}$), capable of maintaining the system dynamics (1), contained in Ω_c, can be computed online by solving the linear programming problem (LP) as described in [16] or offline from the solution of the following multiparametric programming problem (mp-LP) [23]:

$$\begin{gathered} \min_{v} \quad c^T v + d^T x, \\ \textit{subject to:} \quad Dv \leq W + Ex, \end{gathered} \tag{2}$$

where

$$\begin{aligned} c^T &= \begin{bmatrix} 0 & 1 \end{bmatrix}, \\ D &= \begin{bmatrix} G_c B & -\rho_c \\ V & 0 \end{bmatrix}, \\ v &= \begin{bmatrix} u \\ \varepsilon \end{bmatrix}, \\ W &= \begin{bmatrix} 0 \\ \varphi \end{bmatrix}, \\ E &= \begin{bmatrix} GA \\ 0 \end{bmatrix}, \end{aligned} \tag{3}$$

where $0 < \varepsilon \leq \lambda$ is the contraction rate to be minimized at each time step, $u(k)$ is the control action to be computed, and $x(k)$ is set of states contained inside Ω_c. The expression $G_c(Ax(k) + Bu(k)) \leq \varepsilon\rho$ represents a convex polyhedron in the space $\mathbb{R}^{n+m}$, and $\mathfrak{U} = \{u : Vu \leq \varphi\}$ is a convex polytope that represents the constraints in the control variable.

In the design of controllers under constraints, the solution of the mp-LP (problem (2)) results in a PWA state feedback control law over the polyhedral regions in the space of parameters $x(k)$ as follows [24]:

(1) The set Ω_c (controlled invariant polyhedral) is partitioned into N_q different polyhedral regions:

$$R_j = \{x \in \mathbb{R}^n \mid P_j x \leq b_j\} \quad j = 1, \cdots, N_q, \tag{4}$$

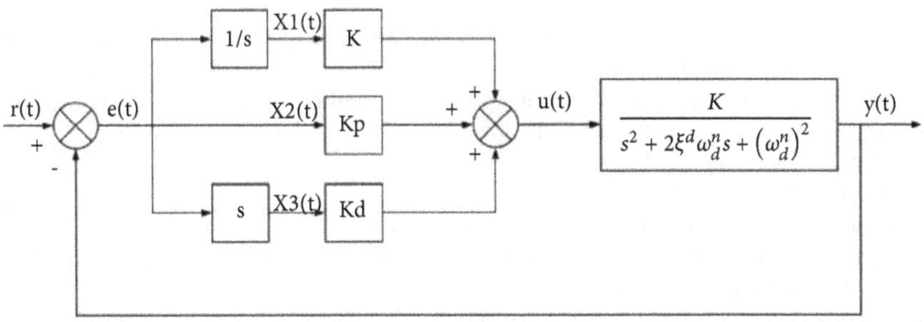

FIGURE 1: System controlled by a PID.

(2) The optimal solution $u^*(x(k)) : \Omega_c \longrightarrow \mathbb{R}^m$ is a PWA function over R_j:

$$u^*(x(k)) = F^j x(k) + g^j \quad for\ x(k) \in R_j. \tag{5}$$

As the system is in the state space form it is possible to find the largest λ-contractive invariant set and, in sequence, the parameters of the control law are computed which maintain the dynamics of the states within the λ-contractive invariant set. In order to associate with the PID controller, we will call this "the tuning step", because we find the controller's parameters that guarantee positive invariance and λ-contractivity. That is, by using this process we will be able to find a PID control law, PWA, that allows the controller to synthesize control actions capable of controlling the process under constraints.

2.1. Tuning of Gain-Scheduling PID Control Design (mp_PID). Based on formulation that allows the reorganization of a second-order systems in state space form, described in [20], whose states are the tracking error, integral of the error, and derivative of the error, we propose the tuning of a type of gain-scheduling PID controller by using the PWA state feedback control law computed by multiparametric linear programming, described in (2).

2.1.1. PID Controllers. Consider now the system presented in (1) is described by

$$\frac{Y(s)}{U(s)} = \frac{K}{s^2 + 2\xi^{ol}\omega_n^{ol} s + (\omega_n^{ol})^2}. \tag{6}$$

Because the external setpoint does not affect the controller design, we assume $r(t) = 0$ for the system of Figure 1 [20].

The relation $y(t) = -e(t)$ is valid for the standard regulation problem. Thus, it is possible to place the system in function of the error and control signal, as presented in [20]

$$\left[s^2 + 2\xi^{ol}\omega_n^{ol} s + (\omega_n^{ol})^2\right] E(s) = -KU(s), \tag{7}$$

$$\Longrightarrow \ddot{e} + 2\xi^{ol}\omega_n^{ol}\dot{e} + (\omega_n^{ol})^2 e = -Ku. \tag{8}$$

Use the following definitions:

$$x_1 = \int_0^t e(t)\,dt,$$
$$x_2 = e(t), \tag{9}$$
$$x_3 = \frac{de(t)}{dt}.$$

Equation (8) can be rewritten as

$$\ddot{x}_3 + 2\xi^{ol}\omega_n^{ol}\dot{x}_3 + (\omega_n^{ol})^2 x_2 = -Ku. \tag{10}$$

Considering the state space, the formulation becomes [20]

$$\begin{bmatrix}\dot{x}_1\\ \dot{x}_2\\ \dot{x}_3\end{bmatrix} = \begin{bmatrix}0 & 1 & 0\\ 0 & 0 & 1\\ 0 & -(\omega_n^{ol})^2 & -2\xi^{ol}\omega_n^{ol}\end{bmatrix}\begin{bmatrix}x_1\\ x_2\\ x_3\end{bmatrix} + \begin{bmatrix}0\\ 0\\ -K\end{bmatrix}u. \tag{11}$$

In this case, we assume the reference signal equal to zero and because the system is regulatory it is organized in a way that the states tend to zero and tend to eliminate the disturbance. When applied in electrical vehicle motor control, we intend to make changes in the reference, so some considerations must be realized:

(i) The external reference does not affect the controller design.

(ii) It is possible to work step-type references in two ways, by using model illustrated in Figure 1 or by forcing changes in the operational point, so the new system's reference is forced to be at the origin likewise linearizing a nonlinear system in the operational point.

Concerning the direct change of reference, it is possible to verify that as the system is stable in closed loop, it tends to converge to the reference. However, in this case, only the error and derivative of the error states will converge to zero and the integral of the error will become a value that maintains the necessary control action to force the output to zero, in the same way conventional PID does.

The second way to control the system is using the linearization idea and it can be observed in Figure 2.

1: Convert the continuous system into a state space system, where the states are the errors and the input is the control action, according to equation (11);
2: Compute the maximal $\lambda-$ contractive controlled invariant set contained in the set of constraints, which must be reexplained in terms of the new state representation;
3: Solve the mp-LP problem described in equation (2);
4: Compute the integral of error, error, and derivative of error;
5: Identify which polyhedral region the computed state $x(k)$ belongs to.
6: Use the affine control law to control the system with $F^j = [Ki_j\ Kp_j\ Kd_j]^T$ and g^j being the mp_PID parameters, corresponding to the j-th polyhedral region;
7: If the control routine is not interrupted, return to step 4.

ALGORITHM 1

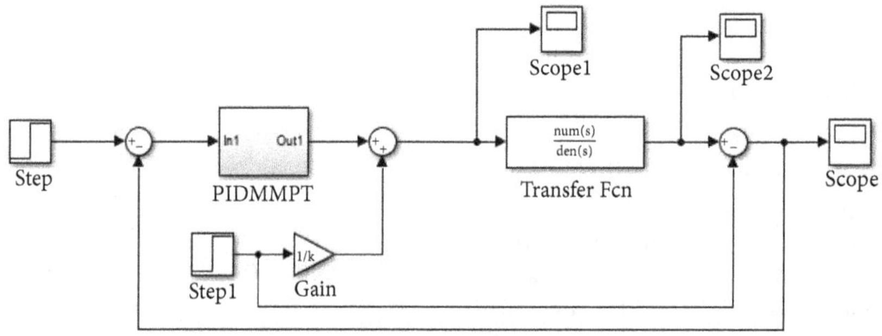

FIGURE 2: System controlled by a PID.

Figure 2 considers the reference equal to zero (step) and, to make changes in reference, we are changing the operational point using step 1. The concept is very simple; by changing the reference (step 1) in this system, the operational point will change. Thus, the zero will be shifted; i.e., the entered reference will be recognized as an instantaneous error disturbance on the system's states. The system was designed to force the error, integral of the error, and the derivative of the error to the origin, so the system will work to stabilize states in this "new zero", which is the operational point. In this way, the system's output will be forced toward the reference using the same considerations as depicted in Figure 2. In this scheme when there is a reference change the invariant set moves along with it; i.e., if the error limit is equal to 1, for example, and the reference is 5, the output will be limited between the values 6 and 4. However, if the reference is changed to 4, the output will be limited to the interval between 3 and 5. This is particularly useful when it is desirable to constrain the error around a variable reference. However, it is needed to be careful not to vary the reference above the limits of the invariant set and not to exceed the physical limitations of the system. When the operator wants to limit the output between 4 and 6 and use the operating points 4 and 5, because of this, the constraints on the magnitude of the error would be within ±1 in the first example and in the second one between 2 and 0 ($error = -y$). This means that, for these examples, it would be necessary to modify the system's constraints by making a set of constraints for each operational point.

Given the main considerations about the state space system, it is necessary to tune and use the controller. In this case the concept of invariant sets will be used to find an optimum tuning for a PWA PID control law, which we call mp_PID. Thus, the λ-contractive controlled invariant set is computed by using the algorithms proposed by [22]; then, mp-LP problem (2) is solved to compute the parameters F^j and g^j of the affine law $u(x) = F^j x + g^j$, where F_i multiplies the state $X = [x_1\ x_2\ x_3]$ corresponding to integral of the error, the error, and the derivative of the error. This means that the parameters of the proposed PID controller are defined by $F^j = [Ki_j\ Kp_j\ Kd_j]$ plus affine term g^j which is associated with a polyhedral region R_j, forming a gain-scheduling PID control (mp_PID). Such approach can be obtained from the implementation of Algorithm 1.

As described in Algorithm 1, the proposed PID control design is performed from a sequence of steps, where at first the system must be rearranged in state space form, so that the input is the control action and the states are the integral of the error, error, and derivative of the error, according to (11). Then, the new system is discretized with the desired sampling period and, then given the new state space extended system, the maximal $\lambda-$contractive controlled invariant set is computed as well as the multiparametric linear programming problem (mp-LP) (2). The solution of the mp-LP problem represents a PWA control law over polyhedral regions R_j associated with a PID controller. At the end of these steps, the control law is inserted into the control loop, where the error tends to zero and the constraints are not violated as long as the

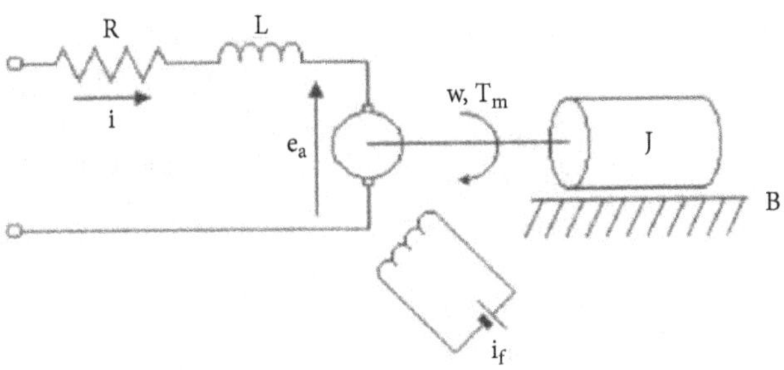

FIGURE 3: DC motor with separated winding field with constant current ($i_f(t)$) [25].

initial state is inside the λ−contractive controlled invariant set.

3. DC Motors Overview

After the definition of the PID tuning method the application must be studied in order to have mp_PID applied to it, so in this section we present an overview of electric vehicle DC motors.

Recently, electric vehicles are gaining popularity among the population resulting in more demand for these types of car. Electric vehicles are efficient and need less maintenance than fuel-based cars and they do not pollute the environment [15]. However, the electric car depends heavily on battery systems that are finite and have fewer storage capabilities. Therefore, strategies to reduce and use more efficiently the energy stored in the batteries are needed.

The consuming of the electric vehicles depends heavily on the used motor type and adopted control strategy. Actually, DC motors and induction motors are being proposed to be used in the electric vehicle industry, with DC motor type being a good candidate to be used in electric vehicle applications [26, 27]. DC motors are efficient, presenting high reliability and easy maintenance. They are easy to control resulting in smooth acceleration and efficient battery usage [7, 8]. Although brushed DC motor suffers from the brush maintenance, the aforementioned advantages turn the DC motor to be applicable for use in electric cars.

DC motors have the advantage of being easy to control resulting in several control techniques for velocity control using PID type control. Techniques using metaheuristic PID [28, 29], multivariate PID [30], genetic algorithm method with PID [12], and Adaptive PID Dynamic, Fuzzy, and Neurofuzzy Controller [19] were proposed to control DC motors.

DC motors have different configurations as compound, shunt, series, permanent magnetic DC and separated field winding where each of them has advantages and disadvantages. DC motors in series configuration are the type of motors that can be used in electric cars due to their instantaneous torque and smooth acceleration. However, these types of motors need a minimal load not to be damaged, which is a condition that occurs in electric car applications. One way to limit the damage in the DC motor series control is to limit the maximum velocity in the adopted control strategy [31]. DC motors with separated field are the most versatile because they allow controlling torque and velocity separately using the armature current and field current separately with more options for control applications. In this paper, DC motor with separated field was chosen to test the control algorithm because it allows more restrictions options to be tested by the proposed control algorithm.

In sequence, we present the DC motor modelling with excited separately field and afterwards we present the numerical examples controlling the motor and testing the performance of the tuned controllers.

A DC motor controlled by current armature with independent field is depicted in Figure 3.

The torque induced (T_m) in the motor is given by

$$T_m(t) = J\frac{dw(t)}{dt} + Bw(t), \tag{12}$$

$$T_m(t) = K_a i(t), \tag{13}$$

where $i(t)$ is the armature current, J is the inertia moment, w is the angular velocity, B is the friction coefficient, and K_a is the torque constant.

The armature induced voltage ($e_a(t)$) with constant field current (i_f) is given by

$$e_a(t) = K_b w(t), \tag{14}$$

where K_b is the velocity constant.

As the DC motor has separated field, the voltage applied ($v(t)$) in the armature windings is given by

$$v(t) = L\frac{di(t)}{dt} + Ri(t) + e_a(t), \tag{15}$$

where L is the armature inductance, and R is the armature resistance.

The following second-order transfer function resulted:

$$\frac{W(s)}{V(s)} = \frac{K_a}{(Js+B)(Ls+R) + K_a K_b}. \tag{16}$$

4. Numerical Examples

(i) The parameters of a motor suitable for use in electric cars are determined.

TABLE 1: Motor parameters.

V_n	48 V
R	0.1 Ω
L	0.005 H
K_b	0.004 $V.s/rad$
K_a	0.0036 $N.m/A$
J	0.1 $N.m^2$
B	0.05 $N.m.s/rad$

(ii) The mp_PID controllers are tuned.

(iii) The proposed tuning algorithm is performed using different parameters and, then, the results are compared.

(iv) The influence of each parameter on the final process performance is analyzed.

4.1. DC Motor Parameters. At first, in order to control an electric vehicle DC motor we specify its parameters (the motor) in such a way that it is able to move the vehicle. The voltage armature, resistance and inductance armatures, moment of inertia, velocity constant, torque constant, and friction coefficient motor parameters presented in Table 1 were used to test the proposed control algorithm performance.

Replacing the motor parameters in (17),

$$\frac{W(s)}{V(s)} = \frac{0.0036}{(s+0.05)(0.005s+0.1)+1.44e^{-5}}, \quad (17)$$

then, the simplified transfer function of the used motor becomes

$$G(s) = \frac{7.2}{s^2+20.5s+10.03}. \quad (18)$$

In the model of (18) the output is the angular velocity (rad/s) in the range of $\pm 34.56 rad/s$ and the input is the voltage that will vary in the range of $\pm 48V$. Note that as this type of motor allows two-direction movement, we will have both positive and negative voltages and positive and negative speeds considered in the controller constraints.

4.2. Tuning of the Controllers mp_PID. This section presents the general tuning process of the mp_PID controllers. Some important aspects to emphasize are described as follows:

(i) The output constraints can be transformed into constraints on the error, especially when assuming the operating point at zero. The output constraints for the electric vehicle DC motor case will be the maximum speed allowed at a voltage of 48V.

(ii) The control signal constraints must be limited to $\pm 48V$, since we need to limit motor damage from overcurrent as the armature current is directly dependent on the applied load and armature voltage, so the algorithm is able to optimize the system within this set of constraints, unlike other techniques which only insert constraint on control action without taking in account the constraints at tuning phase resulting in lower dynamic performance of the vehicle DC motor.

(iii) The integral and derivative of the error values will be appropriately chosen; i.e., they will be used as a tuning parameter and the influence of these constraints will be evaluated in order to emphasize how the final performance of the system is changed, in aspects such as overshooting and stabilization time.

(iv) The main tuning parameter of the mp_PID controller is the value of λ. It is related to the contractivity and will also be an observed parameter.

(v) In a specific case we observe the difference between varying the operating point and varying the reference.

4.3. Set of Tests. Based on these considerations, tests are performed varying the constraints on the integral of the error, derivative of the error, and the λ parameter. After that, a comparison between changing the reference and changing the operating point will be made.

4.3.1. Test 1: Integral of the Error Constraints Change Effect. In this first test the objective is to observe the influence, in system's performance, of integral of the error constraints change. Despite the direct relationship between error and the output of the motor (angular velocity), the integral of the error is an internal state of the system in state space and influences the motor dynamic performance but not necessarily forces physically the system outside its limits. Therefore, by hypothesis we assume that we can freely modify these parameters. In this way, the test conditions presented in Table 2 were chosen. In these test conditions, the integral of the error limits was varied and the derivative of the error limits and λ were maintained constant.

The system's outputs and the control actions for test condition 1 are depicted in Figures 5 and 6. Also, Figure 4 presents the polyhedron formed by the first test case (test 1).

As seen, Figure 4 presents the polyhedron formed by the first test case (test 1), where the default value is 34.56 so $MaxIe = 1$ means 1 times 34.56 and $MaxDe = 5$ means 5 times 34.56 and this concept is applicable to all the others figures. Because this polyhedron changes its shape for each test in all tests we present only the quantity of sets generated within the invariant set that in this case are 10, 8, and 10. The number of regions within the invariant set reveals the computational cost difference (in Algorithm 1 we need to search the region to find which PID controller parameters will be used to compute the control signal). In this case, as observed, the quantity is very similar; for a reference, there may be cases in which the amount of computed regions may be greater than 400, which in this case points out that the control signal computation will be inexpensive.

Figure 5 shows the output, i.e., the angular velocity in the 3 performed tests. We observe that, with lower constraint values in the integral of the error signal, the system tends to converge faster; on the other hand, with more relaxed restriction in the integral of the error signal, the system tends to stabilize slower. Another important factor to be analyzed is

TABLE 2: Table with the 3 tests, varying integral of error.

Simulations	λ	Integral of the Error Limits	Derivative of the Error Limits
1	0.999	±34.56	±172.80
2	0.999	±172.80	±172.80
3	0.999	±1728	±172.80

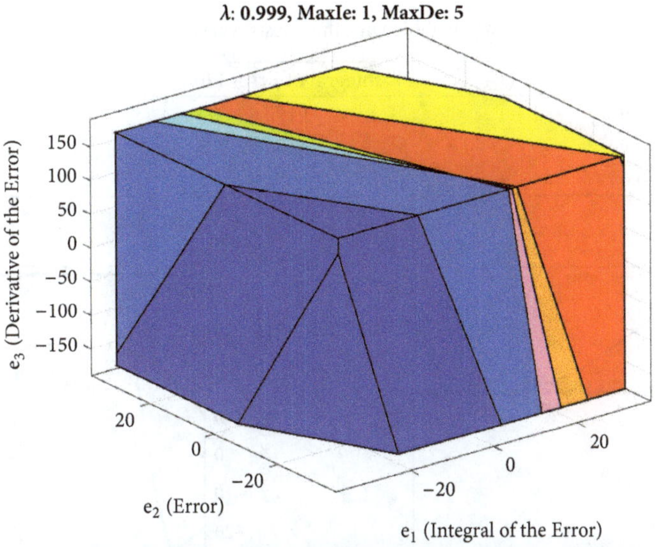

FIGURE 4: Polyhedron test 1 for test condition 1.

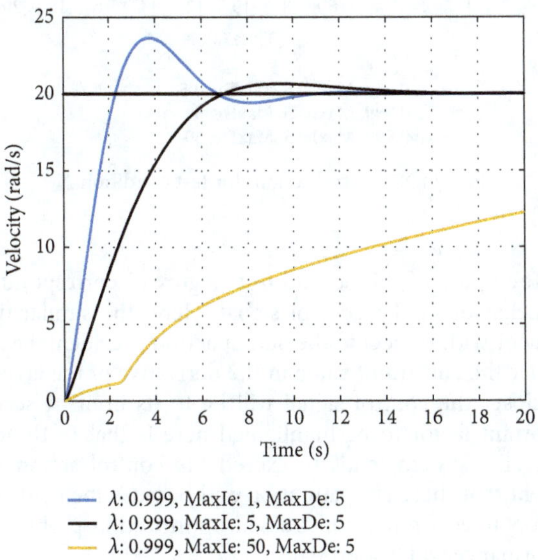

FIGURE 5: System's outputs for test condition 1.

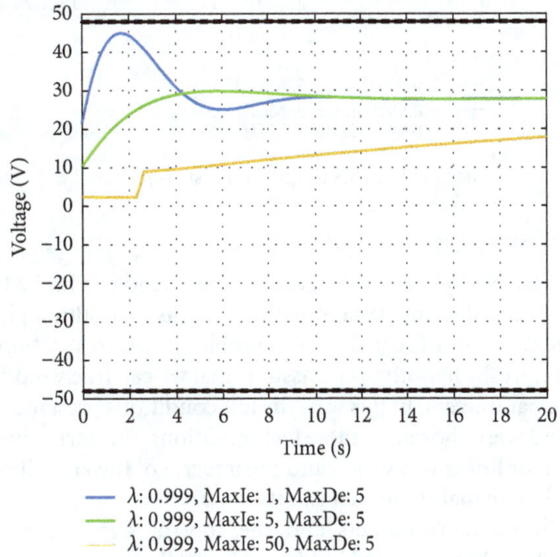

FIGURE 6: Control actions for test condition 1.

that the lower the restriction value in the integral of the error is, the greater the overshooting system tends to present.

Figure 6 corresponds to the behavior of the control signal. Here, the similarity in behavior with respect to the output signal is evident; i.e., the smaller the restriction value in the integral of the error is, the closer to the limits the signal will be, which in this case is ±48.

4.3.2. Test 2: Derivative of the Error Constraints Parameters Change Effect. Similarly to the first test, in this second test our objective is to observe the influence in the performance of the system, when the derivative of the error constraints is changed. This also happened to the integral of the error constraint, the derivative of the error is an internal state of the system in state space and influences the performance but not

Table 3: Table with the 3 tests, varying derivative of error.

Simulations	λ	Integral of the Error Limits	Derivative of the Error Limits
1	0.999	±172.80	±17.28
2	0.999	±172.80	±172.80
3	0.999	±172.80	±1728.0

Table 4: Table with the 3 tests, varying λ.

Simulations	λ	Integral of Error Limits	Derivative of Error Limits
1	0.99	±172.80	±172.80
2	0.90	±172.80	±172.80
3	0.60	±172.80	±172.80

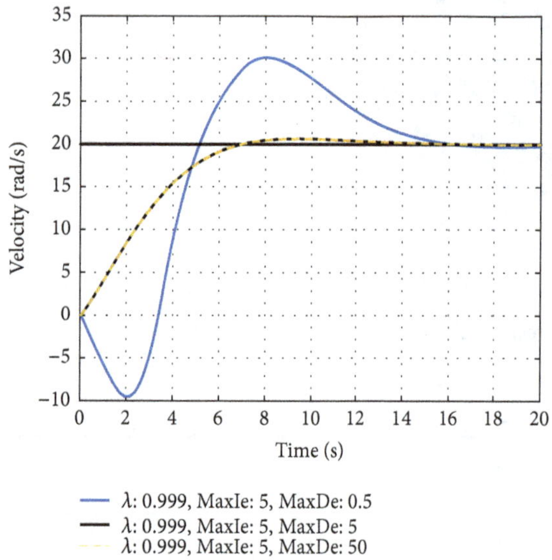

Figure 7: System's outputs for test condition 2.

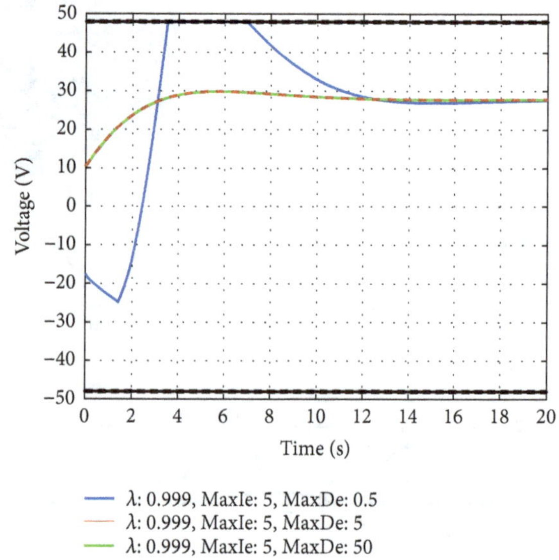

Figure 8: Control actions for test condition 2.

necessarily physically forces the system outside its limits (we are not considering variation effects in this case, although in some cases they are included in problems with restrictions). Therefore, by hypothesis we assume that we can freely modify these parameters. In this way, the test conditions presented in Table 3 were chosen. In these test conditions, the derivative of the error limits was varied and the integral of the error limits and λ were maintained constant.

The system's outputs and the control signals for test condition 2 are depicted in Figures 7 and 8.

Figure 5 shows the output, i.e., the angular velocity in the 3 performed tests. We observe that as the derivative of the error limits increases the system begins to present nonminimum phase and it is more difficult to stabilize it in smaller times; this means that in the case in which the limits are relaxed the integral of the error values can be larger and consequently the system tends to have a better transient.

In these tests the quantity of sets generated within the controlled invariant set is 8, 8, and 12. Similarly to the previous test we emphasize that it will be very inexpensive to compute the control action.

The figure of this second test (Figure 6) corresponds to the behavior of the control signal. Here, the similarity in behavior with respect to the output action is evident; i.e., the smaller the constraint value in the derivative of the error is, the closer the control signal will be to its limit. A second important factor to be highlighted here is that in the case where the system tends to exceed the control action it is evident that there is a saturation and because the controller was obtained from a constrained optimization problem its performance will be optimal.

4.3.3. Test 3: Effect of Changing λ. Unlike the first 2 tests, this one will not present the effect in constraints, but in the tuning parameter (λ) of the optimization problem. What should be highlighted here is that this parameter is associated with the contraction rate, which makes the invariant set smaller at each iteration and it is related to the system's convergence speed. In this way, the test conditions presented in Table 4 were chosen. In these test conditions, λ was varied and the integral and derivative of the error limits were maintained constant.

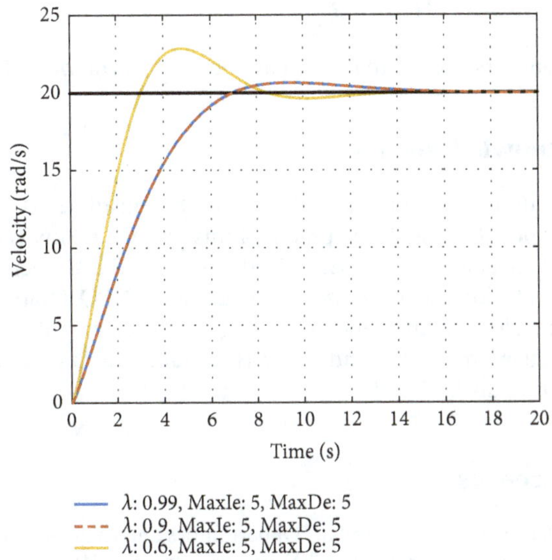

Figure 9: System's output for test condition 3.

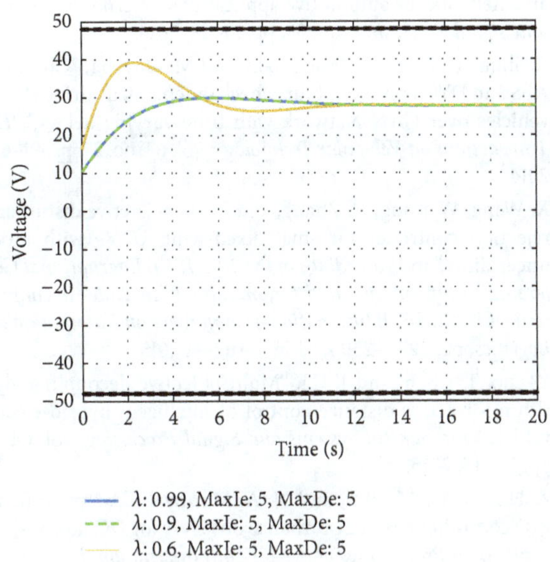

Figure 10: Control actions for test condition 3.

The system's outputs and the control actions for the test condition 3 are depicted in Figures 9 and 10.

This third test resulted as the quantities of sets within the invariant sets 8, 14, and 12. Also, Figure 9 shows the output, i.e., the angular velocity in the 3 performed tests. We observe that the lower the λ value is the more aggressive the system's behavior is, forcing the stabilization to happen before, but reflecting in greater overshooting. This also means that by increasing the value of this parameter we have less aggressive systems, but with slower responses.

The last figure of this third test (Figure 10) corresponds to the behavior of the control signal. Here, the similarity to the output signal is evident; i.e., the smaller the value of λ is the closer to its limits it is.

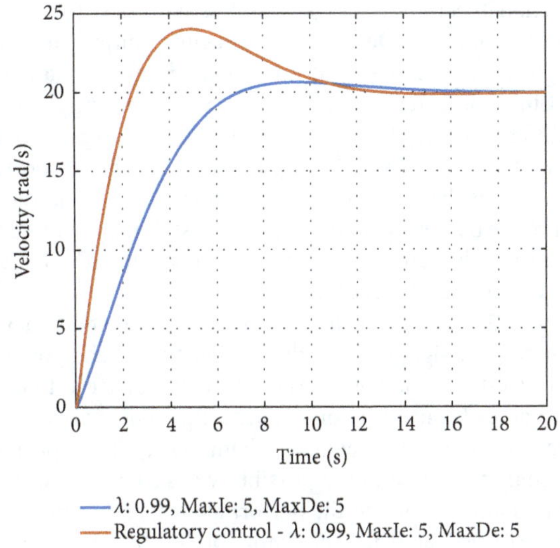

Figure 11: System's output.

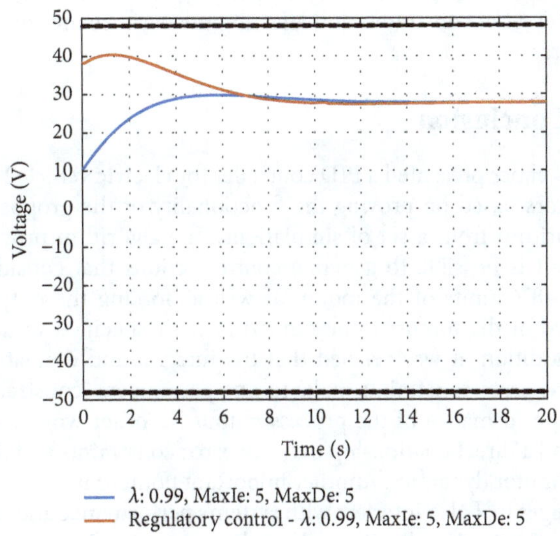

Figure 12: Control action.

4.3.4. Test 4: Reference Changes versus Setpoint Changes. The first three tests aimed to show the results related to the controllers tuning when the systems are subject to reference variations (step). This means that the controlled invariant set remained unchanged as well as the operating point of the system. In this fourth test, the objective is to present the changes related to the operating point. Note that there is a possibility of differences in behaviors, because although it has the same amount of variation, this test directly affects the activated regions in the control process, which consequently changes the response of the system. Thus, the test condition is to change the setpoint by 20 units and compare it with the changed operating point (setpoint 1) by 20.

The system's outputs and the control actions for the test condition 4 are depicted in Figures 11 and 12.

Figure 11 shows the output, i.e., the angular velocity in the 3 performed tests. We observe that the system's response, which had its setpoint changed, was less aggressive (observing the overshooting). Despite the presented output, it is not enough to emphasize that in any and every change of operating point the response will behave accordingly, since this difference is related to the fact that different regions of the invariant set are activated in both cases. Another important fact to note is that, depending on the stabilization criteria, we observe the operating point change case stabilizing before the response to the setpoint change case; this is mainly due to a faster reaction of this response.

Figure 12 of the fourth test corresponds to the behavior of the control signal. Here, the similarity in behavior with respect to the output signal is evident. We observe that the initial value of control action in operating point change case is greater than that of the reference change case; this impacts on the initial response and the possibility of faster stabilization. Note that additionally to the performed tests other possibilities were investigated, i.e., the change of the objective function to minimize the integral of the error and the derivative of the error (as alternatives to changing constraints). However, the results were not satisfactory, impacting directly on the size of the set of constraints and affecting the performance of the process.

5. Conclusion

This paper presented a PID controller for electric vehicle DC motors based on proving the functionality of the proposed algorithm from a set of simulations. The algorithm proved that it is possible to generate control actions that consider the 48V limits of the motor, as well as forcing the output speed of the motors to be limited to their specified values. In addition, it was verified that the integral and derivative of the error constraints make tuning parameters "constrain" the performance of the process output. In other words, we found a direct relationship between error constraints and the DC motor dynamics. Another important point to note is that changes in λ also interfere with system's performance and the output behavior also depends on how the error is changed. In this second case, we verify the behavior of the system to reference and setpoint variations.

In conclusion, control of electric vehicle DC motors under constraints and tuning controllers adjusting the behavior of the system's output is possible by using the proposed algorithm. In this sense, important improvements can be proposed. The first one would be to use a methodology to automatically tune parameters, since there are no tuning rules for these constraints (in the integral of the error and in the derivative of the error). A second improvement would be to extend the tests by verifying that the behavior presented to the DC motors, i.e., verified, can be extended to other processes.

6. Future Works

Although DC motor with excited separately field can be used in an electric vehicle, new types of motors as permanent magnet synchronous motor (PMSM) are being used recently [11, 32, 33]. The usage of the proposed algorithm in this type of motors will be investigated.

Conflicts of Interest

The authors declare that they have no conflicts of interest.

Acknowledgments

The authors would like to mention the institutions that contributed to the development of this project: UFRN (Federal University of Rio Grande do Norte), DCA (Department of Computation and Automation), CNPQ (National Council for Scientific and Technological Development), UnP (Potiguar University), and LAUT (Laboratory of Automation in Petroleum).

References

[1] S. Chu and A. Majumdar, "Opportunities and challenges for a sustainable energy future," *Nature*, vol. 488, no. 7411, pp. 294–303, 2012.

[2] X. Zhu, H. Zhang, D. Cao, and Z. Fang, "Robust control of integrated motor-transmission powertrain system over controller area network for automotive applications," *Mechanical Systems and Signal Processing*, vol. 58, pp. 15–28, 2014.

[3] Z. Shuai, H. Zhang, J. Wang, J. Li, and M. Ouyang, "Combined AFS and DYC control of four-wheel-independent-drive electric vehicles over CAN Network with time-varying delays," *IEEE Transactions on Vehicular Technology*, vol. 63, no. 2, pp. 591–602, 2014.

[4] X. Wang, W. Kong, D. Zhang, and L. Shen, "Active disturbance rejection controller for small fixed-wing UAVs with model uncertainty," in *Proceedings of the 2015 IEEE International Conference on Information and Automation, ICIA 2015 - In conjunction with 2015 IEEE International Conference on Automation and Logistics*, pp. 2299–2304, China, August 2015.

[5] Y. Luo, T. Chen, and K. Li, "Multi-objective decoupling algorithm for active distance control of intelligent hybrid electric vehicle," *Mechanical Systems and Signal Processing*, vol. 64-65, pp. 29–45, 2015.

[6] Z. Has, A. H. Muslim, and N. A. Mardiyah, "Adaptive-fuzzy-PID controller based disturbance observer for DC motor speed control," in *Proceedings of the 2017 4th International Conference on Electrical Engineering, Computer Science and Informatics (EECSI)*, pp. 1–6, Yogyakarta, September 2017.

[7] N. Matsui, "Sensorless PM brushless DC motor drives," *IEEE Transactions on Industrial Electronics*, vol. 43, no. 2, pp. 300–308, 1996.

[8] J. S. Ko, J. H. Lee, S. K. Chung, and M. J. Youn, "A robust digital position control of brushless DC motor with dead beat load torque observer," *IEEE Transactions on Industrial Electronics*, vol. 40, no. 5, pp. 512–520, 1993.

[9] H. Jeongheon and R. Skelton, "An LMI optimization approach for structured linear controllers," in *Proceedings of the 42nd IEEE International Conference on Decision and Control*, pp. 5143–5148, Maui, HI, USA, 2003.

[10] A. Kalangadan, N. Priya, and T. K. S. Kumar, "PI, PID controller design for interval systems using frequency response model matching technique," in *Proceedings of the International Conference on Control, Communication and Computing India, ICCC 2015*, pp. 119–124, India, November 2015.

[11] S. Hu, Z. Liang, W. Zhang, and X. He, "Research on the integration of hybrid energy storage system and dual three-phase PMSM Drive in EV," *IEEE Transactions on Industrial Electronics*, vol. 65, no. 8, pp. 6602–6611, 2018.

[12] M. Jaiswal, M. Phadnis, and H. O. D. Ex, "Speed control of DC motor using genetic algorithm based PID controller," *International Journal of Advanced Research in Computer Science and Software Engineering*, vol. 3, no. 7, pp. 2277-128, 2013.

[13] E. Vinodh Kumar and J. Jerome, "LQR based optimal tuning of PID controller for trajectory tracking of magnetic levitation system," in *Proceedings of the 2013 International Conference on Design and Manufacturing, IConDM 2013*, pp. 254–264, India, July 2013.

[14] M. Kvasnica, P. Grieder, M. Baotić, and M. Morari, "Multi-parametric toolbox (MPT)," *Lecture Notes in Computer Science (including subseries Lecture Notes in Artificial Intelligence and Lecture Notes in Bioinformatics): Preface*, vol. 2993, pp. 448–462, 2004.

[15] S.-M. Lu, "A review of high-efficiency motors: specification, policy, and technology," *Renewable & Sustainable Energy Reviews*, vol. 59, pp. 1–12, 2016.

[16] F. Blanchini, "Ultimate boundedness control for uncertain discrete-time systems via set-induced Lyapunov functions," *Institute of Electrical and Electronics Engineers Transactions on Automatic Control*, vol. 39, no. 2, pp. 428–433, 1994.

[17] A. D. O. d. S. Dantas, C. E. T. Dorea, and A. F. O. d. A. Dantas, "Projeto de controladores para sistemas lineares sujeitos a restricões utilizando análise de agrupamento de dados," *Simpósio Brasileiro de Automação Inteligente - SBAI*, 2015.

[18] S. Das, I. Pan, K. Halder, S. Das, and A. Gupta, "LQR based improved discrete PID controller design via optimum selection of weighting matrices using fractional order integral performance index," *Applied Mathematical Modelling: Simulation and Computation for Engineering and Environmental Systems*, vol. 37, no. 6, pp. 4253–4268, 2013.

[19] A. K. Hassan, M. S. Saraya, M. S. Elksasy, and F. F. Areed, "Brushless DC motor speed control using PID controller, fuzzy controller, and neuro fuzzy controller," *International Journal of Computer Applications*, vol. 180, no. 30, pp. 47–52, 2018.

[20] J.-B. He, Q.-G. Wang, and T.-H. Lee, "PI/PID controller tuning via LQR approach," *Chemical Engineering Science*, vol. 55, no. 13, pp. 2429–2439, 2000.

[21] F. Blanchini and S. Miani, *Set-theoretic methods in control*, Systems & Control: Foundations & Applications, Birkhauser Boston, Inc., Boston, MA, USA, 2008.

[22] C. E. Dórea and J. C. Hennet, "(a, b)-invariant polyhedral sets of linear discrete-time systems," *Journal of Optimization Theory and Applications*, vol. 35, pp. 521–542, 1999.

[23] G. P. Weinkeller, J. L. Salles, and T. F. B. Filho, "Controle preditivo via programação multiparamétrica aplicado no modelo cinemático de uma cadeira de rodas robótica," in *Proceedings of the DINCON 2013—Conferência Brasileira de Dinâmica, Controle e Aplicações*, 2008.

[24] A. Bemporad, F. Borrelli, and M. Morari, "Model predictive control based on linear programming—the explicit solution," *IEEE Transactions on Automatic Control*, vol. 47, no. 12, pp. 1974–1985, 2002.

[25] A. A. R. Coelho and L. S. Coelho, "Identificação de sistemas dinâmicos lineares," *Didáctica (Editora da UFSC)*, 2004.

[26] S. Lixin, "Electric vehicle development: the past, present & future," in *Proceedings of the 2009 3rd International Conference on Power Electronics Systems and Applications, PESA 2009*, China, May 2009.

[27] C. Chan, "The past, present and future of electric vehicle development," in *Proceedings of the IEEE 1999 International Conference on Power Electronics and Drive Systems. PEDS'99 (Cat. No.99TH8475)*, pp. 11–13 vol.1, Hong Kong, July 1999.

[28] D. Puangdownreong, "Optimal PID controller design for DC motor speed control system with tracking and regulating constrained optimization via cuckoo search," *Journal of Electrical Engineering & Technology*, vol. 13, no. 1, pp. 460–467, 2018.

[29] P. Payakkawan, K. Klomkarn, and P. Sooraksa, "Dual-line PID controller based on PSO for speed control of DC motors," in *Proceedings of the 2009 9th International Symposium on Communications and Information Technology, ISCIT 2009*, pp. 134–139, Republic of Korea, September 2009.

[30] B. Gasbaoui and A. Nasri, "A novel multi-drive electric vehicle system control based on multi-input multi-output PID controller," *Serbian Journal of Electrical Engineering*, vol. 9, no. 2, pp. 279–291, 2012.

[31] Z. Bitar, S. Al Jabi, and I. Khamis, "Modeling and simulation of series DC motors in electric car," in *Proceedings of the International Conference on Technologies and Materials for Renewable Energy, Environment and Sustainability, TMREES 2014 - EUMISD*, pp. 460–470, Lebanon, April 2014.

[32] R. Xiao, Z. Wang, H. Zhang, J. Shen, and Z. Chen, "A Novel Adaptive Control of PMSM for Electric Vehicle," in *Proceedings of the 2017 IEEE Vehicle Power and Propulsion Conference (VPPC)*, pp. 1–8, Belfort, France, December 2017.

[33] C. Shi, Y. Tang, and A. Khaligh, "A three-phase integrated onboard charger for plug-in electric vehicles," *IEEE Transactions on Power Electronics*, vol. 33, no. 6, pp. 4716–4725, 2018.

16

Neural Network Predictive Control for Autonomous Underwater Vehicle with Input Delay

Jiemei Zhao

School of Mathematics and Computer Science, Wuhan Polytechnic University, Wuhan 430023, China

Correspondence should be addressed to Jiemei Zhao; jiemeizhao@163.com

Academic Editor: Jongrae Kim

A path tracking controller is designed for an autonomous underwater vehicle (AUV) with input delay based on neural network (NN) predictive control algorithm. To compensate for the time-delay in control system and realize the purpose of path tracking, a predictive control algorithm is proposed. An NN is used to estimate the nonlinear uncertainty of AUV induced by hydrodynamic coefficients and the coupling of the surge, sway, and yaw angular velocity. By Lyapunov theorem, stability analysis is also given. Simulation results show the effectiveness of the proposed control strategy.

1. Introduction

With the rapid development of demands for the resources, countries around the world have attached importance to the exploration and application of the marine resources. Autonomous underwater vehicle (AUV) is a mobile carrier which is small in size and convenient in controllability as the special equipment for resource exploration, environmental monitoring, and ocean investigation. It has the ability for long-time navigating and great-weight carrying and satisfies the different demands of the fields of military science and economics (see [1–3] and references therein).

In recent years, control problems of AUV such as setpoint stabilization, trajectory tracking control, and path tracking control have been actively considered by many researchers. Based on nonlinear control theory, several control methods have been proposed, such as sliding mode control [4–6], adaptive control [7–11], and predictive control [12–15]. However, a common problem of the above literatures is that the time-delays are not taken into account. In practical systems, time-delays are unavoidable in information acquisition and transmission. Time-delay phenomenon is often a source of instability and poor performance [16–18]. From this point of view, considerable amount of attention has been paid to the problem of stabilization and control of time-delay systems. Predictive control is a good method with the ability to handle constraints and time-delays [19–23]. Now, it has become one of the most popular control methodologies no matter in theory or the reality (see [24–27]). The NN predictive control for nonlinear dynamic systems with input delay was studied in [24], but the considered predictive model is required for linear ones and this condition is removed in this paper. The predictor-based control algorithm for an uncertain input delay Euler-Lagrange system was studied in [26], but the controller is an iteration form. To overcome the problem of input delay in Euler-Lagrange dynamical systems directly, a predictor with uncertain system dynamics was proposed in [27]. Recently, predictive control has been applied in many kinds of practical systems [28–31]. Up to now, only a few papers have considered this problem because of its complexity. Paper [32] addressed the control problem with input delay and synthesized a robust controller for underwater vehicles which requires only knowledge of mass matrix. The region tracking problem for AUV with input delay based on predictive control was studied in [33], but it assumes that all the states are known in advance. Therefore, it is a very challenging and significant work to investigate the path tracking control of AUV with input delay.

In this paper, a novel controller is investigated for path tracking control of AUV with input delay. Because of the hydrodynamic coefficients and the surge, sway, and yaw angular velocity coupling, an NN is used to identify the nonlinear part of AUV at first. Then predictive control algorithm is employed to compensate for the delay produced in input channel. The proposed predictive model is a nonlinear model. Stability of the closed-loop system is guaranteed based on Lyapunov stability theory. Finally, a simulation example is presented to show the effectiveness of the proposed control strategy.

The remainder of this paper is organized as follows. The problem of path tracking for AUV is formulated in Section 2. Section 3 is devoted to identification of AUV system by NN. Stability analysis for the boundness of error state and NN weight estimation error are also performed. The predictor and the corresponding control are derived in Section 4. The problems of dealing with the time-delay and stability analysis are illustrated in Section 5. Section 6 validates the feasibility and performance of the proposed control law by simulation experiment. Some conclusions are given in Section 7.

2. Problem Formulation

In the horizontal plane, a 3-DOF AUV with input delay can be modeled as

$$\begin{gathered}\dot{\eta} = J(\eta)\nu, \\ M\dot{\nu} + C(\nu)\nu + D(\nu)\nu + g(\eta) = \overline{\tau}(t-d), \\ h = \eta,\end{gathered} \tag{1}$$

where $\eta = [x\ y\ \psi]^T$ denotes the vehicle location and orientation in the earth-fixed frame. The vector $\nu = [u\ \ v\ \ r]^T$ is the velocities expressed in the body-fixed frame. $M = M_{RB} + M_A$ is the inertia matrix of rigid body M_{RB} with added mass M_A. The matrix $C(\nu)$ is skew symmetrical and it denoted the Coriolis and centripetal forces. Linear and quadratic damping forces are considered in the total hydrodynamic damping matrix $D(\nu)$. The vector $g(\eta)$ is the combined gravitational and buoyancy forces in the body-fixed frame. $\overline{\tau}$ is the input of the system and the vector of the forces and moments on AUV induced by the input and fins. d is a known constant time-delay. h is the output of the system. The kinematic transformation matrix transformation from the body-fixed frame to earth-fixed frame is denoted by $J(\eta)$, and

$$J(\eta) = R(\psi) = \begin{bmatrix} \cos\psi & -\sin\psi & 0 \\ \sin\psi & \cos\psi & 0 \\ 0 & 0 & 1 \end{bmatrix}. \tag{2}$$

Let

$$\begin{gathered}\xi_1 = \eta, \\ \xi_2 = J(\eta)\nu.\end{gathered} \tag{3}$$

Then system (1) changed to

$$\begin{aligned}\dot{\xi}_1 &= \xi_2, \\ \dot{\xi}_2 &= f(\xi_1,\xi_2) + J(\xi_1)M^{-1}\overline{\tau}(t-d), \\ y &= \xi_1,\end{aligned} \tag{4}$$

where nonlinear uncertain function

$$\begin{aligned} f(\xi_1,\xi_2) &= \dot{J}(\xi_1)J^{-1}(\xi_1)\xi_2 + J(\xi_1) \\ &\quad\cdot M^{-1}\left[-C\left(J^{-1}(\xi_1)\xi_2\right)\right. \\ &\quad\left.-D\left(J^{-1}(\xi_1)\xi_2\right)J^{-1}(\xi_1)\xi_2 - g(\xi_1)\right]. \end{aligned} \tag{5}$$

The objective of this paper is that the output y of system (4) tracks a desired trajectory η_d, with all internal signals and control commands remaining bounded. For this purpose, we make the following assumption.

Assumption 1. The desired trajectory vector $\zeta_d = [\eta_d\ \dot{\eta}_d\ \ddot{\eta}_d]^T$ is available for measurement, and η_d and $\dot{\eta}_d$ are bounded.

3. Identification of AUV System

There are two steps to design the output feedback controller for AUV with input delay. First, an NN is designed to identify system (4). Then we will use predictive control algorithm to compensate for the delay that presents in communication channel of AUV.

Let

$$\begin{gathered} A = \begin{bmatrix} 0 & I_3 \\ A_1 & A_2 \end{bmatrix}, \\ B = \begin{bmatrix} 0 \\ I_3 \end{bmatrix}, \\ C = [I_3\ \ 0]^T, \end{gathered} \tag{6}$$

where I_3 denotes the identity matrix; matrices A_1 and A_2 are parameters that can be chosen such that matrix A is stable. Then there exist symmetric positive definite matrices P and Q such that Lyapunov matrix equations $A^TP + PA = -Q$ hold.

Hence, system (4) can be expressed as

$$\begin{gathered}\dot{\xi} = A\xi + B\left[f_0(\xi) + \tau(t-d)\right], \\ y = C^T\xi,\end{gathered} \tag{7}$$

where $\xi = [\xi_1^T\ \ \xi_2^T]^T$, $\tau(t-d) = J(\xi_1)M^{-1}\overline{\tau}(t-d)$, and $f_0(\xi) = f(\xi) - A_1\xi_1 - A_2\xi_2$.

According to the approximation of NN, there exists a bounded reconstruction error $\varepsilon(\|\varepsilon\| \le \overline{\varepsilon})$ and an ideal weight W such that system (4) is described by

$$\begin{gathered}\dot{\xi} = A\xi + B\left[W^T\Phi(\xi) + \tau(t-d) + \varepsilon\right], \\ y = C^T\xi,\end{gathered} \tag{8}$$

where W is the ideal NN weight and $\|W\| \le \overline{W}$ ($\overline{W}$ is a positive constant). The sigmoid function $\Phi(Z) = [\Phi_1(Z), \Phi_2(Z), \cdots, \Phi_n(Z)]^T \in R^n$ is differentiable with respect to x and $\|\Phi(\cdot)\| \le \overline{\Phi}$ holds with a positive constant $\overline{\Phi}$.

Then the NN of system (8) can be written as

$$\begin{aligned}\dot{\widehat{\xi}} &= A\widehat{\xi} + B\left[\widehat{W}^T\Phi\left(\widehat{\xi}\right) + \tau(t-d)\right],\\ \widehat{y} &= C^T\widehat{\xi},\end{aligned} \tag{9}$$

where $\widehat{\xi}$ is a state vector of NN and $\widehat{W}$ is a synaptic weight matrix. The sigmoid function $\Phi(x) = a/(1+e^{-bx}) + c, \forall a, b \in R^+, c \in R$, where a, b, and the real number c are the bound, the slope, and the bias of sigmoidal curvature, respectively.

Let estimation error $\widetilde{\xi} = \xi - \widehat{\xi}$, output error $\widetilde{y} = y - \widehat{y}$, and NN weight error $\widetilde{W} = W - \widehat{W}$. Using (8) and (9), we have

$$\begin{aligned}\dot{\widetilde{\xi}} &= A\widetilde{\xi} + B\left[W^T\Phi(\xi) - \widehat{W}^T\Phi\left(\widehat{\xi}\right) + \varepsilon\right],\\ \widetilde{y} &= C^T\widetilde{\xi}.\end{aligned} \tag{10}$$

Next, a main result will be given. In the following, $\lambda_{\mathrm{m}}(P)$ and $\lambda_{\mathrm{M}}(P)$ denote the minimum and maximum eigenvalue of corresponding matrix P.

Theorem 2. *Consider system (1) with the identification model (9) and conditions (19). Let the NN weight update law be provided by*

$$\dot{\widehat{W}} = \Gamma\left(\widetilde{y}^T\Phi\left(\widehat{\xi}\right) - \mu\widehat{W}\right), \tag{11}$$

in which $\Gamma = \Gamma^T > 0$ is the learning parameter and μ is a constant. Then the estimation error $\widetilde{\xi}$ and neural network weight error $\widetilde{W}$ are uniformly ultimately bounded (UUB).

Proof. Consider a Lyapunov function defined by

$$V = \frac{1}{2}\widetilde{\xi}^T P\widetilde{\xi} + \frac{1}{2}\operatorname{tr}\left(\widetilde{W}^T\Gamma^{-1}\widetilde{W}\right). \tag{12}$$

Calculate the derivative (12) along (10) and (11); we have

$$\begin{aligned}\dot{V} = &-\frac{1}{2}\widetilde{\xi}^T Q\widetilde{\xi} + \widetilde{\xi}^T PB\left[W^T\Phi(\xi) - \widehat{W}^T\Phi\left(\widehat{\xi}\right) + \varepsilon\right]\\ &+ \operatorname{tr}\left(\widetilde{W}^T\Gamma^{-1}\dot{\widetilde{W}}\right).\end{aligned} \tag{13}$$

From (11), it follows that

$$\begin{aligned}\dot{V} = &-\frac{1}{2}\widetilde{\xi}^T Q\widetilde{\xi} + \widetilde{\xi}^T PB\left[W^T\Phi(\xi) - \widehat{W}^T\Phi\left(\widehat{\xi}\right) + \varepsilon\right]\\ &- \widetilde{y}\widetilde{W}^T\Phi\left(\widehat{\xi}\right) + \mu\operatorname{tr}\left(\widetilde{W}^T\widehat{W}\right).\end{aligned} \tag{14}$$

From the definition of $\widetilde{W}$, we obtain the following equation:

$$\operatorname{tr}\left(\widetilde{W}^T\widehat{W}\right) = \operatorname{tr}\left(\widetilde{W}^T\left(W - \widetilde{W}\right)\right) = W^T\left\|\widetilde{W}\right\| - \left\|\widetilde{W}\right\|^2. \tag{15}$$

Moreover, there exist three positive constants α_1, α_2, and α_3 such that

$$\left\|W^T\Phi(\xi) - \widehat{W}^T\Phi\left(\widehat{\xi}\right) + \varepsilon\right\| \le \alpha_1\left\|\widetilde{\xi}\right\| + \alpha_2\left\|\widetilde{W}\right\| + \alpha_3. \tag{16}$$

So

$$\begin{aligned}\dot{V} \le &-\left(\frac{1}{2}\lambda_{\mathrm{m}}(Q) - \alpha_1\|PB\|\right)\left\|\widetilde{\xi}\right\|^2\\ &+ \left(\alpha_2\|PB\| + \overline{\Phi}\right)\left\|\widetilde{W}\right\|\left\|\widetilde{\xi}\right\| + \alpha_3\|PB\|\left\|\widetilde{\xi}\right\|\\ &+ \mu\overline{W}\left\|\widetilde{W}\right\| - \mu\left\|\widetilde{W}\right\|^2.\end{aligned} \tag{17}$$

According to Young's inequality $ab \le (a^2 + \kappa^2 b^2)/2\kappa$ with $\kappa > 0$; then there are $\kappa_0 > 0$, $\kappa_1 > 0$ and $\kappa_2 > 0$ such that

$$\begin{aligned}\left\|\widetilde{W}\right\|\left\|\widetilde{\xi}\right\| &\le \frac{\left\|\widetilde{\xi}\right\|}{2\kappa_0} + \frac{\kappa_0}{2}\left\|\widetilde{W}\right\|^2,\\ \alpha_3\|PB\|\left\|\widetilde{\xi}\right\| &\le \frac{\left(\alpha_3\|PB\|\right)^2}{2\kappa_1} + \frac{\kappa_1}{2}\left\|\widetilde{\xi}\right\|^2,\\ \|W\|\left\|\widetilde{W}\right\| &\le \frac{\|W\|^2}{2\kappa_2} + \frac{\kappa_2}{2}\left\|\widetilde{W}\right\|^2.\end{aligned} \tag{18}$$

Assume that

$$\begin{aligned}R_1 &= \frac{1}{2}\lambda_{\mathrm{m}}(Q) - \alpha_1\|PB\| - \frac{1}{2\kappa_0} - \frac{1}{2}\kappa_1 > 0,\\ R_2 &= \mu - \frac{1}{2}\mu\kappa_2 - \frac{1}{2}\kappa_0 > 0.\end{aligned} \tag{19}$$

Then

$$\dot{V} \le -R_1\left\|\widetilde{\xi}\right\|^2 - R_2\left\|\widetilde{W}\right\|^2 + R_3 \le -\eta V + R_3, \tag{20}$$

where

$$\begin{aligned}\eta &= \min\left\{\frac{R_1}{\lambda_{\mathrm{M}}(P)}, R_2\right\},\\ R_3 &= \frac{\alpha_3^2\|PB\|^2}{2\kappa_1} + \frac{\|W\|^2}{2\kappa_2}.\end{aligned} \tag{21}$$

Thus, estimator error $\widetilde{\xi}$ and NN weight error $\widetilde{W}$ are UUB. □

4. Predictive Control

Input delay (measurement delay and computational delay can be represented by input delay) is a source of instability, which is frequently encountered in the practical systems. For achieving tracking performance, a predictive controller is proposed to compensate for the time-delay present in AUV. Figure 1 is the control structure diagram of AUV system (1).

In fact, the NN weight $\widehat{W}$ stores the dynamical system information. Based on the structure of NN in (9), an online predictor is proposed. For improving the accuracy of path tracking effectively, the nonlinear prediction model is employed here. Now, let the predictor of system (8) be

$$\begin{aligned}\dot{\xi}_p(t+d \mid t) = &A\xi_p(t+d \mid t)\\ &+ B\left[\widehat{W}^T\Phi\left(\xi_p(t+d \mid t)\right) + \tau(t)\right],\\ y_p(t+d \mid t) = &C^T\xi_p(t+d \mid t),\end{aligned} \tag{22}$$

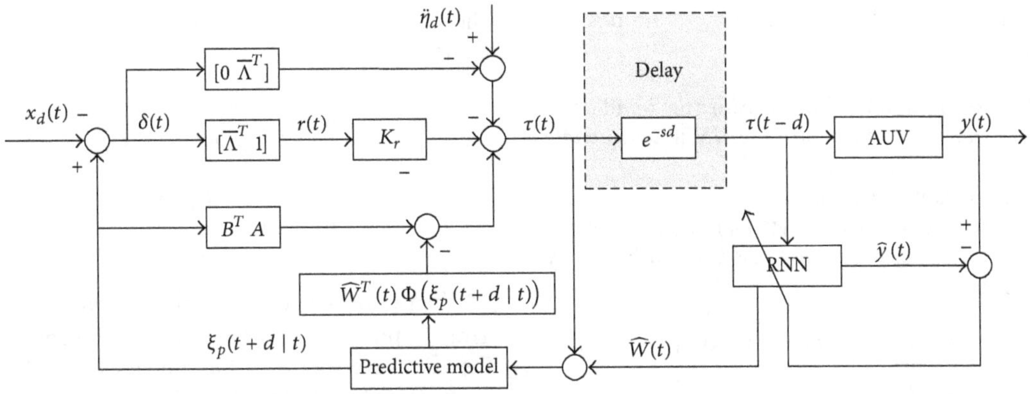

FIGURE 1: Control structure of AUV system (1).

where $\xi_p(t+d \mid t)$ and $y_p(t+d \mid t)$ are the prediction state and output of system (8) with the initial condition $\xi_p(d \mid 0) = \xi(0)$.

If prediction model (22) is precise, then $\xi_p(t+d \mid t) = \xi(t+d)$. This mean that ξ ahead of time d can be predicted via $\xi_p(t+d \mid t)$ in prediction model. Therefore, the difficulty in controlling time-delay plant can be overcome. However, due to the modeling errors, in prediction model (22) errors exist inevitably. Now, define a predictor error as $e(t) = \xi(t+d) - \xi_p(t+d \mid t)$. It follows from (8) and (22) that

$$\begin{aligned}\dot{e}(t) &= Ae(t) + B\left[W^T\Phi(\xi(t+d)) - \widehat{W}^T(t)\Phi\left(\xi_p(t+d \mid t)\right) + \varepsilon(t+d)\right],\\ y_e(t) &= C^T e(t).\end{aligned} \quad (23)$$

Next, we will prove that the predictor error (23) is bounded. Define an error vector as

$$\delta(t) = \xi_p(t+d \mid t) - \eta_d(t). \quad (24)$$

Define a filtered error as

$$r(t) = \Lambda^T\delta(t) = \left[\overline{\Lambda}^T\ \ 1\right]\delta(t), \quad (25)$$

where $\overline{\Lambda} = [\lambda_1\ \lambda_2]^T$ is an appropriately chosen coefficient vector such that $\delta(t) \to 0$ exponentially as $r(t) \to 0$. Then, using (22), the filtered error can be written as

$$\begin{aligned}\dot{r}(t) &= \Lambda^T\dot{\delta}(t) = \Lambda^T\left[A\xi_p(t+d \mid t) + B\left(\widehat{W}^T\Phi\left(\xi_p(t+d \mid t)\right) + \tau(t)\right)\right]\\ &= \left[0\ \ \overline{\Lambda}^T\right]\delta(t) - \ddot{\eta}_d + B^T A\xi_p(t+d \mid t) + \widehat{W}^T\Phi\left(\xi_p(t+d \mid t)\right) + \tau(t).\end{aligned} \quad (26)$$

Now choose $K_r > 0$ and let

$$\begin{aligned}\tau(t) &= -\left[0\ \ \overline{\Lambda}^T\right]\delta(t) + \ddot{\eta}_d - B^T A\xi_p(t+d \mid t) - \widehat{W}^T\Phi\left(\xi_p(t+d \mid t)\right) - K_r r(t)\\ &= -\left[0\ \ \overline{\Lambda}^T\right]\delta(t) + \ddot{\eta}_d - B^T A\delta(t) - B^T A\eta_d(t) - \widehat{W}^T\Phi\left(\xi_p(t+d \mid t)\right) - K_r r(t).\end{aligned} \quad (27)$$

That is, a control input of AUV is

$$\begin{aligned}\overline{\tau}(t) &= MJ^{-1}(\xi_1)\left(-\left[0\ \ \overline{\Lambda}^T\right]\delta(t) + \ddot{\eta}_d - B^T A\xi_p(t+d \mid t) - \widehat{W}^T\Phi\left(\xi_p(t+d \mid t)\right) - K_r r(t)\right) = MJ^{-1}(\xi_1)\left(-\left[0\ \ \overline{\Lambda}^T\right]\delta(t) + \ddot{\eta}_d\right.\\ &\quad \left.- B^T A\delta(t) - B^T Ax_d(t) - \widehat{W}^T\Phi\left(\xi_p(t+d \mid t)\right) - K_r r(t)\right).\end{aligned} \quad (28)$$

Note that the predictor state $\xi_p(t+d \mid t)$ and the associated error $r(t)$ are used in AUV controller (28); the NN approximation term $\widehat{W}^T(t)\Phi(\xi_p(t+d \mid t))$ from (22) is employed to accommodate the unknown nonlinearity. Therefore, the stability of the closed-loop system can be guaranteed.

From (28), (26) becomes

$$\dot{r}(t) = -K_r r(t). \quad (29)$$

5. Stability Analysis

Assume that the parameters are chosen such that

$$\begin{aligned}R_4 &= R_2 + \frac{1}{2}\kappa_3 > 0,\\ R_5 &= \frac{1}{2}\lambda_m(Q)\frac{1}{2\kappa_3} - \frac{1}{2}\kappa_4 - \beta_2 > 0,\\ R_6 &= -K_r > 0,\end{aligned} \quad (30)$$

where R_2 is defined as in Theorem 2 and κ_3, κ_4 are positive constants that can be chosen.

Theorem 3 (let Assumption 1 hold). *Consider the input delay AUV system (1) under condition (30), the NN weight update law (11), and controller (28). Then*

(1) all the closed-loop signals are UUB;

(2) the path tracking error $\omega(t) = y(t+d) - \eta_d(t)$ converges to a neighborhood of the origin, whose size can be adjusted by control parameters.

Proof. Consider a Lyapunov function defined by

$$\overline{V} = \frac{1}{2}\tilde{\xi}^T P\tilde{\xi} + \frac{1}{2}\widetilde{W}^T\Gamma^{-1}\widetilde{W} + \frac{1}{2}e^T Pe + \frac{1}{2}r^2 := V_1 + V_2 + V_3 + V_4. \tag{31}$$

The derivative of V_1 and V_2 can be deduced following the proof of Theorem 2. Thus we have

$$\dot{V}_3 = \frac{1}{2}e^T\left(A^T P + PA\right)e + e^T PB\left[W^T\Phi(\xi(t+d)) - \widehat{W}^T\Phi\left(\xi_p(t+d\mid t)\right) + \varepsilon(t+d)\right]. \tag{32}$$

In fact, via Taylor series expansion, there exist positive constants β_1, β_2, and β_3 such that

$$\begin{aligned}&\left\|W^T\Phi(\xi(t+d)) - \widehat{W}^T\Phi\left(\xi_p(t+d\mid t)\right)\right\| \\ &= \left\|\widetilde{W}\Phi\left(\xi_p(t+d\mid t)\right) + W^T\left[\Phi(\xi(t+d)) - \Phi\left(\xi_p(t+d\mid t)\right)\right]\right\| \\ &\le \beta_1\left\|\widetilde{W}\right\| + \beta_2\|e\| + \beta_3.\end{aligned} \tag{33}$$

So

$$\dot{V}_3 \le -\frac{1}{2}Q_m\|e\|^2 + \|e\|\left(\beta_1\left\|\widetilde{W}\right\| + \beta_2\|e\| + \beta_3 + \bar{\varepsilon}\right). \tag{34}$$

By using Young's inequality, there exist positive numbers κ_3 and κ_4 such that

$$\begin{aligned}\|e\|\left\|\widetilde{W}\right\| &\le \frac{\|e\|^2}{2\kappa_3} + \frac{\kappa_3}{2}\left\|\widetilde{W}\right\|^2, \\ (\beta_3+\bar{\varepsilon})\|e\| &\le \frac{(\beta_3+\bar{\varepsilon})^2}{2\kappa_4} + \frac{\kappa_4}{2}\|e\|^2,\end{aligned} \tag{35}$$

$$\dot{V}_3 \le -\left(\frac{1}{2}Q_m - \frac{1}{2\kappa_3} - \frac{1}{2}\kappa_4 - \beta_2\right)\|e\|^2 + \frac{1}{2}\kappa_3\left\|\widetilde{W}\right\| + \frac{\left(\beta_4+\bar{d}+\bar{\varepsilon}\right)^2}{2\kappa_4}, \tag{36}$$

$$\dot{V}_4 = r\dot{r} = -K_r\|r\|^2. \tag{37}$$

Therefore,

$$\begin{aligned}\dot{\overline{V}} &\le -R_1\left\|\tilde{\xi}\right\|^2 - R_4\left\|\widetilde{W}\right\|^2 - R_5\|e\|^2 - R_6\|r\|^2 + R_7 \\ &\le -\bar{\eta}\overline{V} + R_7,\end{aligned} \tag{38}$$

where R_1 is defined as in (19), and $\bar{\eta}$ is positive constant defined by

$$\begin{aligned}\bar{\eta} &= \min\left\{\frac{R_1}{\lambda_{\max}(P)}, R_4, \frac{R_5}{\lambda_{\max}(P)}, R_6\right\}, \\ R_7 &= R_3 + \frac{\left(\beta_4+\bar{d}+\bar{\varepsilon}\right)^2}{2\kappa_4}.\end{aligned} \tag{39}$$

Then according to Lyapunov theorem, error $\tilde{\xi}$, NN weight error $\widetilde{W}$, predictor error e, and filtered error r are all UUB. The control error δ is thus bounded based on (24) and Assumption 1. Therefore, the NN weights $\widehat{W}$ and $\xi_p(t+d\mid t)$ are bounded.

Finally, the boundedness of path tracking error $\omega(t)$ will be proved. Since

$$\begin{aligned}\omega(t) &= y(t+d) - \eta_d(t) \\ &= \xi_1(t+d) - \xi_{p1}(t+d\mid t) + \xi_{p1}(t+d\mid t) - y_d = e_1(t) + \delta_1(t),\end{aligned} \tag{40}$$

then

$$\begin{aligned}\lim_{t\to\infty}\|\omega(t)\| &= \lim_{t\to\infty}\|e_1(t)\| + \lim_{t\to\infty}\|\delta_1(t)\| \\ &\le \lim_{t\to\infty}\|e(t)\| + \lim_{t\to\infty}\|\delta(t)\|.\end{aligned} \tag{41}$$

Therefore, the tracking error $\omega(t)$ is bounded because $e(t)$, $\delta(t)$, and $\xi(t)$ are bounded. □

Remark 4. Compared with [20–33], there are three advantages. Firstly, output feedback is considered in this paper. Secondly, the nonlinear prediction model is employed to improve the accuracy of predictive control. Finally, time-delay is considered in path tracking control of AUV which has more real significance.

6. Simulation Analysis

Example 5. The simplified dynamics model of INFANTE AUV [2] in the horizontal plane with input delay is adopted as follows in this paper:

$$\begin{aligned}\dot{x} &= u\cos(\psi) - v\sin(\psi), \\ \dot{y} &= u\sin(\psi) + v\cos(\psi), \\ \dot{\psi} &= r, \\ 0 &= m_v\dot{v} + m_{ur}ur + d_v, \\ \Gamma &= m_r\dot{r} + d_r,\end{aligned} \tag{42}$$

where x, y, and ψ are the surge position, sway position, and yaw angle in the body-fixed frame, and u, v, and r denote surge, sway, and yaw velocities, respectively. Γ is the yaw moment. The symbol I_z denotes the moment of inertia of the AUV, $N_{\{\cdot\}}$ is nonlinear hydrodynamic damping, and

$$
\begin{aligned}
m_v &= m - Y_{\dot{v}}, \\
m_{ur} &= m - Y_r, \\
d_v &= -Y_v uv - Y_{v|v|} v |v|, \\
m_r &= I_z - N_{\dot{r}}, \\
d_r &= -N_v uv - N_{v|v|} v |v| - N_r ur.
\end{aligned} \tag{43}
$$

Since the considered model (42) is in the horizontal plane and has no disturbance, we can assume that $u = 1m/s$ in principle. In the following, the desired path is $x_d(t) = 20\sin 2\pi t/200$, $y_d(t) = 20 - 20\cos 2\pi t/200$, and $\psi_d(t) = 2\pi t/200$.

The model parameters of INFANTE AUV are as follows:

$$
\begin{aligned}
m &= 2234.5kg, \\
I_z &= 2000Nm^2, \\
X_{\dot{u}} &= -142kg, \\
N_{\dot{r}} &= -1350Nm^2, \\
Y_{\dot{v}} &= -1715kg, \\
Y_v &= -346kg/m, \\
Y_r &= 435kg, \\
N_v &= -686kg, \\
N_r &= -1427kgm, \\
Y_{v|v|} &= -667kg/m, \\
N_{v|v|} &= 443kg.
\end{aligned} \tag{44}
$$

NN parameters are selected as follows: $\Phi(x) = 1/(1 + e^{-ax})$, where $a = 0.5$, $\mu = 0.3$, $\Gamma = 6$.

The delay constant $d = 2$. Other parameters in controller are $A_1 = -3I_3$, $A_2 = -2I_3$, $\lambda_1 = 2$, $\lambda_2 = 2$, and $K_r = 8$.

The initial position and the surge speed of the AUV are $(0, 20)$ and $0m/s$, respectively. The simulation results are shown in Figures 2–4. The path tracking errors in x, y, and ψ are given in Figure 2. The control forces in x and y and the control torque of yaw ψ are given in Figure 3. From these simulation figures, we can see that the tracking performance is unsatisfying at the beginning of simulation; this is because the controller performs mainly depending on the adaptive control. The good tracking of position is obtained by the proposed adaptive NN predictive controller by and by. Figure 4 is the path tracking in horizontal plane. From Figure 4 we can see that AUV can realize tracking control smoothly and converge to the desired trajectory.

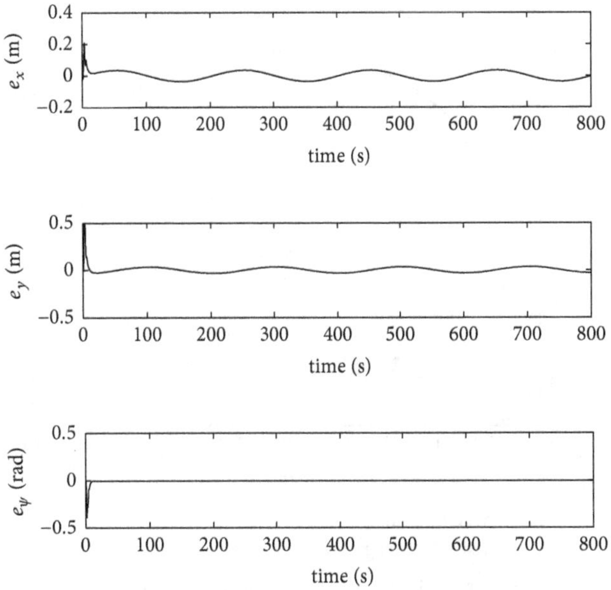

FIGURE 2: Path tracking errors.

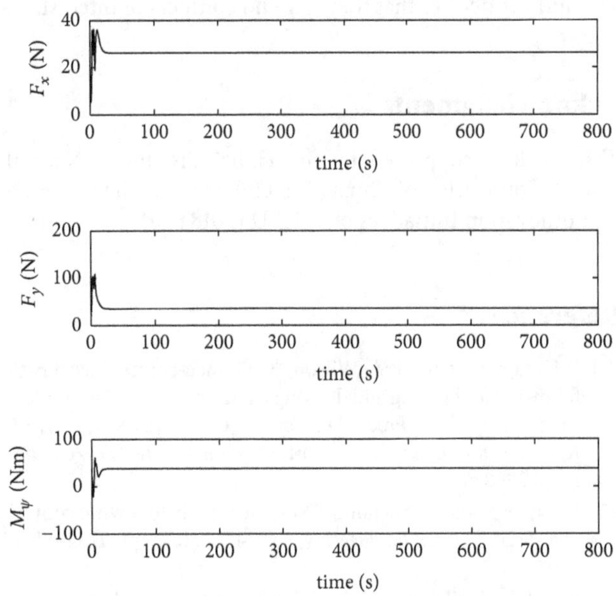

FIGURE 3: Control forces in x and y and control torque of yaw.

7. Conclusion

This paper investigates the path tracking problem for an AUV with input delay. Based on predictive and adaptive NN control theory, a predictive controller is given. The output feedback control algorithm is employed here. The NN is used to estimate the dynamic uncertain nonlinear function induced by hydrodynamic coefficients and coupling of the surge, sway, and yaw angular velocity. The predictive control is introduced to compensate the input delay present in AUV. The stability of the controller was analyzed by Lyapunov theorem. Simulation results showed that the proposed controller performs well with stability.

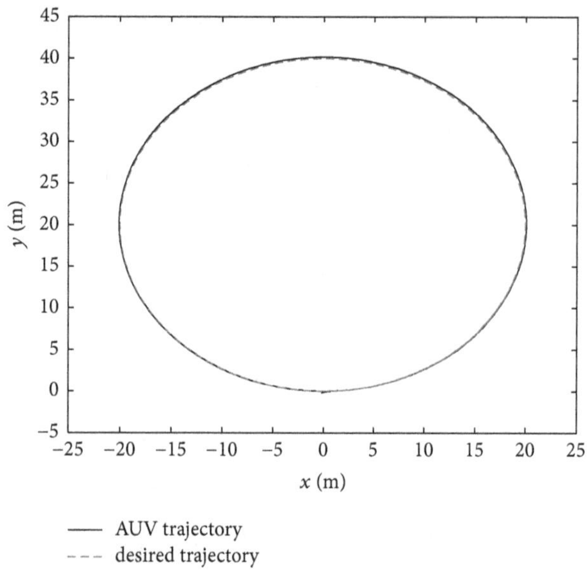

FIGURE 4: Path tracking response in xy plane.

Conflicts of Interest

The authors declare that they have no conflicts of interest.

Acknowledgments

This work is supported by the Hubei Provincial Natural Science Foundation of China (2016CFB273) and the Research and Innovation Initiatives of WHPU (2018Y20).

References

[1] P. Jantapremjit and P. A. Wilson, "Guidance-control based path following for homing and docking using an autonomous underwater vehicle," in *Proceedings of the OCEANS'08 MTS/IEEE Kobe-Techno-Ocean'08 - Voyage toward the Future, OTO'08*, jpn, April 2008.

[2] L. Lapierre and D. Soetanto, "Nonlinear path-following control of an AUV," *Ocean Engineering*, vol. 34, no. 11-12, pp. 1734–1744, 2007.

[3] A. Adhami-Mirhosseini, M. J. Yazdanpanah, and A. P. Aguiar, "Automatic bottom-following for underwater robotic vehicles," *Automatica*, vol. 50, no. 8, pp. 2155–2162, 2014.

[4] M. Kim, H. Joe, J. Kim, and S.-c. Yu, "Integral sliding mode controller for precise manoeuvring of autonomous underwater vehicle in the presence of unknown environmental disturbances," *International Journal of Control*, vol. 88, no. 10, pp. 2055–2065, 2015.

[5] R. Cui, X. Zhang, and D. Cui, "Adaptive sliding-mode attitude control for autonomous underwater vehicles with input nonlinearities," *Ocean Engineering*, vol. 123, pp. 45–54, 2016.

[6] Y. Wang, L. Gu, M. Gao, and K. Zhu, "Multivariable output feedback adaptive terminal sliding mode control for underwater vehicles," *Asian Journal of Control*, vol. 18, no. 1, pp. 247–265, 2016.

[7] X. Qi, "Adaptive coordinated tracking control of multiple autonomous underwater vehicles," *Ocean Engineering*, vol. 91, pp. 84–90, 2014.

[8] R. Cui, C. Yang, Y. Li, and S. Sharma, "Adaptive Neural Network Control of AUVs With Control Input Nonlinearities Using Reinforcement Learning," *IEEE Transactions on Systems, Man, and Cybernetics: Systems*, vol. 47, no. 6, pp. 1019–1029, 2017.

[9] B. B. Miao, T. S. Li, and W. L. Luo, "A DSC and MLP based robust adaptive NN tracking control for underwater vehicle," *Neurocomputing*, vol. 111, pp. 184–189, 2013.

[10] Y.-C. Liu, S.-Y. Liu, and N. Wang, "Fully-tuned fuzzy neural network based robust adaptive tracking control of unmanned underwater vehicle with thruster dynamics," *Neurocomputing*, vol. 196, pp. 1–13, 2016.

[11] J. Ghommam and M. Saad, "Backstepping-based cooperative and adaptive tracking control design for a group of underactuated AUVs in horizontal plan," *International Journal of Control*, vol. 87, no. 5, pp. 1076–1093, 2014.

[12] C. Shen, B. Buckham, and Y. Shi, "Modified C/GMRES Algorithm for Fast Nonlinear Model Predictive Tracking Control of AUVs," *IEEE Transactions on Control Systems Technology*, vol. 25, no. 5, pp. 1896–1904, 2017.

[13] J. Gao, A. A. Proctor, Y. Shi, and C. Bradley, "Hierarchical Model Predictive Image-Based Visual Servoing of Underwater Vehicles with Adaptive Neural Network Dynamic Control," *IEEE Transactions on Cybernetics*, vol. 46, no. 10, pp. 2323–2334, 2016.

[14] J. Sun and J. Chen, "Networked predictive control for systems with unknown or partially known delay," *IET Control Theory & Applications*, vol. 8, no. 18, pp. 2282–2288, 2014.

[15] S. Li and G.-P. Liu, "Networked predictive control for nonlinear systems with stochastic disturbances in the presence of data losses," *Neurocomputing*, vol. 194, pp. 56–64, 2016.

[16] J. Zhao, L. Zhang, and X. Qi, "A necessary and sufficient condition for stabilization of switched descriptor time-delay systems under arbitrary switching," *Asian Journal of Control*, vol. 18, no. 1, pp. 266–272, 2016.

[17] S. Sirouspour and A. Shahdi, "Model predictive control for transparent teleoperation under communication time delay," *IEEE Transactions on Robotics*, vol. 22, no. 6, pp. 1131–1145, 2006.

[18] Y. Sheng, Y. Shen, and M. Zhu, "Delay-dependent global exponential stability for delayed recurrent neural networks," *IEEE Transactions on Neural Networks and Learning Systems*, 2016.

[19] O. M. Smith, "A controller to overcome deadtime," *ISA J*, vol. 6, p. 2833, 1959.

[20] C. F. Caruntu and C. Lazar, "Network delay predictive compensation based on time-delay modelling as disturbance," *International Journal of Control*, vol. 87, no. 10, pp. 2012–2026, 2014.

[21] M. Messadi, A. Mellit, K. Kemih, and M. Ghanes, "Predictive control of a chaotic permanent magnet synchronous generator in a wind turbine system," *Chinese Physics B*, vol. 24, no. 1, Article ID 010502, 2015.

[22] S. H. HosseinNia and M. Lundh, "A general robust MPC design for the state-space model: application to paper machine process," *Asian Journal of Control*, vol. 18, no. 5, pp. 1891–1907, 2016.

[23] Y. Ding, Z. Xu, J. Zhao, K. Wang, and Z. Shao, "Embedded MPC controller based on interior-point method with convergence depth control," *Asian Journal of Control*, vol. 18, no. 6, pp. 2064–2077, 2016.

[24] J.-Q. Huang and F. L. Lewis, "Neural-network predictive control for nonlinear dynamic systems with time-delay," *IEEE Transactions on Neural Networks and Learning Systems*, vol. 14, no. 2, pp. 377–389, 2003.

[25] B. An and G. Liu, "Model-based predictive controller design for a class of nonlinear networked systems with communication delays and data loss," *Chinese Physics B*, vol. 23, no. 8, p. 080202, 2014.

[26] N. Sharma, C. M. Gregory, and W. E. Dixon, "Predictor-based compensation for electromechanical delay during neuromuscular electrical stimulation," *IEEE Transactions on Neural Systems and Rehabilitation Engineering*, vol. 19, no. 6, pp. 601–611, 2011.

[27] N. Fischer, A. Dani, N. Sharma, and W. E. Dixon, "Saturated control of an uncertain nonlinear system with input delay," *Automatica*, vol. 49, no. 6, pp. 1741–1747, 2013.

[28] T. Wang, H. Gao, and J. Qiu, "A combined adaptive neural network and nonlinear model predictive control for multirate networked industrial process control," *IEEE Transactions on Neural Networks and Learning Systems*, vol. 27, no. 2, pp. 416–425, 2016.

[29] P. Dai, Z. Gong, and G. Guo, "Performance analysis of a modular multilevel converter drive system with model predictive control," *International Journal of Control and Automation*, vol. 9, no. 1, pp. 387–398, 2016.

[30] C. Wei, J. Luo, H. Dai, Z. Yin, W. Ma, and J. Yuan, "Globally robust explicit model predictive control of constrained systems exploiting SVM-based approximation," *International Journal of Robust and Nonlinear Control*, vol. 27, no. 16, pp. 3000–3027, 2017.

[31] M. Parvez Akter, S. Mekhilef, N. Mei Lin Tan, and H. Akagi, "Modified Model Predictive Control of a Bidirectional AC-DC Converter Based on Lyapunov Function for Energy Storage Systems," *IEEE Transactions on Industrial Electronics*, vol. 63, no. 2, pp. 704–715, 2016.

[32] R. Prasanth Kumar, A. Dasgupta, and C. S. Kumar, "Robust trajectory control of underwater vehicles using time delay control law," *Ocean Engineering*, vol. 34, no. 5-6, pp. 842–849, 2007.

[33] K. Mukherjee, I. N. Kar, and R. K. P. Bhatt, "Region tracking based control of an autonomous underwater vehicle with input delay," *Ocean Engineering*, vol. 99, pp. 107–114, 2015.

17

Circular Formation Control of Multiagent Systems with Any Preset Phase Arrangement

Lina Jin,[1] **Shuanghe Yu**,[2] **and Dongxu Ren**[2]

[1]*School of Computer and Communication Engineering, Liaoning Shihua University, Fushun, China*
[2]*College of Marine Electrical Engineering, Dalian Maritime University, Dalian, China*

Correspondence should be addressed to Shuanghe Yu; shuanghe@dlmu.edu.cn

Academic Editor: Yongji Wang

This paper deals with the circular formation control problem of multiagent systems for achieving any preset phase distribution. The control problem is decomposed into two parts: the first is to drive all the agents to a circle which either needs a target or not and the other is to arrange them in positions distributed on the circle according to the preset relative phases. The first part is solved by designing a circular motion control law to push the agents to approach a rotating transformed trajectory, and the other is settled using a phase-distributed protocol to decide the agents' positioning on the circle, where the ring topology is adopted such that each agent can only sense the relative positions of its neighboring two agents that are immediately in front of or behind it. The stability of the closed-loop system is analyzed, and the performance of the proposed controller is verified through simulations.

1. Introduction

Imitating the collective behaviors that occur in nature, the distributed control of multiagent systems (MAS), such as multiple autonomous underwater vehicles (AUVs) and unmanned aerial vehicles (UAVs) [1–3], has attracted a great deal of attention in control and robotic communities [4, 5] and has been extensively explored with different settings, including consensus [6], formation control [7, 8], flocking [9], distributed sensor networks [10], rendezvous [11], and source seeking [12, 13], through coordinating multiple autonomous mobile agents.

As one of these fundamental problems, the pattern-forming problem has attracted a considerable amount of research interest, where the agents are required to cooperatively generate and maintain the desired geometric patterns to perform various teamwork tasks. Formation patterns are typically limited to a point (rendezvous), line (flocking), or circle. The circular formation is a design method for steering the agents to orbit around a target along a common circle, which provides a simple geometric shape to collect data with a desired spatial and temporal distribution. In the community of systems and control, research efforts have been devoted to the circle formation problem for multiagent systems modeled as single or double integrators [14–18] and unicycles [19–26] under different communication topologies. Circular motion has been studied in the scenario of cyclic pursuit with ring topology [14–23]. A collective circular motion is addressed with a jointly connected communication condition [24, 25]. Under all-to-all communication condition, the phase potentials are used for uniform phase arrangement of particles along a circle [26]. In the aforementioned works, all the agents can enclose a fixed or moving target with position, distance, or bearing measurements in an equally circular distribution manner. However, for some special robotic application occasions, uniform distribution is unable to meet the practical demands; for example, AUV formation detects the concentration of oil pollution and UAV formation performs special escort missions in a nonuniform distribution [16]. Only a few works have presented the distributed control laws for a group of agents to formulate any given phase arrangement on a circle, and it should be noted that the agents are restricted to move in the one-dimensional space of a circle [27, 28].

The problem of circular formation of multiple agents with any preset phase arrangement in the two-dimensional space is addressed in this paper. The contributions can be

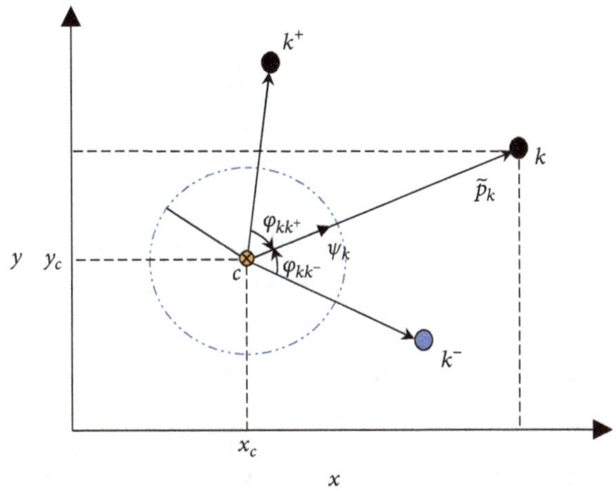

FIGURE 1: Agent, target, and the graphical view of notations.

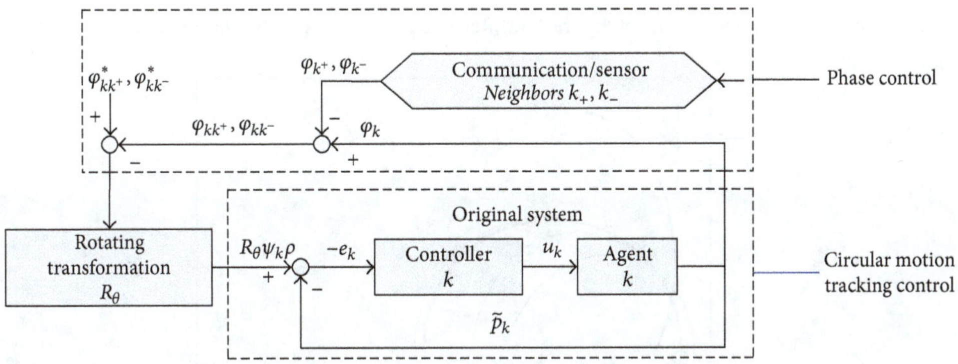

FIGURE 2: Feedback control design scheme of circular formation.

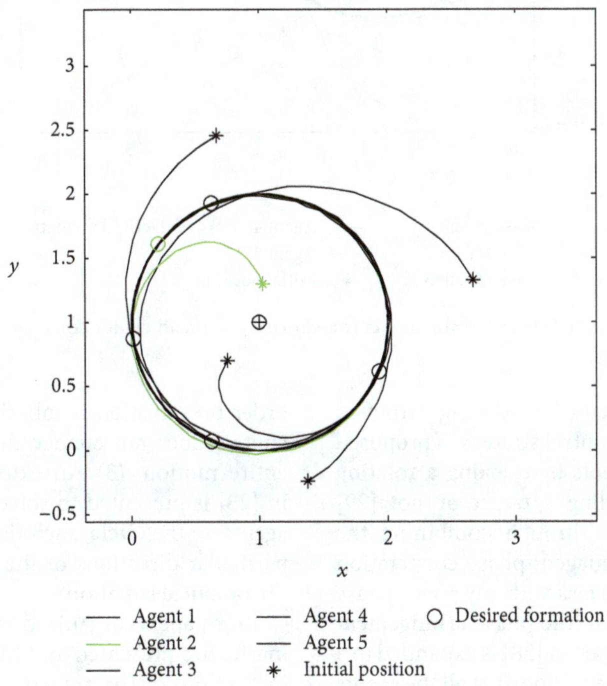

FIGURE 3: Circular formation trajectories $p_k(t)$ with a fixed target.

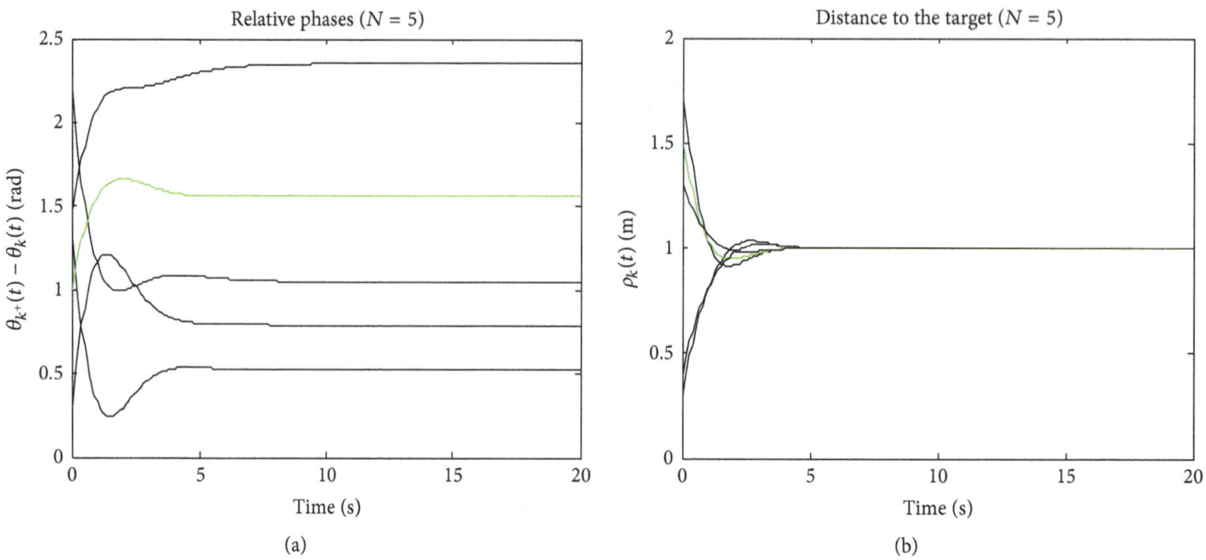

Figure 4: Trajectories with time of Figure 3. (a) Relative phases $\varphi_{k^+k}(t)$. (b) Distances to the target.

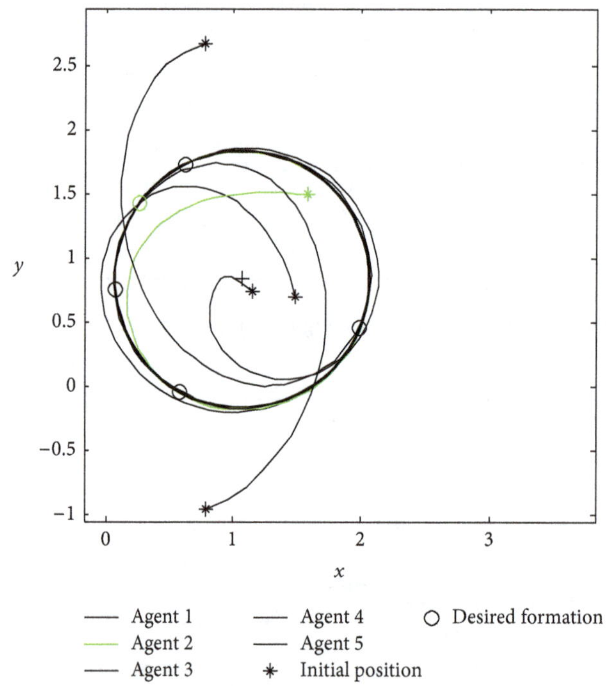

Figure 5: Circular formation trajectories $p_k(t)$ with center consensus.

summarized as follows. (1) Through introducing a rotated affine transformation, a tracking control strategy is proposed to achieve circular motion of agents by tracking a rotating matrix, where two cases of circling a target or not [29] are, respectively, considered. (2) Through combining the above control strategy with a multiagent phase cooperation mechanism, the circular formation task with any preset phase arrangement is implemented. Then, the phase arrangement algorithm without circle forming part in [28] is expanded to a two-dimensional model; that is, the positions of all the agents can be initialized out of the circle instead of being initialized on the assumed given circle. Furthermore, the advantage of order preservation is inherited because the circular motion control does not change the phase distribution during the entire motion. (3) An extension of the phase control law in [28] is presented to solve the positioning problem of the agents on the circle, such that some agents are located in the particular directions of the surrounded target according to the practical situation.

The paper is organized as follows. Some necessary preliminaries are presented and the control problem is formulated in Section 2. The tracking control law of circular motion is designed and analyzed in Section 3. Section 4 combines the above circular tracking control algorithm with a phase

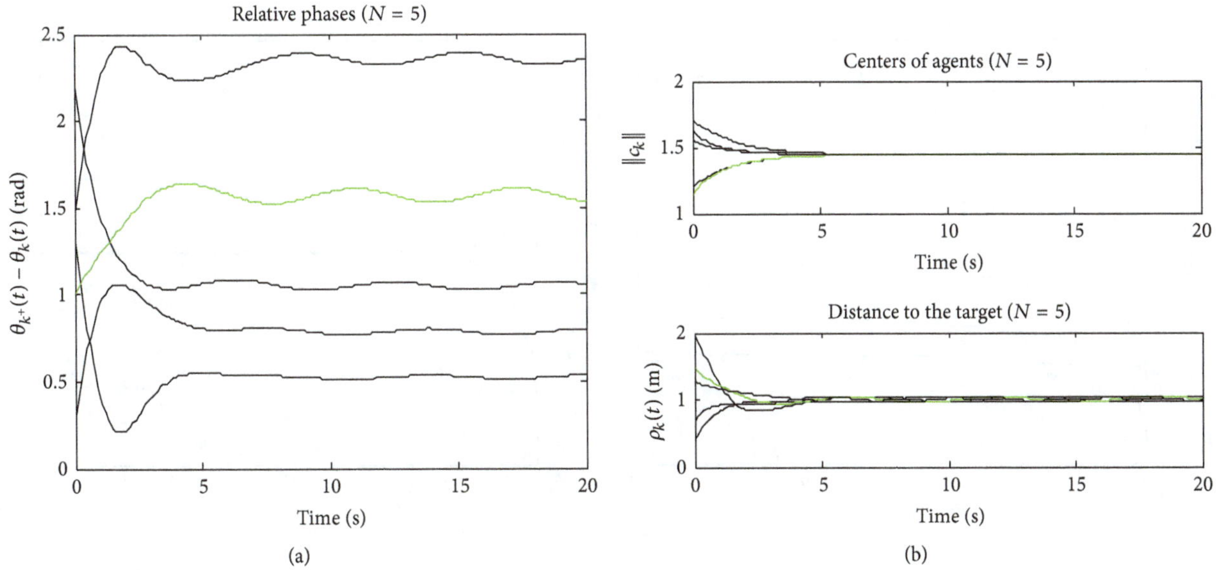

FIGURE 6: Trajectories with time of Figure 5. (a) Relative phases $\varphi_{k^+k}(t)$. (b) Consensus center and distances to it.

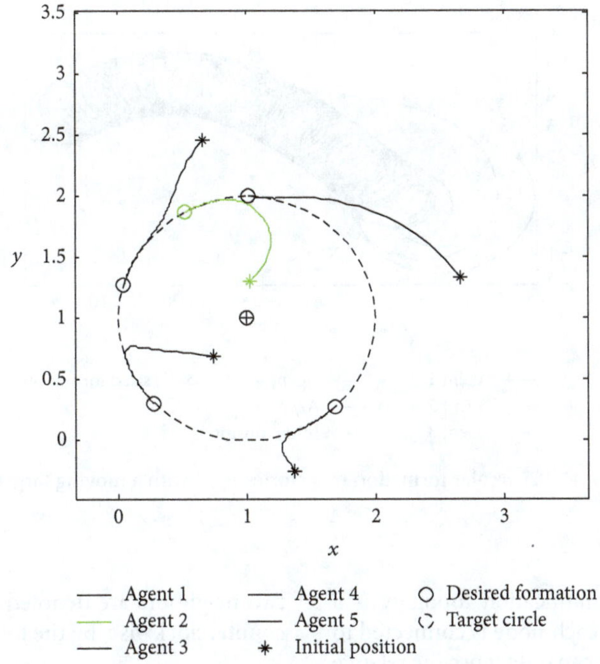

FIGURE 7: Circular formation trajectories $p_k(t)$ with $\varphi_d = \pi/2$.

arrangement control law to achieve any phase arrangement along the circle. Simulation results given in Section 5 validate the strategy.

2. Preliminaries and Problem Formulation

2.1. Model of the Agents. Consider N ($N > 2$) single-integrator-modeled agents moving in the plane

$$\dot{p}_k(t) = u_k(t), \quad k = 1, \ldots, N, \tag{1}$$

where $p_k(t) \in \mathbb{R}^2$ and $u_k(t) \in \mathbb{R}^2$ denote the position and the control input of agent k, respectively. $P(t) = [p_1(t), \ldots, p_N(t)]^T$ denotes the stacked column vector. As shown in Figure 1, for a target with position $c \in \mathbb{R}^2$, $\tilde{p}_k(t) = p_k(t) - c$ denotes the relative displacement between agent k in (1) and the target, ρ denotes the radius of the desired circle, and $\psi_k \in \mathbb{R}^2$ is a unit vector on the line passing through agent k and the target; that is,

$$\psi_k(t) = \frac{p_k(t) - c_k(t)}{\|p_k(t) - c_k(t)\|}. \tag{2}$$

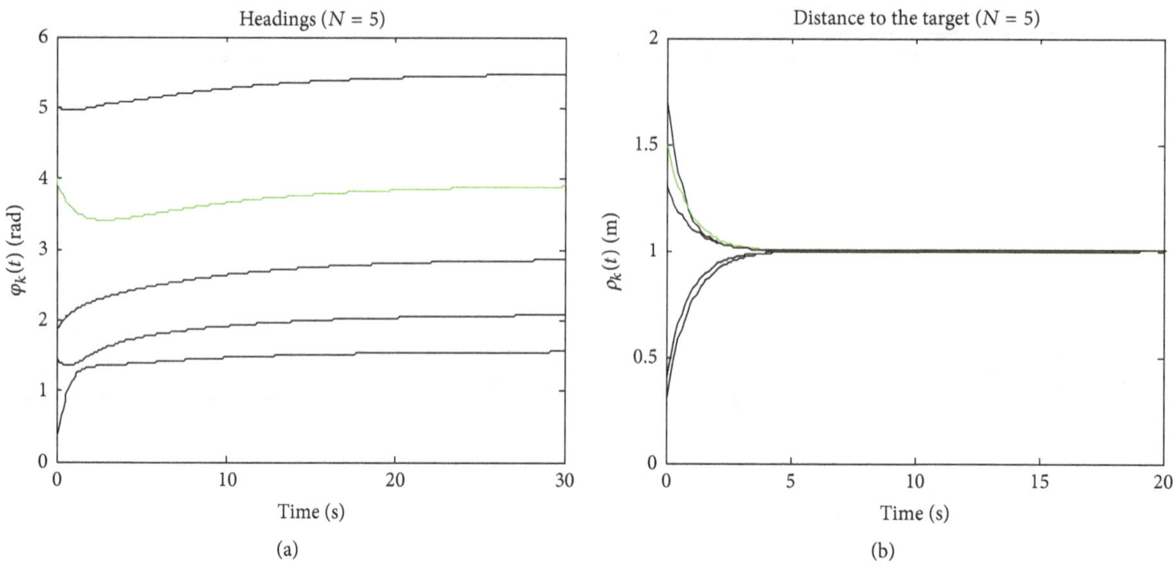

FIGURE 8: Trajectories with time of Figure 7. (a) Phases $\varphi_k(t)$ of agents. (b) Distances to the target.

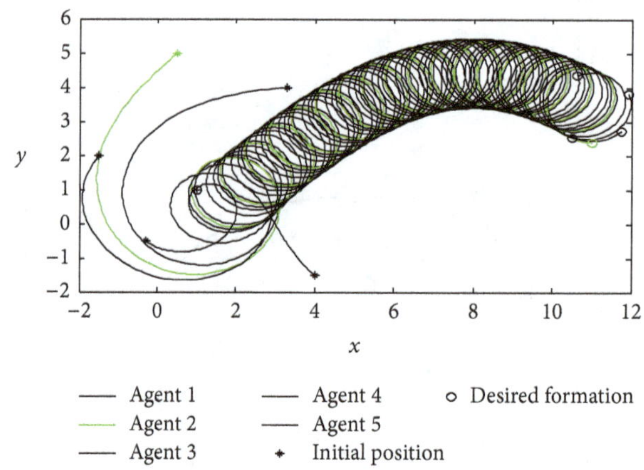

FIGURE 9: Circular formation trajectories $p_k(t)$ with a moving target.

2.2. Interaction Graphs. The communication topology is a connected ring topology in which each node is connected to two neighboring nodes. Each agent can only sense the relative positions of its two neighboring agents that are immediately in front of or behind it. Then, the graph describing the neighbor relationships is a weighted communication topology graph (undirected ring) $\mathcal{G} = (\mathcal{I}, \mathcal{E}, \mathcal{A})$, where $\mathcal{I} = \{1, 2, \ldots, N\}$ is the set of vertices, $\mathcal{E} = \{(1,2), (2,3), \ldots, (N-1, N), (N,1)\} \subseteq \mathcal{I} \times \mathcal{I}$ is the set of edges, and $\mathcal{A} = [a_{kj}]$ is the $N \times N$ adjacency matrix given by $a_{kj} \neq 0$ if $(k, j) \in \mathcal{E}$ and $a_{kj} = 0$ otherwise. Both the adjacency and the Laplacian matrices of the ring topology are circulant; that is, they are completely defined by their first row; each subsequent row of the circulant matrix is the previous row shifted one position to the right with the first entry equal to the last entry of the previous row.

For a node $k \in \mathcal{I}$, the set of adjacent nodes is denoted by $\mathcal{N}_k$, which is called the neighbors of agent k. Agent k's two neighbors are denoted by k^+ and k^-, which are labeled counterclockwise, by the following two rules:

$$k^+ = \begin{cases} k+1 & \text{when } k = 1, 2, \ldots, N-1 \\ 1 & \text{when } k = N, \end{cases}$$
$$k^- = \begin{cases} N & \text{when } k = 1 \\ k-1 & \text{when } k = 2, 3, \ldots, N. \end{cases} \quad (3)$$

Furthermore, as shown in Figure 1, the relative phases (bearing measurements to target) from agent k to its immediate counterclockwise and clockwise neighboring agents are, respectively, denoted by variables φ_{kk^+} and φ_{kk^-} as follows:

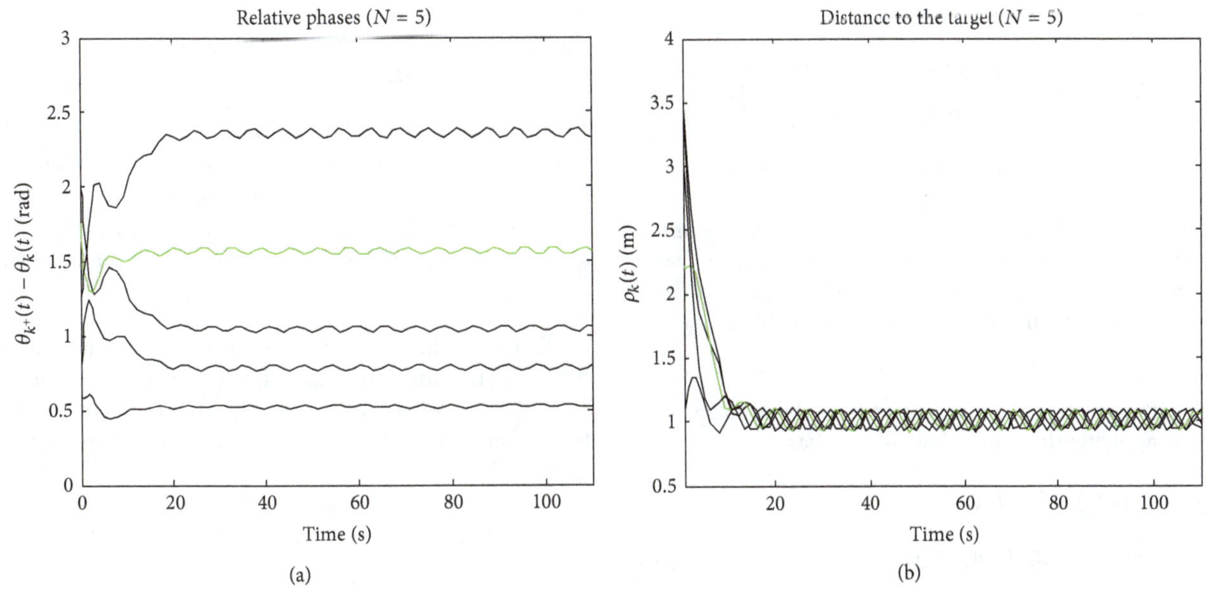

FIGURE 10: Trajectories with time of Figure 9. (a) Relative phases $\varphi_{k^+k}(t)$. (b) Distances to the target.

$$\varphi_{kk^+} = \begin{cases} \varphi_k - \varphi_{k^+} & \text{when } k = 1, 2, \ldots, N-1 \\ PV(\varphi_k - \varphi_{k^+}) \triangleq [(\varphi_k - \varphi_{k^+}) \bmod 2\pi] = \varphi_k - \varphi_{k^+} - 2\pi & \text{when } k = N, \end{cases}$$
$$\varphi_{kk^-} = \begin{cases} \varphi_k - \varphi_{k^-} & \text{when } k = 2, 3, \ldots, N \\ PV(\varphi_k - \varphi_{k^-}) \triangleq [(\varphi_k - \varphi_{k^-}) \bmod 2\pi] = \varphi_k - \varphi_{k^-} + 2\pi & \text{when } k = 1. \end{cases} \tag{4}$$

Moreover, $\sum_{k=1}^{N} |\varphi_{kk^+}(t)| \equiv 2\pi$ and $\sum_{k=1}^{N} |\varphi_{kk^-}(t)| \equiv 2\pi$ always hold. The desired relative phases are denoted by $\varphi_{kk^+}^*$ and $\varphi_{kk^-}^*$. $\sum_{k=1}^{N} |\varphi_{kk^+}^*(t)| \equiv 2\pi$ and $\sum_{k=1}^{N} |\varphi_{kk^-}^*(t)| \equiv 2\pi$ always hold as well.

Assumption 1. No two agents initially occupy the same phase, and the initial phases $\varphi_k(t_0)$ satisfy

$$0 \le \varphi_1(t_0) < \varphi_2(t_0) < \cdots < \varphi_k(t_0) < \varphi_{k+1}(t_0) < \cdots < \varphi_N(t_0) < 2\pi. \tag{5}$$

2.3. Control Objectives. The main objective is to design a distributed control law $u_k(t)$ such that the agents in (1) are formulated to a circular motion in any preset desired pattern (not limited to uniform distribution) along the circle; moreover, the agents' spatial ordering can be preserved. The structure of feedback control design of circular formation is shown in Figure 2.

3. Circular Motion Control Design

In this section, a trajectory tracking control strategy is designed to stabilize an agent to a circular motion, as illustrated in the *circular motion tracking control* part of Figure 2.

3.1. Circling a Fixed Target. In this case, the control objective can be expressed as

$$\lim_{t\to\infty} \tilde{p}_k(t) = R_{\theta_k}(t)\,\psi_k(t_0)\,\rho, \tag{6}$$

where $R_{\theta_k}(t)$ is the rotation matrix associated with $\theta_k(t)$ as

$$R_{\theta_k}(t) = \begin{pmatrix} \cos\theta_k(t) & -\sin\theta_k(t) \\ \sin\theta_k(t) & \cos\theta_k(t) \end{pmatrix}. \tag{7}$$

The right side of (6) represents a trajectory rotating on a circle with the radius ρ and rotating direction decided by $R_{\theta_k}(t)$, which can be considered as the reference trajectory for the agent state. The tracking error is defined by $e_k(t) = \tilde{p}_k(t) - R_{\theta_k}(t)\psi_k(t_0)\rho$, and the control law is given by

$$u_k(t) = -\beta\left(\tilde{p}_k(t) - R_{\theta_k}(t)\,\psi_k(t_0)\,\rho\right) + \dot{R}_{\theta_k}(t)\,\psi_k(t_0)\,\rho, \tag{8}$$

where $\beta > 0$ is the control parameter and

$$\dot{R}_{\theta_k}(t) = \dot{\theta}_k(t)\begin{pmatrix} -\sin\theta_k(t) & -\cos\theta_k(t) \\ \cos\theta_k(t) & -\sin\theta_k(t) \end{pmatrix}. \quad (9)$$

Theorem 2. *Consider a once differentiable function $\theta_k(t)$ with a bounded first time derivative; under the control law (8), $\tilde{p}_k(t)$ asymptotically converges to $R_{\theta_k}(t)\psi_k(t_0)\rho$ for any initial condition $p_k(t_0)$, meaning that agent k is enforced to converge to a circular motion of center c and radius ρ by tracking a rotating trajectory with direction determined by the sign of $\dot{\theta}_k(t)$.*

Proof. Differentiating the tracking error $e_k(t)$ with respect to time and applying the control law (8), one has

$$\begin{aligned} \dot{e}_k(t) &= \dot{p}_k(t) - \dot{R}_{\theta_k}(t)\psi_k(t_0)\rho \\ &= u_k(t) - \dot{R}_{\theta_k}(t)\psi_k(t_0)\rho \\ &= -\beta\left(\tilde{p}_k(t) - R_{\theta_k}(t)\psi_k(t_0)\rho\right) = -\beta e_k(t). \end{aligned} \quad (10)$$

Therefore, the tracking error $e_k(t)$ asymptotically converges to zero. Consider the given $\tilde{p}_k(t)$ and $\dot{\theta}_k(t)$; the closed-loop dynamics of the original system will formulate a circular motion with tracking a rotating trajectory. □

Remark 3. From both sides of (6), we note that the initial direction of $\tilde{p}_k(t)$ and $R_{\theta_k}(t)\psi_k(t_0)\rho$ is the same. Therefore, if a uniform angular velocity $\dot{\theta}_k(t) = \omega \neq 0$ is adopted, a circular motion with initial phase arrangement can be achieved.

3.2. Circling without Target. Instead of circling a target, now we consider the problem of how the agents are formulated to circular formation with circling the resulting coincident center. A protocol for center consensus can be firstly designed as

$$\dot{c}_k(t) = -k_c \sum_{j\in\mathcal{N}_k(t)} a_{kj}(t)\left(c_k(t) - c_j(t)\right), \quad (11)$$

where $k_c > 0$. Equation (11) can also be written in a vector form as

$$\dot{C}(t) = -k_c L(t) \otimes I_2 C(t), \quad (12)$$

where $C(t) = [c_1(t), \ldots, c_N(t)]^T$ and $L(t)$, which represents the communication topology, is bounded and piecewise continuous in time.

Lemma 4 (see [29]). *Suppose that $\mathcal{G}(t) = (\mathcal{I}, \mathcal{E}(t), \mathcal{A}(t))$ is uniformly connected and $L(t)$ is bounded and piecewise continuous in time. Then, the equilibrium set $\overline{C} = \{C \in \mathbb{R}^{2N} : c_k = c_j\ \forall k, j \in \mathcal{I}\}$ of (12) is uniformly exponentially stable. Furthermore, the states converge to a center point in $\overline{C} = [\bar{c}, \ldots, \bar{c}]^T$.*

Consider the center $c_k(t)$ to be a time-varying variable here; the tracking error is redefined by $e_k(t) = \tilde{p}_k(t) - R_{\theta_k}(t)\psi_k(t)\rho$, where time-varying $c_k(t)$ replaces the fixed target c in $\tilde{p}_k(t)$ and $\psi_k(t)$. Then, the error dynamics can be expressed as follows:

$$\begin{aligned} \dot{e}_k(t) &= \dot{p}_k(t) - \dot{c}_k(t) - \dot{R}_{\theta_k}(t)\psi_k(t)\rho \\ &\quad - R_{\theta_k}(t)\dot{\psi}_k(t)\rho \\ &= u_k(t) - \dot{c}_k(t) - \dot{R}_{\theta_k}(t)\psi_k(t)\rho \\ &\quad - R_{\theta_k}(t)\dot{\psi}_k(t)\rho. \end{aligned} \quad (13)$$

And a circular motion control law for the agents can be given as

$$\begin{aligned} u_k(t) &= -\beta\left(\left(p_k(t) - c_k(t)\right) - R_{\theta_k}(t)\psi_k(t)\rho\right) \\ &\quad + \dot{R}_{\theta_k}(t)\psi_k(t)\rho + R_{\theta_k}(t)\dot{\psi}_k(t)\rho + \dot{c}_k(t). \end{aligned} \quad (14)$$

Theorem 5. *Consider a once differentiable function $\theta_k(t)$ with bounded first time derivative; under the center consensus law (12) and the control law (14), $\bar{e}_k(t)$ converges to zero for any initial conditions $p_k(t_0)$ and $c_k(t_0)$; that is, all agents are simultaneously enforced to converge to a common circle of center $\bar{c}$ and radius ρ by tracking a rotating transformed trajectory with direction determined by the sign of $\dot{\theta}_k(t)$.*

Proof. Lemma 4 confirms the existence of $\bar{c}$ such that $\lim_{t\to\infty} c_k(t) = \bar{c}$ for all $k \in \mathcal{I}$. Based on the existence of $\bar{c}$, a new tracking error $\bar{e}_k(t)$ is defined as $\bar{e}_k(t) = (p_k(t) - \bar{c}) - R_{\theta_k}(t)\overline{\psi}_k\rho$, where $\bar{c}$ replaces $c_k(t)$ in $\overline{\psi}_k$. Differentiating the tracking error $\bar{e}_k(t)$ with respect to time and applying the control law (14), one has

$$\begin{aligned} \dot{\bar{e}}_k(t) &= \dot{p}_k(t) - \dot{R}_{\theta_k}(t)\overline{\psi}_k\rho = u_k(t) - \dot{R}_{\theta_k}(t)\psi_k(t_0)\rho \\ &= -\beta\bar{e}_k(t) \underbrace{-\beta e_{c_k}(t) - e_{\psi_k}(t)\left(\beta R_{\theta_k}(t)\rho + \dot{R}_{\theta_k}(t)\rho\right) + R_{\theta_k}(t)\dot{\psi}_k(t)\rho + \dot{c}_k(t)}_{w(t)\triangleq} = -\beta\bar{e}_k(t) + w(t), \end{aligned} \quad (15)$$

where $e_{c_k}(t) = \bar{c} - c_k(t)$ and $e_{\psi_k}(t) = \overline{\psi}_k - \psi(t)$. Thus, the overall error dynamics can be written in a vector form as

$$\dot{\overline{E}}(t) = -\beta\overline{E}(t) + W(t), \quad (16)$$

where $\overline{E}(t) = [\bar{e}_1(t), \ldots, \bar{e}_N(t)]^T$ and $W(t) = [w_1(t), \ldots, w_N(t)]^T$. Note that (16) and the dynamics of $W(t)$ form a cascade system. It is noted from Lemma 4 that $e_{c_k}(t)$, $e_{\psi_k}(t)$, $\dot{\psi}_k$, and $\dot{c}_k(t)$ converge to zero; that is, $W(t) = 0$, as $t \to \infty$. Thus, $\bar{e}_k(t)$ will converge to zero asymptotically according to $\dot{\bar{e}}_k(t) = -\beta \bar{e}_k(t)$. □

Remark 6. Note that the center consensus subsystem and the circle control subsystem form a cascade system such that the center consensus subsystem does not rely on the position information. Thus, before the system starts, each agent is initialized at any random center instead of a real target, and all centers can reach a common center point far faster than that of the formation control.

4. Any Preset Phase Arrangement along a Circle

The previous control law does not take into consideration the phase configuration problem among agents. In order to stabilize the agents to a circular formation with any preset phase distribution, the circular motion control law (8) or (14) must include a cooperative part for arranging phases. Through the analysis of control law (8) or (14), the rotation matrix $R_{\theta_k}(t)$ is found as a breakthrough point. The phase $\varphi_k(t)$ of agent k can be arranged by controlling the rotation angle $\theta_k(t)$ of $R_{\theta_k}(t)$ with the relative phases φ_{kk^+} and φ_{kk^-}. Thus, a control scheme is designed as shown in the *phase control* part of Figure 2.

Firstly, a phase control law has been proposed in [21] as

$$\begin{aligned}\dot{\varphi}_k(t) = &-\frac{\left|\varphi^*_{kk^-}\right|}{\left|\varphi^*_{kk^+}\right| + \left|\varphi^*_{kk^-}\right|}\varphi_{kk^+}(t) \\ &- \frac{\left|\varphi^*_{kk^+}\right|}{\left|\varphi^*_{kk^+}\right| + \left|\varphi^*_{kk^-}\right|}\varphi_{kk^-}(t) \quad k = 1, 2, \ldots, N.\end{aligned} \tag{17}$$

With the stacked column vector $\Phi(t) = [\varphi_1(t), \varphi_2(t), \ldots, \varphi_N(t)]^T$, (17) can be written in a vector form as

$$\dot{\Phi}(t) = -L^* \Phi(t), \tag{18}$$

$$L^* = \begin{bmatrix} 1 & -\frac{\left|\varphi^*_{1N}\right|}{\left|\varphi^*_{12}\right| + \left|\varphi^*_{1N}\right|} & 0 & \cdots & 0 & -\frac{\left|\varphi^*_{12}\right|}{\left|\varphi^*_{12}\right| + \left|\varphi^*_{1N}\right|} \\ -\frac{\left|\varphi^*_{23}\right|}{\left|\varphi^*_{23}\right| + \left|\varphi^*_{21}\right|} & 1 & -\frac{\left|\varphi^*_{21}\right|}{\left|\varphi^*_{23}\right| + \left|\varphi^*_{21}\right|} & \cdots & 0 & 0 \\ 0 & -\frac{\left|\varphi^*_{34}\right|}{\left|\varphi^*_{34}\right| + \left|\varphi^*_{32}\right|} & 1 & \cdots & 0 & 0 \\ \vdots & \vdots & \vdots & \vdots & \vdots & \vdots \\ 0 & 0 & 0 & \cdots & 1 & -\frac{\left|\varphi^*_{(N-1)(N-2)}\right|}{\left|\varphi^*_{(N-1)N}\right| + \left|\varphi^*_{(N-1)(N-2)}\right|} \\ -\frac{\left|\varphi^*_{N1}\right|}{\left|\varphi^*_{N1}\right| + \left|\varphi^*_{N(N-1)}\right|} & 0 & 0 & \cdots & -\frac{\left|\varphi^*_{N(N-1)}\right|}{\left|\varphi^*_{N1}\right| + \left|\varphi^*_{N(N-1)}\right|} & 1 \end{bmatrix}. \tag{19}$$

Lemma 7 (see [21]). *Given any admissible circular formation characterized by $\varphi^*_{kk^+}$ and $\varphi^*_{kk^-}$, the circular formation problem is solved with order preservation under the proposed control law (17).*

The above phase control law (17) can drive the phase $\varphi_k(t)$ of agent k to move towards its waypoint that is determined completely by its two neighbors' relative phases and the desired relative phases $\varphi^*_{kk^+}$ and $\varphi^*_{kk^-}$. Instead of a fixed circular formation, in order to keep the phase arrangement always in a circular motion, a new waypoint based phase control law can be equivalently redesigned in a rotating manner as

$$\begin{aligned}\dot{\varphi}_k(t) = \omega_0 + \alpha\Bigg(&-\frac{\left|\varphi^*_{kk^-}\right|}{\left|\varphi^*_{kk^+}\right| + \left|\varphi^*_{kk^-}\right|}\varphi_{kk^+}(t) \\ &- \frac{\left|\varphi^*_{kk^+}\right|}{\left|\varphi^*_{kk^+}\right| + \left|\varphi^*_{kk^-}\right|}\varphi_{kk^-}(t)\Bigg),\end{aligned} \tag{20}$$

where ω_0 is a constant and $\alpha > 0$ is a control parameter. One feature of the proposed control law (20) is that it guarantees that the spatial ordering of the agents is preserved throughout the system's evolution in the initial condition (5), and thus no collision takes place during the process of circular formation. In addition, it is noted that the relationship between the phase $\varphi_k(t)$ and the rotation angle $\theta_k(t)$ is

$$\theta_k(t) = \varphi_k(t) - \varphi_k(t_0). \tag{21}$$

Then, combined with the phase control, the control law (8) can be redesigned as

$$u_k(t) = -\beta\left(\tilde{p}_k(t) - R_{\theta_k}(t)\psi_k(t_0)\rho\right) + \dot{\varphi}_k(t)R_{\theta_k}(t)\psi_k(t_0)\rho, \tag{22}$$

where $\dot{\varphi}_k(t)$ is given in (20).

Theorem 8. *Let $\mathscr{G}$ be a ring topology graph and L^* be its corresponding Laplacian matrix defined by (19). Consider model (1) with any initial conditions $p_k(t_0)$ and satisfying Assumption 1; under the circular motion control law (22) and phase control law (20), the agents are all enforced to converge to a circular formation of center c and radius ρ with phase arrangement by presetting $\varphi^*_{kk^+}$ and $\varphi^*_{kk^-}$, for $k = 1, 2, \ldots, N$. Simultaneously, the agents' spatial ordering is preserved.*

Proof. The proof of the circular motion control part is analyzed in Theorem 2. The convergence proof of the phase control part is a consequence of Theorem 1 of [21]. So, the proof of this theorem is omitted here. □

In some practical situations, the uneven distribution circular formation needs some particular agents to observe the target from particular directions, such as agent 1. In this case, by introducing a phase reference φ_d, the phase control law for this agent can be redesigned as

$$\dot{\varphi}_1(t) = \omega_0 + \alpha\left(-\frac{|\varphi^*_{1N}|}{|\varphi^*_{12}| + |\varphi^*_{1N}|}\varphi_{12}(t) - \frac{|\varphi^*_{12}|}{|\varphi^*_{12}| + |\varphi^*_{1N}|}\varphi_{1N}(t)\right) + d\sin(\varphi_d - \varphi_1), \tag{23}$$

where $d > 0$ is a constant. This yields the following extension of Theorem 8.

Corollary 9. *Under the same conditions of Theorem 8, with the circular motion control law (22) and phase control laws (20) and (23), all the agents will be enforced to converge to a circular formation as described in Theorem 8 and the phases φ_k of agent k can be steered to their desired phase φ^*_k for $k = 2, \ldots, N$ by referring to the phase $\varphi_1 = \varphi_d$ of agent 1.*

Proof. By defining $y_k(t) = \varphi_{k^+k}(t)$, the stability of the control system can be studied by Lyapunov function as

$$V = \frac{1}{2}y^T y + d(1 - \cos(\varphi_d - \varphi_1)), \tag{24}$$

where $y = [y_1, y_2, \ldots, y_N]^T$. Differentiating V yields

$$\dot{V} = -\alpha y^T L^* y + d\sin(\varphi_d - \varphi_1)(\dot{\varphi}_d - \dot{\varphi}_1). \tag{25}$$

The control law (23) with $\dot{\varphi}_d = \omega_0$ for agent 1, (20) for agents $k = 2, \ldots, N$, and u_k given by (22) result in $\dot{V} = -\alpha y^T L^* y \le 0$ and $L^*\Phi(t) = 0$ because the matrix L^* has rank $N - 1$. The relation

$$d\sin(\varphi_d - \varphi_1) = 0 \tag{26}$$

implies that $\varphi_1 = \varphi_d$ or $\varphi_d + \pi$. Therefore, the relative equilibrium is $\varphi_1 = \varphi_d$ which minimizes the potential V. □

It is noted that, in the case $\omega_0 = 0$, Corollary 9 proves that the agents are stabilized to a fixed circular formation with agent 1's phase $\varphi_1 = \varphi_d$, so that the other agents are localized with respect to agent 1's position. In addition, the result of Theorem 8 and Corollary 9 can also be applied to the time-varying center case; that is, the control law (14) can also be combined with phase control law (23).

5. Simulation Results

This section presents the simulation of five agents ($N = 5$) modeled by (1), governed by the proposed control strategies above. The desired circular formation is defined by radius $\rho = 1$ and relative headings are arranged as $\pi/6, \pi/4, \pi/3, \pi/2, 3\pi/4$. The simulation results with $k_c = \alpha = \beta = \omega_0 = d = 1$ are, respectively, demonstrated in the following several scenarios. (1) Figures 3 and 4 show that all the agents circumnavigate the fixed target $c = (1, 1)$ under the circular motion control law (22) and the phase control law (20). (2) While there is no fixed target, the control laws (14) and (20) formulate a circle circumnavigating the resulting coincident center by the consensus law (11) as shown in Figures 5 and 6. (3) In Figures 7 and 8, a fixed circular formation for positioning agent 1 at $\varphi_d = \pi/2$ is formulated by the control laws (20), (22), and (23). (4) All the agents circumnavigate a slow moving target under the control laws (14) and (20) as shown in Figures 9 and 10, where $c(t) = (1 + (1/10)t, 1 + 4.5\sin(2\pi/300)t)^T$.

The First Scenario. See Figures 3 and 4.

The Second Scenario. See Figures 5 and 6.

The Third Scenario. See Figures 7 and 8.

The Fourth Scenario. See Figures 9 and 10.

All the above simulation results show that the group of agents can asymptotically converge to the desired circular formation under the proposed control law. In particular, the figures clearly demonstrate that the agents preserve their orderings in the entire motion.

6. Conclusion

A new control strategy is presented to formulate a desired circular formation with any preset phase distribution. The problem is solved in two parts: circular motion control and phase control. The circular motion control part is designed to circumnavigate a fixed target. Particularly, with a center consensus method, no target case is solved by considering the center $c_k(t)$ to be a time-varying variable. Through combining the circular motion control law with a phase control law, the limitation on a one-dimensional space of circular motion in [21] is released by introducing the radius control of circular motion. Furthermore, the positioning problem of agents on the circle is solved for some application situations. Future works will focus on extending the proposed strategy to agents with more practical dynamics, including some additional constraints to ensure obstacle avoidance. In addition, it is of interest to consider the extension of the proposed strategy to three-dimensional space.

Conflicts of Interest

The authors declare that they have no conflicts of interest.

Acknowledgments

This work was supported by the Scientific Research Fund of Liaoning Provincial Education Department (L2016011), Liaoning Provincial Natural Science Foundation (201601321), and Talent Scientific Research Fund of LSHU (2016XJJ-033) of China.

References

[1] Z. Peng, D. Wang, Y. Shi, H. Wang, and W. Wang, "Containment control of networked autonomous underwater vehicles with model uncertainty and ocean disturbances guided by multiple leaders," *Information Sciences*, vol. 316, pp. 163–179, 2015.

[2] B. Das, B. Subudhi, and B. B. Pati, "Co-operative control coordination of a team of underwater vehicles with communication constraints," *Transactions of the Institute of Measurement and Control*, vol. 38, no. 4, pp. 463–481, 2016.

[3] X. Dong, B. Yu, Z. Shi, and Y. Zhong, "Time-varying formation control for unmanned aerial vehicles: theories and applications," *IEEE Transactions on Control Systems Technology*, vol. 23, no. 1, pp. 340–348, 2015.

[4] R. Olfati-Saber, J. A. Fax, and R. M. Murray, "Consensus and cooperation in networked multi-agent systems," *Proceedings of the IEEE*, vol. 95, no. 1, pp. 215–233, 2007.

[5] N. E. Leonard, D. A. Paley, F. Lekien, R. Sepulchre, D. M. Fratantoni, and R. E. Davis, "Collective motion, sensor networks, and ocean sampling," *Proceedings of the IEEE*, vol. 95, no. 1, pp. 48–74, 2007.

[6] R. Olfati-Saber and R. M. Murray, "Consensus problems in networks of agents with switching topology and time-delays," *Institute of Electrical and Electronics Engineers Transactions on Automatic Control*, vol. 49, no. 9, pp. 1520–1533, 2004.

[7] D. A. Paley, N. E. Leonard, and R. Sepulchre, "Stabilization of symmetric formations to motion around convex loops," *Systems & Control Letters*, vol. 57, no. 3, pp. 209–215, 2008.

[8] M. Ou, H. Du, and S. Li, "Finite-time formation control of multiple nonholonomic mobile robots," *International Journal of Robust and Nonlinear Control*, vol. 24, no. 1, pp. 140–165, 2014.

[9] R. Olfati-Saber, "Flocking for multi-agent dynamic systems: algorithms and theory," *Institute of Electrical and Electronics Engineers Transactions on Automatic Control*, vol. 51, no. 3, pp. 401–420, 2006.

[10] P. Ogren, E. Fiorelli, and N. E. Leonard, "Cooperative control of mobile sensor networks: adaptive gradient climbing in a distributed environment," *IEEE Transactions on Automatic Control*, vol. 49, no. 8, pp. 1292–1302, 2004.

[11] D. V. Dimarogonas and K. J. Kyriakopoulos, "On the rendezvous problem for multiple nonholonomic agents," *Institute of Electrical and Electronics Engineers Transactions on Automatic Control*, vol. 52, no. 5, pp. 916–922, 2007.

[12] W. Wu and F. Zhang, "A speeding-up and slowing-down strategy for distributed source seeking with robustness analysis," *IEEE Transactions on Control of Network Systems*, vol. 3, no. 3, pp. 231–240, 2016.

[13] S. Li, R. Kong, and Y. Guo, "Cooperative distributed source seeking by multiple robots: Algorithms and experiments," *IEEE/ASME Transactions on Mechatronics*, vol. 19, no. 6, pp. 1810–1820, 2014.

[14] J.-C. Juang, "On the formation patterns under generalized cyclic pursuit," *Institute of Electrical and Electronics Engineers Transactions on Automatic Control*, vol. 58, no. 9, pp. 2401–2405, 2013.

[15] L. Ma and N. Hovakimyan, "Vision-based cyclic pursuit for cooperative target tracking," *Journal of Guidance, Control, and Dynamics*, vol. 36, no. 2, pp. 617–622, 2013.

[16] C. Song, L. Liu, and G. Feng, "Coverage control for mobile sensor networks with input saturation," *Unmanned Systems*, vol. 4, no. 1, pp. 15–21, 2016.

[17] Y. J. Shi, R. Li, and K. L. Teo, "Cooperative enclosing control for multiple moving targets by a group of agents," *International Journal of Control*, vol. 88, no. 1, pp. 80–89, 2015.

[18] I. Shames, S. Dasgupta, B. Fidan, and B. D. Anderson, "Circumnavigation using distance measurements under slow drift," *IEEE Transactions on Automatic Control*, vol. 57, no. 4, pp. 889–903, 2012.

[19] L. Briñón-Arranz, L. Schenato, and A. Seuret, "Distributed source seeking via a circular formation of agents under communication constraints," *IEEE Transactions on Control of Network Systems*, vol. 3, no. 2, pp. 104–115, 2016.

[20] J. A. Marshall, M. E. Broucke, and B. A. Francis, "Formations of vehicles in cyclic pursuit," *IEEE Transactions on Automatic Control*, vol. 49, no. 11, pp. 1963–1974, 2004.

[21] S. Daingade, A. Sinha, A. V. Borkar, and H. Arya, "A variant of cyclic pursuit for target tracking applications: theory and implementation," *Autonomous Robots*, vol. 40, no. 4, pp. 669–686, 2016.

[22] R. Zheng, Y. Liu, and D. Sun, "Enclosing a target by nonholonomic mobile robots with bearing-only measurements," *Automatica*, vol. 53, pp. 400–407, 2015.

[23] A. Jain and D. Ghose, "Collective circular motion in synchronized and balanced formations with second-order rotational dynamics," *Communications in Nonlinear Science and Numerical Simulation*, vol. 54, pp. 156–173, 2018.

[24] Z. Chen and H.-T. Zhang, "No-beacon collective circular motion of jointly connected multi-agents," *Automatica*, vol. 47, no. 9, pp. 1929–1937, 2011.

[25] Z. Chen and H.-T. Zhang, "A remark on collective circular motion of heterogeneous multi-agents," *Automatica*, vol. 49, no. 5, pp. 1236–1241, 2013.

[26] R. Sepulchre, D. A. Paley, and N. E. Leonard, "Stabilization of planar collective motion: all-to-all communication," *IEEE Transactions on Automatic Control*, vol. 52, no. 5, pp. 811–824, 2007.

[27] C. Wang, G. Xie, and M. Cao, "Forming circle formations of anonymous mobile agents with order preservation," *IEEE Transactions on Automatic Control*, vol. 58, no. 12, pp. 3248–3254, 2013.

[28] C. Wang, G. Xie, and M. Cao, "Controlling anonymous mobile agents with unidirectional locomotion to form formations on a circle," *Automatica*, vol. 50, no. 4, pp. 1100–1108, 2014.

[29] L. Moreau, "Stability of continuous-time distributed consensus algorithms," in *Proceedings of the 43rd IEEE Conference on Decision and Control (CDC '04)*, pp. 3998–4003, December 2004.

18

Safety-Guaranteed Course Control of Air Cushion Vehicle with Dynamic Safe Space Constraint

Mingyu Fu, Shuang Gao, and Chenglong Wang

College of Automation, Harbin Engineering University, Harbin 150001, China

Correspondence should be addressed to Shuang Gao; gaussian@hrbeu.edu.cn

Academic Editor: Sing Kiong Nguang

This paper develops a safety-guaranteed course controller for air cushion vehicle (ACV). As the safety criterion, the unique safety limit of ACV including turn rate (TR) and sideslip angle (SA) changes with the speed when ACV is turning. To be more intuitive to show the change of safety limit and more convenient for safety monitoring and control, dynamic safe space of ACV is proposed. If the work point is within the safe space during the manual operation or automatic control, the sailing of ACV is safe. Then, the safety-guaranteed controller is designed to keep TR and SA within the safe space during the course control process based on the dynamic safe space constraint, terminal sliding mode control, and adaptive mechanism. The adaptive mechanism can effectively estimate the system uncertainty and external disturbances online without the requirement of their upper bounds. The proposed controller guarantees the convergence of tracking error. Simulations are implemented to demonstrate the efficacy of the designed controller.

1. Introduction

An air cushion vehicle (ACV, as shown in Figure 1) is supported totally by its air cushion, with a flexible skirt system around its periphery to seal the cushion air [1]. The ACV is able to run at high speed over shallow water, rapids, ice, and swamp where no other craft can go. These "special abilities" have attracted many military and civil users with particular mission requirements. The study about the safety-guaranteed motion control of ACV moving with high velocities is meaningful to reduce the burden of drivers.

From a detailed review of the available literatures about the motion control of ACV such as Antonio Pedro Aguiar [2], Hebertt Sira-Ramirez [3], Tanaka K [4], Han B [5], Aranda [6], Zhao Jingbo [7], Morales R [8], Gerasimos G. Rigatos[9], and Shojaei Khoshnam [10, 11], the designs only concern the success of control task and the improvement of control performance. And the safety of ACV which is the necessary precondition to complete the control task is neglected by them.

The turn rate (TR) and sideslip angle (SA) play key roles for the safety of ACV in the high-speed moving process [12, 13]. If the SA exceeds the angle of drift which corresponds to the maximum of hydrodynamic forces, the behavior of hovercraft will be nonstable [14]. The dangers caused by the TR and SA include stern-kickoff, plough-in, and great heeling [12]. Hence, safety limit of TR and SA must be strictly obeyed in the high-speed moving process to ensure safe maneuvers of ACV [13]. For instance, safety limit of an ACV in [13] shows that if speed is 35 knots, the limits of TR and SA are $1.2^{\circ}/s$ and 2°. If speed is 14 knots, the limits of TR and SA are $3.5^{\circ}/s$ and 20°. The safety limit of TR and SA has been considered in the course controller and trajectory controller by Mingyu Fu [1, 15]. However, the constant safety limit is used in them. In practice, the safety limit changes with the speed of ACV. Hence in this paper, the changed safety limit is considered. And a dynamic safe space is established based on the relationship of the speed and the safety limit.

Moreover, sliding mode control (SMC) has attracted significant amount of interest due to its fast global convergence, simplicity of implementation, and high robustness to external disturbances [16, 17]. Compared to the conventional linear sliding mode (LSM) control, terminal sliding mode (TSM) control has superior properties such as fast and finite-time convergence and smaller steady-state tracking errors [18–22]. So a safety-guaranteed course controller is designed based

FIGURE 1: A 3D model of hovercraft. Photo is from the international cooperation project described in the Acknowledgment.

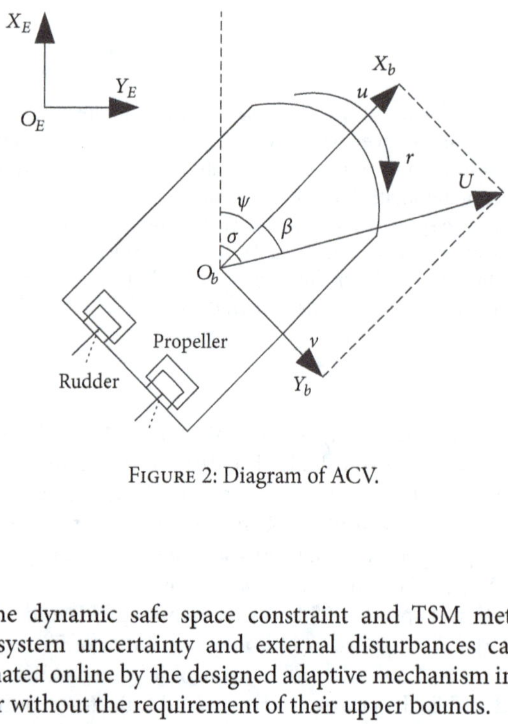

FIGURE 2: Diagram of ACV.

on the dynamic safe space constraint and TSM method. The system uncertainty and external disturbances can be estimated online by the designed adaptive mechanism in this paper without the requirement of their upper bounds.

Remark 1. The control task in this paper is chosen as course control. It is an important and common control task for ACV due to the difference with heading control as shown in Figure 2. But it needs to be noted that the dynamic safe space constraint method is suitable to other control task of ACV for safety.

2. Problem Formulation

2.1. ACV Model Description. The three degrees of freedom (DOF) model is

$$\begin{aligned} \dot{\eta} &= R(\psi)\,\nu \\ M\dot{\nu} + f(\nu) &= \tau_c + \tau_w + d(t) \end{aligned} \tag{1}$$

where

$$R(\psi) = \begin{bmatrix} \cos\psi & -\sin\psi & 0 \\ \sin\psi & \cos\psi & 0 \\ 0 & 0 & 1 \end{bmatrix} \tag{2}$$

M is the inertia matrix of ACV, $\eta = (\xi, \eta, \psi)^T$ and $\nu = (u, v, r)^T$ are position/Euler angles and velocities, $d(t) = (d_u(t), d_v(t), d_r(t))^T$ is the bounded input disturbance, $\|d(t)\| < d_0$, where $d_0 > 0$, $\tau_c = (\tau_{cu}, 0, \tau_{cr})^T$ is the forces and moments provide by the actuators, and $\tau_w = (\tau_{wu}, \tau_{wv}, \tau_{wr})^T$ is the external environment forces written as

$$\begin{aligned} F_{ua} &= -0.5\rho_a V_a^2 C_{ua} S_{PP} \\ F_{va} &= -0.5\rho_a V_a^2 C_{va} S_{LP} \\ M_{ra} &= -0.5\rho_a V_a^2 C_{ra} S_{HP} H_{hov} + F_{va} x_a + F_{ua} y_a \\ F_m &= \rho_a V_a Q \\ F_{wm} &= \frac{C_{wm} p_c^2 B_c}{\rho_w g} \\ F_{sk} &= 0.5\rho_w V_a^2 C_{sk} \left(\frac{h}{l_{sk}}\right)^{-0.34} l_{sk} S_c^{0.5} \\ &\quad + \left(2.8167 \left(\frac{p_c}{l_c}\right)^{-0.259} - 1\right) F_{wm} \\ \tau_{wu} &= F_{ua} - F_m \cos\beta - F_{wm} \cos\beta - F_{sk} \cos\beta \\ \tau_{wv} &= F_{va} - F_m \sin\beta - F_{wm} \sin\beta - F_{sk} \sin\beta \\ \tau_{wr} &= M_{ra} - F_m x_m \sin\beta + F_m y_m \cos\beta \\ &\quad - F_{wm} x_{wm} \sin\beta + F_{wm} y_{wm} \cos\beta \\ &\quad - F_{sk} x_{sk} \sin\beta + F_{sk} y_{sk} \cos\beta \end{aligned} \tag{3}$$

where the suffixes ua, va, and ra mean aerodynamic profile drags in three DOF, respectively. The suffixes m, wm, and sk mean wave-making drag, air momentum drag, and skirt drag, respectively. l_c and B_c denote cushion length and beam, S_{PP}, S_{LP}, and S_{HP} are positive, lateral, and horizontal projection areas, S_c means cushion area, β means drift angle, h is the average clearance for air leakage in static hovering mode, l_{sk} is the total peripheral length of the skirts, H_{hov} is the height of the hull, ρ_a and ρ_w are air and water density, and (x_a, y_a), (x_m, y_m),(x_{wm}, y_{wm}), and (x_{sk}, y_{sk}) are the coordinates of these acting points. V_a can be obtained by

$$\begin{aligned} V_a &= \sqrt{[u + V_w \cos(\beta_w - \psi)]^2 + [v + V_w \sin(\beta_w - \psi)]^2} \\ \beta_a &= \tan^{-1}\left[\frac{v + V_w \sin(\beta_w - \psi)}{u + V_w \cos(\beta_w - \psi)}\right] \end{aligned} \tag{4}$$

in which V_w and β_w are absolute wind speed and direction. For more details can see [1, 17, 23].

Considering the modeling errors and the variations of parameters, M and $f(\nu)$ can be rewritten as follows:

$$M = M_0 + \Delta M$$
$$f(\nu) = f_0(\nu) + \Delta f(\nu) \tag{5}$$

where

$$M_0 = \begin{bmatrix} m & 0 & 0 \\ 0 & m & 0 \\ 0 & 0 & I_z \end{bmatrix} \tag{6}$$

$$f_0(\nu) = \begin{bmatrix} -mvr \\ mur \\ 0 \end{bmatrix} \tag{7}$$

Substituting (5) into (3) yields

$$M_0 \dot{\nu} + f_0(\nu) = \tau_c + \tau_w + \delta(t) \tag{8}$$

where $\delta(t) = (\delta_1(t), \delta_2(t), \delta_3(t))^T$ is defined as follows:

$$\delta(t) = -\Delta M \dot{\nu} - \Delta f(\nu) + d(t) \tag{9}$$

Then the following assumptions are made.

Assumption 2. The norm of inertia matrix M is upper bounded by a positive number a_0.

$$\|M\| < a_0 \tag{10}$$

Assumption 3. The vector $f(\nu)$ is upper bounded by a positive function

$$\|f(\nu)\| < \rho_0 + \rho_1 \|\nu\| + \rho_2 \|\dot{\nu}\|^2 \tag{11}$$

where ρ_0, ρ_1, and ρ_2 are positive numbers.

Then the system uncertainty will be bounded in the following form:

$$\|\delta(t)\| < b_0 + b_1 \|\nu\| + b_2 \|\dot{\nu}\|^2 \tag{12}$$

This bounded property has been used by some researchers of [24–26].

2.2. Dynamic Safe Space of ACV

Lemma 4. *Assuming that ε is any positive value, then the open interval $(a-\varepsilon, a+\varepsilon)$ is ε neighborhood of a, write for $N(a,\varepsilon)$, that is,*

$$N(a,\varepsilon) = \{x \mid a-\varepsilon < x < a+\varepsilon\} \tag{13}$$

Definition 5. From Lemma 4, we define the following expression:

$$S(a,\varepsilon_1,\varepsilon_2) = \{(x,y) \mid x \in (a-\varepsilon_1, a+\varepsilon_1),\ y \in (a-\varepsilon_2, a+\varepsilon_2)\} \tag{14}$$

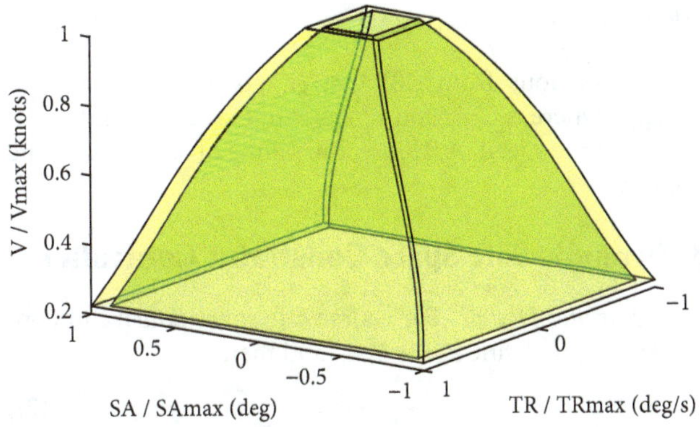

FIGURE 3: The relationship of safe space, green space, and buffer space.

Defining the interval of speed of ACV as ϕ_u, for any $u_i \in \phi_u$, the safety limit values of TR and SA are written as $\varepsilon_{u_i rsafe}$ and $\varepsilon_{u_i \beta safe}$, and then the safety area of ACV at the speed u_i is

$$S(0, \varepsilon_{u_i rsafe}, \varepsilon_{u_i \beta safe}) \tag{15}$$

Then the sailing is safe if the following condition is satisfied.

$$(r_{u_i}, \beta_{u_i}) \in S(0, \varepsilon_{u_i rsafe}, \varepsilon_{u_i \beta safe}) \tag{16}$$

The green limit values of TR and SA at the speed of $u_i \in \phi_u$ are defined as $\varepsilon_{u_i rgs}$ and $\varepsilon_{u_i \beta gs}$ which satisfy $0 \le \varepsilon_{u_i rgs} < \varepsilon_{u_i rsafe}$ and $0 \le \varepsilon_{u_i \beta gs} < \varepsilon_{u_i \beta safe}$.

From Definition 5, we know that

$$S(0, \varepsilon_{u_i rgs}, \varepsilon_{u_i \beta gs}) \subset S(0, \varepsilon_{u_i rsafe}, \varepsilon_{u_i \beta safe}) \tag{17}$$

Then the dynamic safe space $\chi_{\phi_u safe}$ is obtained as follows:

$$\chi_{\phi_u safe} = S(0, \varepsilon_{rsafe}, \varepsilon_{\beta safe}) = \bigcup_{u_i \in \phi_u} S(0, \varepsilon_{u_i rsafe}, \varepsilon_{u_i \beta safe}) \tag{18}$$

The dynamic green space $\chi_{\phi_u gs}$ is defined as follows:

$$\chi_{\phi_u gs} = S(0, \varepsilon_{rgs}, \varepsilon_{\beta gs}) = \bigcup_{u_i \in \phi_u} S(0, \varepsilon_{u_i rgs}, \varepsilon_{u_i \beta gs}) \tag{19}$$

Then the safe buffer space $\chi_{\phi_u bs}$ is gotten by

$$\chi_{\phi_u bs} = \chi_{\phi_u safe} - \chi_{\phi_u gs} \tag{20}$$

The relationships of $\chi_{\phi_u gs}$, $\chi_{\phi_u safe}$, and $\chi_{\phi_u bs}$ are shown in Figure 3. If the speed of ACV exceeds V_{max}, the motion of turning is not allowed. If the speed is lower than $0.2V_{max}$, the TR and SA are not required to be limited. The inside space of green boundary is green space. The inside space of yellow boundary is safe space. The space between green boundary and yellow boundary is buffer space.

Theorem 6. *For any $u_k \in \phi_u$, if $(r_{u_k}, \beta_{u_k}) \in \chi_{\phi_u gs}$ or $(r_{u_k}, \beta_{u_k}) \in \chi_{\phi_u bs}$, the sailing of ACV is safe.*

Proof. $(r_{u_k}, \beta_{u_k}) \in \chi_{\phi_u gs}$ or $(r_{u_k}, \beta_{u_k}) \in \chi_{\phi_u bs}$ means $(r_{u_k}, \beta_{u_k}) \in \chi_{\phi_u safe}$.

It is obvious from (18) that $(r_{u_k}, \beta_{u_k}) \in S(0, \varepsilon_{u_k rsafe}, \varepsilon_{u_k \beta safe})$, where $\varepsilon_{u_k rsafe}$ and $\varepsilon_{u_k \beta safe}$ are the safety limit of TR and SA at the speed u_k. Hence, the sailing of ACV is safe and stability. □

3. Dynamic Safe Space Constraint Controller

3.1. Controller Design. For course safe space constraint controller design, define $e_1 = \sigma - \sigma_d$ and then

$$e_2 = \dot{e}_1 = \dot{\sigma} - \dot{\sigma}_d = r + \dot{\beta} - \dot{\sigma}_d \tag{21}$$

$$\dot{e}_2 = \dot{r} + \ddot{\beta} - \ddot{\sigma}_d = \frac{1}{I_z}\left(\tau_{cr} + \tau_{wr} + \delta_3(t)\right) + \ddot{\beta} - \ddot{\sigma}_d \tag{22}$$

Choose the sliding mode surface and approaching law as follows based on the fast terminal sliding mode approach:

$$s = e_1 + \frac{1}{\alpha_1} e_1^{g/h} + \frac{1}{\alpha_2} e_2^{p/q} \tag{23}$$

$$\dot{s} = \left(-\phi s - \gamma s^{m/n}\right) e_2^{p/q-1} \tag{24}$$

where $\alpha_1, \alpha_2 \in R^+, p, q, g, h \in N$ are odd numbers and $1 < p/q < 2, g/h > p/q$. $\phi \in R^+, \gamma \in R^+, m, n \in N$ are odd numbers, and $0 < m/n < 1$.

After considering the safe space constraint, the control law is given by

$$\begin{aligned}\tau_{cr} = {} & -I_z\left(Y - \ddot{\psi}_d\right) - \tau_{wr} - k_1 s \xi_{gs} \\ & - \left(\widehat{b}_{02} + \widehat{b}_{12}\|r\| + \widehat{b}_{22}\|\dot{r}\|^2\right)\end{aligned} \tag{25}$$

where

$$\begin{aligned}Y = {} & \frac{\alpha_2 q}{p}\left(\phi s + \gamma s^{m/n}\right) \\ & + \frac{\alpha_2 q}{p} e_2^{2-(p/q)}\left(1 + \frac{g}{\alpha_1 h} e_1^{(g/h)-1}\right)\end{aligned}$$

$$\begin{aligned}\xi_{gs} = {} & k_r f_{lock}\left(|r| - \delta_{u_i rgs}\right) \\ & + k_\beta f_{lock}\left(|\beta| - \delta_{u_i \beta gs}\right)\end{aligned} \tag{26}$$

$$f_{lock}(x) = \begin{cases} x, & x > 0 \\ 0, & x \le 0 \end{cases}$$

and $k_1 > 0,\ k_r > 0,\ k_\beta > 0$.

The adaptive laws are

$$\begin{aligned}\dot{\widehat{b}}_{02} &= x_0 \|s\| \frac{p}{I_z \beta q} e_2^{p/q-1} \\ \dot{\widehat{b}}_{12} &= x_1 \|s\| \frac{p}{I_z \beta q} e_2^{p/q-1} \|r\| \\ \dot{\widehat{b}}_{22} &= x_2 \|s\| \frac{p}{I_z \beta q} e_2^{p/q-1} \|\dot{r}\|^2\end{aligned} \tag{27}$$

where x_0, x_1, and x_2 are arbitrary positive constants.

3.2. Stability Analysis

Theorem 1. *If the control law (25), the sliding mode surface (23), approaching law (24), and adaptive laws (27) are applied to ACV, the controller is stable and the course error will converge to zero in finite time.*

Proof. The following Lyapunov function is defined:

$$V = \frac{1}{2}s^2 + \frac{1}{2}\sum_{i=0}^{2} x_i^{-1}\widetilde{b}_{i2}^2 \tag{28}$$

where $\widetilde{b}_{i2} = b_{i2} - \widehat{b}_{i2},\ i \in \{0, 1, 2\}$.

Then

$$\begin{aligned}\dot{V} &= s\dot{s} - \sum_{i=0}^{2} x_i^{-1}\widetilde{b}_{i2}\dot{\widehat{b}}_{i2} \\ &= s\dot{e}_1 + \frac{g}{\alpha_1 h} s e_1^{g/h-1}\dot{e}_1 - \sum_{i=0}^{2} x_i^{-1}\widetilde{b}_{i2}\dot{\widehat{b}}_{i2} \\ &\quad + \frac{s p e_2^{p/q-1}}{\alpha_2 q}\left(\frac{1}{I_z}\left(\tau_{cr} + \tau_{wr} + \delta_3(t)\right) + \ddot{\beta} - \ddot{\sigma}_d\right)\end{aligned} \tag{29}$$

Substituting (25) into (29) yields

$$\begin{aligned}\dot{V} &= \frac{s p e_2^{p/q-1}}{I_z \alpha_2 q}\left(\delta_3(t) - \left(\widehat{b}_{02} + \widehat{b}_{12}\|r\| + \widehat{b}_{22}\|\dot{r}\|^2\right)\right) \\ &\quad - \dot{V}_1 - \dot{V}_2 - \sum_{i=0}^{2} x_i^{-1}\widetilde{b}_{i2}\dot{\widehat{b}}_{i2} \\ &\le \frac{\|s\| p}{I_z \alpha_2 q} e_2^{p/q-1}\|\delta_3(t)\| \\ &\quad - \frac{\|s\| p}{I_z \alpha_2 q} e_2^{p/q-1}\left(\widehat{b}_{02} + \widehat{b}_{12}\|r\| + \widehat{b}_{22}\|\dot{r}\|^2\right) - \dot{V}_1 \\ &\quad - \dot{V}_2 - \sum_{i=0}^{2} x_i^{-1}\widetilde{b}_{i2}\dot{\widehat{b}}_{i2}\end{aligned} \tag{30}$$

where

$$\dot{V}_1 = e_2^{p/q-1}\left(\phi s^2 + \gamma s^{m/n+1}\right) \tag{31}$$

$$\dot{V}_2 = k_1 \frac{s^2 p}{I_z \alpha_2 q} e_2^{p/q-1}\xi_{gs} \tag{32}$$

Using (27), we rewrite (30) as

$$\begin{aligned}\dot{V} &\le -\dot{V}_1 - \dot{V}_2 + \frac{\|s\| p}{I_z \alpha_2 q} e_2^{p/q-1}\|\delta_3(t)\| \\ &\quad - \frac{\|s\| p}{I_z \alpha_2 q} e_2^{p/q-1}\left(\widehat{b}_{02} + \widehat{b}_{12}\|r\| + \widehat{b}_{22}\|\dot{r}\|^2\right) \\ &\quad - \frac{\|s\| p}{I_z \alpha_2 q} e_2^{p/q-1}\left(\widetilde{b}_{02} + \widetilde{b}_{12}\|r\| + \widetilde{b}_{22}\|\dot{r}\|^2\right)\end{aligned} \tag{33}$$

Table 1: Main particulars of hovercraft.

m(kg)	40000	l_{sk}(m)	65
I_z(kgm^2)	1.8×10^6	B_c(m)	8.9
S_{PP}(m^2)	45	l_c(m)	23.6
S_{LP}(m^2)	93	h(m)	1
S_{HP}(m^2)	260	H_{hov}(m)	5.9
Q(m^3/s)	140.8	S_c(m^2)	212
V_w(knots)	10	β_w(deg)	45

From $\widetilde{b}_i = b_i - \widehat{b}_i$, $i \in \{0, 1, 2\}$, we have

$$\begin{aligned} \dot{V} &\leq -\dot{V}_1 - \dot{V}_2 + \frac{\|s\| p}{I_z \alpha_2 q} e_2^{p/q-1} \|\delta_3(t)\| \\ &\quad - \frac{\|s\| p}{I_z \alpha_2 q} e_2^{p/q-1} \left(b_{02} + b_{12} \|r\| + b_{22} \|\dot{r}\|^2 \right) \\ &\leq -\dot{V}_1 - \dot{V}_2 - \dot{V}_3 \end{aligned} \tag{34}$$

From [27], we know that $\dot{V}_1 \geq 0$. It is obvious from (26), (31)-(32), and (34) that $\dot{V} \leq 0$. Hence, the closed-loop system is stable.

Integrating both sides of (34), we have

$$V(t) - V(0) = -\int_0^t \left(\dot{V}_1 + \dot{V}_2 + \dot{V}_3 \right) d\tau \tag{35}$$

Then

$$V(t) = V(0) - \int_0^t \left(\dot{V}_1 + \dot{V}_2 + \dot{V}_3 \right) d\tau \leq V(0) < \infty \tag{36}$$

We can see that

$$\begin{aligned} V(0) &= V(t) + \int_0^t \left(\dot{V}_1 + \dot{V}_2 + \dot{V}_3 \right) d\tau \\ &\geq \int_0^t \left(\dot{V}_1 + \dot{V}_2 + \dot{V}_3 \right) d\tau \end{aligned} \tag{37}$$

Therefore

$$\lim_{t \to \infty} \int_0^t \left(\dot{V}_1 + \dot{V}_2 + \dot{V}_3 \right) d\tau \leq V(0) < \infty \tag{38}$$

According to Barbalat's Lemma, we have

$$\begin{aligned} \lim_{t \to \infty} s &= 0, \\ \lim_{t \to \infty} e_2 &= 0 \end{aligned} \tag{39}$$

From (23), we have

$$\lim_{t \to \infty} e_1 = 0 \tag{40}$$

This concludes the proof. □

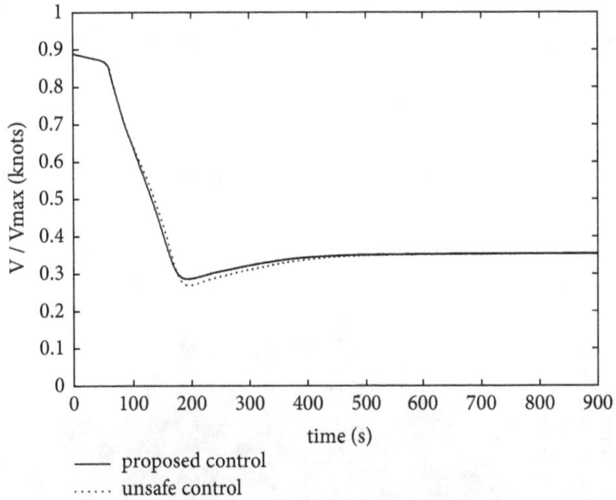

Figure 4: The speed of ACV during the control process.

4. Simulations

Simulations are implemented to verify the effectiveness and superiority of the proposed controller. In simulations, the main particulars of ACV are shown in Table 1.

The initial course angle is 0°, and the desired course angle is −55°. The bounded input disturbance is chosen as $d_r(t) = 1100\sin(0.02t)$.

You can see from Figures 4 and 6–8 that the proposed dynamic safe space can effectively reflect the variations of safety limit following the speed. The course controller is verified to make the error converge to zero from Figure 5. From the comparison with general unsafe controller without considering the safety limit of ACV in Figures 6–8, the proposed safety-guaranteed course controller can effectively keep TR and SA in the safe space during the course control process.

5. Conclusion

In this paper, dynamic safe space of ACV is proposed to be more intuitive to show the change of safety limit and more convenient for safety monitoring and control. For the safety of ACV, a safety-guaranteed course controller is designed based on the dynamic safe space constraint. In the controller, the adaptive mechanism is designed to estimate the system uncertainty and external disturbances online without the requirement of their upper bounds. The proposed controller

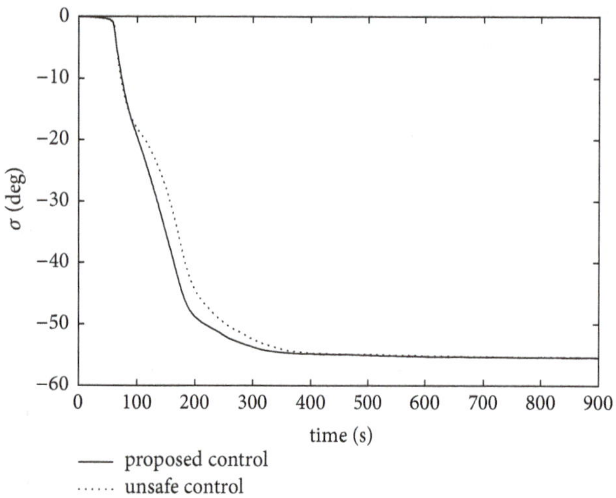

FIGURE 5: The course angle of ACV.

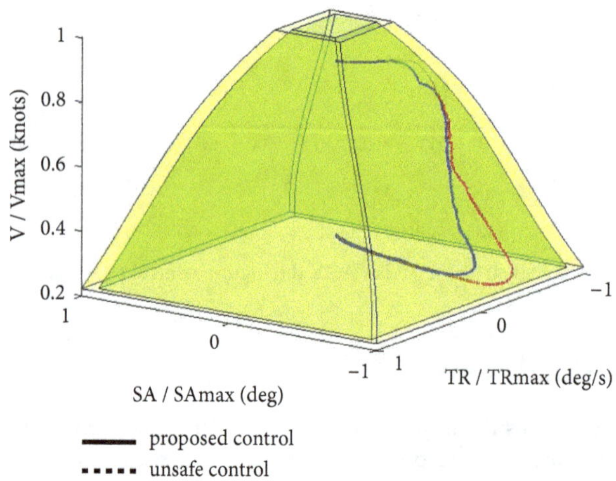

FIGURE 6: The variations of TR and SA.

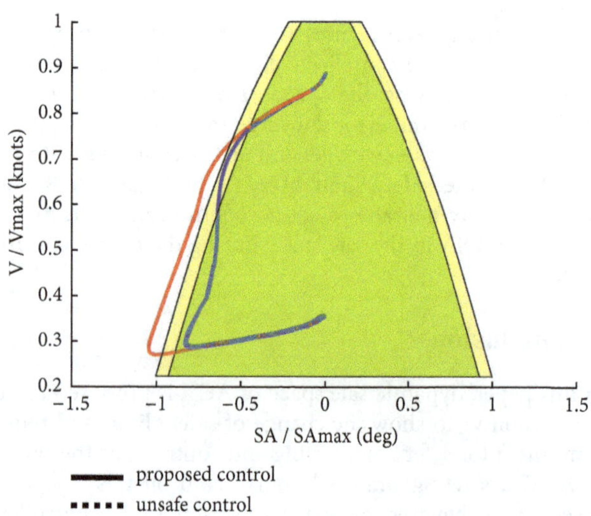

FIGURE 7: The variation of SA.

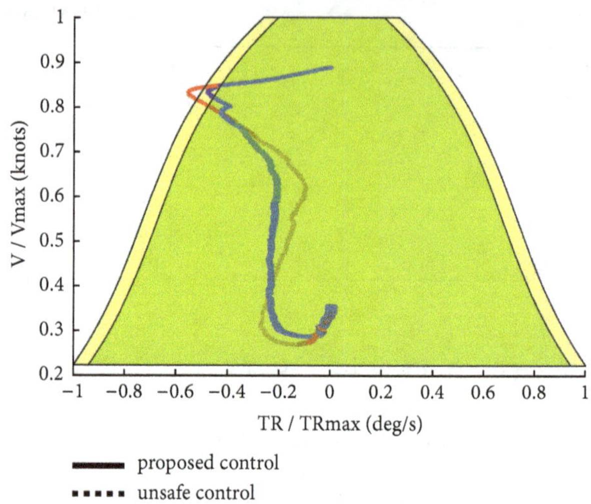

FIGURE 8: The variation of TR.

guarantees the safe space constraint and the convergence of the tracking error.

Conflicts of Interest

The authors declare that there are no conflicts of interest regarding the publication of this paper.

Acknowledgments

The supports from the National Natural Science Foundation of China (Grant no. 51309062) and the project "Research on Maneuverability of High Speed Hovercraft" (Project no. 2007DFR80320) are gratefully acknowledged.

References

[1] M. Fu, S. Gao, and C. Wang, "Safety-Guaranteed Trajectory Tracking Control for the Underactuated Hovercraft with State and Input Constraints," *Mathematical Problems in Engineering*, vol. 2017, Article ID 9452920, 12 pages, 2017.

[2] A. Aguiar, L. Cremean, and J. Hespanha, "Position tracking for a nonlinear underactuated hovercraft: controller design and experimental results," in *Proceedings of the 42nd IEEE International Conference on Decision and Control*, pp. 3858–3863, Maui, HI, USA, 2003.

[3] H. Sira-Ramírez, "Dynamic second-order sliding mode control of the hovercraft vessel," *IEEE Transactions on Control Systems Technology*, vol. 10, no. 6, pp. 860–865, 2002.

[4] K. Tanaka, M. Iwasaki, and O. Wang H, "Switching control of an R/Chovercraft: stabilization and smooth switching," *IEEE Transactions on Systems Man & Cybernetics Part B Cybernetics A Publication of the IEEE Systems Man & Cybernetics Society*, vol. 31, no. 6, pp. 853–863, 2001.

[5] B. Han and G. Zhao, "Course-keeping control of underactuated hovercraft," *Journal of Marine Science and Application*, vol. 3, no. 1, pp. 24–27, 2004.

[6] J. Aranda, "A Control for tracking and stabilization of an underactuated nonlinear RC hovercraft," *International Federation of Automatic Control*, 2006.

[7] J. Zhao and J. Pang, "Trajectory control of underactuated hovercraft," in *Proceedings of the 8th World Congress on Intelligent Control and Automation (WCICA '10)*, vol. 20, pp. 3904–3907, July 2010.

[8] R. Morales, H. Sira-Ramírez, and J. A. Somolinos, "Linear active disturbance rejection control of the hovercraft vessel model," *Ocean Engineering*, vol. 96, pp. 100–108, 2015.

[9] G. G. Rigatos and G. V. Raffo, "Input-output linearizing control of the underactuated hovercraft using the derivative-free nonlinear kalman filter," *Unmanned Systems*, vol. 03, no. 02, pp. 127–142, 2015.

[10] K. Shojaei, "Trajectory tracking control of autonomous underactuated hovercraft vehicles with limited torque," in *Proceedings of the 2nd RSI/ISM International Conference on Robotics and Mechatronics (ICRoM '14)*, pp. 468–473, October 2014.

[11] K. Shojaei, "Neural adaptive robust control of underactuated marine surface vehicles with input saturation," *Applied Ocean Research*, vol. 53, pp. 267–278, 2015.

[12] F. Hua, "Analysis and consideration on safety of all-lift hovercraft," *Ship & Boat*, 2008.

[13] M. Tao and W. Chengjie, *Hovercraftperformance and skirt-cushion system dynamics design*, NationalDefence Industry Press, 2012.

[14] G. Zilman, "Mathematical simula- tion of hovercraft maneuvering," in *Proceedings of the the Twenty-Third American Towing tank Conference*, pp. 373–382, 1993.

[15] M. Fu, F. Ding, and X. Shi, "Research on the course and turn-rate coordinated control for an air cushion vehicle," in *Proceedings of the IEEE International Conference on Mechatronics and Automation, ICMA '07*, pp. 2424–2428, China, August 2007.

[16] T. Elmokadem, M. Zribi, and K. Youcef-Toumi, "Trajectory tracking sliding mode control of underactuated AUVs," *Nonlinear Dynamics*, vol. 84, no. 2, pp. 1079–1091, 2016.

[17] M. Fu, S. Gao, C. Wang, and M. Li, "Human-centered Automatic Tracking System for Underactuated Hovercraft Based on Adaptive Chattering-free Full-order Terminal Sliding Mode Control," *IEEE Access*, 2018.

[18] B. Patre, P. Londhe, L. Waghmare, and S. Mohan, "Disturbance estimator based non-singular fast fuzzy terminal sliding mode control of an autonomous underwater vehicle," *Ocean Engineering*, vol. 159, pp. 372–387, 2018.

[19] L. Qiao and W. Zhang, "Double-Loop Integral Terminal Sliding Mode Tracking Control for UUVs With Adaptive Dynamic Compensation of Uncertainties and Disturbances," *IEEE Journal of Oceanic Engineering*, pp. 1–25.

[20] L. Qiao and W. Zhang, "Adaptive second-order fast nonsingular terminal sliding mode tracking control for fully actuated autonomous underwater vehicles," *IEEE Xplore: IEEE Journal of Oceanic Engineering*, pp. 1–23.

[21] X. Liu, M. Zhang, and F. Yao, "Adaptive fault tolerant control and thruster fault reconstruction for autonomous underwater vehicle," *Ocean Engineering*, vol. 155, pp. 10–23, 2018.

[22] L. Qiao and W. Zhang, "Adaptive non-singular integral terminal sliding mode tracking control for autonomous underwater vehicles," *IET Control Theory & Applications*, vol. 11, no. 8, pp. 1293–1306, 2017.

[23] L. Yun and A. Bliault, *Theory & Design of Air Cushion Craft*, Arnold, 2000.

[24] M. B. R. Neila and D. Tarak, "Terminal sliding mode control for robotic manipulators with uncertaindynamics and bounded disturbance," in *Proceedings of theof the 10th International Conference on Sciences and Techniques*, pp. 1641–1652, 2009.

[25] T. Madani, B. Daachi, and K. Djouani, "Non-singular terminal sliding mode controller: application to an actuated exoskeleton," *Mechatronics*, vol. 33, pp. 136–145, 2016.

[26] R. K. Patnaik and P. K. Dash, "Fast adaptive finite-time terminal sliding mode power control for the rotor side converter of the DFIG based wind energy conversion system," *Sustainable Energy, Grids and Networks*, vol. 1, pp. 63–84, 2015.

[27] L. Sheng-Bo I, L. Ke-Qiang I, J. Wang Q et al., "Nonsingular and Fast Terminal Sliding Mode Control Method," *Information & Control*, vol. 38, no. 1, pp. 1–8, 2009.

19

Estimating Remaining Useful Life for Degrading Systems with Large Fluctuations

Dang-Bo Du, Jian-Xun Zhang, Zhi-Jie Zhou, Xiao-Sheng Si, and Chang-Hua Hu

High-Tech Institute of Xi'an, Xi'an, Shaanxi 710025, China

Correspondence should be addressed to Chang-Hua Hu; hch_reu@sina.com

Academic Editor: Darong Huang

Remaining useful life (RUL) prediction method based on degradation trajectory has been one of the most important parts in prognostics and health management (PHM). In the conventional model, the degradation data are usually used for degradation modeling directly. In engineering practice, the degradation of many systems presents a volatile situation, that is, fluctuation. In fact, the volatility of degradation data shows the stability of system, so it could be used to reflect the performance of system. As such, this paper proposes a new degradation model for RUL estimation based on the volatility of degradation data. Firstly the degradation data are decomposed into trend items and random items, which are defined as a stochastic process. Then the standard deviation of the stochastic process is defined as another performance variable because standard deviation reflects the system performance. Finally the Wiener process and the normal stochastic process are used to model the trend items and random items separately, and then the probability density function (PDF) of the RUL is obtained via a redefined failure threshold function that combines the trend items and the standard deviation of the random items. Two practical case studies demonstrate that, compared with traditional approaches, the proposed model can deal with the degradation data with many fluctuations better and can get a more reasonable result which is convenient for maintenance decision.

1. Introduction

With the development of industrial system, the technology of intelligent vehicle has become more and more mature, and its safety and reliability have become the key factor which restricts its development. An intelligent driving system of a vehicle is a complicated system, which usually needs to monitor the running state, position, and environment [1]. The navigation system is one of its monitoring systems. It can provide information about the position, direction, speed, and acceleration. With the development of intelligent driving technology, navigation system is more and more important. As a key part of the navigation system, the performance of the gyroscope will directly affect the performance of the whole system, and then influence the safety of the entire vehicle system. Therefore, it is necessary to forecast its remaining useful life for better maintenance or health management strategies, using the available condition monitoring information [2, 3].

Generally, the current methods for estimating RUL can usually be divided into two types: physics of failure based methods and data driven methods. Physics of failure based methods rely on the physics of the failure mechanisms. However, with advances in design and production technologies, engineering systems become more complex and large-scale, so it is difficult to obtain the physical failure mechanisms in advance. In contrast, data driven methods achieve RUL estimation via data fitting mainly including machine learning and statistics-based approaches. Therefore, data driven methods attempt to derive models directly from collected CM data and life data and have gained much attention in variety of industrial assets [4, 5]. Considering that the life data may not be available for cost-expensive or highly reliable products, we mainly focus on the statistical data driven approaches for RUL estimation in this paper.

Under a recognized definition of lifetime, that is, first hitting time (FHT) [6, 7], many RUL estimation approaches have been reported in literature [8–10]. References [6, 11] provide a detailed overview of the degradation modeling and RUL predicting method. It can be founded that these models mainly concern the system degrading with a soothing wave.

However, some degradation processes have high dynamics and the observed degradation data exhibit many fluctuations. In this case, it is difficult to model those degradation data for estimating the RUL via traditional approaches [12, 13], while the fluctuation of degradation can reflect the stability of the system. With the deterioration of the system, the stability of the system may worsen gradually, which causes the fluctuation's degree of the degradation data to usually increase over time. Such large fluctuating characteristics can be described by a stochastic degradation process with time-varying mean and variance. Hence, in order to simplify the expression, we use "large fluctuation" to describe the characteristics of these degradation data with time-varying mean and variance. Therefore, the issue of how to model those data with many fluctuations is of practical significance.

In this paper, it is assumed that the degradation process of the system is a stochastic process with time-dependent expectation and standard deviation at each time. Considering that the fluctuation can be represented by a stochastic degradation of standard deviation, which is a new key performance factor, then the whole system RUL can be estimated by considering the stochastic degradation data and the fluctuation. The model parameters can be estimated by maximum likelihood estimation method. Finally, a numbering validation method and INS gyro study are provided to show the superiority of our approach.

The remaining parts are organized as follows. In Section 2, the problem of failure prognosis is formulated and defined. In Section 3, a degradation model for prognosis is developed based on the degradation data and the standard deviation. Section 4 provides two case studies to illustrate the application and usefulness of the developed model. Section 5 draws the main conclusions.

2. Problem Formulation

In this section, based on the fluctuation and the stochastic degradation data, a new model for RUL estimation is proposed. Firstly, we introduce the variables used in this paper.

2.1. Notations. The notations that will be used in this paper are listed as follows:

t_i: time of the ith CM point (can be irregularly spaced)

ξ: failure threshold

$X(t)$: random variable representing the degradation at time instant t

x_i: degradation observation at t_i

$\mathbf{X}_{0:i}$: history of degradation observations for the system up to t_i

$R(t)$: trend item

$C(t)$: fluctuation item

μ_c: expectation of $C(t)$

σ_c: standard deviation of $C(t)$

$\psi(\cdot)$: failure function

$\sigma_c(t)$: standard deviation function at the time instant t

T: lifetime

$\boldsymbol{\vartheta}$, $\boldsymbol{\lambda}_\mu$, $\boldsymbol{\lambda}_\sigma$, $\boldsymbol{\lambda}_R$, Θ: parameters column vector

$f_C(\cdot)$: PDF of $C(t)$

$g_R(t_i, \boldsymbol{\lambda}_R)$: function of t_i and $\boldsymbol{\lambda}_R$

$g_\mu(t, \boldsymbol{\lambda}_\mu)$: function of t_i and $\boldsymbol{\lambda}_\mu$

$g_\sigma(t, \boldsymbol{\lambda}_\sigma)$: function of t_i and $\boldsymbol{\lambda}_\sigma$

$T_{\text{RUL}}(t)$: RUL of the system at time instant t

$L(\cdot)$: log-likelihood function

t_{ch}: time instant of the change-point

n_{ch}: number of the observation data at time instant t_{ch}

ϕ: initial degradation

θ, σ: drift and diffusion parameters

$B(t)$: standard Brownian motion

$E(\cdot)$: expectation operator

$\varepsilon(t_i)$: individual difference random variable.

2.2. Problem Formulation of the New Approach for RUL Estimation. In this paper, we model the degradation process of the system as a stochastic process as $\{X(t),\ t \geq 0\}$, where $X(t)$ is the degradation quantity a time t. Therefore, the lifetime T of the system can be defined as follows:

$$T = \inf\{t : X(t) > \xi;\ t > 0\}, \tag{1}$$

where ξ is the failure threshold.

From (1), it is important to establish the model of stochastic degradation process. In general, a stochastic degradation process can be represented as

$$X(t) = R(t) + C(t), \quad t \geq 0, \tag{2}$$

where $R(t)$ is the trend term, $C(t)$ is the stochastic fluctuating term, and both of them change over time.

It is assumed that $C(t)$ is the complex stochastic process with the expectation μ_c and standard deviation σ_c. Note that σ_c determines the fluctuation of the stochastic degradation process. Thus, jointly modeling the standard deviation and trend items is necessary for RUL estimation. Generally, there are two methods to model the standard deviation: the first method is that the standard deviation is regarded as an independent failure factor, but the difficulty lies in how to set the failure threshold; the second method is to formulate a new failure function $\psi(\cdot)$ which establishes the linkage between the standard deviation σ_c and the trend $R(t)$. Accordingly, the lifetime of the system can be defined as follows:

$$T = \inf\{t : \psi(R(t), \sigma_c(t), \xi, \boldsymbol{\vartheta}) > 0;\ t > 0\}, \tag{3}$$

where $\psi(\cdot)$ is the failure function, $\sigma_c(t)$ is the standard deviation, which is the function of the time instant t, and $\boldsymbol{\vartheta}$ is the parameters column vector.

Based on the estimated lifetime T, the RUL at time t_i is obtained as

$$T_{\mathrm{RUL}}(t_i) = T - t_i, \quad (4)$$

where t_i is the time at the i CM points. In this way, the fluctuation in the degradation process could be incorporated into the RUL estimation.

From (3), we can observe that the fluctuation of the degradation data is taken into account in estimating the lifetime. Then the next section is to model the trend term $R(t)$ and the stochastic fluctuating term $C(t)$.

3. Degradation Modeling and RUL Estimation

In this section we describe how to model the degradation process and estimate the parameter based on the observed degradation data in detail and then how to obtain the RUL of the system via the proposed approach.

3.1. Degradation Modeling. Based on the problem formulation of our approach, the main algorithm formulation for degradation modeling and RUL estimation is outlined as follows.

First, we define the stochastic disturbance term $C(t)$ as a function of μ_c and σ_c:

$$C(t_i) = f_c(\mu_c, \sigma_c), \quad (5)$$

where both the expectation μ_c and the standard deviation σ_c are time-dependent and formulated as

$$\begin{aligned} \mu_c(t) &= g_\mu(t, \boldsymbol{\lambda}_\mu), \\ \sigma_c(t) &= g_\sigma(t, \boldsymbol{\lambda}_\sigma). \end{aligned} \quad (6)$$

As a result, the PDF of $C(t)$ can be represented as

$$f_C(c; \mu_c, \sigma_c) = f_C(c; g_\mu(t, \boldsymbol{\lambda}_\mu), g_\sigma(t, \boldsymbol{\lambda}_\sigma)). \quad (7)$$

Suppose the trend $R(t)$ is also a function of t as follows:

$$R(t) = g_R(t, \boldsymbol{\lambda}_R). \quad (8)$$

Then the degradation process defined in (2) can be further described as follows:

$$X(t) = g_R(t, \boldsymbol{\lambda}_R) + f_c(g_\mu(t, \boldsymbol{\lambda}_\mu), g_\sigma(t, \boldsymbol{\lambda}_\sigma)). \quad (9)$$

In practice, the degradation process is often discretely monitored at time $0 = t_0 < t_1 < \cdots < t_k$ and let $x_k = X(t_k)$ denote the degradation observation at time t_k. Then, the set of the degradation observations up to t_k is represented by $X_{1:k} = \{x_0, x_1, \ldots, x_k\}$. In this paper, we can utilize the history of the degradation observations $X_{1:k}$ to evaluate the parameters $\boldsymbol{\lambda}_R$ $\boldsymbol{\lambda}_\mu$ $\boldsymbol{\lambda}_\sigma$ in (4). The details of algorithm for parameter estimation are summarized in Appendix A.

3.2. Decomposing Trend Items and Fluctuation Items. As discussed in Section 3.1, the degradation process is decomposed into the trend item and fluctuation item at first. Nowadays the common methods of trend extraction, such as the average slope method, finite difference method, LPF (Low-Pass Filter) method, and least square fit method, need the form of the trend to be defined in advance, which leads to difficulty in applying those methods to the degradation signals with the unknown trend. Because the EMD (Empirical Mode Decomposition) method is an adaptive trend extraction method [14–16], which has been widely used for trend extraction, then, in our paper, the EMD method is adopted for trend extraction.

3.2.1. Estimation Parameters $\boldsymbol{\lambda}_\mu$, $\boldsymbol{\lambda}_\sigma$ of Fluctuation Items. Suppose $\mathbf{R}_{0:i} = \{r_0, r_1, \ldots, r_i\}$ are the trend items obtained by EMD method; let $Y_i = X_i - R_i$, so, as discussed in Section 3.1, $\mathbf{Y}_{0:i}$ denotes the observed data of the stochastic fluctuating process $C(t)$.

Since the stochastic disturbance $C(t)$ is defined as a normal random process, so it could be concluded that the mean of stochastic fluctuating process is $\boldsymbol{\lambda}_\mu = 0$ based on the good property of the normal random process. And it can be obtained as follows:

$$C(t) \sim N(0, \sigma_c^2), \quad (10)$$

where σ_c is defined as a function of t.

In engineering practice, many systems have two stages of degradation. In the first stage, the trend of the degradation is not obvious, and the fluctuation is stable. In the second stage, the trend of the degradation has an obvious increase, and the fluctuation becomes higher over time. So, it is reasonable to adopt a two-stage model to describe the degrading systems with large fluctuations. The two-stage model can be expressed by [17]

$$\sigma_c(t_i) = \begin{cases} \boldsymbol{\lambda}_\sigma & t_i \le t_{\mathrm{ch}} \\ g_\sigma(t_i, \boldsymbol{\lambda}_\sigma) & t_i > t_{\mathrm{ch}}, \end{cases} \quad (11)$$

where t_{ch} is the time of the change-point. It is noted that the change-point assumed can be easily found and here we regard it as known information [17]. When $t_i \le t_{\mathrm{ch}}$, $\sigma_c = \boldsymbol{\lambda}_\sigma$, $\boldsymbol{\lambda}_\sigma$ is the fixed value; when $t_i > t_{\mathrm{ch}}$, $\sigma_c(t_i) = g_\sigma(t_i, \boldsymbol{\lambda}_\sigma)$, which is the function of time t_i.

When $t_i \le t_{\mathrm{ch}}$, $\sigma_c = \lambda_\sigma$, λ_σ is a constant. λ_σ could be evaluated by the maximum likelihood estimation based on the property of normal distribution [18] as follows:

$$\lambda_\sigma = \sqrt{\sum_{i=1}^{n_{\mathrm{ch}}} \frac{\left(Y(t_i) - \overline{Y}_{\mathrm{ch}}\right)}{n_{\mathrm{ch}}}}. \quad (12)$$

In (12), n_{ch} denotes the number of the observed data at time t_{ch} and $\overline{Y}_{\mathrm{ch}}$ is the arithmetic average of the observed data before t_{ch}.

When $t_i > t_{\mathrm{ch}}$, σ_c is the function of t_i. In order to simplify the modeling, it is assumed that $\sigma_c(t_i)$ is a linear function of time t_i, expressed as

$$g_\sigma(t_i, \boldsymbol{\lambda}_\sigma) = \lambda'_\sigma t_i + \lambda''_\sigma. \quad (13)$$

Based on the maximum likelihood estimation method, we can easily obtain the log-likelihood function L. Taking the partial derivatives of the log-likelihood function to all parameters, we have

$$\begin{aligned}\frac{\partial L}{\partial \lambda'} &= \sum_{i=0}^{k}\left\{-\frac{t_i}{\lambda'_\sigma t_i + \lambda''_\sigma} + \frac{Y^2(t_i)\,t_i}{(\lambda'_\sigma t_i + \lambda''_\sigma)^3}\right\} = 0\\ \frac{\partial L}{\partial \lambda''_\sigma} &= \sum_{i=0}^{k}\left\{-\frac{1}{\lambda'_\sigma t_i + \lambda''_\sigma} + \frac{Y^2(t_i)}{(\lambda'_\sigma t_i + \lambda''_\sigma)^3}\right\} = 0.\end{aligned} \tag{14}$$

It is observed from (14) that the analytic solution of those parameters via maximum likelihood estimation could not be obtained. To be solvable, the numerical method is adopted to estimate λ'_σ, λ''_σ, which is summarized in Appendix C.

3.2.2. Modeling the Trend Items. Because the systems in the same batch may have different degradation paths, we adopt the random variable $\varepsilon(t_i)$ to describe such individual difference. As such, the real trend item can be represented as

$$\mathbf{R}_{0:i} = \widetilde{\mathbf{R}}_{0:i} + \boldsymbol{\varepsilon}_{0:i}, \tag{15}$$

where $\widetilde{\mathbf{R}}_{0:i}$ is the trend item from initial time 0 to time instant t_i and $\boldsymbol{\varepsilon}_{0:i}$ is the random variable reflecting individual difference.

From the above formulation, it could be concluded that $R(t_i)$ is also random due to individual difference. So we could model the trend items by statistics-based data driven methods. Considering that the trend items may not be a monotonic process, we utilize Wiener process to describe this random trend process [18].

In order to simplify modeling, only the linear degradation model or the model that could be converted into linear form will be discussed in this paper. In general, a linear Wiener-process-based degradation model can be represented as follows [19]:

$$R(t_i) = \phi + \theta t_i + \sigma B(t_i), \tag{16}$$

where ϕ is the initial degradation, θ and σ are the drift and diffusion parameters, respectively, and $B(t_i)$ denotes the standard Brownian Movement (BM), which represents the stochastic dynamics. We assume that θ is the stochastic coefficient while σ and ϕ are deterministic. And we further assume that when $t_0 = 0$, $x_0 = 0$; thus $\phi = 0$.

Define $Z_i = R_i - R_{i-1} = \theta(t_i - t_{i-1}) + \sigma B_i - \sigma B_{i-1}$. Then the joint sampling distribution $f(Z_1, Z_2, \ldots, Z_k \mid \theta)$ can be calculated as

$$\begin{aligned}f(Z_1, Z_2, \ldots, Z_k \mid \theta) &= \frac{1}{\prod_{i=1}^{k}\sqrt{2\pi\sigma^2(t_i - t_{i-1})}} \cdot \exp\left[-\sum_{j=1}^{k}\frac{(Z_i - Z_{i-1} - \theta(t_i - t_{i-1}))^2}{2\sigma^2(t_i - t_{i-1})}\right]\\ &= \frac{1}{\prod_{i=1}^{k}\sqrt{2\pi\sigma^2(t_i - t_{i-1})}} \cdot \exp\left[-\sum_{j=1}^{k}\frac{(Z_i - \theta(t_i - t_{i-1}))^2}{2\sigma^2(t_i - t_{i-1})}\right].\end{aligned} \tag{17}$$

In Bayesian framework, it is assumed that the prior distribution of θ follows $N(\mu_0, \sigma_0)$. Thus, we can obtain the following [20]:

$$\begin{aligned}f(\theta \mid Z_1, \ldots, Z_k) &\propto f(Z_1, Z_2, \ldots, Z_k \mid \theta)\,p(\theta)\\ &\propto \exp\left[-\sum_{j=1}^{k}\frac{(Z_i - \theta(t_i - t_{i-1}))^2}{2\sigma^2(t_i - t_{i-1})}\right] \cdot \exp\left(-\frac{(\theta - \mu_0)^2}{2\sigma_0^2}\right) \propto \exp\left[-\frac{(\theta - \mu_{\theta,k})^2}{2\sigma_{\theta,k}^2}\right].\end{aligned} \tag{18}$$

Due to the property of the normal distribution, we can obtain the posterior estimate of θ as follows:

$$P(\theta \mid Z_1, \ldots, Z_k) = \frac{1}{\sqrt{2\pi\sigma_{\theta,k}^2}}\exp\left[-\frac{(\theta - \mu_{\theta,k})^2}{2\sigma_{\theta,k}^2}\right] \tag{19}$$

with

$$\begin{aligned}\mu_{\theta,k} &= \frac{(\mu_0\sigma^2 + x_k\sigma_0^2)}{(t_k\sigma_0^2 + \sigma^2)}\\ \sigma_{\theta,k}^2 &= \frac{\sigma^2\sigma_0^2}{(t_k\sigma_0^2 + \sigma^2)}.\end{aligned} \tag{20}$$

It is obvious that when a new observation is available, the posterior estimate of θ can be easily updated. $\Theta = [\sigma^2, \mu_0, \sigma_0^2]$ is used to denote the unknown parameters. Actually, we can also estimate those unknown parameters in Θ via EM algorithm, which provides a possible framework for estimating the parameters involving hidden variables [19, 21, 22]. For example, let $\Theta_k = [\sigma_k^2, \mu_{0,k}, \sigma_{0,k}^2]$ denote the estimated Θ based on the observed data $\mathbf{Z}_{0:k}$. Then, Θ_k can be obtained as follows:

$$\begin{aligned}\sigma_k^{2(i+1)} &= \frac{1}{k}\sum_{j=1}^{k}\frac{(Z_j - Z_{j-1})^2 - 2\mu_{\theta,k}(t_j - t_{j-1})(Z_j - Z_{j-1}) + (t_j - t_{j-1})^2(\mu_{\theta,k}^2 + \sigma_{\theta,k}^2)}{(t_j - t_{j-1})}\\ \widehat{\mu}_{0,k}^{(i+1)} &= \mu_{\theta,k}\\ \widehat{\sigma}_{0,k}^{2(i+1)} &= \sigma_{\theta,k}^2.\end{aligned} \tag{21}$$

The details of the derivation are summarized in Appendix D.

3.3. Joint Model for RUL Estimation. If X is a normal distribution, then

$$P\{X < \mu + 3\sigma,\ X \sim N(\mu, \sigma)\} = 0.9987. \quad (22)$$

It is well-known that $[\mu - 3\sigma, \mu + 3\sigma]$ is frequently used to be the confidence interval of normal distributed random variable. Therefore, we assume that if $R(t)+3\sigma_c$ does not reach the failure threshold ξ, it is reasonable that the degradation data $X(t)$ does not reach the failure threshold ξ. For those highly critical systems, the maintenance after failure is too expensive or the consequence of failure is disastrous. Thus, for those systems, it is essential that the conservative method for PHM be adopted to avoid unexpected failure. Therefore, the failure function is defined as

$$\psi(R(t), \sigma_t, \vartheta) = R(t) + 3\sigma_c. \quad (23)$$

Then, as we discussed before, the failure function can be further represented as follows:

$$\begin{aligned}\psi(R(t), R_t, \vartheta) &= R(t) + 3\sigma_c \\ &= \phi + \theta t_i + \sigma B(t_i) + 3(\lambda'_\sigma t_i + \lambda'').\end{aligned} \quad (24)$$

If t_{end} is used to denote the failure time, the following result can be obtained directly:

$$\phi + \theta t_{\text{end}} + \sigma B(t_{\text{end}}) + 3(\lambda'_\sigma t_{\text{end}} + \lambda'') = \xi. \quad (25)$$

If t_{now} is used to denote the current time, we have

$$\begin{aligned}&\phi + \theta t_{\text{now}} + \sigma B(t_{\text{now}}) + 3(\lambda'_\sigma t_{\text{now}} + \lambda'') \\ &\quad = R(t_{\text{now}}) + 3\sigma_c(t_{\text{now}}).\end{aligned} \quad (26)$$

where $\sigma_c(t_{\text{now}}) = \lambda'_\sigma t_{\text{now}} + \lambda''$.

Solving the equations in (25) and (26), we can get

$$\begin{aligned}&\theta(t_{\text{end}} - t_{\text{now}}) + \sigma(B(t_{\text{end}}) - B(t_{\text{now}})) \\ &\quad + 3\lambda'_\sigma(t_{\text{end}} - t_{\text{now}}) \\ &\quad = \xi - R(t_{\text{now}}) - 3(\lambda'_\sigma t_{\text{now}} + \lambda'').\end{aligned} \quad (27)$$

Based on the definition of RUL in (1), the RUL estimated at time t_j could be represented as

$$T_{\text{RUL}}(t_j) = t_{\text{end}}(t_j) - t_{\text{now}}(t_j). \quad (28)$$

Further, utilizing the property of Wiener process, the PDF of the RUL estimation at time t_j can be obtained:

$$\begin{aligned}f(T_k \mid \theta, \mathbf{X}_{1:k}) &= \frac{\xi - R(t_k) - 3(\lambda'_\sigma t_k + \lambda'')}{\sqrt{2\pi T_k^3 \sigma^2}} \\ &\quad \cdot \exp\left(-\frac{(\xi - R(t_k) - 3(\lambda'_\sigma t_k + \lambda'') - \theta T_k - 3\lambda'_\sigma T_k)^2}{2\sigma^2 T_k}\right).\end{aligned} \quad (29)$$

Recall that θ follows $N(\mu_{\theta,k}, \sigma_{\theta,k})$; Theorem 1 in [20] can be used to estimate the PDF of the RUL as

$$\begin{aligned}f(T_k \mid \theta, \mathbf{X}_{1:k}) &= \frac{\xi - b(t_k) - 3(\lambda'_\sigma t_k + \lambda'')}{\sqrt{2\pi T_k^3 (\sigma^2 + \sigma_{\theta,k}^2 t_k)}} \\ &\quad \cdot \exp\left(-\frac{(\xi - b(t_k) - 3(\lambda'_\sigma t_k + \lambda'') - \mu_\theta T_k - 3\lambda'_\sigma T_k)^2}{2T_k(\sigma^2 + \sigma_{\theta,k}^2 T_k)}\right).\end{aligned} \quad (30)$$

Clearly, there are five steps of RUL estimation in our approach, which are summarized below.

3.4. A Procedure of Proposed Approach for RUL Estimation. As discussed in the previous subsection, the procedure of the approach proposed in this paper for RUL estimation is summarized as follows.

Step 1. Through EMD method, the degradation data $\mathbf{X}_{0:i} = \{x_0, x_1, \ldots, x_i\}$ are divided into the trend items $\mathbf{R}_{0:i} = \{r_0, r_1, \ldots, r_i\}$ and the fluctuation items $\mathbf{Y}_{0:i} = \{y_0, y_1, \ldots, y_i\}$ when the monitored observations $\mathbf{X}_{0:i} = \{x_0, x_1, \ldots, x_i\}$ are available.

Step 2. The trend items $\mathbf{R}_{0:i} = \{r_0, r_1, \ldots, r_i\}$ and the fluctuation items $\mathbf{Y}_{0:i} = \{y_0, y_1, \ldots, y_i\}$ are modeled by Wiener process and a normal stochastic process separately; the details of algorithm and parameter estimation are summarized in Section 3.2.1 and Appendices B, C, and D.

Step 3. According to the failure function defined in Section 3.2.2 and the model proposed in Step 2, the PDF of the estimated RUL can be obtained.

Step 4. When the monitored observation X_{i+1} is available, let $t = t + 1$ and go to Step 1. Otherwise, go to Step 5.

Step 5. Using the PDF of the estimated RUL, the system's performance can be evaluated.

4. Case Studies

In this section, two practical cases for gyros in the inertial navigation system (INS) are provided to illustrate the application of our model and compare the performance of our model with traditional models in [12, 23].

4.1. Case Description. As discussed before, gyro plays a key role in intelligent vehicle. With the working hours increase, gyro may cause a kind of degradation that makes gyro drift. Usually, this gyro can be compensated using a specific compensating model. However, when gyro drift degrades to a predefined threshold, the gyro's performance can not satisfy the vehicle INS system requirement. The gyro's drift can not be monitored directly, and a performance indicator is used instead.

In the following, two practical cases are provided to illustrate the adaptability and rationality of our approach for RUL estimation.

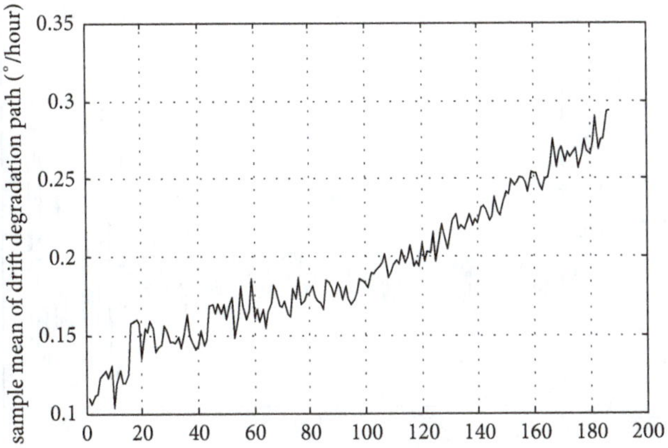

FIGURE 1: The sample mean of the gyro drift.

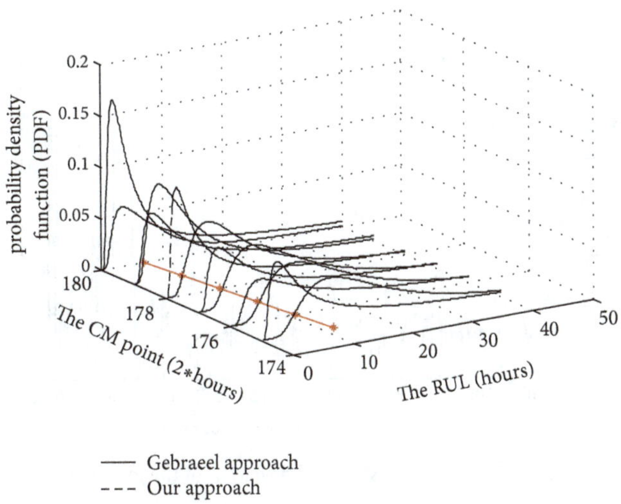

FIGURE 2: Illustration of the RUL distributions at last 6 CM points (* means actual RUL).

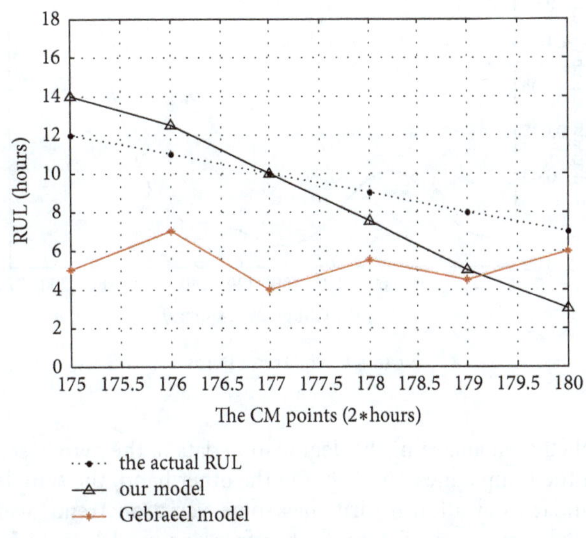

FIGURE 3: The estimated mean RUL and the actual RUL at last 6 CM points.

4.2. Two Practical Case Studies

4.2.1. Case 1: The Fluctuations of Degradation Data Are Stable. In this section, we utilize the degradation data from INS to illustrate, where its operating time is 374 hours and the interval is 2 hours. The collected data are shown in Figure 1.

In the practice of the INS health monitoring, it is usually required that the sample mean of the drift measurement along the sense axis should not exceed 0.30 (°/hour), which is predetermined at the design stage and is strictly enforced in practice since an INS is a critical device used in the navigated space system. From Figure 1, it could be observed that the degradation data is much similar to the linear case. As for this practical case, the distribution of the RUL could be obtained at each CM point by the proposed method. At the same time, the approach in [12] is adopted for comparison. Obviously, from Figures 2 and 3 our approach can reduce the uncertainty of the RUL estimation. At the last 6 points, it is observed that the estimated RUL of our approach is more accurate than the model in [12] from Figure 3. Compared with the actual RUL relying on the definition of FHT, the estimated RUL of our approach is more conservative, which shows the availability of our model.

Through the above comparisons, it is shown that our approach can deal with the degradation as well as the traditional approach and could decrease the influence of the fluctuations in the observed degradation signals.

4.2.2. Case 2: The Fluctuations of Degradation Data Are Diverged

(1) Description and Collection of the Degradation Data. With the development of technology, the stability of the gyro becomes better and better. In this case, the sample mean of the drift does not have obvious increase within a short time. Figure 4 shows another monitored INS in certain space system with test time of 420 hours; data of 210 points of drift coefficients were collected with regular CM intervals of 2 hours in the field condition. It is obvious that those data do not have obvious increase trend, which leads to the

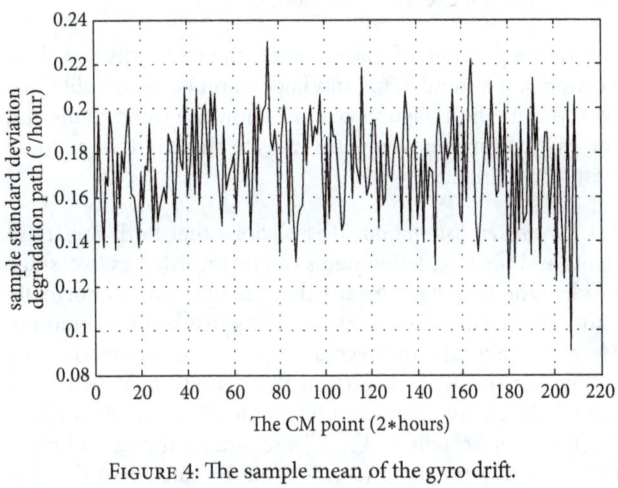

FIGURE 4: The sample mean of the gyro drift.

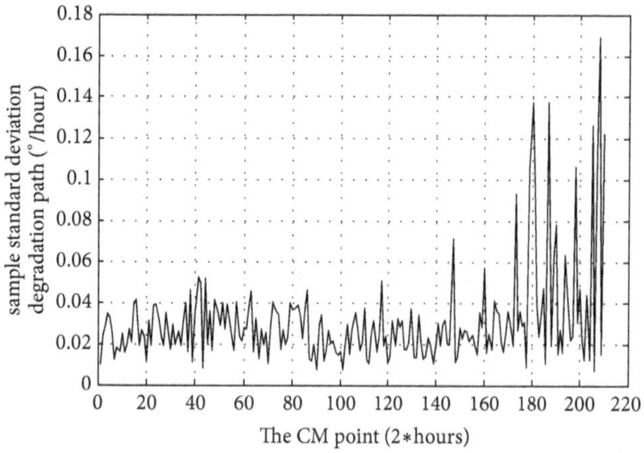

FIGURE 5: The sample standard deviation of the gyro drift.

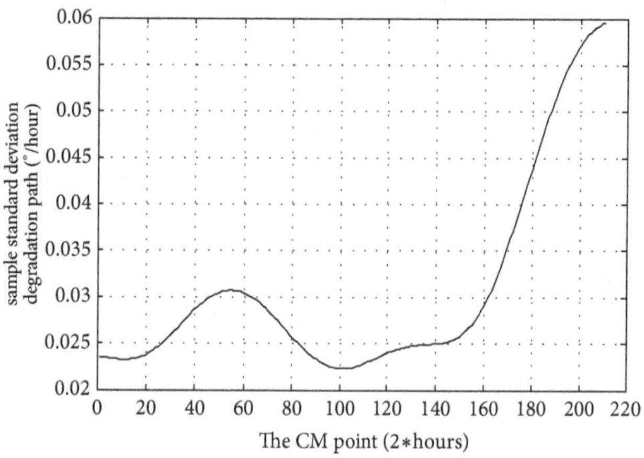

FIGURE 6: The trend items.

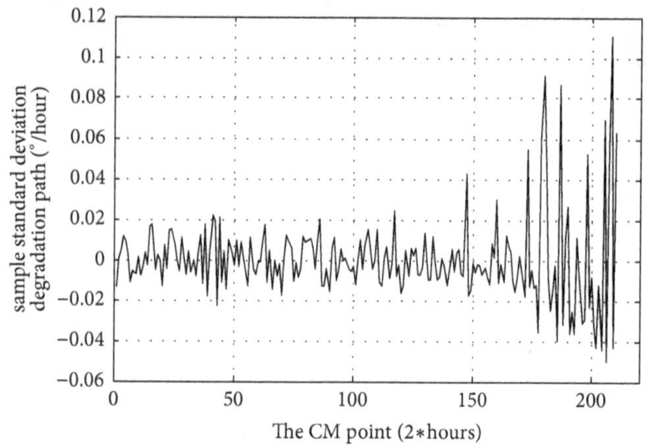

FIGURE 7: The fluctuation items.

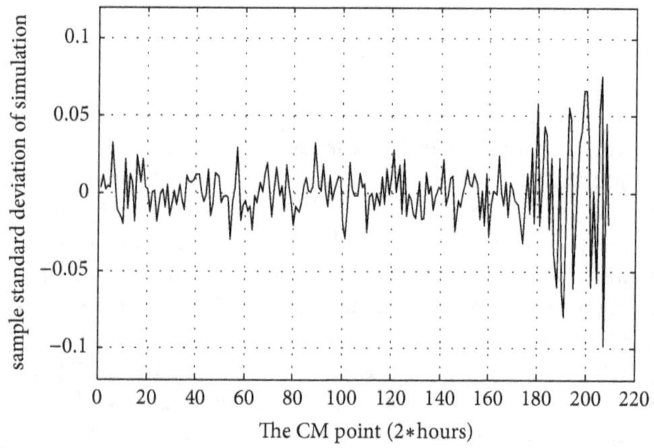

FIGURE 8: Simulation based on the parameter evaluations.

difficulty in analyzing the degradation data of the gyro based on the sample mean of drift. On the other hand, the sample standard deviation of drift has some variation trend over time, as shown in Figure 5. Therefore our model could be used for RUL estimation via the data of the drift sample standard deviation which usually required that the sample standard deviation of the drift measurement along the sense axis should not exceed 0.15 (°/hour).

(2) Decomposition of Degradation Data. As discussed in Section 3, the trend items and fluctuation items are obtained by EMD method. Figures 6 and 7 show the trend items and the fluctuation items which illustrate the changes of these terms over time.

(3) Parameter Estimation. It is obvious that both the trend items and the fluctuation items can be regarded as two-stage model. The first stage means that the gyro works normally and the second stage means that the gyros begin degrading. Here we only take the second stage into consideration for estimating the RUL. Based on the definition of two-stage model, the change-point is at the 150th CM point. As we have discussed in Section 3, λ'_σ, λ''_σ are almost the same before the 150th CM point via (18). It could be calculated that the random items follow $N(0, 0.0103)$ in first stage, where $t_{\text{RUL}} = T_{\text{Rated}} - t_{\text{now}}$. Then the model discussed in Section 3 is used to evaluate and update the parameters, which are used to estimate the RUL.

(4) Verification of Parameter Estimation. In order to illustrate the accuracy of the estimated λ'_σ and λ''_σ, a numerical simulation is adopted. Figure 8 shows that the simulation is produced based on the evaluations of λ'_σ and λ''_σ, which are used to illustrate the accuracy of estimation for parameters λ'_σ and λ''_σ. Comparing Figure 8 with Figure 7, it is noted that the increases of the two fluctuation items are not the same when the degradation path begins stepping into the next stage, which is caused by the lack of the observed data. From Appendix C, it is noted that Std_i could be used to describe the change of standard deviation. So comparing Std_i of the simulation with the real fluctuation items, it is obvious that both of their paths are close, as shown in Figure 9. Thus, it could be concluded that evaluation of parameters is accurate.

(5) Comparative Results. Figure 10 shows the PDF of the estimated RUL at the last 8 CM points. From Figure 10, it is obvious that our approach can deal with the degradation with many fluctuations and get an explicit result, which is

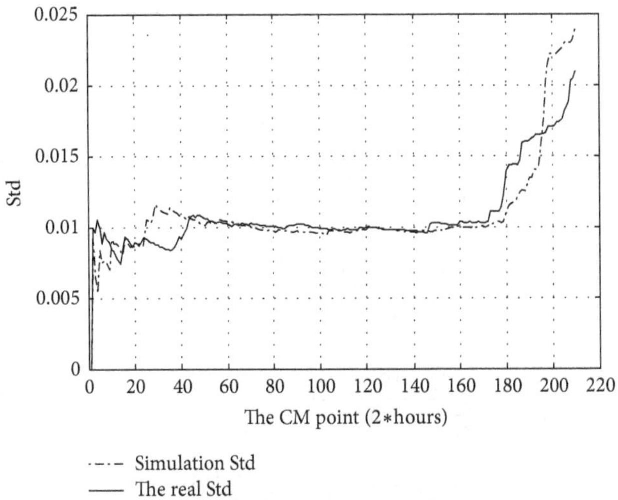

FIGURE 9: Comparison of the real Std and the simulation Std.

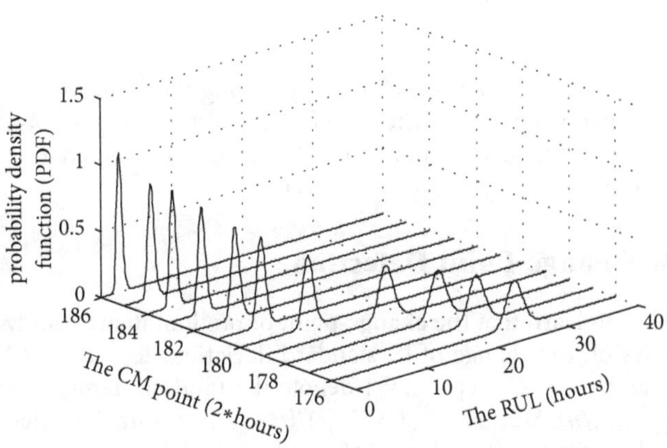

FIGURE 10: Illustration of the RUL distributions at last 10 CM points.

convenient for maintenance decision. From Figure 7, it is noted that the degradation data reach the failure threshold at the 205th CM point, but the result of our model shows that the gyro's failure occurs at about 187th CM point, which reflects that our approach is more conservative than the traditional approach, since the fluctuation is regarded as another performance variable for evaluating the performance of the gyro. Even through our approach is more conservative, it can be safer when the result of the RUL estimations is used for PHM of safety-critical systems such as gyros, whose maintenance cost after failure is too expensive and the consequence of failure are disastrous. And our approach can obtain the regular PDF of RUL which could be convenient for maintenance decision. It is shown in Figure 11 that the traditional approaches [8, 9] could not obtain regular PDF of RUL.

Comparing the results of case 1 and case 2, it is obvious that our approach can not only deal with the degradation with many fluctuations, but also get more reasonable result. Overall, these case studies imply that our approach is an effective model for RUL estimation, especially when the degradation data experience strong fluctuation.

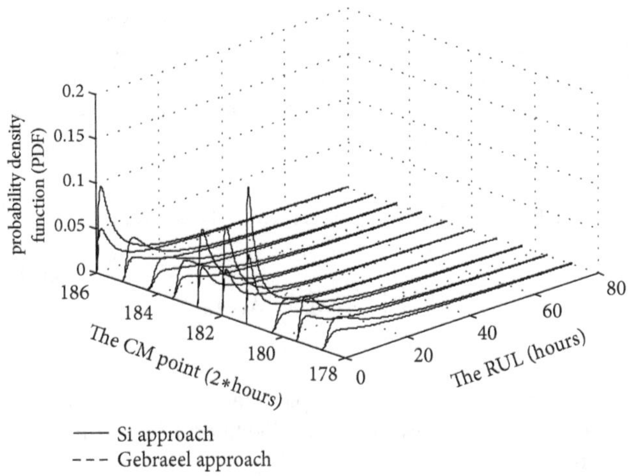

FIGURE 11: Illustration of the RUL distributions at last 8 CM points.

5. Conclusion

In this paper, we study how to estimate the RUL of dynamic degrading systems with high fluctuating degradation data. Specifically, we present a stochastic process-based degradation model with a time-varying standard deviation function to describe the degradation process of the system. Considering the fact that the fluctuation reflects the stability of the system, we define the fluctuation as an indicator of the system's health condition in addition to directly using the observed degradation data and further using the standard deviation to represent the degree of fluctuation. Then, based on the influence of the standard deviation, we define a failure threshold based method to estimate the marginal RUL distributions. And then the joint RUL distribution is formulated based on the correlation between two performance variables. Finally, through a practical case, the feasibility and applicability of the presented model are validated.

In this paper, we are concerned with a situation where the system is degrading with high fluctuation. Then the collected degradation data are difficult to model for evaluating the RUL of system by traditional approaches. In order to solve the above problem, we redefine the failure function that is depended on two aspects including the degradation trend and the fluctuations. So we decompose the degradation data into trend items and fluctuation items firstly, and then model the two items separately. Finally we evaluate the PDFs of RUL based on the failure function that has been defined above. Through the two practical cases, it is obvious that our model can deal with this problem well and get the reasonable result used for PHM.

Appendix

A. Formulation of Parameter Estimation

In order to estimate the parameters in (7), the maximum likelihood estimation method is adopted. Here two cases are considered.

Case 1 (the functional form of $g_R(t, \boldsymbol{\lambda}_R)$ is already known). Because the functional form of $g_R(t, \boldsymbol{\lambda}_R)$ is known, the probability density function of X_i can be defined as $f_{X(t)}(x; \boldsymbol{\lambda}_\mu, \boldsymbol{\lambda}_\sigma, \boldsymbol{\lambda}_R, t)$. Therefore, the likelihood function for $\mathbf{X}_{1:i}$ can be written as

$$\begin{aligned} L\left(\boldsymbol{\lambda}_\mu, \boldsymbol{\lambda}_\sigma, \boldsymbol{\lambda}_R\right) &= f\left(x_0, x_1, \ldots, x_k \mid \boldsymbol{\lambda}_\mu, \boldsymbol{\lambda}_\sigma, \boldsymbol{\lambda}_R\right) \\ &= \prod_{i=1}^{k} f_{X_i}\left(x_i; \boldsymbol{\lambda}_\mu, \boldsymbol{\lambda}_\sigma, \boldsymbol{\lambda}_R, t\right). \end{aligned} \tag{A.1}$$

In order to estimate those parameters in (A.1), we take the partial derivatives of the log-likelihood function of all parameters:

$$\begin{aligned} \frac{\partial L}{\partial \boldsymbol{\lambda}_\mu} &= \sum_{i=0}^{k} \frac{1}{f_{X_i}\left(x_i; \boldsymbol{\lambda}_\mu, \boldsymbol{\lambda}_\sigma, \boldsymbol{\lambda}_R, t_i\right)} \cdot \frac{\partial f_{X_i}\left(x_i; \boldsymbol{\lambda}_\mu, \boldsymbol{\lambda}_\sigma, \boldsymbol{\lambda}_R, t_i\right)}{\partial \boldsymbol{\lambda}_\mu} = 0 \\ \frac{\partial L}{\partial \boldsymbol{\lambda}_\sigma} &= \sum_{i=0}^{k} \frac{1}{f_{X_i}\left(x_i; \boldsymbol{\lambda}_\mu, \boldsymbol{\lambda}_\sigma, \boldsymbol{\lambda}_b, t_i\right)} \cdot \frac{\partial f_{X_i}\left(x_i; \boldsymbol{\lambda}_\mu, \boldsymbol{\lambda}_\sigma, \boldsymbol{\lambda}_R, t_i\right)}{\partial \boldsymbol{\lambda}_\sigma} = 0 \\ \frac{\partial L}{\partial \boldsymbol{\lambda}_R} &= \sum_{i=0}^{k} \frac{1}{f_{X_i}\left(x_i; \boldsymbol{\lambda}_\mu, \boldsymbol{\lambda}_\sigma, \boldsymbol{\lambda}_R, t_i\right)} \cdot \frac{\partial f_{X_i}\left(x_i; \boldsymbol{\lambda}_\mu, \boldsymbol{\lambda}_\sigma, \boldsymbol{\lambda}_R, t_i\right)}{\partial \boldsymbol{\lambda}_R} = 0. \end{aligned} \tag{A.2}$$

Solving the above system of equations in (A.2) numerically, we can obtain the estimations of $\boldsymbol{\lambda}_\sigma$, $\boldsymbol{\lambda}_R$, $\boldsymbol{\lambda}_\mu$.

Case 2 (the functional form of $g_R(t_i, \boldsymbol{\lambda}_R)$ is unknown). In the practical case, the degradation data used to be nonlinear and different system may be in the different degradation path so that it is difficult to model the data by the same simple model, which makes the functional form of $g_R(t_i, \boldsymbol{\lambda}_R)$ unknown. For this reason, we extract the trend items $\mathbf{R}_{0:i} = \{r_0, r_1, \ldots, r_i\}$ from degradation data, which reflect the variation tendency of degradation. Let $Y_i = X_i - R_i$, so it could be assumed that Y_i denotes the value of stochastic disturbance $c(t)$ at time t_i; then the likelihood function $L(\boldsymbol{\lambda}_\mu, \boldsymbol{\lambda}_\sigma)$ can be represented as

$$\begin{aligned} L\left(\boldsymbol{\lambda}_\mu, \boldsymbol{\lambda}_\sigma\right) &= f\left(y_0, y_1, \ldots, y_k \mid \boldsymbol{\lambda}_\mu, \boldsymbol{\lambda}_\sigma\right) \\ &= \prod_{i=1}^{k} f_{Y_i}\left(y_i; \boldsymbol{\lambda}_\mu, \boldsymbol{\lambda}_\sigma, t\right). \end{aligned} \tag{A.3}$$

Taking its partial derivatives of all parameters, we have

$$\begin{aligned} \frac{\partial L}{\partial \boldsymbol{\lambda}_\mu} &= \sum_{i=0}^{k} \frac{\partial f_{C_i}\left(c_i; \mu_c, \sigma_c\right)}{\partial \boldsymbol{\lambda}_\mu} \\ &= \sum_{i=0}^{k} \frac{f'_{C\mu_c}\left(c; g_\mu\left(t_i, \boldsymbol{\lambda}_\mu\right), g_\sigma\left(t_i, \boldsymbol{\lambda}_\sigma\right)\right)}{f_C\left(c; g_\mu\left(t_i, \boldsymbol{\lambda}_\mu\right), g_\sigma\left(t_i, \boldsymbol{\lambda}_\sigma\right)\right)} \cdot \frac{\partial g_\mu\left(t_i, \boldsymbol{\lambda}_\mu\right)}{\partial \boldsymbol{\lambda}_\mu} = 0 \\ \frac{\partial L}{\partial \boldsymbol{\lambda}_\sigma} &= \sum_{i=0}^{k} \frac{\partial f_{C_i}\left(c_i; \mu_c, \sigma_c\right)}{\partial \boldsymbol{\lambda}_\sigma} \\ &= \sum_{i=0}^{k} \frac{f'_{C\sigma_c}\left(c; g_\mu\left(t_i, \boldsymbol{\lambda}_\mu\right), g_\sigma\left(t_i, \boldsymbol{\lambda}_\sigma\right)\right)}{f_C\left(c; g_\mu\left(t_i, \boldsymbol{\lambda}_\mu\right), g_\sigma\left(t_i, \boldsymbol{\lambda}_\sigma\right)\right)} \cdot \frac{\partial g_\sigma\left(t_i, \boldsymbol{\lambda}_\sigma\right)}{\partial \boldsymbol{\lambda}_\sigma} = 0. \end{aligned} \tag{A.4}$$

$\boldsymbol{\lambda}_\sigma$, $\boldsymbol{\lambda}_R$, $\boldsymbol{\lambda}_\mu$ can be obtained by solving the above system of equations in (A.4). Then the functional form of $g_R(t_i, \boldsymbol{\lambda}_R)$ can be obtained by modeling the history of degradation observations $X_{1:k} = \{x_1, x_2, \ldots, x_k\}$.

B. Change-Point Detection

We can learn that the change-point of random items mainly relys on the change of its standard deviation. It is assumed that $\mathbf{X}_{0:n} = \{x_0, x_1, \ldots, x_n\}$ denote the random items, $\overline{x} = \sum_{i=0}^{n}(x_i/n)$, $\mathrm{Std}_n = \sum_{i=0}^{n}((x_i - \overline{x})/(n-1))$. It is obvious that Std_n can reflect the change of standard deviation. Therefore we propose a simple method of change-point detection. The procedure for change-point detection can be summarized as follows.

Step 1. Calculate Std_i, via the observed data $\mathbf{X}_{0:i}$.

Step 2. Define $\overline{S}_i = \sum_{j=i-m}^{i}(\mathrm{Std}_j/i)$, where m denotes the length of the moving window.

Step 3. Compare $\overline{S}_{i-1}$ and Std_i; if $\overline{S}_{i-2} < \mathrm{Std}_{i-2}$, Std_{i-1}, and Std_i, then the $(i-2)$th CM point is change-point; else $i = i+1$, and repeat the Steps 2 and 3.

C. Parameter Estimation in (17)

It is obvious that the analytic solution of parameters λ'_σ, λ''_σ could not be calculated via the maximum likelihood estimation method. Fortunately, there are only two parameters that need to be evaluated, so the numerical method can be adopted. Generally the common numerical method is needed to set the initial value. We mainly discuss how to set the initial values. It is noted that Std_i discussed in Appendix B can reflect the trend of standard deviation σ_c and is less than the real standard deviation σ_c. There is a simulation case to illustrate the proposed point. We simulate a stochastic process with two stages: one is a normal distribution with standard deviation $\sigma_c = 1$; another is a normal stochastic process with standard deviation $\sigma_c = 0.1t - 14$, where the time of change-point is at 150. Calculate Std_i of the stochastic process and compare Std_i and the real standard deviation σ_c shown in Figure 9. From Figure 9, it is obvious that the trend of Std_i is approximate with the real standard deviation. So Std_i can be used to set the initial value of the parameters λ'_σ and λ''_σ. And if the requirement of real-time is high, Std_i can be used as obtain the approximate solution of λ'_σ and λ''_σ.

D. Parameter Estimation of Θ

Due to $p(Z_{1:k}, \theta \mid \Theta_k) = p(Z_{1:k} \mid \theta, \Theta_k) + p(\theta \mid \Theta_k)$, we can obtain

$$\begin{aligned} &\ln p(\mathbf{Z}_{1:k}, \theta \mid \Theta_k) \\ &= \ln p(\mathbf{Z}_{1:k} \mid \theta, \Theta_k) + \ln p(\theta \mid \Theta_k) \\ &= -\frac{k+1}{2}\ln 2\pi - \frac{1}{2}\sum_{j=1}^{k}\ln\left(t_j - t_{j-1}\right) - \frac{k}{2}\ln\sigma_k^2 \\ &\quad - \sum_{j=1}^{k}\frac{\left(z_j - z_{j-1} - \theta\left(t_j - t_{j-1}\right)\right)^2}{2\sigma_k^2\left(t_j - t_{j-1}\right)} - \frac{1}{2}\ln\sigma_{0,k}^2 \\ &\quad - \frac{\left(\theta - \mu_{0,k}\right)^2}{2\sigma_{0,k}^2}. \end{aligned} \tag{D.1}$$

Take the expectation operator on both sides of (D.1), and we can obtain

$$\begin{aligned} E_\theta\left[\ln p(\mathbf{Z}_{1:k}, \theta \mid \Theta_k)\right] &= -\frac{k+1}{2}\ln 2\pi - \frac{1}{2}\sum_{j=1}^{k}\ln\left(t_j - t_{j-1}\right) - \frac{k}{2}\ln\sigma_k^2 \\ &\quad - \sum_{j=1}^{k}\frac{\left(z_j - z_{j-1}\right)^2 - 2\mu_{\theta,k}\left(t_j - t_{j-1}\right)\left(z_j - z_{j-1}\right) + \left(t_j - t_{j-1}\right)^2\left(\mu_{\theta,k}^2 + \sigma_{\theta,k}^2\right)}{2\sigma_k^2\left(t_j - t_{j-1}\right)} - \frac{1}{2}\ln\sigma_{0,k}^2 \\ &\quad - \frac{\mu_{\theta,k}^2 + \sigma_{\theta,k}^2 - 2\mu_{\theta,k}\mu_{0,k} + \mu_{0,k}^2}{2\sigma_{0,k}^2}. \end{aligned} \tag{D.2}$$

Given $\partial E_\theta[\ln p(\mathbf{Z}_{1:k}, \theta \mid \Theta_k)]/\partial\Theta_k = 0$, then the unknown parameter $\Theta_k = [\sigma_k^2, \mu_{0,k}, \sigma_{0,k}^2]$ can be calculated as

$$\begin{aligned} \sigma_k^{2(i+1)} &= \frac{1}{k}\sum_{j=1}^{k}\frac{\left(z_j - z_{j-1}\right)^2 - 2\mu_{\theta,k}\left(t_j - t_{j-1}\right)\left(z_j - z_{j-1}\right) + \left(t_j - t_{j-1}\right)^2\left(\mu_{\theta,k}^2 + \sigma_{\theta,k}^2\right)}{\left(t_j - t_{j-1}\right)} \\ \widehat{\mu}_{0,k}^{(i+1)} &= \mu_{\theta,k} \\ \widehat{\sigma}_{0,k}^{2(i+1)} &= \sigma_{\theta,k}^2. \end{aligned} \tag{D.3}$$

Conflicts of Interest

The authors declare that they have no conflicts of interest.

Acknowledgments

This work was partially supported by the NSFC under Grants 61573365, 61374126, 61473094, 61673311, 61603398, and 61773386 and the Yong Elite Scientists Sponsorship Program (TESS) of China Association for Science and Technology (CAST) under Grant 2016QNRC001.

References

[1] H. Darong, K. Lanyan, C. Xiaoyan, Z. Ling, and M. Bo, "Fault diagnosis for the motor drive system of urban transit based on improved Hidden Markov Model," *Microelectronics Reliability*, vol. 82, pp. 179–189, 2018.

[2] Z. J. Zhou, G. Y. Hu, and B. C. Zhang, "A model for hidden behavior prediction of complex systems based on belief rule base and power set," *IEEE Transactions on Systems Man Cybernetics Systems*, vol. 99, pp. 1-7, 2017.

[3] D. Huang, C. Chen, G. Sun, and L. Zhao, "Recognition and Diagnosis Method of Objective Entropy Weight for Power Transformer Fault," *Dianli Xitong Zidonghua/Automation of Electric Power Systems*, vol. 41, no. 12, pp. 206–211, 2017.

[4] M. G. Pecht, *Prognostics and Health Management of Electronics*, John Wiley, New Jersey, USA, 2008.

[5] D.-R. Huang, C.-S. Chen, G.-X. Sun, L. Zhao, and B. Mi, "Linear Discriminant Analysis and Back Propagation Neural Network Cooperative Diagnosis Method for Multiple Faults of Complex Equipment Bearings," *Acta Armamentarii*, vol. 38, no. 8, pp. 1649–1657, 2017.

[6] M. Mesbahul Alam and K. Suzuki, "Lifetime estimation using only failure information from warranty database," *IEEE Transactions on Reliability*, vol. 58, no. 4, pp. 573–582, 2009.

[7] X.-S. Si, W. Wang, C.-H. Hu, and D.-H. Zhou, "Remaining useful life estimation—a review on the statistical data driven approaches," *European Journal of Operational Research*, vol. 213, no. 1, pp. 1–14, 2011.

[8] X.-S. Si, W. Wang, C.-H. Hu, M.-Y. Chen, and D.-H. Zhou, "A Wiener-process-based degradation model with a recursive filter algorithm for remaining useful life estimation," *Mechanical Systems and Signal Processing*, vol. 35, no. 1-2, pp. 219–237, 2013.

[9] N. Z. Gebraeel, M. A. Lawley, R. Li, and J. K. Ryan, "Residual-life distributions from component degradation signals: a Bayesian approach," *IIE Transactions*, vol. 37, no. 6, pp. 543–557, 2005.

[10] S.-T. Tseng, J. Tang, and I.-H. Ku, "Determination of burn-in parameters and residual life for highly reliable products," *Naval Research Logistics (NRL)*, vol. 50, no. 1, pp. 1–14, 2003.

[11] C. Hu, Z. Zhou, J. Zhang, and X. Si, "A survey on life prediction of equipment," *Chinese Journal of Aeronautics*, vol. 28, pp. 25–33, 2015.

[12] C.-Y. Peng and S.-T. Tseng, "Mis-specification analysis of linear degradation models," *IEEE Transactions on Reliability*, vol. 58, no. 3, pp. 444–455, 2009.

[13] H. Darong, T. Jianping, and Z. Ling, "A fault diagnosis method of power systems based on gray system theory," *Mathematical Problems in Engineering*, vol. 2015, Article ID 971257, 2015.

[14] N. E. Huang, Z. Shen, S. R. Long et al., "The empirical mode decomposition and the Hilbert spectrum for nonlinear and non-stationary time series analysis," *Proceedings A*, vol. 454, pp. 903–995, 1998.

[15] R. Yan and R. X. Gao, "A tour of the tour of the Hilbert-Huang transform: An empirical tool for signal analysis," *IEEE Instrumentation & Measurement Magazine*, vol. 10, no. 5, pp. 40–45, 2007.

[16] G. Rilling, P. Flandrin, and P. Goncalves, *On empitical mode decomposition and its algorithms*, France, 2003.

[17] W. Wang, "A two-stage prognosis model in condition based maintenance," *European Journal of Operational Research*, vol. 182, no. 3, pp. 1177–1187, 2007.

[18] E. T. Lee and J. W. Wang, *Statistical methods for survival data analysis*, Wiley Series in Probability and Statistics, John Wiley & Sons, Inc., Fourth edition, 2013.

[19] X.-L. Meng and D. B. Rubin, "Maximum likelihood esumation via the ECM algorithm:a general framework," *Biometrika*, vol. 80, no. 2, pp. 267–278, 1993.

[20] X.-S. Si, W. Wang, M.-Y. Chen, C.-H. Hu, and D.-H. Zhou, "A degradation path-dependent approach for remaining useful life estimation with an exact and closed-form solution," *European Journal of Operational Research*, vol. 226, no. 1, pp. 53–66, 2013.

[21] C. F. J. Wu, "On the convergence properties of the EM algorithm," *The Annals of Statistics*, vol. 11, no. 1, pp. 95–103, 1983.

[22] T. B. Schön, A. Wills, and B. Ninness, "System identification of nonlinear state-space models," *Automatica*, vol. 47, no. 1, pp. 39–49, 2011.

[23] W. Wang, M. Carr, W. Xu, and K. Kobbacy, "A model for residual life prediction based on Brownian motion with an adaptive drift," *Microelectronics Reliability*, vol. 51, no. 2, pp. 285–293, 2011.

20

A New Robust Method for Mobile Robot Multifloor Navigation in Distributed Life Science Laboratories

Ali A. Abdulla,[1,2] Hui Liu,[1] Norbert Stoll,[3] and Kerstin Thurow[1]

[1] *Center for Life Science Automation (CELISCA), University of Rostock, 18119 Rostock, Germany*
[2] *College of Engineering, University of Mosul, Mosul, Iraq*
[3] *Institute of Automation, University of Rostock, 18119 Rostock, Germany*

Correspondence should be addressed to Ali A. Abdulla; ali.abdulla@celisca.de

Academic Editor: Yongji Wang

A new robust method is proposed for multifloor navigation in distributed Life Science Laboratories. This method proposes a solution for many technical issues including (a) mapping and localization with ceiling landmarks and a StarGazer module for achieving an accurate and low cost multifloor navigation system, (b) a new method for path planning to navigate across multiple floor environments called backbone method and embedded transportation management system, (c) elevator environment handler with the necessary procedures to interact with the elevator presenting a new approach for elevator entry button and internal buttons detection, and (d) communication system to get an expandable network; this method utilizes a TCP/IP network for the communication. Many experiments in real Life Science Laboratories proved the efficient performance of the developed multifloor navigation system in life science environment.

1. Introduction

In recent years, the growing interest towards automated system development caused a huge change in life science working processes [1, 2]. Laboratories robots are commonly used in automation processes [3, 4]. Different types of laboratories robots have been developed for transportation tasks including stationary robot [5, 6] and mobile robot [7, 8]. Many technical issues have to be handled for an effective mobile robot navigation system including localization [9, 10], communication [11, 12], path planning [13, 14], elevator aspects [15, 16], collision avoidance [17, 18], and multifloor navigation [19]. Liu et al. developed a new intelligent transportation system based on mobile robot H20 for lab ware transportation between different laboratories in single floor [20, 21]. This system uses a Floyd-Genetic algorithm for path planning and a blind grasping method for arm manipulation. Life Science Laboratories are often distributed on different floors; thus suitable methods for multifloor navigation are required. This will also lead to more and higher flexibility and will open the way to a complete automation of complex laboratories in life sciences. This paper will present an innovative multifloor mobile robot navigation system for transportation tasks.

A survey of the actual literature will clarify the working method and demonstrate advantages/disadvantages of existing solutions. Yu et al. applied a computer vision for elevator button detection and recognition. The detection is divided into button panel (BP) detection and button recognition. A method called multiple partial model has been used to enhance the BP detection and combines structural inference, Hough transform, and multisymbol recognition to propose a method for button recognition. The experiments show a detection success rate of 80.6% for external button and 87.3% for internal buttons detection. It is impossible to detect the button when the detection of the BP has failed [22]. Baek and Lee used stereovision in mobile robot navigation to propose an elevator entrance door detection algorithm. This method includes three parts as follows: (a) image capturing and filtering, (b) implementation of an algorithm to extract the elevator door, and (c) detection of the size of the elevator. In this algorithm, the detection of the door depends on the camera angle, which has to be more than 50 degrees to ensure an easy detection of the elevator entrance [23].

Kang et al. proposed a new strategy for mobile robot navigation in multifloor environment. Digital image processing and template matching neural network have been used to find the outside elevator entry button, outside direction indicator, inside current floor, and destination floor button. This strategy works only for the elevator recognition without dealing with button pressing issues. Thus the robot cannot work autonomously [24]. Chae et al. utilized a group of sensors: laser range finder, two dead reckoning sensors, and passive artificial landmarks to build a mapping cart. The mapping cart is used to remove the human effort when building a large scale map. The method was applied in indoor T-city environments. The results show that this method is convenient and efficient in building a large scale map [25]. Ul-Haque and Prassler analyzed the performance of low cost sensors, called StarGazer module (SGM), in comparison with other sensors. The use of this technique leads to an increased detection of the position and angular accuracy, precision, reliability, repeatability, and robustness [26]. Troniak et al. proposed a multifloor mobile robot transportation system that integrates the stereovision, monocular vision, and laser range finder to sense the environment. The map is built based on an SLAM algorithm and is then used to find the plan to the destination. The template matching algorithm is used to find the elevator buttons. The accuracy of the elevator button detection from the outside reaches 85% and 63% from the inside. This system requires a long time to detect the elevator entrance button (4.6 s); also the robot needs a long time to enter the elevator which may cause it to miss the elevator. Thus human assistance is still required [27].

Although a few researches have focused on the multifloor navigation issue there is still the need to develop a new method with more features such as high speed, low cost, high precision, robustness, immunity to light effects, and easy expansion. In this study a smart passive landmark reader called SGM (Hagisonic Company, Korea) is adopted for localization purposes [28]. This module can meet all above-mentioned features but is used for one floor navigation only [26]. The main contribution of this work is the development of a new multifloor navigation system (mapping, localization, and path planning) taking into consideration the elevator environment handler system for laboratory transportation tasks. The organization of the paper is as follows: Section 2 describes the system architecture. The multifloor navigation system with relative map, localization, innovative path planning method, elevator aspects, and communication socket is described in Section 3. Analyzing the experimental results for the evaluation of the system is demonstrated in Section 4. Section 5 discusses the conclusions and describes the work related to a given problem.

2. System Structure

The architecture of the system is shown in Figure 1. It consists mainly of four blocks: Multifloor System (MFS), Robot Remote Center (RRC), Robot Motion Center (RMC), and mobile robot.

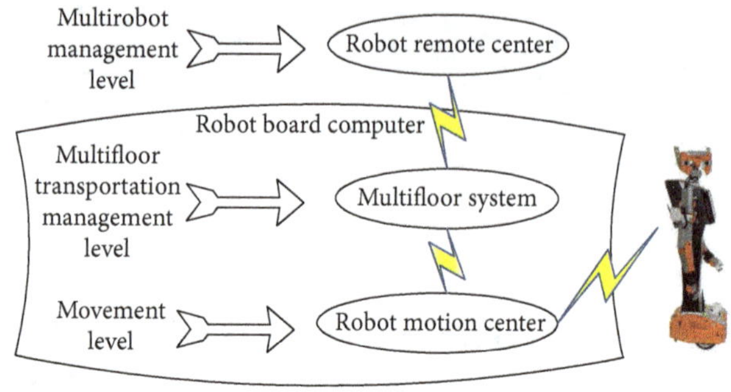

FIGURE 1: System structure.

The presented system can be explained as follows: (a) RRC is the highest level control. It manages transportation tasks by observing robots' status and then forwards incoming transportation task to the MFS of the appropriate robot. (b) The MFS is the transportation management level, which is installed on the H20 mobile robots. It communicates with both RRC and RMC. The MFS is responsible for mapping, localization, path planning, and automated door control. (c) The RMC is the movement execution level, which executes the incoming movement orders from the MFS by controlling the hardware modules of the H20 mobile robot. It also reads the mobile robot data including images stream, ultrasonic measurement distance sensor, infrared measurement distance sensors, and power status. Finally, the mobile robot represents the hardware level of the mobile robot system. The TCP/IP client/server architecture has been utilized to communicate between these controls levels. Figure 2 shows the data flow between RRC, MFS, and RMC in transportation task as follows: (a) the RRC sends the transportation order with requested start and end position. MFS replies to RRC with transportation status, which includes transportation start, transportation finished, and problem in transportation due to wrong points or paths. The MFS sends the robot position in multifloor environment as X, Y, angle, and ID and the power status of each battery periodically as well. (b) MFS sends orders to RMC including move straight, rotate, vision landmark reader data request, ultrasonic distance reading request, and power status request. When the RMC receives the MFS orders it reports with movement and rotation acknowledgment. Besides that, the RMC replies with the requested data information including current landmark reader data (X, Y, angle, and ID), the power data and status for both robot batteries, and ultrasonic distance measurement data.

3. Multifloor Navigation System

In mobile robot navigation systems, many issues have to be handled: mapping, localization, path planning, elevator, and communication systems. In this section, the mapping and localization methods will be proposed. The steps for path planning and elevator aspects will be described. Finally, the integration into the general automated mobile robot transportation system will be clarified.

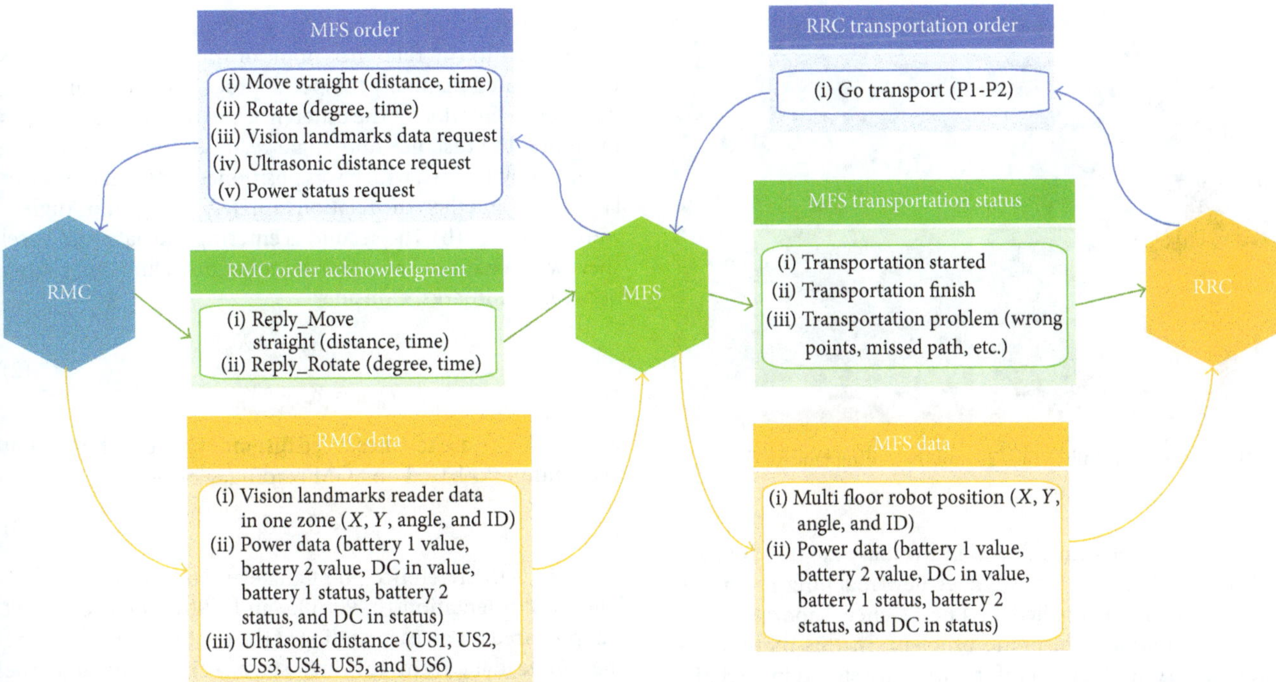

FIGURE 2: Data flow between MFS, RRC, and RMC.

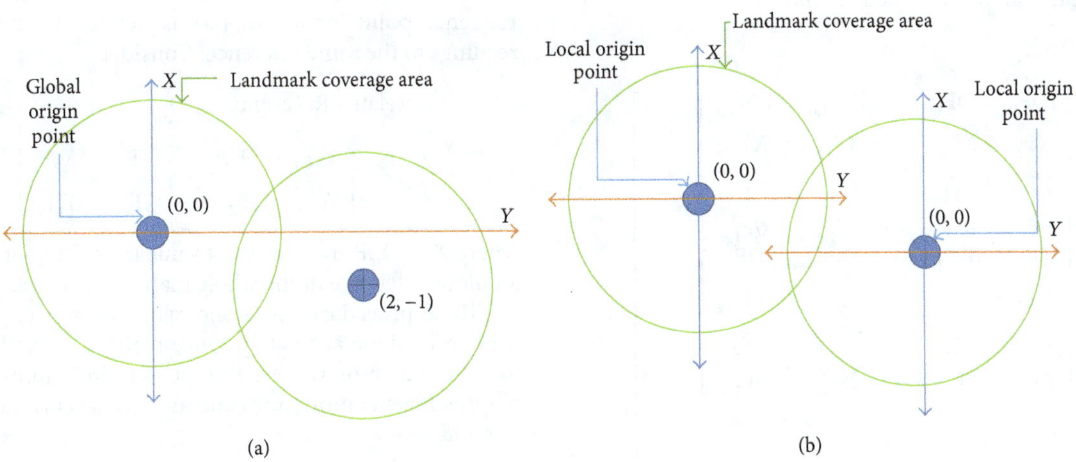

FIGURE 3: SGM modes. (a) Map mode. (b) Alone mode.

3.1. Mapping. In multifloor transportation systems certain features are required like high speed, easy coverage of a large area, low costs, and low sensitivity to light effects. The SGM is used to achieve the mentioned conditions, which consist of an infrared camera, a group of infrared LED projectors, an internal flash memory, and a digital image processing board. The SGM internal work procedures can be explained as projecting an infrared light, capturing the reflected IR images from passive landmarks, and analysis of the IR images to find the ID, angle, X, and Y. This module can work in alone mode or map mode. Figure 3(a) shows the map mode, which has only one global origin point while in alone mode every landmark has a local origin point as clarified in Figure 3(b). Figure 4 shows the installed SGM at the top of the H20 mobile robot and the installed ceiling passive landmark in CELISCA.

Despite the advantages of the SGM, it cannot deal in mapping mode with multifloor environments for two reasons: (a) SGM can build only one map at the same time and (b) it cannot build a map inside the elevator due to height limitation (SGM cannot read a position correctly when the height of landmark to the SGM is less than 1.1 m).

The system was applied using SGM in mapping mode, which stores the complete map inside the SGM and this module gives the position in the building map to the robot and, according to the transportation task, the robot moves in a generated path [20]. This method works for single floor

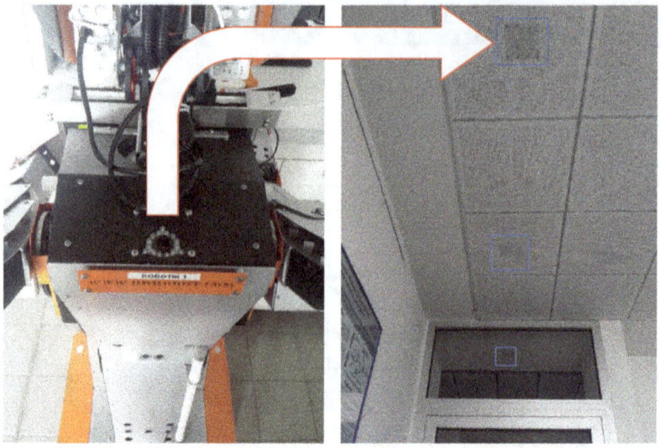

FIGURE 4: SGM installed on H20 and the ceiling landmarks [20].

and requires a long time due to individual map building and path generation for each robot. A new mapping method is proposed and was applied in Life Science Laboratories to solve the multifloor mapping problem. In this method the landmarks ID and related information are stored in a relative map that is used with a navigation program to reach the required destination based on the planning path. Relative map structure can be presented as follows:

$$\text{RelativeMap} = \begin{bmatrix} \begin{bmatrix} IDF_{1,1} & IDF_{1,2} & \cdots & IDF_{1,n-1} & IDF_{1,n} \\ XF_{1,1} & XF_{1,2} & \cdots & XF_{1,n-1} & XF_{1,n} \\ YF_{1,1} & YF_{1,2} & \cdots & YF_{1,n-1} & YF_{1,n} \\ \theta F_{1,1} & \theta F_{1,2} & \cdots & \theta F_{1,n-1} & \theta F_{1,n} \end{bmatrix} \\ \begin{bmatrix} IDF_{2,1} & IDF_{2,2} & \cdots & IDF_{2,n-1} & IDF_{2,n} \\ XF_{2,1} & XF_{2,2} & \cdots & XF_{2,n-1} & XF_{2,n} \\ YF_{2,1} & YF_{2,2} & \cdots & YF_{2,n-1} & YF_{2,n} \\ \theta F_{2,1} & \theta F_{2,2} & \cdots & \theta F_{2,n-1} & \theta F_{2,n} \end{bmatrix} \\ \vdots, \\ \vdots, \\ \vdots, \\ \begin{bmatrix} IDF_{m-1,1} & IDF_{m-1,2} & \cdots & IDF_{m-1,n-1} & IDF_{m-1,n} \\ XF_{m-1,1} & XF_{m-1,2} & \cdots & XF_{m-1,n-1} & XF_{m-1,n} \\ YF_{m-1,1} & YF_{m-1,2} & \cdots & YF_{m-1,n-1} & YF_{m-1,n} \\ \theta F_{m-1,1} & \theta F_{m-1,2} & \cdots & \theta F_{m-1,n-1} & \theta F_{m-1,n} \end{bmatrix} \\ \begin{bmatrix} IDF_{m,1} & IDF_{m,2} & \cdots & IDF_{m,n-1} & IDF_{m,n} \\ XF_{m,1} & XF_{m,2} & \cdots & XF_{m,n-1} & XF_{m,n} \\ YF_{m,1} & YF_{m,2} & \cdots & YF_{m,n-1} & YF_{m,n} \\ \theta F_{m,1} & \theta F_{m,2} & \cdots & \theta F_{m,n-1} & \theta F_{m,n} \end{bmatrix} \end{bmatrix}. \quad (1)$$

From (1), the map consists of multilayers each one representing one floor. Each layer contains the map elements which are landmark ID number (IDF), X position (XF) according to start point, Y position (YF) according to the same reference point, and finally the angle (θF) which is also related to the reference point of the current floor. This map is created manually by special procedures: (a) collecting data: representation of the difference between two landmarks including the distance and angle relation. This step is done using a moving cart, which was positioned between every two landmarks and then the relations as ID, X_{ID}, Y_{ID}, and Angle_{ID} were recorded. (b) The second is entering the data into excel sheet with a required mathematic equation to find the relation between landmarks. Consider

$$X_{n+1} = X_n + \Delta X_n, \quad Y_{n+1} = Y_n + \Delta Y_n, \quad (2)$$

where ΔX, ΔY are the measured distance between the current two landmarks based on SGM reading. Consider

$$\theta_{n+1} = \theta_r - \theta_{\text{LM}}, \quad (3)$$

where θ_r is the reference point angle and θ_{LM} is the installed landmark orientation (0, 90, 180, and −90). The excel sheet was prepared for an easy finding of the relations after entering the whole data from the first stage. (c) Coding into the MF system: the relative positions for each landmark (ID, X_relative, Y_relative, and Angle_relative) were taken from the excel sheet and added to the floor layer. A unique reference point for all floors was defined to unify all the readings to the same reference. Consider

$$\text{RelativeReference} = \begin{bmatrix} X_r F_1 & X_r F_2 & \cdots & X_r F_{n-1} & X_r F_n \\ Y_r F_1 & Y_r F_2 & \cdots & Y_r F_{n-1} & Y_r F_n \end{bmatrix}, \quad (4)$$

where X_rF, Y_rF are a constant value for each floor to extract a unique reference in the whole map.

These procedures are done manually and require some time to be done accurately. Once realized it can be applied to N number of robots; thus it definitely minimizes the map/path generation time compared to the currently applied method.

3.2. Localization. Localization is considered a main key for robot's automation. It can be defined as estimating the absolute or relative position information. A simple and efficient localization method is required in laboratories of life science automation. The need to create a new localization method appeared after developing the relative map to localize the mobile robot in multifloor environments, while the earlier system depends completely on SGM in mapping and localization [20]. The new localization method is based on running the SGM in alone mode as HEX reader. In working mode, the localization method is used to translate the StarGazer reading into a useful information based on relative map using map matching method as clarified in Algorithm 1. This method starts by acquiring the raw data from the StarGazer reader and extracting the useful information about the position in the current landmark zone followed by finding the current floor based on ID numbers. Next step is the orientation translation to determine the correct robot coordinates and the

Input: SGM Raw data in alone mode $(\mathrm{ID}_{\mathrm{SGM}}, X_{\mathrm{SGM}}, Y_{\mathrm{SGM}}, \theta_{\mathrm{SGM}})$
Input: Relative Map $[(\mathrm{ID}, X_{\mathrm{Re}}, Y_{\mathrm{Re}}, \theta_{\mathrm{Re}})_{1,1}, \ldots, (\mathrm{ID}, X_{\mathrm{Re}}, Y_{\mathrm{Re}}, \theta_{\mathrm{Re}})_{m,n}]$
m: Number of Floor
n: Number of landmark for each floor
CF: Current Floor
CLP: Current Landmark Position

For $i = 1 : m$ //Floor Selection
    For $j = 1 : n$ //landmark Selection
        If $\mathrm{ID}_{\mathrm{SGM}} = \mathrm{RelativeMap}[i, 1, j]$ //search about matched Landmark ID
            $\mathrm{CF} = i$
            $\mathrm{CLP} = j$
        End IF
    End For
End For
Apply (5) to correct the landmark orientation $(x_{\mathrm{SGO}}, y_{\mathrm{SGO}}, \theta_{\mathrm{G}})$
Apply (6) as following: $X_{\mathrm{G}} = X\mathrm{F}_{\mathrm{CF,CLP}} + X_{\mathrm{Re}}\mathrm{F}_{\mathrm{CF}} + x_{\mathrm{SGO}}$
Apply (7) as following: $Y_{\mathrm{G}} = Y\mathrm{F}_{\mathrm{CF,CLP}} + Y_{\mathrm{Re}}\mathrm{F}_{\mathrm{CF}} + y_{\mathrm{SGO}}$
Return robot position in global map $(X_{\mathrm{G}}, Y_{\mathrm{G}}, \theta_{\mathrm{G}}, \mathrm{CF})$

ALGORITHM 1: Localization method in multifloor environment.

global angle based on the predefined landmark orientation (θ_{R}) in relative map. Consider

$$x_{\mathrm{SGO}} = \begin{cases} x_{\mathrm{SG}}, & \theta_{\mathrm{R}} = 0 \\ -x_{\mathrm{SG}}, & \theta_{\mathrm{R}} = 180 \\ y_{\mathrm{SG}}, & \theta_{\mathrm{R}} = 90 \\ -y_{\mathrm{SG}}, & \theta_{\mathrm{R}} = -90, \end{cases}$$

$$y_{\mathrm{SGO}} = \begin{cases} y_{\mathrm{SG}}, & \theta_{\mathrm{R}} = 0 \\ -y_{\mathrm{SG}}, & \theta_{\mathrm{R}} = 180 \\ -x_{\mathrm{SG}}, & \theta_{\mathrm{R}} = 90 \\ x_{\mathrm{SG}}, & \theta_{\mathrm{R}} = -90, \end{cases} \tag{5}$$

$$\theta_{\mathrm{G}} = \begin{cases} \theta_{\mathrm{SG}}, & \theta_{\mathrm{R}} = 0 \\ \theta_{\mathrm{SG}} - 180, & \theta_{\mathrm{R}} = 180 \\ \theta_{\mathrm{SG}} - 90, & \theta_{\mathrm{R}} = 90 \\ \theta_{\mathrm{SG}} + 90, & \theta_{\mathrm{R}} = -90. \end{cases}$$

Finally x, y StarGazer readings are translated into real coordinates in a global multifloor map:

$$x_{\mathrm{G}} = x_{\mathrm{SGO}} + x_{\mathrm{R}} + x_{\mathrm{FR}}, \tag{6}$$

$$y_{\mathrm{G}} = y_{\mathrm{SGO}} + y_{\mathrm{R}} + y_{\mathrm{FR}}, \tag{7}$$

where x_{R}, y_{R} are relative values from relative map and x_{FR}, y_{FR} are multifloor reference for the current floor, which is used as a multifloor unique reference.

The map matching method is coded in a way that easily allows an expansion to m number of floors and n number of landmarks. If any ceiling landmark cannot be recognized correctly during the mobile robot's movement using the built multifloor map, the corresponding mobile robot finds and presents the missing landmark ID based on the last landmark and the next landmark. This landmark error-handling approach can guarantee a high-successful rate during robot movement.

3.3. Path Planning. Path planning is an important task to achieve an efficient navigation system. Path planning can be defined as a motion control from start point to the goal point with collision avoidance taking the shortest path to the destination. Many algorithms have been proposed and developed for this purpose [29–32]. Compared to the standard ones, the path planning issue has more challenges for the mobile robot transportation in Life Science Laboratories as follows: since the proposed path planning method will be applied in the real robot transportation system, the real-time planning results are desired. However, there are so many combinations of the transportation starting positions and destinations, so the expected path planning method should be able to combine the dynamic path combinations fast and flexibly. For this purpose, in this study a new path planning method (named as backbone method) is presented, which is comprised of an essential path for each floor and a group of flexible subpaths for all laboratory stations. The corresponding Internal Transportation Management System (ITMS) software has also been developed by adopting the proposed backbone method.

3.3.1. Internal Transportation Management System (ITMS). ITMS is a transportation management system inside the MFS. It has been coded based on multiple thread technique to increase the efficiency and the ability to access the MFS at any time. The management system shown in Figure 5 divides the incoming transporting task into small tasks and keeps monitoring the execution of these tasks continuously. It also informs the Robot Remote Center (RRC) about the current

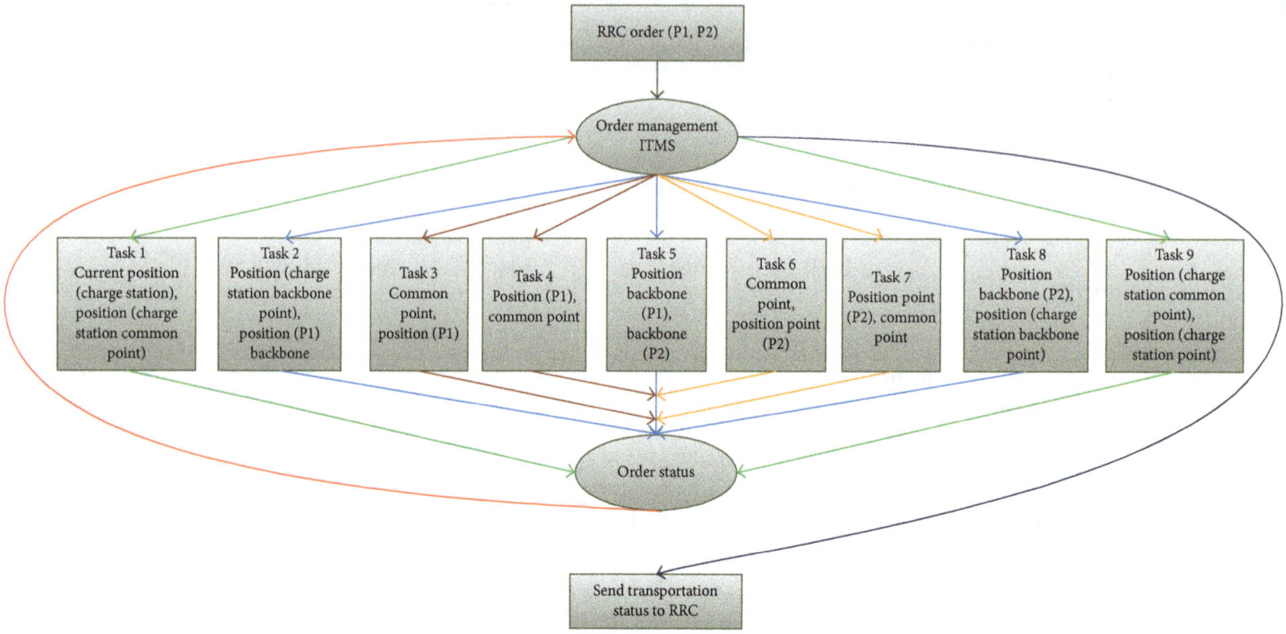

FIGURE 5: Internal transportation management system.

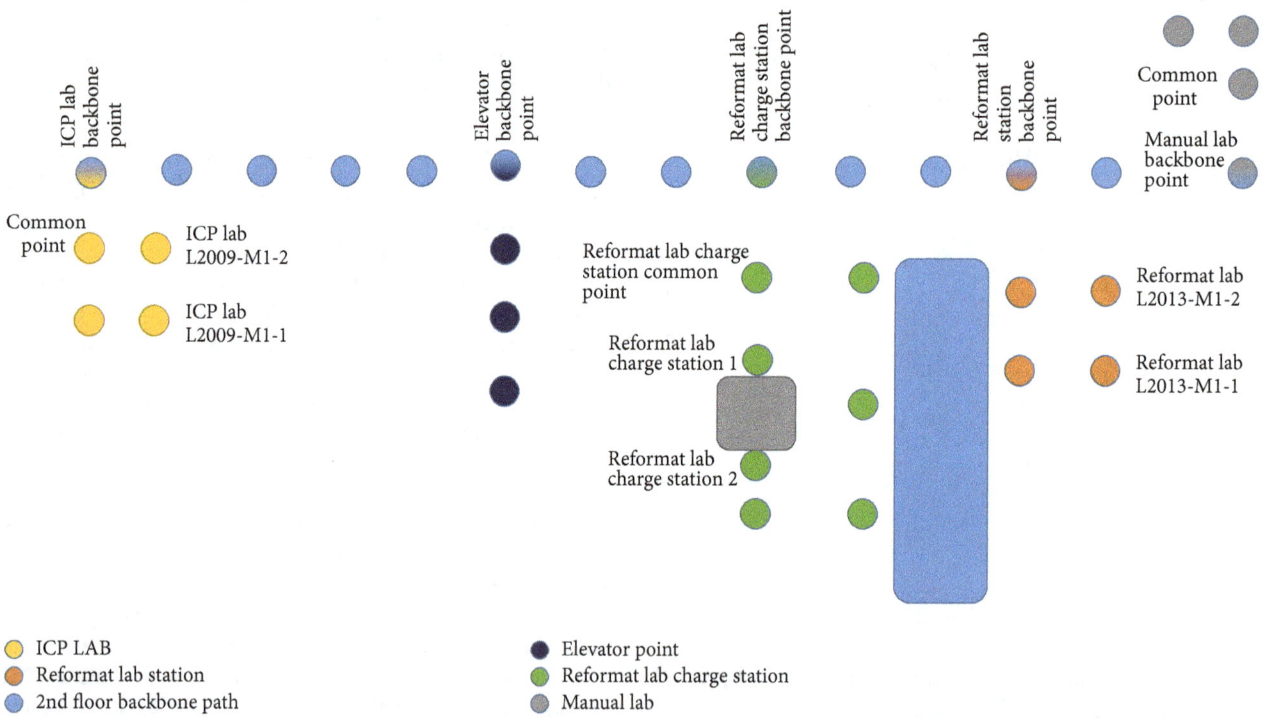

FIGURE 6: Backbone waypoint for 2nd floor in CELISCA.

transportation state (robot busy, transportation start, RAKM problem, wrong path/points parameters, and transportation finished). The ITMS chooses the right path and the points, sends it to the movement core, and waits till the end of this task. Then another path is loaded until the whole transportation task is done. When there is a problem on any task the ITMC stops other tasks immediately and informs the RRC.

3.3.2. The Backbone Method. This method is based on building a main path for each floor. This path has multiple key points leading to subpaths that complete the transportation path as shown in Figure 6. This method decreases the number of paths required to cover a large map with many stations (multiple source and multiple destination points). In addition to that, this method makes the path creation easier than normal static path. The number of paths required in the static path method is $N * N - 1$ where N is the number of stations. The total number of path required in the backbone method is (backbone path + $N/2$). For a floor with six transportation station points with two charge station points, result was 56 paths in static method and 5 for the backbone path method. Table 1 shows the reduction of the required paths by the backbone method especially in multifloor environment. When the robot receives the transportation task from RRC, the ITMS divides the transportation task into small tasks and is starting the execution of these tasks as follows: (a) search inside the created paths to find the way from the charge station to the point near the backbone path. (b) Load the current floor essential path (backbone) and according to the source station position either move directly to the grasping station key point or move through multiple paths (current path towards elevator, elevator path, and grasping floor essential path to the grasping station key point) if the grasping station is on another floor. (c) Use the current grasping subpath to reach the grasping station and execute the required operation. (d) Repeat (b) and (c) taking into consideration that the movement is towards placing station. (e) Load the current floor backbone path and either move till reaching the charge key point on the path occurs or move through (current path towards elevator, elevator path, and charging floor essential path to the charging station key point) in case the charge station is on a different floor. (f) Use the charging path to reach the charge station and wait for the new transportation task.

TABLE 1: Static method and backbone method.

Number of transportation station points	Normal paths method	Backbone path method
8 (2nd) floors	56	5
12 (2nd + 3rd) floors	132	9

3.4. Elevator Aspects. The elevator should be used to move across levels in multifloor environment. As a first step to handle the elevator, the robot moves towards the elevator door based on ceiling landmarks until reaching a calculated point which brings the robot in a right position to press the button and enter the elevator at the same time as clarified in Figure 7(d). After button pressing the mobile robot continuously checks the door to enter the elevator when the door is open. As for the next step, the robot localizes himself using the installed landmark inside the elevator. Next, the robot moves to a suitable point near the control panel to recognize the destination floor button and control the elevator door correctly. The developed methods for buttons detection were divided into outside and inside elevator as follows.

3.4.1. The Elevator Entry Button. Since the entry button and its panel are made from the same reflecting material, the detection of the button is difficult to realize. Thus, a landmark has been installed around the entry button with a specific shape and color to enable an easy and applicable recognition process for mobile robot. The elevator entry button detection process can be explained as follows: (a) initialize the Kinect sensor with the required frame rate, image resolution, and depth resolution. (b) Capture an image from Kinect RGB sensor. (c) Use an RGB channel filter to remove the selected band of colors. (d) Search for connected pixels (object) inside the image. (e) Select candidates based on specific width and height. (f) Check if the chosen candidates have square shape with the specified acceptance distortion and color. (g) Get the depth snapshot from the Kinect sensor; then do the mapping between position in pixel and depth to extract the real coordinates (X, Y, and Z). (h) Calibration between Kinect positions and the arm reference for arm pressing operation is performed.

This method has 88.5% successful rate at a stable lightening condition. But the working environment has different lighting and sunlight conditions which easily affect the detection process with RGB color system. Thus, it reduces the successful rate. This problem has been solved using the HSL color representation which is more stable against the changing of lighting conditions as will be clarified in the experiment section. Figure 7(a) shows the HSL based button detection method.

3.4.2. The Elevator Internal Button. For successful buttons detection, the diameter of the button must be equal to or larger than 1.9 cm and the button label should be placed to the left of the related button based on American Disability Association (ADA) recommendations [33]. In this method, it is important to recognize each button label separately; thus the developed method using a combination of filters includes grayscale conversion that makes the captured image suitable for next stages, stretch contrast to improve the contrast of the image by stretching the domain of intensity, adaptive threshold to choose the best threshold under different light condition for binary image conversion that have been applied. Next, searching for each butto's candidates using specific features (width and height), taking the inverse value of the pixel, and flipping the candidates horizontally (Kinect image stream has a mirror image) to make it suitable for the Optical Character Recognition (OCR) stage, each extracted candidate passes to OCR engine for comparison with the required destination, and finally based on the position of the matching candidate inside the image and depth information, the real coordinates are extracted and translated into robot arm reference. Figure 7(b) demonstrates the internal button detection method.

The proposed detailed robot-elevator interface controlling strategy is demonstrated in Figure 7(c). The elevator handler system has two methods for error handling: (1) in case of failure of the mobile robot to reach the required elevator button pushing area, the position and orientation correction function controls the robot to reach the defined position with high-accuracy orientations. This function checks the robot position after movement and tries to correct the robot position three times till it reaches the defined position

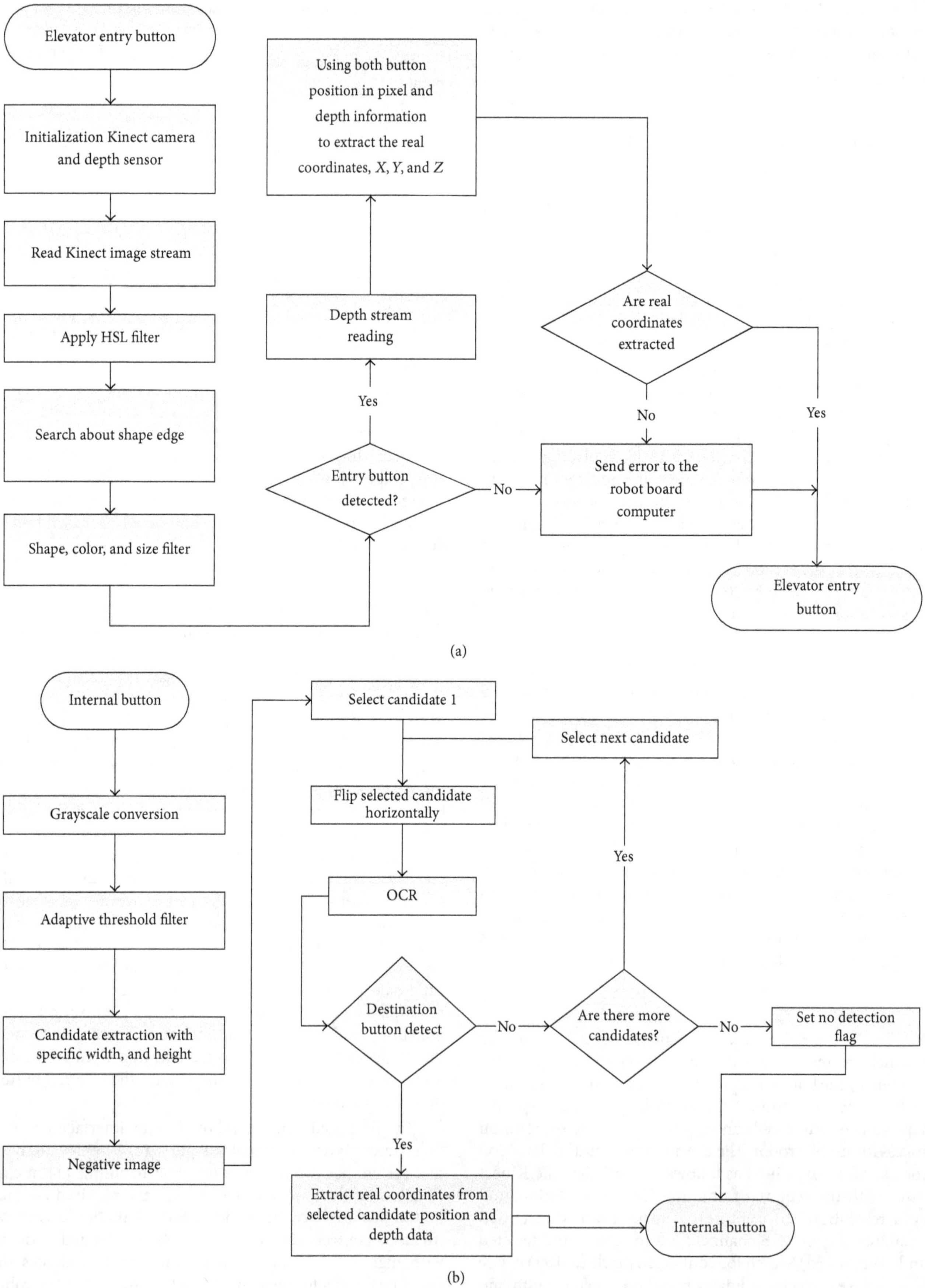

Figure 7: Continued.

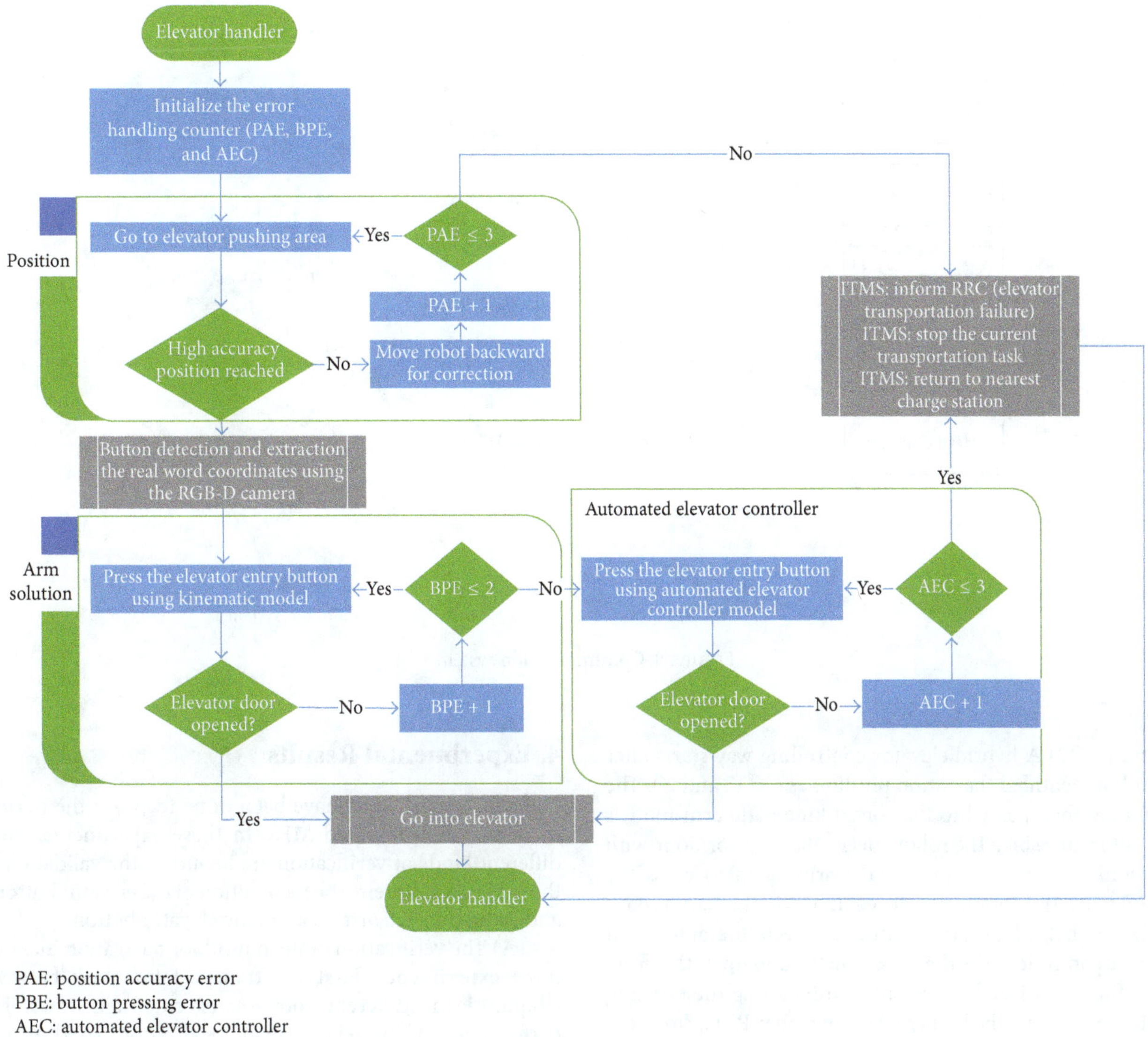

(c)

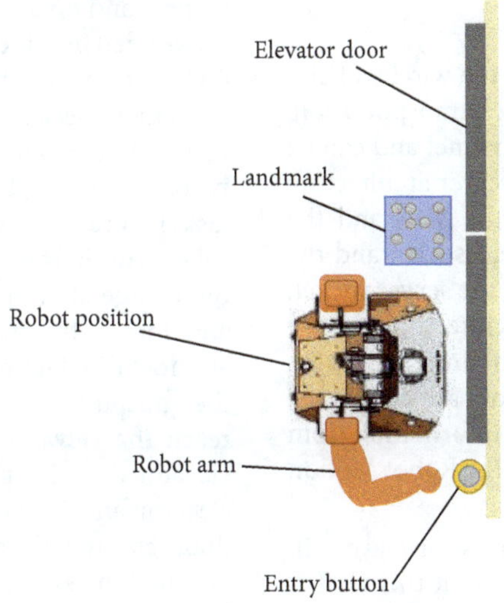

(d)

FIGURE 7: Robot-elevator door controlling case. (a) HSL based entry button detection flow chart. (b) Internal buttons detection flow chart. (c) Button pressing flow chart. (d) Robot and elevator position.

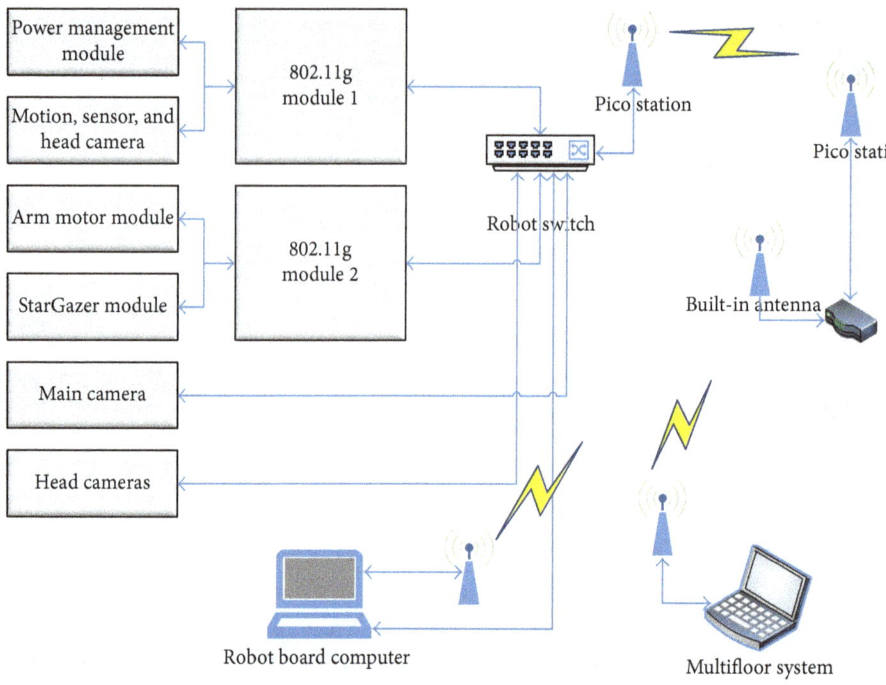

FIGURE 8: Communication system.

accurately. (2) A hybrid elevator controlling way starts after the robot identified the button position as (X, Y, and Z). The values are then passed to the solved kinematic arm models for button pressing. The robot checks the elevator door with ultrasonic sensor and controls the arm again to press the button when the door is still closed. If previous attempts have failed, the hybrid elevator controller selects the automated elevator controller over the socket method to open the door. If the functions listed above failed in handling the elevator, the ITMS informs the higher level controller RRC, stops the current transportation operation, and controls the robot till it reaches the charging station.

3.5. System Communication. In this study the wireless IEEE 802.11g network is utilized for laboratories transportation, which has a suitable width and fast data channel and can be easily expanded for large areas. The client/server architecture is adopted for data exchanging between the MFS and the mobile robot. The MFS represents the client socket and the onboard computer (inside mobile robot) is the server socket. Figure 8 shows the complete network architecture for the mobile robot. The SGM is connected to the internal switch through module 2, while the motion motors are connected through module 1. The MFS reads position information from SGM and sends movement orders through the Robot Motion Center.

A GUI of the MFS communication is displayed in Figure 9(a). From this figure it can be seen that this GUI is used to connect to the server socket, test position readings, and send movement and rotation orders. The server socket is illustrated in Figure 9(b) which shows the connection status and command history.

4. Experimental Results

Different experiments have been done to prove the performance of the presented MFS. In these experiments, three different kinds of verification are included: the validation of the navigation system, the recognition of the elevator buttons, and using the H20 arm to press the elevator button.

(A) The verification of the multifloor navigation includes three experiments. First the transportation task between laboratories in different floor was executed ten times. The performance of the relative map, localization based on relative map, backbone path planning method, communication system, and finally ITMS transportation management system are verified in this experiment. Figure 10 shows the multifloor GUI. Figure 11 clarifies the path created by the robot during its movement between 2nd and 3rd floor including the elevator in real Life Science Laboratories. Figure 12 shows the H20 mobile robot while executing the expected transportation task. Figure contents are explained as follows: in Figure 12(a) robot is initially in charge station; in Figure 12(b) robot leaves the charge station towards the current floor essential path and moves through it till it reaches the grasping station as shown in Figure 12(c); in Figure 12(d) robot is executing the grasping order and returns to the essential path to reach the elevator and in Figure 12(e) it is waiting till the elevator door is opening (see Figure 12(f)); robot enters the elevator and is leaving it after approaching the destination floor as shown in Figures 12(g)–12(i); it passes through destination essential path towards the placing station shown in Figures 12(j) and 12(k). After finishing the transportation task, the mobile robot returns to the charging station through the elevator; the charging station essential path is clarified in Figures 12(l) and 12(m); Figures 12(n) and 12(o) show the

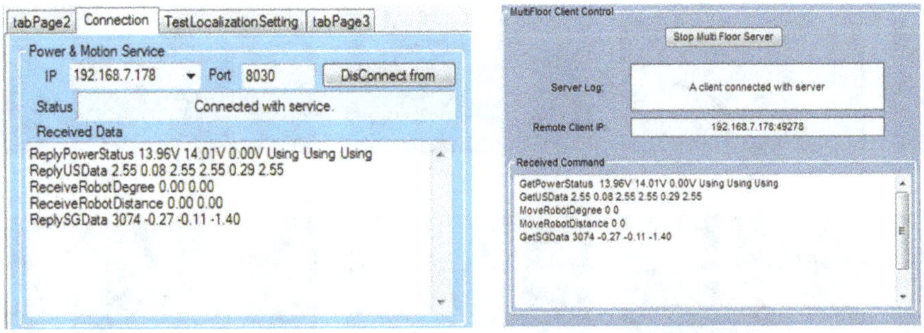

(a) (b)

FIGURE 9: MFS and on-board computer client/server sockets.

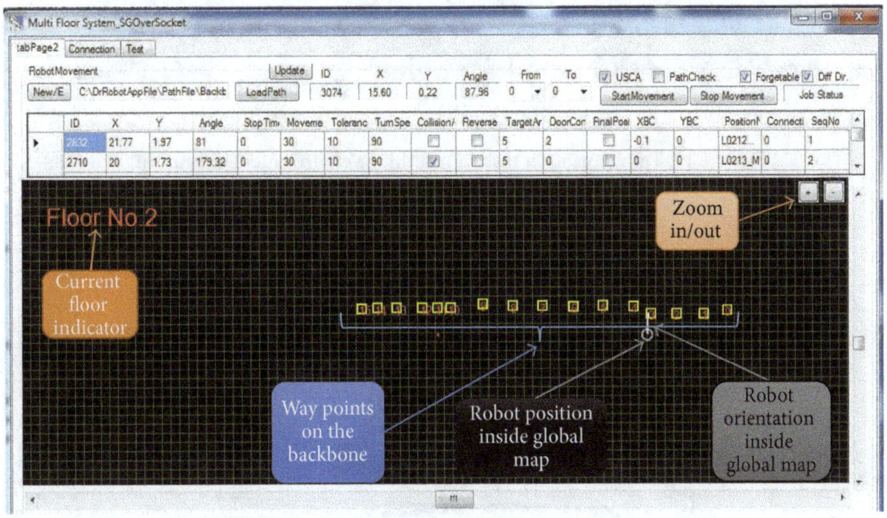

FIGURE 10: Multifloor GUI.

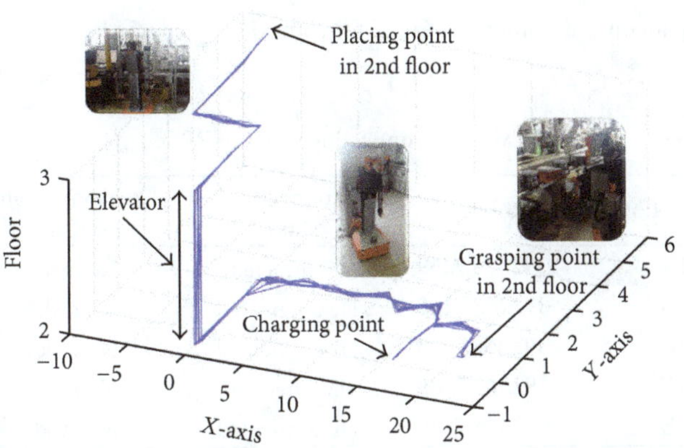

FIGURE 11: The real mobile robot transportation path on multifloor (repeating ten times).

charging procedure including the rotation and movement in backward direction till contacting the charge station. The robot executes the transportation task in multifloor environment efficiently and the internal systems have been verified.

The second experiment has been done to determine the repeatability of the different (grasping, placing, and charging) points in the transportation task. Repeatability is defined as the ability of mobile robot to reach the same position over a period of time while the accuracy is the difference between the reached point and optimal point. This test has been done in one floor for fifty times; each time the robot moves from the charging station towards position 1 for grasping operation, then moves till it reaches position 2 for placing, and returns back to the charging station, as clarified in Figure 13. This experiment has 100% success rate. Figure 14 clarifies the repeatability of the mobile robot for the important stations in the transportation path. In Figure 14(a) we notice that the maximum error reaching the grasping position is about 4 cm. The complicated and narrow area caused a lower mapping accuracy in this position than other tested positions. The repeatability of elevator entry button position had been investigated from the movement of mobile robot in multifloor experiment as clarified in Figure 14(d). The accuracy, mean, and the standard deviation for these points are listed in Table 2.

The last experiment was done for transportation tasks on a single floor to validate the integration of the MFS

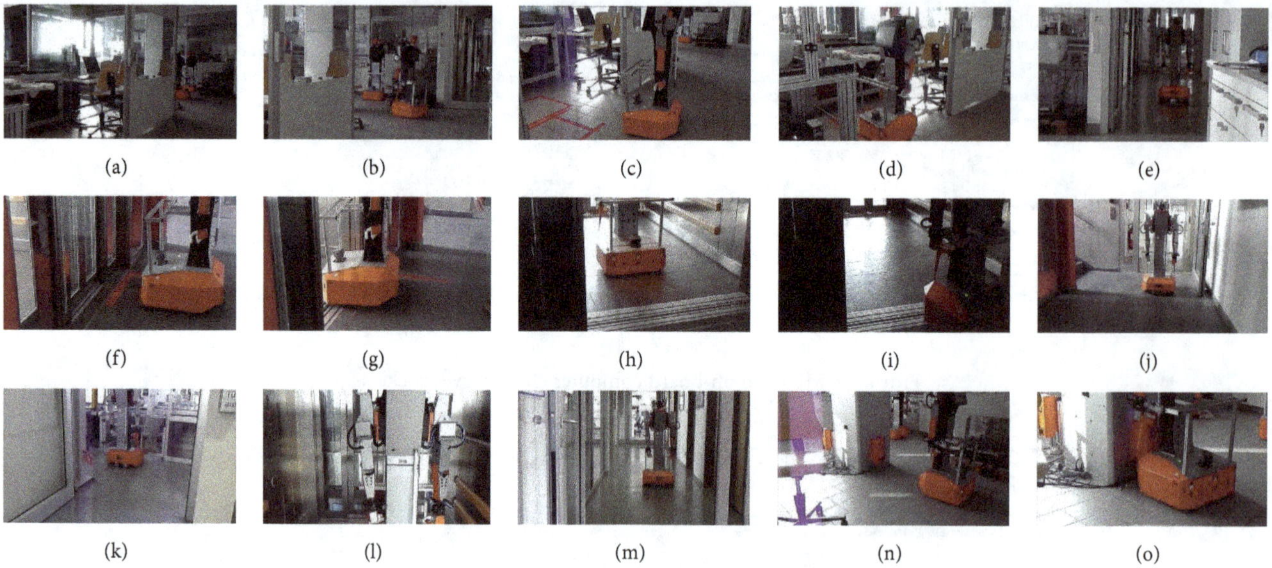

FIGURE 12: Transportation task in multifloor environment execution.

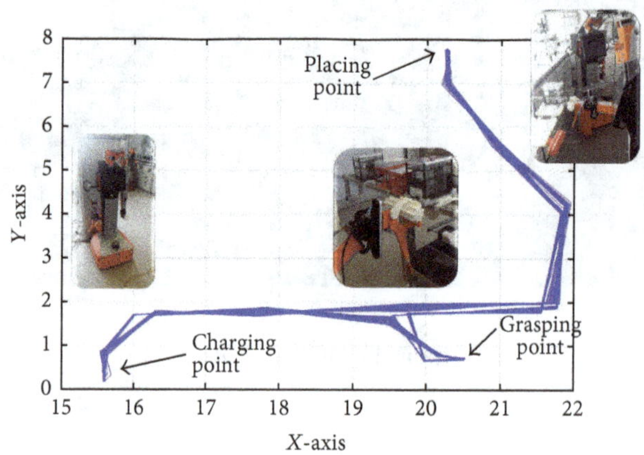

FIGURE 13: The real mobile robot transportation path on single floor (repeating fifty times).

TABLE 2: Accuracy, mean, and standard deviation of important robot position in meter.

	Axis	Accuracy	Mean	SD
Grasping point	X	0.04	20.506	0.008329931
	Y	0.01	0.702	0.011952286
Placing point	X	0.00	20.2682	0.014384
	Y	0.01	7.7676	0.012707
Charging point	X	0.01	15.596	0.006700594
	Y	0.02	0.1998	0.013013337
Elevator entry point	X	0.00	−0.14091	0.009439
	Y	0.01	0.938182	0.015374

TABLE 3: Connecting robot to RRC.

Robot ID	Robot IP	Connection status	Transportation status
1A	192.168.7.159	Connected	No task
2B	192.168.7.168	Disconnected	No task
3C	192.168.7.178	Disconnected	No task
4D	192.168.7.188	Connected	No task
5E	192.168.7.198	Disconnected	No task

with the multirobot management level (RRC). Once the RRC works, the connected robots will be listed as shown in Table 3. The RRC sends the transportation order to one of the available robots that has the highest charging rate to execute the requested task. The MFS of the chosen robot receives this order as raw data (see Figure 15); then the ITMS parses it to receive the grasping/placing position as X, Y, and floor number. The execution of this task starts from the charge station, navigates towards point 1 (Grasping), then is directed to the placing position (point 2), and finally returns to the charge station. This experiment was repeated for ten times with 100% success rate in receiving the order from

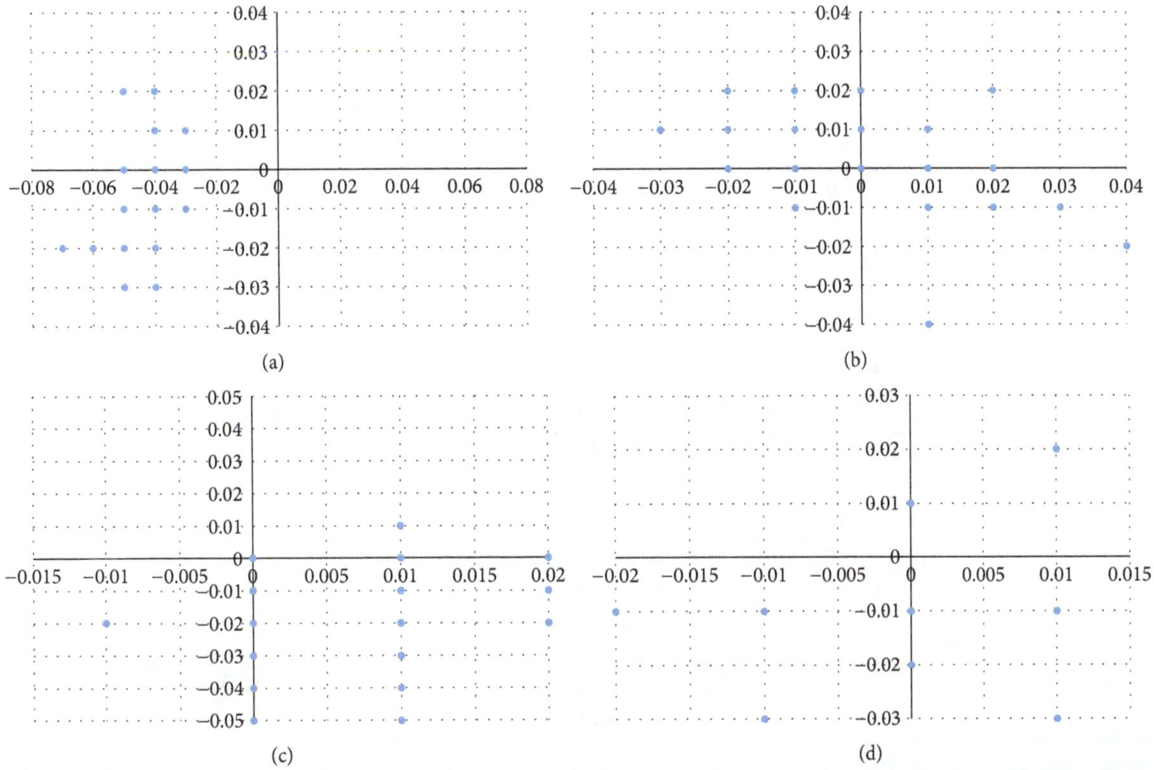

FIGURE 14: Robot movement position error. (a) Grasping position error. (b) Placing position error. (c) Charge station position error. (d) Entry button position error.

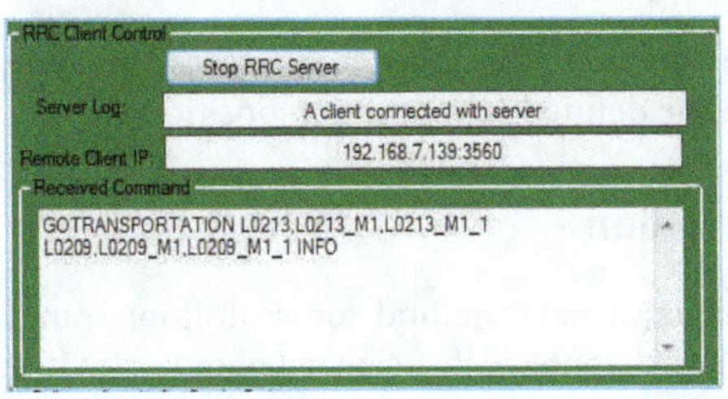

FIGURE 15: Received transportation order on MFS GUI.

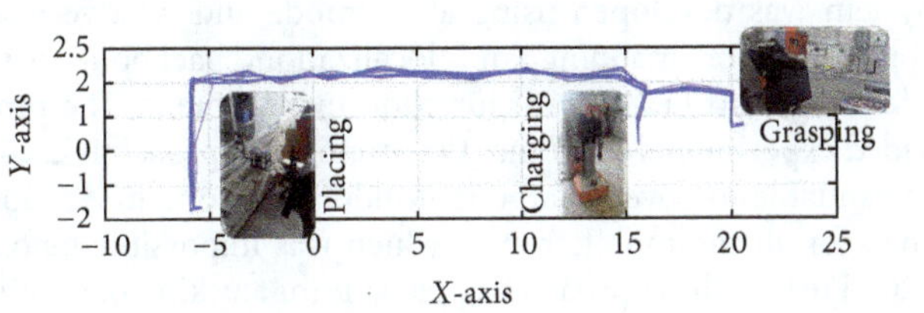

FIGURE 16: The real mobile robot transportation path on single floor.

higher level, controlling the automated doors in the robot path, and navigating to the grasping, placing, and charging position. The generated path of the mobile robot movements is shown in Figure 16. The accuracy and repeatability of the mobile robot in the important points (grasping, placing, and charging points) are demonstrated in Figure 17 while Table 4 lists the accuracy, mean, and standard division of these points.

TABLE 4: Accuracy, mean, and standard deviation of important robot position in meter.

	Axis	Accuracy	Mean	SD
Grasping point	X	0.001	20.55102	0.009717
	Y	0.019	0.729276	0.006221
Placing point	X	0.007	−5.39292	0.011949
	Y	0.009	−1.62057	0.007359
Charging point	X	0.002	15.59156	0.009182
	Y	−0.00768	0.212323	0.022121

(B) The mobile robot with Kinect sensor was moved towards the elevator as shown in Figure 18(a) to validate the elevator entry button detection algorithm. The detection program was started and tests were taken from many distances to detect the button and read the real position according to the mobile robot. The elevator entry button detection experiment has been repeated 100 times for each distance (45 cm, 50 cm, 55 cm, and 60 cm). The system can detect the entry button with high accuracy and speed. The average of the entry button detection success rate from different distances reaches 88.5% using the RGB detection method. Varying lightening condition against the Kinect vision sensor as shown in Figure 18(a) caused errors in detection. The developed method utilizes the HSL color representation instead of RGB representation to stabilize the color detection under different lightening condition. The success rate of the

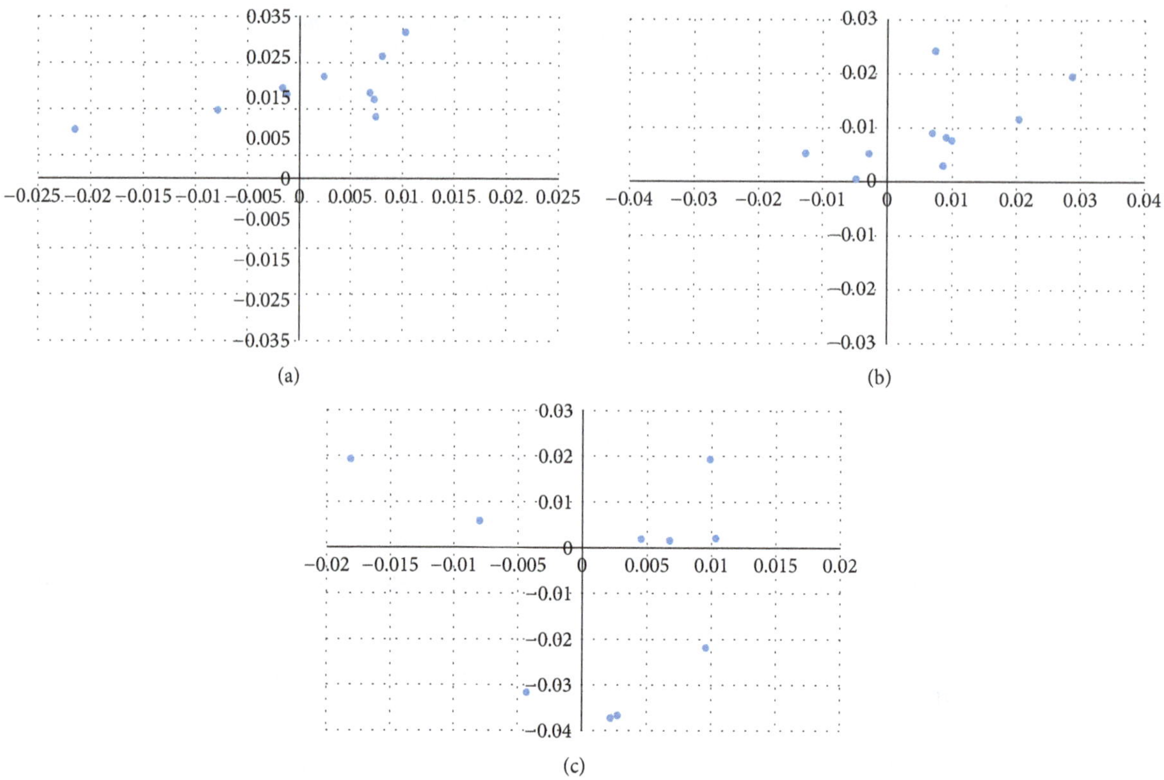

FIGURE 17: Robot movement position error. (a) Grasping position error. (b) Placing position error. (c) Charge station position error.

developed method reaches up to 99%. The contrast results between the HSL and RGB color representation detection methods are shown in Figure 18(b). The depth sensor is also analyzed for both methods; as clearly shown in Figure 18(c) the depth sensor starts to work in 50 cm distance.

The second experiment was executed to evaluate the performance of the internal button detection method. As a first step, the Kinect sensor has been placed in front of the internal buttons panel inside the elevator; then the internal button detection program starts and connects to the Kinect sensor. All buttons were selected one by one as a destination floor to be examined separately. Figure 19 shows the developed GUI for the elevator handler in operation; the left side shows the captured image with the red square drawn around the destination floor button. The detected button position in image coordinates is also shown in the same side, while the other side shows the processed image and the extracted real position. The collected candidates which work as inputs to the OCR engine are shown in Figure 20. This experiment was repeated 600 times to detect the specific button label with its real position. The internal button detection method showed a 98.66% success rate.

(C) The last experiment was performed twenty times to validate the robot arm pressing the elevator entry button as shown in Figure 21. A kinematic model has been built to control the robot arm. The robot position should be chosen correctly to put the entry button in the effective range of robot arm, in the same time allowing the robot to enter the elevator when it is opened. The robot arm succeeds in pressing the button in the defined robot motion positions.

5. Conclusion

In this paper, a new method for multifloor transportation systems in distributed Life Science Laboratories is proposed. To meet all the requirements of life science automation, a ceiling landmark and a StarGazer module sensor are used for indoor localization. An innovative multifloor transportation system was developed using alone mode and relative map for multifloor mapping and localization, backbone path planning, and ITMS transportation management. The provided experiments show that the proposed approach is easily expandable to cover a large multifloor map, one navigation map for all the mobile robots which was impossible before [20]. Further the approach is applicable for any kind of mobile robots and has low costs, has high robustness, is ambient for light effects, and can work under dark conditions, which is not available in other localization systems. A robust high speed method is applied for elevator button detection (entry button and internal buttons). An IEEE 802.11g communication network with TCP/IP protocol is utilized, to establish an expandable communication network.

Competing Interests

The authors declare that they have no competing interests.

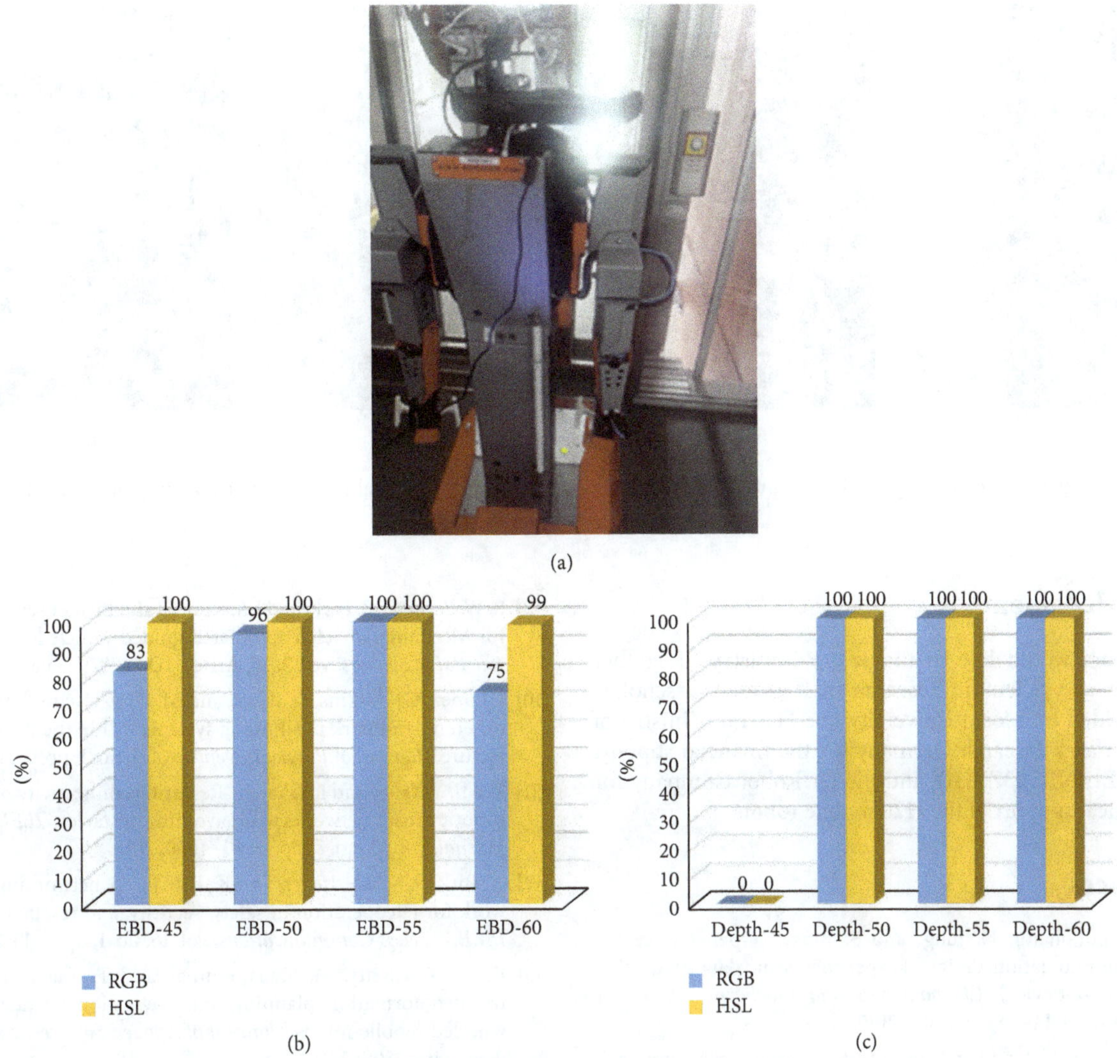

FIGURE 18: Experiment for elevator entry button detection. (a) Experiment environment for elevator entry button detection. (b) Detection success rate for RGB and HSL module. (c) Depth extraction success rate.

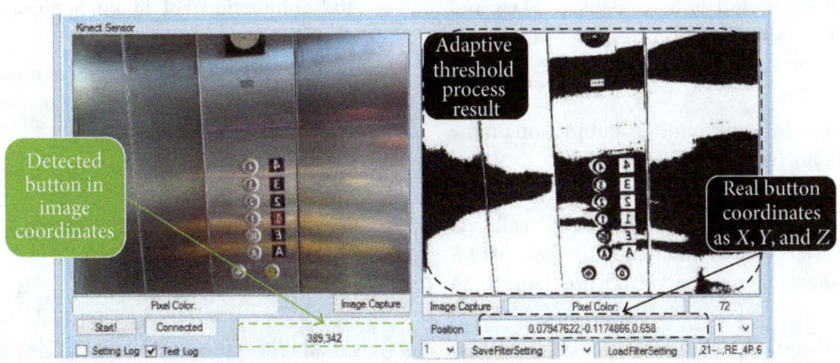

FIGURE 19: Elevator handler for internal buttons.

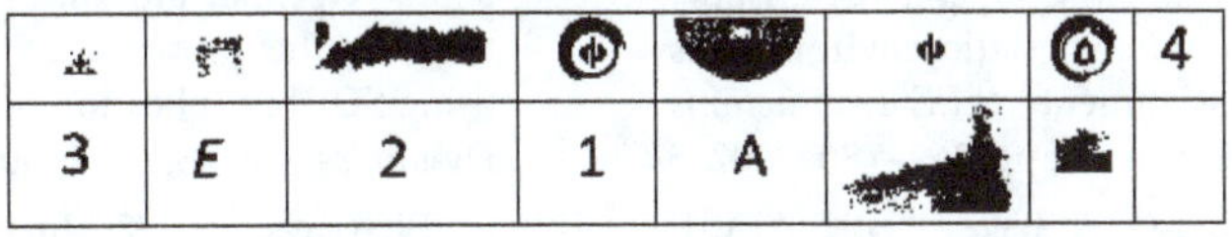

FIGURE 20: Extracted button candidates.

(a) (b) (c) (d)

FIGURE 21: (a) Before the pressing. (b) Button pressing. (c) Robot arm returning to original position after pressing. (d) Ready to enter the elevator.

Acknowledgments

The authors would like to thank the Ministry of Higher Education and Scientific Research in Iraq for the scholarship provided by Mosul University, the Federal Ministry of Education and Research Germany for the financial support (FKZ: 03Z1KN11, 03Z1KI1), and the Dr Robot Company for the technical support of the H20 mobile robots.

References

[1] D. S. Lütjohann, N. Jung, and S. Bräse, "Open source life science automation: design of experiments and data acquisition via 'dial-a-device'," *Chemometrics and Intelligent Laboratory Systems*, vol. 144, pp. 100–107, 2015.

[2] B. Göde, S. Holzmüller-Laue, and K. Thurow, "New approach of standard-based end-to-end process automation in the life science automation," *Chemie-Ingenieur-Technik*, vol. 87, no. 5, pp. 518–536, 2015.

[3] J. G. Petersen, S. A. Bowyer, and F. R. Y. Baena, "Mass and friction optimization for natural motion in hands-on robotic surgery," *IEEE Transactions on Robotics*, vol. 32, no. 1, pp. 201–213, 2016.

[4] T. Chapman, "Lab automation and robotics: automation on the move," *Nature*, vol. 421, no. 6923, pp. 661–666, 2003.

[5] K. Sakaki, H. Esmaeilsabzali, S. Massah et al., "Localized, macromolecular transport for thin, adherent, single cells via an automated, single cell electroporation biomanipulator," *IEEE Transactions on Biomedical Engineering*, vol. 60, no. 11, pp. 3113–3123, 2013.

[6] W. Gecks and S. T. Pedersen, "Robotics—an efficient tool for laboratory automation," *IEEE Transactions on Industry Applications*, vol. 28, no. 4, pp. 938–944, 1992.

[7] N. Miyata, J. Ota, T. Arai, and H. Asama, "Cooperative transport by multiple mobile robots in unknown static environments associated with real-time task assignment," *IEEE Transactions on Robotics and Automation*, vol. 18, no. 5, pp. 769–780, 2002.

[8] J. F. Montgomery, A. H. Fagg, and G. A. Bekey, "USC AFV-I: a behavior-based entry in the 1994 international aerial robotics competition," *IEEE Expert*, vol. 10, no. 2, pp. 16–22, 1995.

[9] S. W. Yoon, S.-B. Park, and J. S. Kim, "Kalman filter sensor fusion for Mecanum wheeled automated guided vehicle localization," *Journal of Sensors*, vol. 2015, Article ID 347379, 7 pages, 2015.

[10] J. López, C. Watkins, D. Pérez, and M. Díaz-Cacho, "Evaluating different landmark positioning systems within the RIDE architecture," *Journal of Physical Agents*, vol. 7, no. 1, pp. 3–11, 2013.

[11] Z. Mi, Y. Yang, and J. Y. Yang, "Restoring connectivity of mobile robotic sensor networks while avoiding obstacles," *IEEE Sensors Journal*, vol. 15, no. 8, pp. 4640–4650, 2015.

[12] A. Birk, S. Schwertfeger, and K. Pathak, "A networking framework for teleoperation in safety, security, and rescue robotics," *IEEE Wireless Communications*, vol. 16, no. 1, pp. 6–13, 2009.

[13] R. M. K. Chetty, M. Singaperumal, and T. Nagarajan, "Distributed formation planning and navigation framework for wheeled mobile robots," *Journal of Applied Sciences*, vol. 11, no. 9, pp. 1501–1509, 2011.

[14] A. Ohya, A. Kosaka, and A. Kak, "Vision-based navigation by a mobile robot with obstacle avoidance using single-camera vision and ultrasonic sensing," *IEEE Transactions on Robotics and Automation*, vol. 14, no. 6, pp. 969–978, 1998.

[15] S.-W. Ma, X. Cui, H.-H. Lee, H.-R. Kim, J.-H. Lee, and H. Kim, "Robust elevator door recognition using LRF and camera," *Journal of Institute of Control, Robotics and Systems*, vol. 18, no. 6, pp. 601–607, 2012.

[16] J.-G. Kang, S.-Y. An, W.-S. Choi, and S.-Y. Oh, "Recognition and path planning strategy for autonomous navigation in the elevator environment," *International Journal of Control, Automation and Systems*, vol. 8, no. 4, pp. 808–821, 2010.

[17] P. K. Mohanty and D. R. Parhi, "A new hybrid optimization algorithm for multiple mobile robots navigation based on the CS-ANFIS approach," *Memetic Computing*, vol. 7, no. 4, pp. 255–273, 2015.

[18] M. H. Korayem, A. Zehfroosh, H. Tourajizadeh, and S. Manteghi, "Optimal motion planning of non-linear dynamic systems in the presence of obstacles and moving boundaries using SDRE: application on cable-suspended robot," *Nonlinear Dynamics*, vol. 76, no. 2, pp. 1423–1441, 2014.

[19] P. Bhargava, S. Krishnamoorthy, A. Shrivastava, A. K. Nakshathri, M. Mah, and A. Agrawala, "Locus: robust and calibration-free indoor localization, tracking and navigation

for multi-story buildings," *Journal of Location Based Services*, vol. 9, no. 3, pp. 187–208, 2015.

[20] H. Liu, N. Stoll, S. Junginger, and K. Thurow, "Mobile robot for life science automation," *International Journal of Advanced Robotic Systems*, vol. 10, pp. 1–14, 2013.

[21] "Dr Robot Inc.: WiFi 802.11 robot, Network-based Robot, robotic, robot kit, humanoid robot, OEM solution," http://www.drrobot.com/products_H20.asp.

[22] X. Yu, L. Dong, Y. Li, and K. E. Hoe, "Lift-button detection and recognition for service robot in buildings," in *Proceedings of the 16th IEEE International Conference on Image Processing (ICIP '09)*, pp. 313–316, IEEE, Cairo, Egypt, November 2009.

[23] J. Baek and M. Lee, "A study on detecting elevator entrance door using stereo vision in multi floor environment," in *Proceedings of the ICROS-SICE International Joint Conference (ICCAS-SICE '09)*, pp. 1370–1373, Fukuoka, Japan, August 2009.

[24] J.-G. Kang, S.-Y. An, and S.-Y. Oh, "Navigation strategy for the service robot in the elevator environment," in *Proceedings of the International Conference on Control, Automation and Systems (ICCAS '07)*, pp. 1092–1097, October 2007.

[25] H. Chae, S.-H. Kim, Y.-C. Lee, and Christiand, "Applications of robot navigation based on artificial landmark in large scale public space," in *Proceedings of the IEEE International Conference on Robotics and Biomimetics (ROBIO '11)*, pp. 721–726, December 2011.

[26] I. Ul-Haque and E. Prassler, "Experimental evaluation of a low-cost mobile robot localization technique for large indoor public environments," in *Proceedings of the Joint 41st International Symposium on Robotics (ISR '10) and 6th German Conference on Robotics (ROBOTIK '10)*, pp. 1–7, Munich, Germany, June 2010.

[27] D. Troniak, J. Sattar, A. Gupta et al., "Charlie rides the elevator—integrating vision, navigation and manipulation towards multi-floor robot locomotion," in *Proceedings of the 10th International Conference on Computer and Robot Vision (CRV '13)*, pp. 1–8, Regina, Canada, May 2013.

[28] HAGISONIC: StarGazer Module, Localization Sensor, http://www.hagisonic.com/.

[29] J. K. Goyal and K. S. Nagla, "A new approach of path planning for mobile robots," in *Proceedings of the 3rd International Conference on Advances in Computing, Communications and Informatics (ICACCI '14)*, pp. 863–867, IEEE, New Delhi, India, September 2014.

[30] B. Park, J. Choi, and W. K. Chung, "An efficient mobile robot path planning using hierarchical roadmap representation in indoor environment," in *Proceedings of the IEEE International Conference on Robotics and Automation (ICRA '12)*, pp. 180–186, Saint Paul, Minn, USA, May 2012.

[31] J. Wang, Y. Sun, Z. Liu, P. Yang, and T. Lin, "Route planning based on floyd algorithm for intelligence transportation system," in *Proceedings of the IEEE International Conference on Integration Technology (ICIT '07)*, pp. 544–546, Shenzhen, China, March 2007.

[32] A. A. Abdulla, H. Liu, N. Stoll, and K. Thurow, "Multi-floor navigation method for mobile robot transportation based on StarGazer sensors in life science automation," in *Proceedings of the IEEE International Instrumentation and Measurement Technology Conference (I2MTC '15)*, pp. 428–433, Pisa, Italy, May 2015.

[33] "Department of Justice ADA Title III Regulation 28 CFR Part 36," 1991, http://www.ada.gov/reg3a.html#Anchor-11540.

21

Rotor Speed Control of a Direct-Driven Permanent Magnet Synchronous Generator-Based Wind Turbine Using Phase-Lag Compensators to Optimize Wind Power Extraction

Ester Hamatwi,[1] Innocent E. Davidson,[2] and Michael N. Gitau[3]

[1] *Department of Electrical, Electronic and Computer Engineering, University of KwaZulu-Natal, University Road, Durban 4000, South Africa*
[2] *Department of Electric Power Engineering, Durban University of Technology, Corner of Botanic Gardens and Steve Biko Roads, Durban 4000, South Africa*
[3] *Department of Electrical, Electronics and Computer Engineering, University of Pretoria, Lynnwood Road, Pretoria 0002, South Africa*

Correspondence should be addressed to Ester Hamatwi; esterhamatwi@gmail.com

Academic Editor: Carlos-Andrés García

Due to the intermittent nature of wind, the wind power output tends to be inconsistent, and hence maximum power point tracking (MPPT) is usually employed to optimize the power extracted from the wind resource at a wide range of wind speeds. This paper deals with the rotor speed control of a 2 MW direct-driven permanent magnet synchronous generator (PMSG) to achieve MPPT. The proportional-integral (PI), proportional-derivative (PD), and proportional-integral-derivative (PID) controllers have widely been employed in MPPT studies owing to their simple structure and simple design procedure. However, there are a number of shortcomings associated with these controllers; the trial-and-error design procedure used to determine the P, I, and D gains presents a possibility for poorly tuned controller gains, which reduces the accuracy and the dynamic performance of the entire control system. Moreover, these controllers' linear nature, constricted operating range, and their sensitivity to changes in machine parameters make them ineffective when applied to nonlinear and uncertain systems. On the other hand, phase-lag compensators are associated with a design procedure that is well defined from fundamental principles as opposed to the aforementioned trial-and-error design procedure. This makes the latter controller type more accurate, although it is not well developed yet, and hence it is the focus of this paper. The simulation results demonstrated the effectiveness of the proposed MPPT controller.

1. Introduction

The increasing energy demand across the globe has led to an enormous interest in the cost-competitive, environmental friendly, and reliable renewable energy (RE) technologies to complement the conventional methods of generating electricity [1, 2]. Wind energy has recently been proven to be technologically mature and fast-growing among all other RE sources due to the growth in the size of commercial wind turbine (WT) designs and the increase in the installed capacity worldwide [2–12]. Modern variable speed wind energy conversion systems (WECSs) are based on doubly fed induction generators (DFIGs) and permanent magnet synchronous generators (PMSGs) [3]. Although the DFIGs are well developed and have previously been widely applied in WECS due to their ability to achieve variable speed control with the reduced cost of converter requirements, the use of slip rings and the protection issues in the event of grid faults are the major drawbacks [8]. As a result, the PMSG-based WECSs have gained more popularity and preference over DFIG-based WECSs owing to their advantages such as high energy density, simple control methodology, high reliability, low maintenance cost, and the self-excitation system [1, 13, 14]. Moreover, direct-driven multipole PMSGs have been adopted to exclude the gearbox, which improves the efficiency of the WT and cuts down the weight of the nacelle

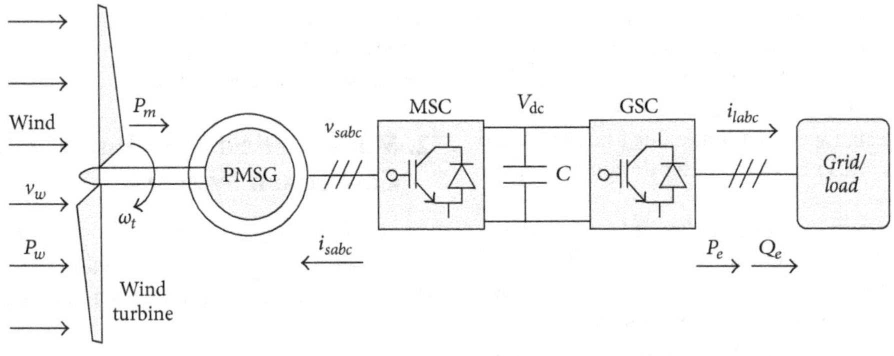

FIGURE 1: The direct-driven PMSG-based WECS [1].

and the operation and maintenance cost of the WT system [3, 13, 15]. The major challenge associated with WECS in general is the intermittent nature of the wind resource leading to an inconsistent wind power output. This inconsistency reduces the system stability and power quality of the electrical grid [6]. Therefore, power electronic converters are required to act as an interfacing medium between WECSs and the grid [16].

This will aid in achieving the grid compliance of the WECS in terms of the voltage level, frequency, active power, and reactive power [17–20]. A direct-driven PSMG-based WECS comprises a machine-side converter (MSC) and a grid-side converter (GSC) interconnected via a DC-link capacitor as shown in Figure 1.

The ability of a variable speed WT to extract optimal power at a wide range of wind speeds makes it possible to be operated in the MPPT mode [1, 12]. Over the past years, various MPPT techniques have been developed, namely, optimal relationship-based (ORB) control, tip speed ratio (TSR) control, optimal torque control (OTC), and Perturb and Observe (P&O)/Hill-Climb Search (HCS) control. Their operation, strengths, and weaknesses have been discussed in [3, 15, 21–23]. In this paper, the MPPT operation is achieved through the use of optimal relationship-based (ORB) control. The ORB control technique has gained interest because of its simplicity, quick response, and enhanced power smoothing capability and it is independent of the wind speed measurement [3]. The ORB control ensures optimal extraction of power from the wind resource by using a precalculated look-up table, and hence the knowledge of the WT's MPPT curve is a prerequisite [3, 24]. On this basis, the amount of power extracted from the wind resource is controlled indirectly by varying the rotor speed ω_r with respect to the change in wind speed v_w. Therefore, optimal aerodynamic efficiency at a given wind speed can be achieved if the MPPT controller allows the generated power to properly track the predefined optimal power curve with variations in wind speed.

The direct (power angle) control and vector-oriented control (VOC) are the well-developed control techniques used to perform the independent control of active and reactive power in grid-connected WECS and they are discussed further in [25–27]. Freire et al. [25] compared the performance and effectiveness of the direct and VOC strategies. It was concluded that although the direct control technique is associated with a fast dynamic response and ease of implementation, the VOC technique has a better performance due to lower current distortions, higher grid power factor, and higher overall efficiency. Moreover, the VOC technique makes it possible to restrain the current stresses that the converter switches may experience during AC faults by limiting the input current to the rated maximum and minimum converter currents [28]. The VOC comprises a dual-loop structure: a slow outer control loop and a fast inner current control loop. The output signals from the outer control loops serve as reference currents for the inner current control loops.

Controllers are mainly incorporated into a system to improve the overall performance and stability of the system. This is done to meet certain design specifications such as percentage overshoot, rise time, settling time, phase margin, and a specified bandwidth [29–33]. The well-developed controllers used for the purpose of meeting the aforementioned design specifications can be classified into two categories:

(i) Proportional-derivative (PD), proportional-integral (PI), and proportional-integral-derivative (PID)

(ii) Phase-lag compensator, phase-lead compensator, and lag-lead compensator

Currently, PI, PD, and PID controllers are widely used in academia and in industrial control systems because they have a simple structure and simple design procedure and they are cost-effective [1, 14, 33–36]. However, the design procedure of PD, PI, and PID controllers involves determining the P, I, and D gains which are usually tuned using a trial-and-error procedure [1, 36]. Therefore, poorly tuned controller gains reduce the accuracy and the dynamic performance of the entire control system. Moreover, these controllers are linear, have a constricted operating range, and are prone to changes in machine parameters, and their design procedure does not take into consideration the initial performance of the uncompensated system [2, 35]. Therefore, control systems comprising PI, PD, and PID controllers are ineffective when applied to nonlinear and uncertain systems [34].

On the other hand, phase-lead, phase-lag, and lag-lead compensators are associated with a design procedure that is well defined from fundamental principles as opposed to the trial-and-error design procedure. Moreover, the initial performance of the uncompensated system is taken into consideration prior to the controllers' modeling and design. The compensators are designed in the frequency domain to achieve the desired design specifications using bode plots and step-response curves. This makes the latter design procedure more accurate.

The literature review has shown that more focus has been put on employing PI controllers in several MPPT control studies for both DFIG-based and PMSG-based WECSs. However, although there have been a few studies on implementing phase-lag compensators in different control applications, particularly in DFIG-based WECS for low voltage ride through control by Wang et al. [37], there is a lack of research studies focusing on the implementation of these compensators in PMSG-based WECSs. Moreover, although nonlinear controllers are becoming more widely popular because they stabilize the nonlinear systems, in this context being WECSs, thereby providing a better performance and stability over a wide range of uncertain wind speed variations, their design procedure tends to be more complicated and they are still gaining their access into the industry and commercial products [38, 39]. Contrariwise, according to Geng and Yang [38], linear controllers have a simpler structure and are easily understood and hence they are widely employed in the industry, of which PD, PI, and PID controllers have been frequently employed as opposed to the phase-lead, phase-lag, and lead-lag compensators which are proposed in this research investigation. Therefore, the main aim of this paper is to close these gaps in literature by implementing phase-lag compensators to control the machine-side converter of a PMSG-based WECS through the use of the dual-loop-based vector-oriented control technique. In this research investigation, the MPPT is achieved by employing the ORB technique to control the rotor speed of a 2 MW direct-driven PMSG-based WECS during wind speeds lower than the WT's rated wind speed. The power electronic converters used in the proposed WECS are of a three-level Neutral Point Clamped (NPC) Voltage Source Converter (VSC) topology.

The research investigation discussed in this paper is motivated by a community-run wind power mini generation pilot project which is implemented in Lüderitz, a coastal town located in the southwestern Namibia, along the Namib Desert [40]. According to an experimental study by Davidson et al. [40], Lüderitz experiences abundant southwestern winds as high as 7.5 m/s in summer and "Berg" winds ranging from 2.8 m/s in winter. This makes it a suitable site for wind power plants. The demonstration site is made up of two 800 W, E230i PMSG-based wind turbines. The generation capacity of the mini power plant is in the range of 250–500 W [40]. This pilot project has brought about social and economic development and public awareness of RE technology in Namibia. Moreover, it is a starting point towards the reduction of the country's dependency on fossil fuels and importing of electricity from neighboring countries. However, this pilot project needs to be expanded to bring about more benefits. Therefore, this research investigation is the starting point of modularly expanding the wind-generation infrastructure in Namibia by thoroughly analyzing and designing the rotor speed controller of a 2 MW PMSG-based WECS to achieve MPPT. Subsequent work will then focus on hardware implementation and testing.

2. Mathematical Modeling of a Wind Energy Conversion System

2.1. Wind Turbine and Drive-Train Mathematical Modeling. The mechanical power extracted from the wind resource by the wind turbine is given by the following equation [3, 4]:

$$P_m = \frac{1}{2}\rho A C_p(\lambda, \beta) v_w{}^3, \tag{1}$$

where ρ is the air density which is equal to 1.225 kg/m^3 at sea level, A is the area swept by the rotor blades [m^2] ($A = \pi r^2$, with r being the radius of the rotor blade in m), v_w is the wind speed upstream of the rotor [m/s], and C_p is the aerodynamic power coefficient dependent on the blade-pitch-angle ϑ [deg] and tip speed ratio λ given by the following equation [3, 4]:

$$\lambda = \frac{r \times \omega_r}{v_w}, \tag{2}$$

where ω_r is the rotor angular speed [rad/s].

2.2. Permanent Magnet Synchronous Generator Mathematical Model. The dynamic modeling of a PMSG is normally carried out in the direct-quadrature (dq) synchronous reference frame with q-axis 90° ahead of d-axis with respect to the direction of rotation [41]. The detailed mathematical modeling and equations describing a PMSG have been discussed in more detail in Kundur [42]. The dynamic equations of the PMSG in the dq synchronous reference frame are given as follows [41, 43]:

$$\begin{aligned} V_{sd} &= R_s i_{sd} + L_{sd}\frac{di_{sd}}{dt} - \omega_e L_{sq} i_{sq}, \\ V_{sq} &= R_s i_{sq} + L_{sq}\frac{di_{sq}}{dt} + \omega_e\left(L_{sd} i_{sd} + \psi_f\right), \end{aligned} \tag{3}$$

where u_{sd} and u_{sq} are the d-axis and q-axis stator voltages [V], respectively; i_{sd} and i_{sq} are the d-axis and q-axis stator currents [A], respectively; L_{sd} and L_{sq} are the d-axis and q-axis inductances [H], respectively; R_s is the stator winding resistance [Ω]; ψ_f is the permanent magnetic flux [Wb]; and ω_e is the electrical rotating speed [rad/s] of the generator.

3. Control System Modeling

3.1. Fundamentals of Control Systems

3.1.1. Gain and Phase Margin. In the frequency domain, the gain and phase margins are the important indicators of system robustness, performance, and stability and hence they are widely used for controller designs. Phase and gain margins are determined by using Bode plots of the system's open-loop transfer functions. A gain margin is referred to as the amount of additional open-loop gain, expressed in decibels (dB),

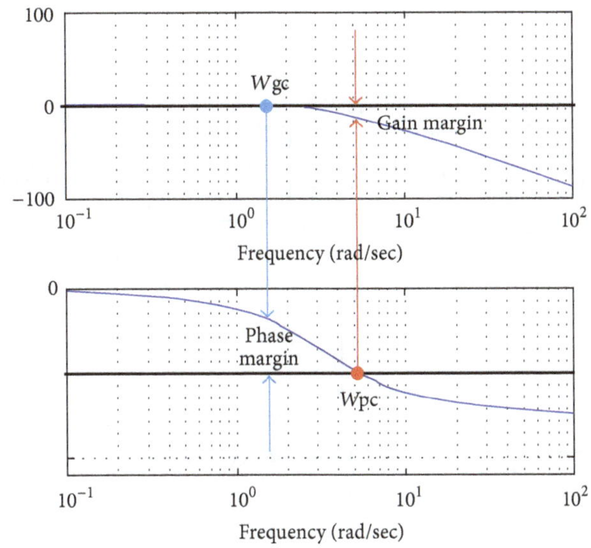

FIGURE 2: Graphical representation of the gain and phase margin and their corresponding crossover frequencies.

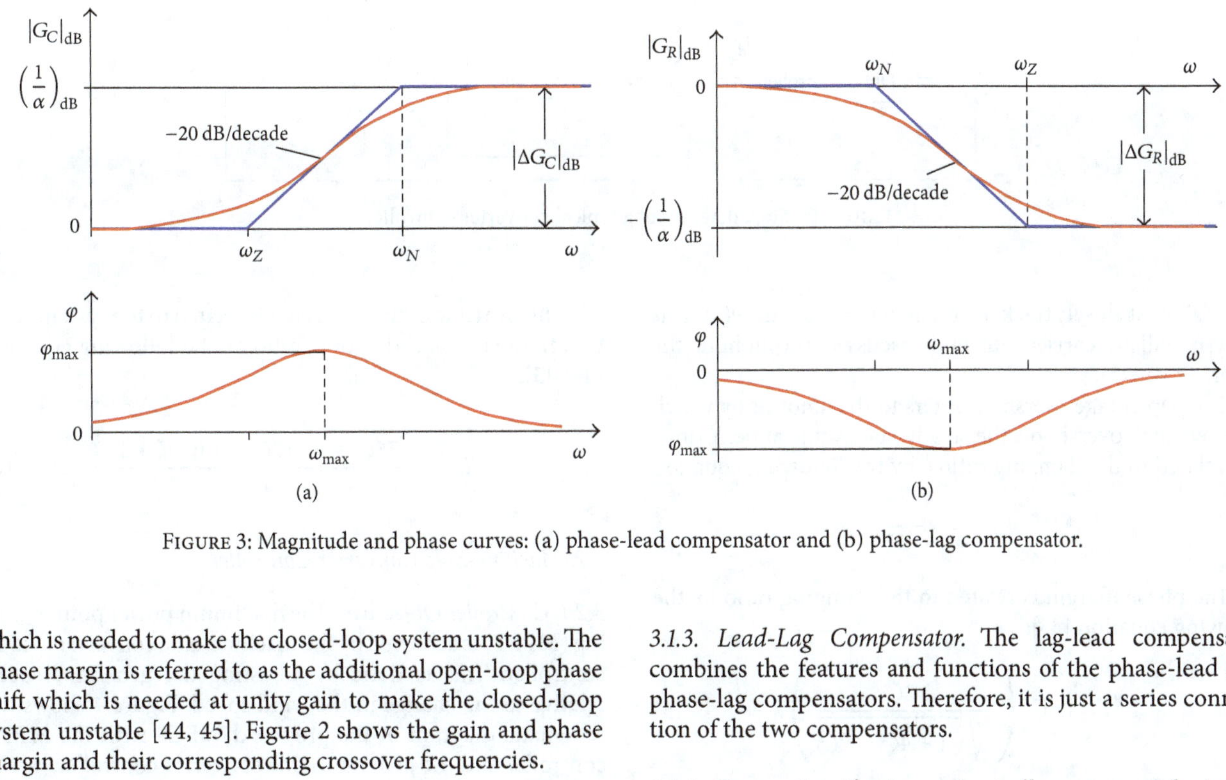

FIGURE 3: Magnitude and phase curves: (a) phase-lead compensator and (b) phase-lag compensator.

which is needed to make the closed-loop system unstable. The phase margin is referred to as the additional open-loop phase shift which is needed at unity gain to make the closed-loop system unstable [44, 45]. Figure 2 shows the gain and phase margin and their corresponding crossover frequencies.

3.1.2. Phase-Lead and Phase-Lag Compensators. The phase-lead compensator improves the performance of the open-loop system by adding a phase boost to the phase curve to obtain the desired phase margin at the desired gain crossover frequency [29], whereas the phase-lag compensator improves the performance of the open-loop system by adding phase attenuation to the phase curve to obtain the desired phase margin at the desired gain crossover frequency [31]. Figures 3(a) and 3(b) show the bode plot indicating the magnitude and phase curves of typical phase-lead and phase-lag compensators, respectively.

3.1.3. Lead-Lag Compensator. The lag-lead compensator combines the features and functions of the phase-lead and phase-lag compensators. Therefore, it is just a series connection of the two compensators.

3.1.4. Design Specifications. Controllers are mainly incorporated into a system to improve the overall performance and stability of the system. This is done to meet certain design specifications such as percentage overshoot, rise time, settling time, phase margin, and a specified bandwidth [29–33]. Therefore, the stability of the proposed controllers is ensured by verifying that the controllers have enabled the desired design specifications to be achieved when they have been incorporated into the entire system. This is analyzed by comparing the measured values to the desired/reference values. These values must correspond to or have a small margin of error, and, in terms of plots, the measured values'

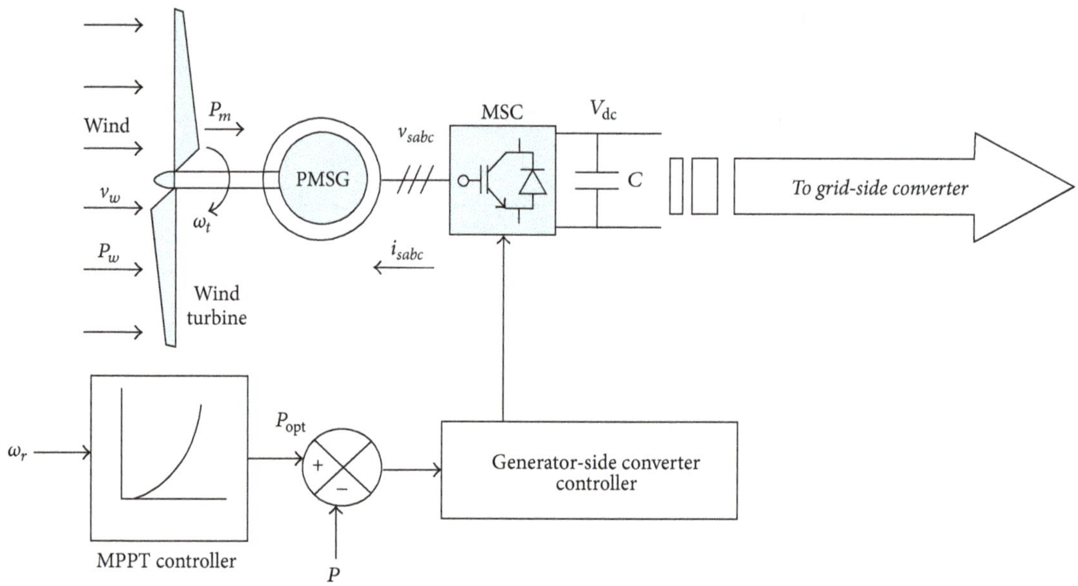

FIGURE 4: Schematic diagram of the PMSG-based WECSs.

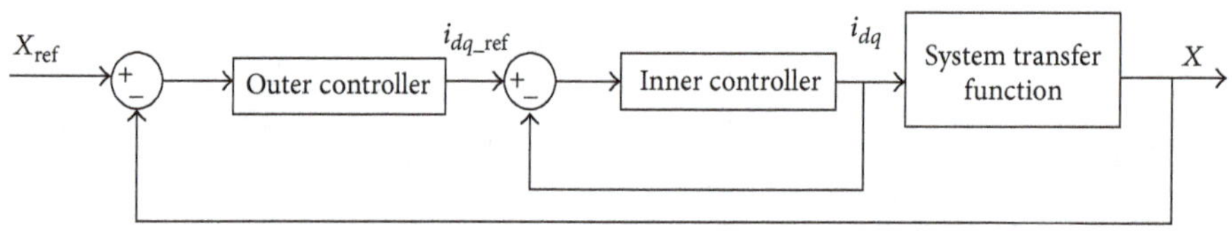

FIGURE 5: Block diagram of a typical converter controller.

curve(s) must closely track the reference values' curve(s). This analysis will be carried out and discussed throughout the paper.

The percentage overshoot refers to the amount by which the waveform overshoots the steady-state value at peak time. It is related to the damping ratio ζ by the following equation [45]:

$$\text{os} = e^{-\pi\zeta/\sqrt{1-\zeta^2}}. \tag{4}$$

The phase margin is related to the damping ratio by the following equation [45]:

$$\text{PM} = \tan^{-1}\left(\frac{2\zeta}{\sqrt{\sqrt{1+4\zeta^2}-2\zeta^2}}\right). \tag{5}$$

The natural frequency of the system is related to the specified bandwidth and damping ratio by the following equation [44, 45]:

$$\omega_n = \frac{\omega_{\text{BW}}}{\sqrt{(1-2\zeta^2)+\sqrt{\zeta^4-4\zeta^2+2}}}. \tag{6}$$

The settling time of the controller is related to the system's natural frequency and damping ratio by the following equation [44, 45]:

$$T_s = \frac{4}{\omega_n \times \zeta}. \tag{7}$$

The rise time of the controller is related to the system's natural frequency and damping ratio by the following equation [44, 45]:

$$T_r = \frac{1.76\zeta^3 - 0.417\zeta^2 + 1.039\zeta + 1}{\omega_n}. \tag{8}$$

3.2. Machine-Side Converter Controller

3.2.1. Controller Objective. The maximum power point tracking (MPPT) is achieved by the machine-side converter (MSC) controller. This is done by adjusting the generator speed relative to the change in the wind speed. Figure 4 shows the schematic diagram of the proposed WECS with the MPPT controller and the generator-side converter controller.

Figure 5 shows a typical control block of a converter station employing a dual-loop structure: a fast inner current control loop and a slower outer control loop.

3.2.2. Derivation of the Transfer Functions: Inner Current Control Loop. Figures 6(a) and 6(b) show the initial and final inner current control loops for the generator-side converter, respectively.

A converter is usually considered as an ideal transformer with a time delay caused by the switching of the converter switches. The delay time is equal to the half of the switching

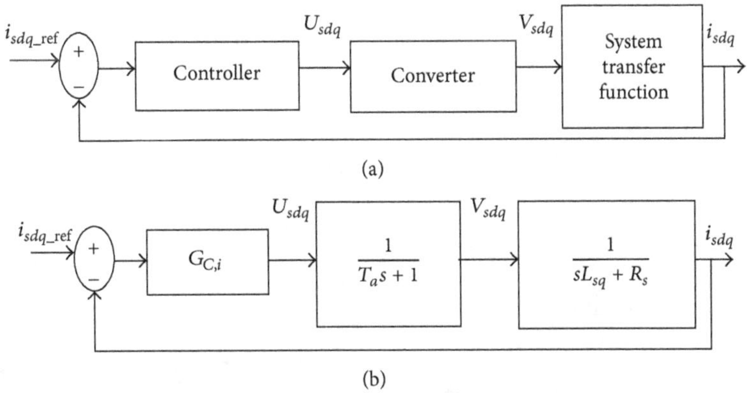

FIGURE 6: Inner current control loop for the generator-side converter: (a) initial control loop and (b) final control loop.

time. The transfer function of the converter block is therefore given by the following equation [46]:

$$Y(s) = \frac{1}{T_a s + 1}, \tag{9}$$

where $T_a = T_{\text{switch}}/2 = 1/2f_{\text{switch}}$ is the time delay in seconds.

The system's behavior is governed by the equations that represent the PMSG's stator voltages in dq synchronous reference frame given in (3). The d-components and q-components of the stator voltage have a speed or frequency induced term $\omega_{se}L_{sq}i_{sq}$ and a speed or frequency and a flux induced term $\omega_{se}(L_{sd}i_{sd} + \psi f)$, respectively, which ensures the cross-coupling between the two axes. Decoupling d-axis and q-axis ensures a good control system performance [10]. Removing the cross-coupling terms gives (10) which represent the system's transfer function in Figure 6(a):

$$V_{sd} = R_s i_{sd} + L_{sd}\frac{di_{sd}}{dt};$$
$$V_{sq} = R_s i_{sq} + L_{sq}\frac{di_{sq}}{dt}. \tag{10}$$

Taking the Laplace transformation of (10) gives (11) relating the output stator current to the stator voltage:

$$\frac{i_{sd}(s)}{V_{sd}(s)} = \frac{i_{sq}(s)}{V_{sq}(s)} = \frac{1}{sL_{sdq} + R_s}. \tag{11}$$

From Figure 6(b), the open-loop transfer function of the inner current control loop is given by the following equation:

$$G_{i,\text{ol}}(s) = \frac{G_{C,i}(s)}{(sTa + 1)\left(sL_{sdq} + R_s\right)}. \tag{12}$$

3.2.3. Design Specifications of the Inner Current Controller. To prevent the ripple generated during the switching of the pulse width modulation (PWM) converter switches from affecting the controller's performance, it is recommended that the inner current controller's bandwidth should be as wide as possible and at least be less than or equal to one-fourth of PWM converter's switching frequency [2]. The choice of the PWM converter's switching frequency mainly depends on the switches' blocking voltage (collector-emitter voltage) and the current carrying capacity. According to [47, 48], the IGBTs are preferred over MOSFETs for blocking voltages above 250 V and they usually operate at low frequencies less than 20 kHz for motor drive/control. In this study, the PWM converter is rated at 3 MVA, above the apparent power rating of the WT given in Table 2.

Considering the wind turbine's rated line-to-line voltage of 690 V as given in Table 2, the switching frequency is taken to be 10 kHz. Therefore, the bandwidth of the inner current control loop is taken to be 2.5 kHz, which is one-fourth of the PWM converter's switching frequency. The percentage overshoot of the inner current control loop is taken to be 25%, since the inner loop is fast, and hence the system can recover from the overshoot faster. Therefore, the damping ratio and phase margin calculated using (4) and (5) are equal to 0.404 and 39.52°, respectively. Furthermore, the settling time and rise time of the inner current controller calculated using (7) and (8) are equal to 0.856 ms and 0.127 ms, respectively.

(1) The Uncompensated Inner Current Control Loop. The bode plot used for the system's stability analysis is plotted using (12) by substituting in the delay time, the synchronous inductance, and stator winding resistance given in Table 2. The delay time calculated using (9) is equal to 50 μs. Therefore, the open-loop transfer function is given by the following equation:

$$G_{I,\text{OL\_uncompen}}(s) = \frac{G_{C,i}(s)}{78 \times 10^{-9}s^2 + 1.573 \times 10^{-3}s + 0.821 \times 10^{-3}}. \tag{13}$$

Figure 7 shows the bode plot of the uncompensated open-loop transfer function of the inner current controller; that is, $G_{C,i}(s) = 1$.

It is observed that the phase margin and bandwidth of the uncompensated system are 88.2° and 635 rad/s, respectively. The bandwidth is too low as compared to the desired value of 1.57×10^4 rad/s (2500 Hz) and hence the system response will be too slow. Furthermore, the phase margin is too high and much closer to 90°.

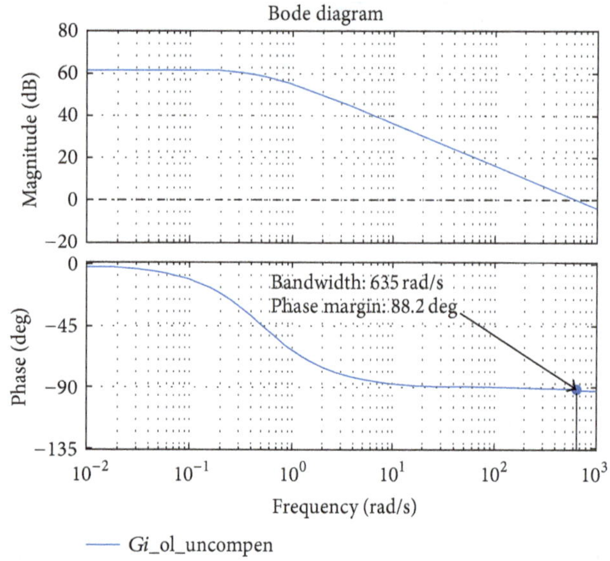

FIGURE 7: Bode plot of the uncompensated open-loop transfer function.

To obtain the desired bandwidth, the system's open-loop transfer function needs to be multiplied by a gain K to raise the magnitude curve up. The gain is determined in such a way that the magnitude curve crosses the 0 dB line at the desired bandwidth. The gain is computed using (14) as follows:

$$\begin{aligned} &\left|G_{I,\mathrm{OL\_uncompen}}(j)\right| \\ &= \frac{K}{78\times10^{-9}j^2+1.573\times10^{-3}j+0.821\times10^{-3}}, \\ &15707.96 \\ &= \frac{K}{\left(78\times10^{-9}j^2+1.573\times10^{-3}j+0.821\times10^{-3}\right)}. \\ &K = 31.12. \end{aligned} \tag{14}$$

Multiplying (13) with the gain K gives (15) which is used for plotting the phase and magnitude curves of the gain-compensated open-loop transfer function:

$$G_{I,\mathrm{OL\_compen\_gain}}(s) = \frac{31.12}{78\times10^{-9}s^2+1.573\times10^{-3}s+0.821\times10^{-3}}. \tag{15}$$

Figure 8 shows the magnitude and phase curves of the gain-compensated open-loop transfer function of the inner current control loop. From Figure 8, it can be observed that

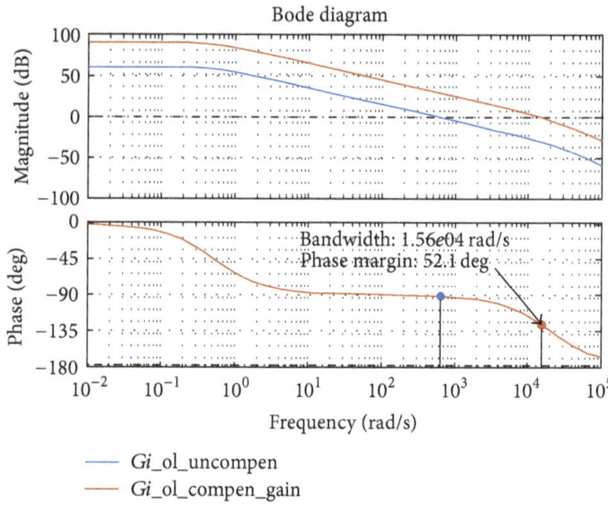

FIGURE 8: Bode plot of the gain-compensated transfer function.

the bandwidth requirement has been met. The phase margin at the desired bandwidth is 52.1°. To obtain the desired phase margin of 39.52°, a phase-lag compensator is needed to shift the phase curve down.

(2) Phase-Lag Compensator Design. The first step of designing a phase-lag compensator is to compute the phase attenuation needed to obtain the desired phase margin. This is done by subtracting the current phase margin from the desired phase margin as follows.

$$\text{Phase attenuation} = \mathrm{PM}_{\mathrm{desired}} - \mathrm{PM}_{\mathrm{current}} = -12.58^\circ.$$

The phase attenuation is used to locate the pole p_c and zero z_c of the phase-lag compensator by calculating the frequency [Hz] where the pole and zero will be located using the following equation [31]:

$$\begin{aligned} f_z &= f_{\mathrm{BW}}\sqrt{\frac{1-\sin(\text{phase-attenuation})}{1+\sin(\text{phase-attenuation})}}, \\ f_p &= f_{\mathrm{BW}}\sqrt{\frac{1+\sin(\text{phase-attenuation})}{1-\sin(\text{phase-attenuation})}}, \end{aligned} \tag{16}$$

where f_{BW} is the bandwidth frequency, which in this study is equal to 2500 Hz. Therefore, f_p and f_z are equal to 3119.39 Hz and 2003.59 Hz, respectively. The angular frequencies ω_p and ω_z where the pole and zero of the phase-lag compensator are located are equal to 19599.7 rad/s and 12588.9 rad/s, respectively. Therefore, the transfer function of the phase-lag compensator is given by the following equation [31]:

$$\begin{aligned} G_{\mathrm{lag}}(s) &= G_{\mathrm{co}}\left(\frac{\tau_{\mathrm{lag}}s+1}{\alpha\tau_{\mathrm{lag}}s+1}\right) = \frac{6.366\times10^{-5}s+1.248}{7.943\times10^{-5}s+1} \\ \text{where; } \alpha &= \frac{\omega_z}{\omega_p} = \frac{19599.71}{12588.93} = 1.557,\ G_{\mathrm{co}} = \sqrt{\alpha} = \sqrt{1.557} = 1.248,\ \tau_{\mathrm{lag}} = \frac{1}{\omega_z} = \frac{1}{19599.71} = 0.05\times10^{-3}. \end{aligned} \tag{17}$$

The phase-lag compensator is applied to the gain-compensated system, and hence multiplying (15) by (17) gives (18) which represents the transfer function of the phase-lag compensated system for the inner current control loop:

$$G_{I,\text{OL\_compen\_lag}}(s) = G_{I,\text{OL\_compen\_gain}}(s) \times G_{\text{lag}}(s),$$
$$G_{I,\text{OL\_compen\_lag}}(s) = \frac{1.97 \times 10^{-3}s + 38.68}{6.25 \times 10^{-12}s^3 + 2.036 \times 10^{-7}s^2 + 1.573 \times 10^{-3}s + 0.821 \times 10^{-3}}. \quad (18)$$

Figure 9 shows the magnitude and phase curves of the phase-lag compensated open-loop transfer function of the inner current control loop.

It is observed that the phase-lag compensator shifted the phase curve down to yield the phase margin of 39.5°. The bandwidth is maintained at 1.57×10^4 rad/s (2500 Hz). Therefore, the phase margin and bandwidth design specifications have been met.

Figures 10(a), 10(b), and 10(c) show the step-response curves indicating the peak response, settling time, and rise time of the uncompensated, gain-compensated, and phase-lag compensated closed-loop transfer function of the inner stator current control loop.

Figure 10(a) shows that the overshoot of the gain-compensated system is 16%, while the overshoot of the phase-lag compensated system is 29.5%. Therefore, the percentage overshoot of the gain-compensated system falls within the desired overshoot of 25%. Figure 10(b) shows that gain K reduced the settling time of the uncompensated system from 6 ms to 0.4 ms. However, the phase-lag compensator reduced the settling time of the gain-compensated system to 0.6 ms, which is within the desired settling time of 0.856 ms.

This illustrates that the settling time design specification of the inner current controller has been achieved. Figure 10(c) shows that gain K reduced the rise time of the uncompensated system from 3.34 ms to 0.0829 ms and this speeds up the system response. However, the phase-lag compensator increased the gain-compensated system's rise time to 0.079 ms which is within the desired rise time of 0.127 ms. This illustrates that the rise time design specification of the inner current controller has also been achieved.

3.2.4. Design Specifications of the Outer Control Loop. The outer control loops are expected to be slower than the inner current control loops. Therefore, the bandwidth of the outer control loops is taken to be 625 Hz (3926.99 rad/s), which is one-fourth of the inner current control loops' switching frequency. Moreover, the slower outer loop needs to have a smaller overshoot as compared to the fast inner current controllers loop to get rid of as many oscillations in the controller as possible, thereby enhancing the stability of the entire system. Therefore, the percentage overshoot is taken to be 5%. Therefore, the damping ratio and phase margin calculated using (4) and (5) are equal to 0.69 and 57.85°, respectively. Furthermore, the settling time and rise time of the inner current controller calculated using (7) and (8) are equal to 1.2 ms and 0.42 ms, respectively.

(1) Derivation of the Transfer Functions: Maximum Power Point Tracking Control Loop. MPPT is achieved when the WT operates at optimal $C_p(C_{p\_\text{opt}})$ and optimal tip speed ratio λ_{opt}. Therefore, the rotor speed ω_r needs to be varied relative to the change in wind speed v_w to keep the tip speed ratio at λ_{opt}. The equation relating the optimal power to the optimal rotor speed ω_{r_opt}, C_{p_opt}, and λ_{opt}, is derived by substituting (2) into (1) as illustrated by the following equation [3]:

$$P_{\text{opt}} = \frac{1}{2}\rho A C_{P_{\text{opt}}} \left(\frac{\omega_{r_{\text{opt}}} \times r}{\lambda_{\text{opt}}} \right)^3. \quad (19)$$

The output power P_{opt} from the MPPT stage serves as an input into the outer active power control loop. The output of the outer controller block is the d-component of the reference current fed into the inner current control loop. Figures 11(a) and 11(b) show the initial and final active power control loops.

Using the dq axis theory, the instantaneous real power and reactive power of the PMSG are described by the following equation [49]:

$$P_{\text{gen}} = \frac{3}{2}\left(V_{sd}i_{sd} + V_{sq}i_{sq}\right);$$
$$Q_{\text{gen}} = \frac{3}{2}\left(V_{sq}i_{sd} - V_{sd}i_{sq}\right). \quad (20)$$

In stator voltage reference frame, d-axis of the synchronous reference frame is aligned with the stator voltage vector. Therefore, the stator voltage only has the d-axis component V_{sd}, while the q-axis component V_{sq} is set to zero. Therefore, (20) becomes

$$P_{\text{gen}} = \frac{3}{2}V_{sd}i_{sd};$$
$$Q_{\text{gen}} = -\frac{3}{2}V_{sd}i_{sq}. \quad (21)$$

Therefore, the system transfer function in Figure 11(a) is given by the following equation:

$$\frac{P_{\text{gen}}}{i_{sd}} = \frac{3}{2}V_{sd}. \quad (22)$$

From Figure 11(b), the open-loop transfer function of the MPPT control loop is given by the following equation:

$$G_{P,\text{ol}}(s) = G_{C,p}(s) \times G_{C,\text{cl}}(s) \times \frac{3}{2}V_{sd}. \quad (23)$$

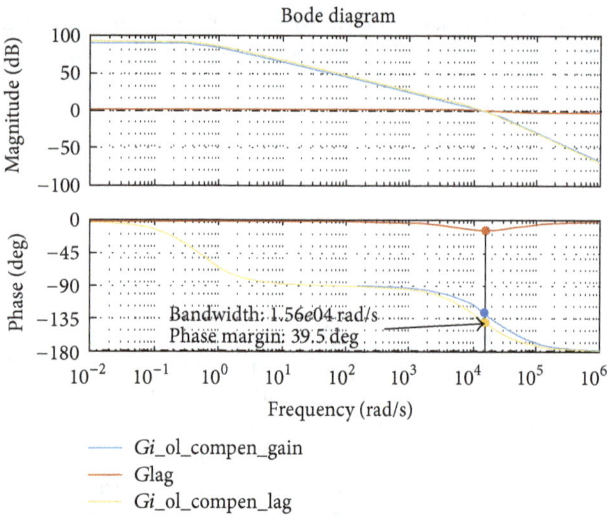

FIGURE 9: Bode plot of the phase-lag compensated transfer function.

(2) The Uncompensated MPPT Control Loop. The open-loop transfer function used to plot the magnitude and phase curves of the outer MPPT control loop is given by (25). It is derived by substituting the converter's voltage of 620.62 V and the closed-loop transfer function $G_{I,\mathrm{CL}}(s)$ of the phase-lag compensated inner current loop given by (24) into (23):

$$G_{I,\mathrm{CL}}(s) = \frac{2.23 \times 10^{-3}s + 43.52}{6.96 \times 10^{-12}s^3 + 2.27 \times 10^{-7}s^2 + 3.99 \times 10^{-3}s + 43.52}, \quad (24)$$

$$G_{P,\mathrm{OL\_uncompen}}(s) = G_{P,\mathrm{ol}}(s) \times G_{I_{\mathrm{CL}}}(s),$$

$$G_{P,\mathrm{OL\_uncompen}}(s) = \frac{\left(4.149s + 8.102 \times 10^4\right) \times G_{C,P}(s)}{1.393 \times 10^{-11}s^3 + 4.55 \times 10^{-7}s^2 + 7.98 \times 10^{-3}s + 87.03}. \quad (25)$$

Figure 12 shows the bode plot of the uncompensated open-loop transfer function of the MPPT control loop. That is, $G_{C,P}(s) = 1$.

From Figure 12, it is observed that the phase margin and bandwidth of the uncompensated system are 1.38° and 5.46×10^5 rad/s, respectively. The bandwidth is much higher than the desired value and the phase margin is very small. To reduce the bandwidth, a phase-lag compensator is needed to drop the magnitude curve down so that it crosses the 0 dB line at the desired bandwidth of 3926.99 rad/s (625 Hz).

(3) Phase-Lag Compensator Design. The first step is to identify the gain magnitude corresponding to the desired bandwidth on the magnitude curve. The gain magnitude at the desired bandwidth is 59.7 dB as shown in Figure 13.

The magnitude at the desired crossover frequency is then used to calculate the attenuation α needed to shift the magnitude curve down. The attenuation is used to locate the pole and zero of the phase-lag compensator and it is calculated using the following equation [31]:

$$\alpha = 10^{\mathrm{gain}/10} = 10^{59.7/10} = 933254.30. \quad (26)$$

The frequencies where the pole p_c and zero z_c will be located are calculated using the following equation:

$$z_c = \omega_{\mathrm{bandwidth}} \times \sqrt{\alpha}, \qquad p_c = \frac{z_c}{\alpha}. \quad (27)$$

Therefore, the pole p_c and zero z_c of the phase-lag compensator with a bandwidth $\omega_{\mathrm{bandwidth}}$ of 3926.99 rad/s and attenuation α of 933254.30 are located at 3.79×10^6 rad/s and 4.065 rad/s, respectively. The transfer function of the phase-lag compensator is given by the following equation [31]:

$$G_{\mathrm{lag}}(s) = \frac{(s/z_c + 1)}{(s/p_c + 1)} = \frac{2.636 \times 10^{-7}s + 1}{0.246s + 1}. \quad (28)$$

Multiplying (25) by (28) gives (29) which represents the open-loop transfer function of the phase-lag compensated system:

$$G_{P,\mathrm{OL\_compen\_lag}}(s) = G_{P,\mathrm{OL\_uncompen}}(s) \times G_{\mathrm{lag}}(s)$$
$$G_{P,\mathrm{OL\_compen\_lag}}(s) = \frac{1.09 \times 10^{-6}s^2 + 4.17s + 8.102 \times 10^4}{3.427 \times 10^{-12}s^4 + 1.118 \times 10^{-7}s^3 + 1.96 \times 10^{-3}s^2 + 21.42s + 87.03}. \quad (29)$$

Figure 14 shows the magnitude and phase curves of the phase-lag compensated open-loop transfer function of the MPPT control loop.

From Figure 14, it is observed that phase-lag compensator shifted the magnitude curve down by 60.6 dB at the bandwidth of 3.93×10^3 rad/s, and hence the bandwidth design specification has been met. The phase margin at the desired bandwidth is 80.6°. Therefore, a second phase-lag compensator is needed to shift the phase curve down to obtain the desired phase margin of 57.85°.

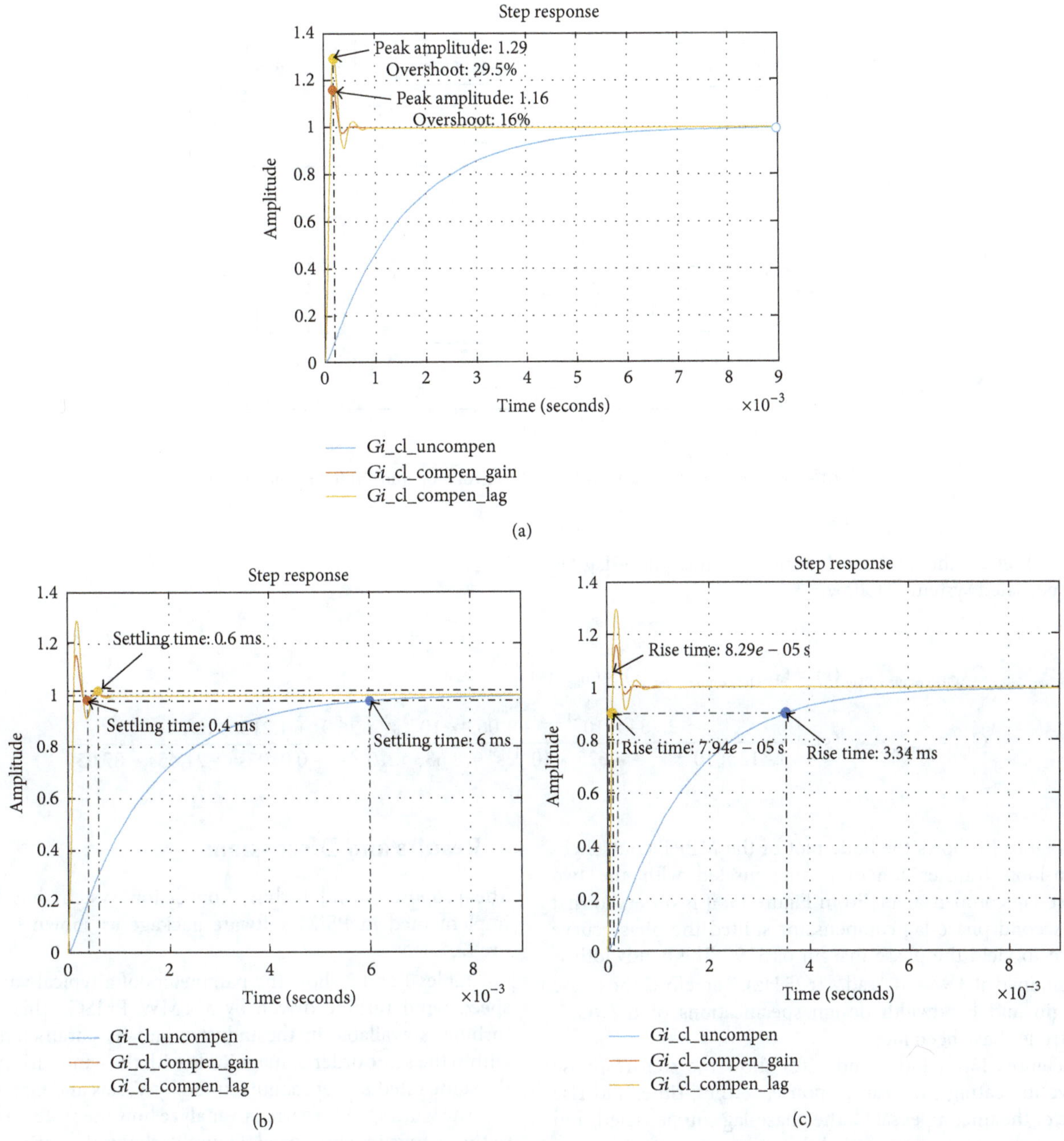

FIGURE 10: Step response of the inner current controller's closed-loop transfer function: (a) peak response, (b) settling time, and (c) rise time.

(4) The Second Phase-Lag Compensator Design. The first step is to compute the phase attenuation needed to obtain the desired phase margin by subtracting the current phase margin from the desired phase margin as follows.

$$\text{Phase attenuation} = \text{PM}_{\text{desired}} - \text{PM}_{\text{current}} = -22.75^{\circ}.$$

Therefore, the frequency [Hz] where the pole p_c and zero z_c calculated using (16) and the phase attenuation are equal to 415.64 Hz and 939.81 Hz, respectively. Moreover, the angular frequencies ω_p and ω_z where the pole and zero of the phase-lag compensator will be located are equal to 2611.54 rad/s and 5905 rad/s, respectively.

The transfer function of the phase-lag compensator is given by the following equation:

$$G_{\text{lag}_2}(s) = G_{\text{co}}\left(\frac{\tau_{\text{lag}}s + 1}{\alpha\tau_{\text{lag}}s + 1}\right) = \frac{0.255 \times 10^{-3}s + 1.504}{0.383 \times 10^{-3}s + 1}. \quad (30)$$

The second phase-lag compensator is applied to the phase-lag compensated system, and hence multiplying (29)

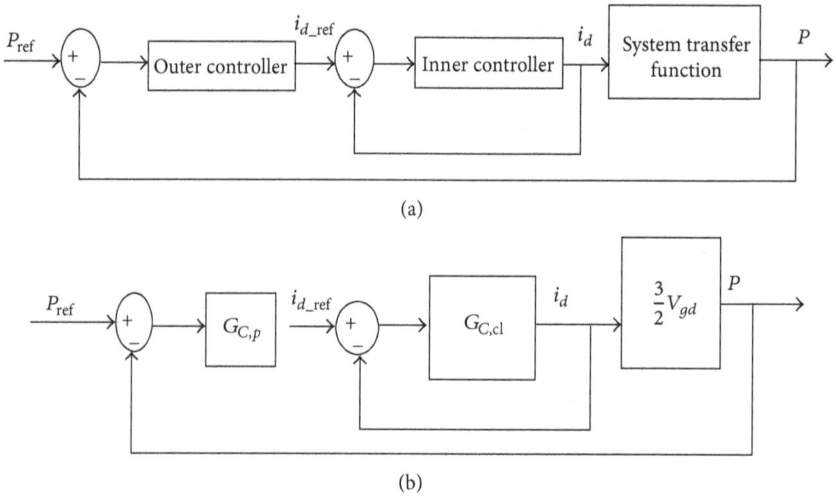

FIGURE 11: MPPT control loop: (a) initial control loop and (b) final control loop.

by (30) gives the transfer function of the phase-lag-lag compensated system as follows:

$$G_{P,OL\_compen\_2lag}(s) = G_{P,OL\_compen\_lag}(s) \times G_{lag_2}(s),$$
$$G_{P,OL\_compen\_2lag}(s) = \frac{2.784 \times 10^{-10}s^3 + 1.063 \times 10^{-3}s^2 + 26.9s + 1.218 \times 10^5}{1.312 \times 10^{-15}s^5 + 4.625 \times 10^{-11}s^4 + 8.633 \times 10^{-7}s^3 + 0.0102s^2 + 21.45s + 87.03}. \quad (31)$$

Figure 15 shows the bode plot of the MPPT controller's open-loop transfer function compensated with the two phase-lag compensators. From Figure 15, it is observed that the second phase-lag compensator shifted the phase curve down to yield the phase margin of 57.9°. The bandwidth is maintained at 3.92×10^3 rad/s (625 Hz). Therefore, the phase margin and bandwidth design specifications of the outer controller have been met.

Figures 16(a), 16(b), and 16(c) show the step-response curve indicating the peak response, settling time, and rise time of the uncompensated, the phase-lag compensated, and second phase-lag compensated closed-loop transfer function of the outer MPPT control loop, respectively.

Figure 16(a) shows that the phase-lag compensators reduced the percentage overshoot of the uncompensated system from 96.5% to 12.1%. Figure 16(b) shows that the first phase-lag compensator increased the settling time of the uncompensated system from 0.593 ms to 0.956 ms. Moreover, the second phase-lag compensator further increased the settling time of the phase-lag compensated system to 1.3 ms, which is close to the desired settling time of 1.2 ms. Figure 16(c) shows that the rise time of the uncompensated system is 1.93 μs. The first phase-lag compensator increased the system's rise time to 0.485 ms, while the second phase-lag compensator increased the rise time to 0.297 ms which is within the desired rise time of 0.42 ms. This shows that the rise time and settling time design specifications of the outer controller have been met.

4. Results and Discussions

The proposed wind turbine conversion system has been implemented in PSIM software package as shown in Figure 17.

Tables 1 and 2 show the parameters of a typical variable speed wind turbine driven by a 2 MW PMSG. This wind turbine is available in the industry and its parameters fall within the same order of magnitude with the wind turbines of the same rated power manufactured by Vestas and Gamesa.

The wind speed is varied to analyze how the system reacts to the change in wind speed. Figure 18 shows the variation in wind speed from the wind turbine's cut-in wind speed (4 m/s) to the rated wind speed (13 m/s) at 0.6 s up until 18 m/s, approaching the WT's cut-out wind speed (see Table 1).

Figures 19(a), 19(b), and 19(c) show the rotor angular speed, the aerodynamic power, and aerodynamic torque, respectively.

Figure 19(a) shows that the rotor angular speed increases as the wind speed increases. Moreover, at the rated wind speed (13 m/s), the measured rotor speed is equal to the rated rotor speed, which is 2.356 rad/s. Figure 19(b) shows that when the wind speed is below the WT's rated wind speed (13 m/s), the power generated from the WT tracks the optimal power curve to ensure MPPT. When the wind speed is equal to the WT's rated wind speed, the power generated is equal to 2 MW which is the WT's rated power. Moreover, as the wind speed increases further above the WT's rated wind

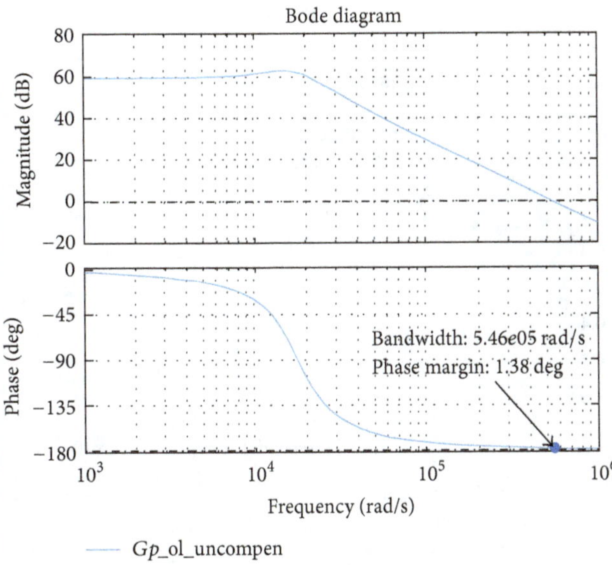

FIGURE 12: Bode plot of the uncompensated transfer function.

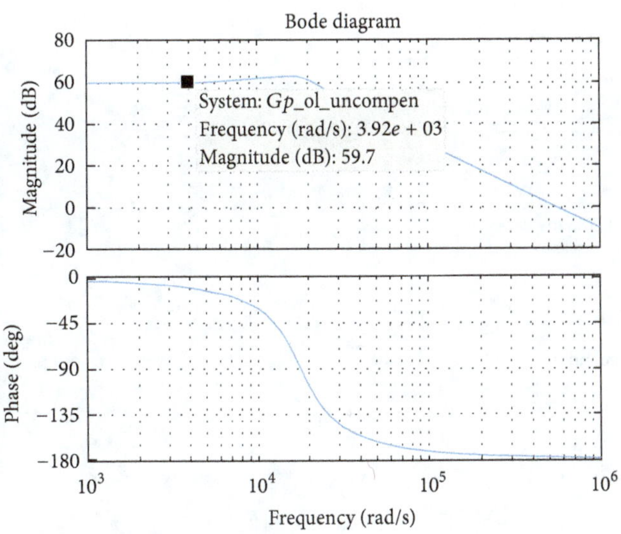

FIGURE 13: Gain magnitude at the desired bandwidth.

TABLE 1: Wind turbine parameters.

Parameters	Values
Rated power (MW)	2
Cut-in wind speed (m/s)	4
Rated wind speed (m/s)	13
Cut-out wind speed (m/s)	25
Number of rotor blades	3
Rotor area (m^2)	4587
Rotor diameter (m)	76.42
Air density (kg/m^2)	1.225
Optimal power coefficient Cp	0.4

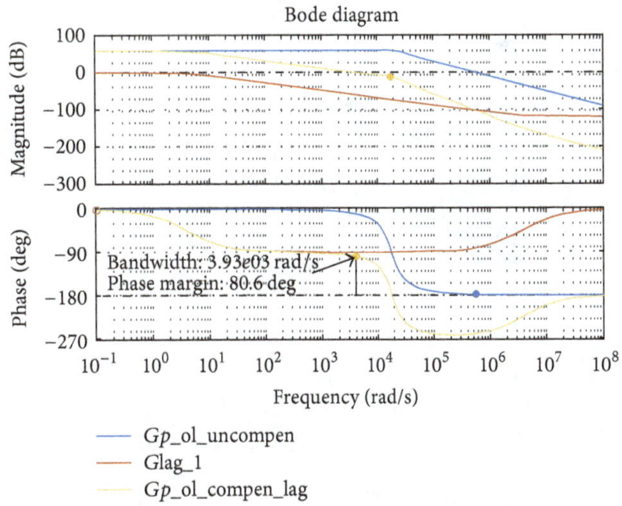

FIGURE 14: Bode plot of the phase-lag compensated transfer function.

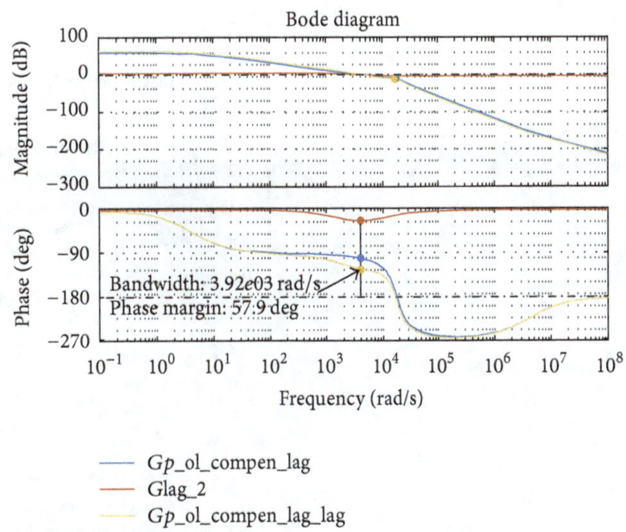

FIGURE 15: Bode plot of the transfer function compensated with the two phase-lag compensators.

TABLE 2: PMSG parameters.

Parameters	Value
Generator type	PMSG
Rated mechanical power, P (MW)	2
Rated apparent power, S (MVA)	2.24
Rated L-L voltage, $V_{L\text{-}L}$ (V)	690 (rms)
Rated power factor, pf	0.89
Rated rotor speed, ω_{rated} (rpm)	22.5
Pole pairs, n_p	26
Rated mechanical torque (kNm)	848.83
Flux linkage, ψ_f (Wb)	4.971
Stator winding resistance, R_s (mΩ)	0.821
Stator d-axis inductance, L_{ds} (mH)	1.573
Stator q-axis inductance, L_{qs} (mH)	1.573

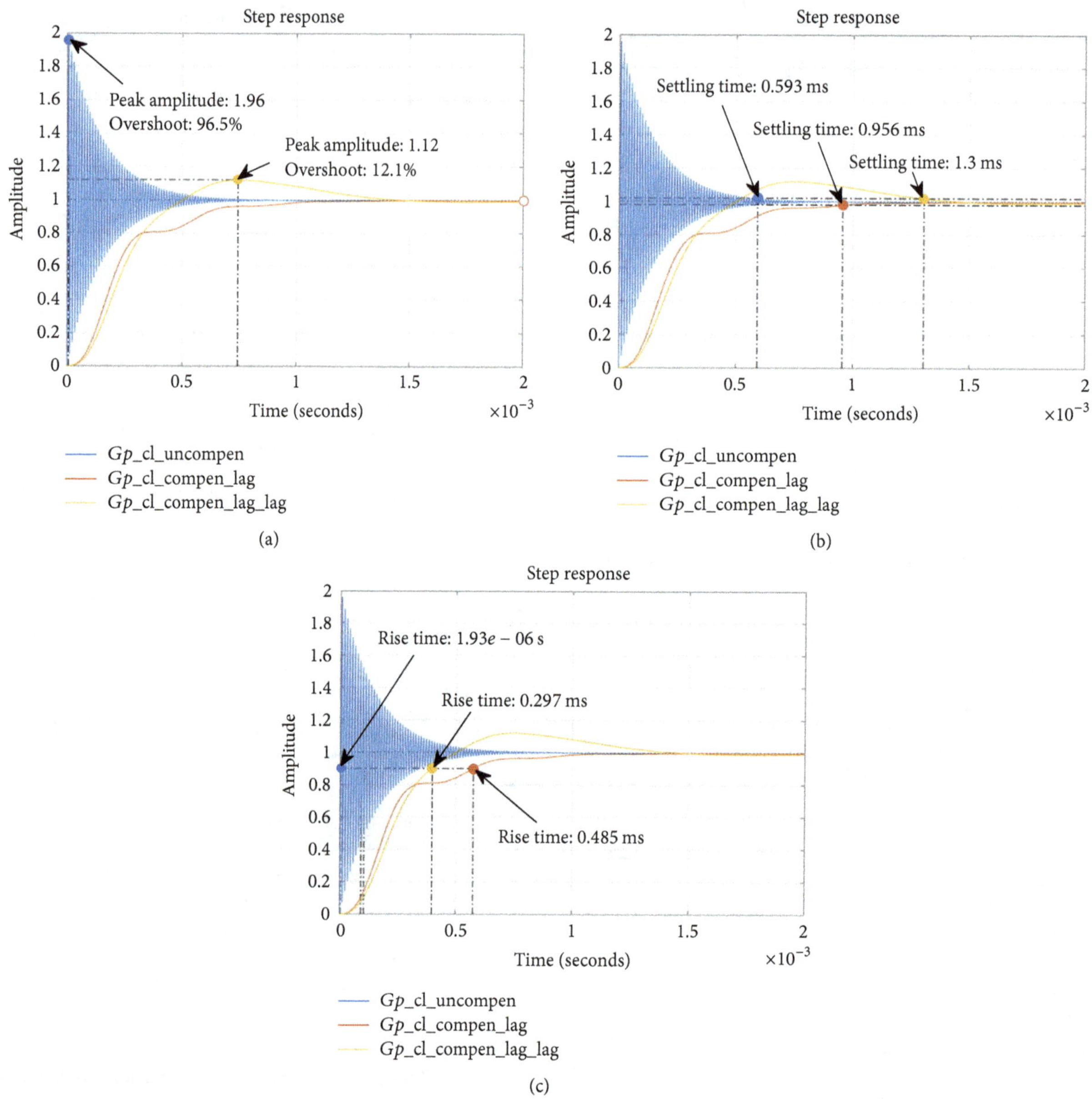

FIGURE 16: Step response of the MPPT controller's closed-loop transfer function: (a) peak response, (b) settling time, and (c) rise time.

speed, the generated power remains equal to the rated power. Figure 19(c) shows that, during the MPPT stage, the generated torque closely tracks the optimal torque curve until the rated WT's rated torque. At the rated power (2 MW), the torque is equal to the WT generator's rated torque of 849 kNm.

5. Conclusions

This research investigation serves as a starting point of modularly expanding the already existing mini wind power plant in Namibia by thoroughly analyzing and designing the rotor speed controller of a 2 MW PMSG-based WECS to achieve MPPT.

The wind turbine system components and the MPPT controller were successfully implemented in PSIM software package. The design and stability analysis of the generator-side converter controller which comprises a phase-lag compensator was carried out in MATLAB software package using bode plots and step-response curves. Based on the simulation results, it was observed that the MPPT controller enabled the power generated from the WT to closely track the optimal power curve to ensure maximum power generation when the wind speed is less than the WT's rated wind speed (13 m/s). It was also observed that when the wind speed is equal to the WT's rated wind speed, the aerodynamic power generated is equal to 2 MW which is the WT's rated power. The MPPT controller also enabled the generated torque to closely track the optimal torque curve when the wind speed was lower than the WT's rated wind speed.

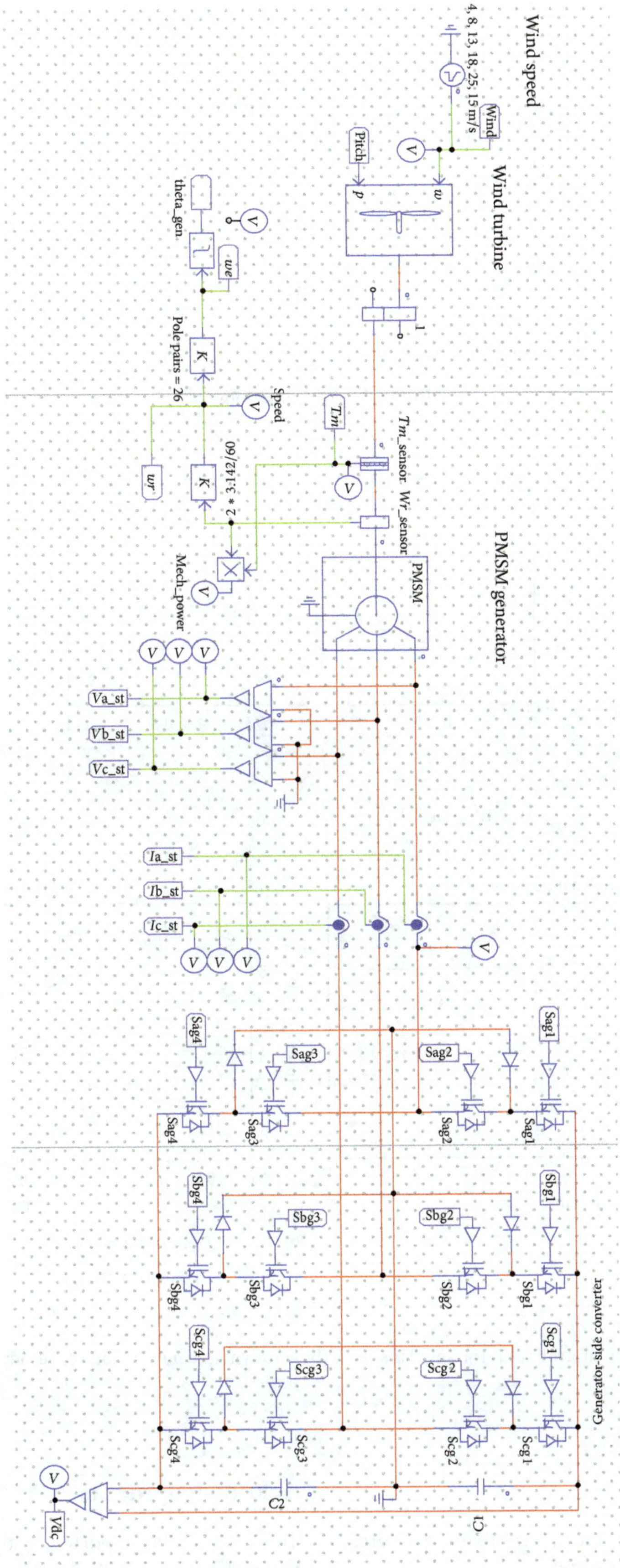

FIGURE 17: Proposed wind energy conversion system.

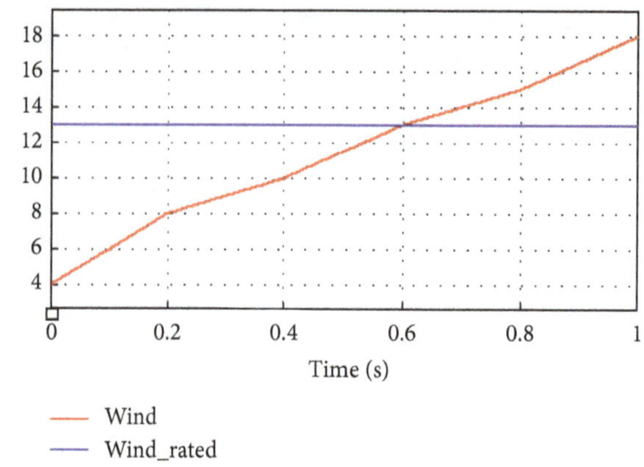

FIGURE 18: Wind speed model.

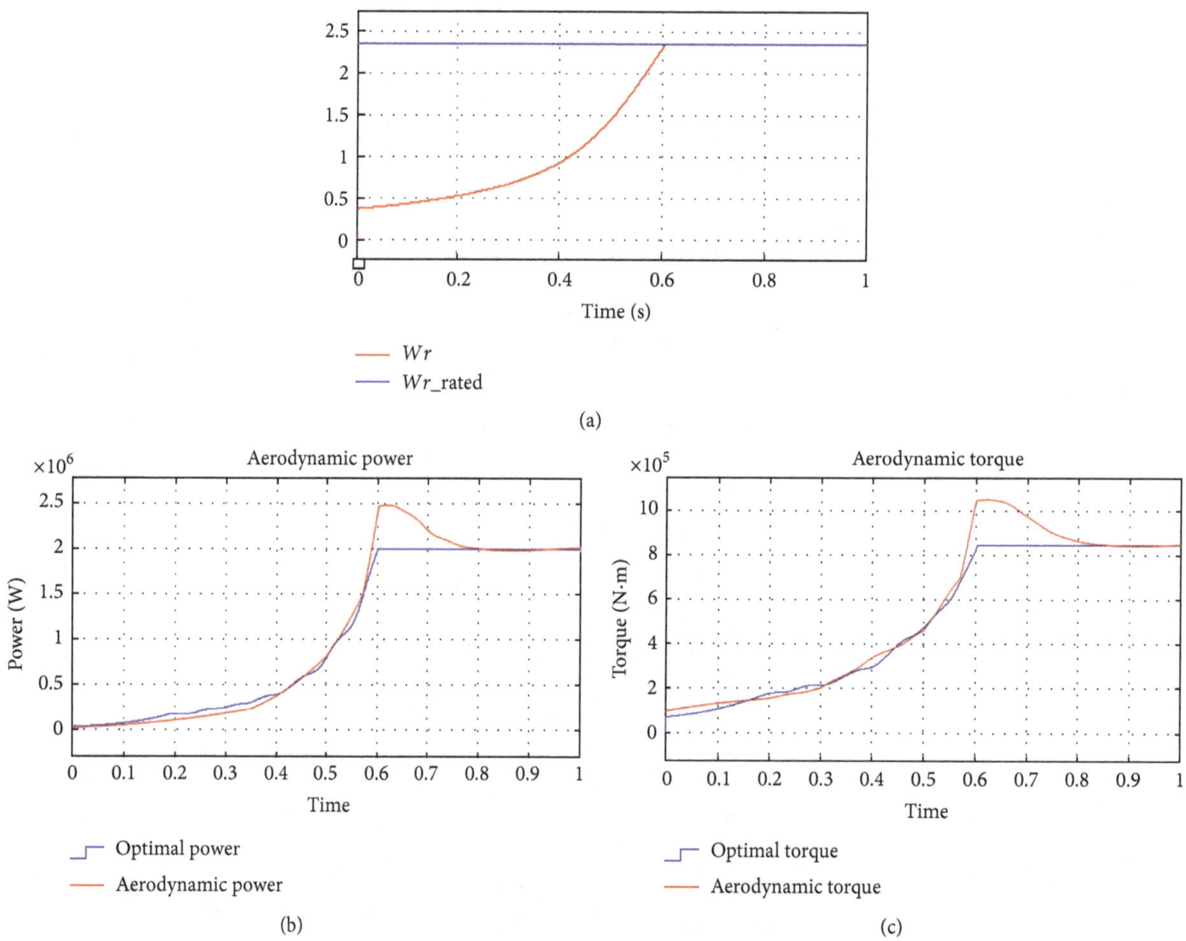

FIGURE 19: MPPT control: (a) rotor angular speed, (b) aerodynamic power, and (c) aerodynamic torque.

Conflicts of Interest

The authors declare that there are no conflicts of interest regarding the publication of this paper.

Acknowledgments

This research is supported by the Eskom Power Plant Engineering Institute (EPPEI) Program.

References

[1] Z. Zhang, Y. Zhao, W. Qiao, and L. Qu, "A discrete-time direct torque control for direct-drive PMSG-based wind energy conversion systems," *IEEE Transactions on Industry Applications*, vol. 51, no. 4, pp. 3504–3514, 2015.

[2] G. M. J. Herbert, S. Iniyan, E. Sreevalsan, and S. Rajapandian, "A review of wind energy technologies," *Renewable and Sustainable Energy Reviews*, vol. 11, no. 6, pp. 1117–1145, 2007.

[3] S. M. Tripathi, A. N. Tiwari, and D. Singh, "Grid-integrated permanent magnet synchronous generator based wind energy conversion systems: a technology review," *Renewable and Sustainable Energy Reviews*, vol. 51, article 4598, pp. 1288–1305, 2015.

[4] A. L. Soriano, W. Yu, and J. J. Rubio, "Modeling and control of wind turbine," *Mathematical Problems in Engineering*, vol. 2013, Article ID 982597, p. 7, 2013.

[5] B. Snyder and M. J. Kaiser, "Ecological and economic cost-benefit analysis of offshore wind energy," *Renewable Energy*, vol. 34, no. 6, pp. 1567–1578, 2009.

[6] Z. Jinhua, L. Yongqian, D. Infield, M. Yuanchi, C. Qunshi, and T. De, "Optimal power dispatch within wind farm based on two approaches to wind turbine classification," *Renewable Energy*, vol. 102, pp. 487–501, 2017.

[7] T. Ackermann and L. Söder, "Wind energy technology and current status: a review," *Renewable and Sustainable Energy Reviews*, vol. 4, no. 4, pp. 315–374, 2000.

[8] P. S. Barendse and P. Pillay, "A doubly-fed induction generator drive for a wind energy conversion system," *Transactions of the South African Institute of Electrical Engineers*, vol. 97, pp. 274–280, 2006.

[9] J. O. Okoronkwo, O. Ojo, and I. Davidson, "Design considerations of the Katsina wind farm in Nigeria," in *Proceedings of the IEEE PES Power Africa Conference*, pp. 251–253, Livingstone, Zambia, June 2016.

[10] A. Edrisian, A. Goudarzi, I. E. Davidson, A. Ahmadi, and G. K. Venayagamoorthy, "Enhancing SCIG-based wind turbine generator performance through reactive power control," in *Proceedings of the Clemson University Power Systems Conference (PSC '15)*, pp. 1–8, Clemson, SC, USA, March 2015.

[11] Y. Zhang, Z. Liu, Z. Yang, and H. Si, "Robust control of wind turbines by using singular perturbation method and linear parameter varying model," *Journal of Control Science and Engineering*, vol. 2016, Article ID 2830736, 2016.

[12] T. Li, A. J. Feng, and L. Zhao, "Neural network compensation control for output power optimization of wind energy conversion system based on data-driven control," *Journal of Control Science and Engineering*, vol. 2012, Article ID 736586, 8 pages, 2012.

[13] M. R. Islam, Y. Guo, and J. Zhu, "A review of offshore wind turbine nacelle: technical challenges, and research and developmental trends," *Renewable and Sustainable Energy Reviews*, vol. 33, pp. 161–176, 2014.

[14] Q. Xu, C. Zhang, L. Zhang, and C. Wang, "Multiobjective optimization of PID controller of PMSM," *Journal of Control Science and Engineering*, vol. 2014, Article ID 471609, 2014.

[15] A. Giallanza, M. Porretto, L. Cannizzaro, and G. Marannano, "Analysis of the maximization of wind turbine energy yield using a continuously variable transmission system," *Renewable Energy*, vol. 102, pp. 481–486, 2017.

[16] J. A. Baroudi, V. Dinavahi, and A. M. Knight, "A review of power converter topologies for wind generators," *Renewable Energy*, vol. 32, no. 14, pp. 2369–2385, 2007.

[17] S. Ravichandran, K. D. R. Pandu, and S. G. Bharathidasan, "Coordinated controller design of grid-connected variable-speed wind energy conversion system with model-based predictive control using response surface methodology," *Electric Power Components and Systems*, vol. 42, no. 11, pp. 1107–1120, 2014.

[18] Z. Chen, J. M. Guerrero, and F. Blaabjerg, "A review of the state of the art of power electronics for wind turbines," *IEEE Transactions on Power Electronics*, vol. 24, no. 8, pp. 1859–1875, 2009.

[19] M. Chinchilla, S. Arnalte, J. C. Burgos, and J. L. Rodríguez, "Power limits of grid-connected modern wind energy systems," *Renewable Energy*, vol. 31, no. 9, pp. 1455–1470, 2006.

[20] A. M. Massoud, S. Ahmed, A. S. Abdel-Khalik, A. A. Elserougi, and K. H. Ahmed, "Control of power converters for emerging applications of power electronics," *Journal of Control Science and Engineering*, vol. 2016, Article ID 9235283, 2 pages, 2016.

[21] Y. Yang, K.-T. Mok, S.-C. Tan, and S. Y. R. Hui, "Nonlinear dynamic power tracking of low-power wind energy conversion system," *IEEE Transactions on Power Electronics*, vol. 30, no. 9, pp. 5223–5236, 2015.

[22] Y. Errami, M. Ouassaid, and M. Maaroufi, "Optimal power control strategy of maximizing wind energy tracking and different operating conditions for permanent magnet synchronous generator wind farm," *Energy Procedia*, vol. 74, pp. 477–490, 2015.

[23] M. Nasiri, J. Milimonfared, and S. H. Fathi, "Modeling, analysis and comparison of TSR and OTC methods for MPPT and power smoothing in permanent magnet synchronous generator-based wind turbines," *Energy Conversion and Management*, vol. 86, pp. 892–900, 2014.

[24] S. Bowes, "A novel harmonic elimination PWM strategy," in *Proceedings of the 7th International Conference on Power Electronics and Variable Speed Drives*, pp. 426–432, London, UK.

[25] N. Freire, J. Estima, and A. Cardoso, "A comparative analysis of PMSG drives based on vector control and direct control techniques for wind turbine applications," *Electrical Review*, vol. 88, no. 1, pp. 184–187, 2012.

[26] L. Zhang, L. Harnefors, and H.-P. Nee, "Power-synchronization control of grid-connected voltage-source converters," *IEEE Transactions on Power Systems*, vol. 25, no. 2, pp. 809–820, 2010.

[27] S. Li, T. A. Haskew, and L. Xu, "Conventional and novel control designs for direct driven PMSG wind turbines," *Electric Power Systems Research*, vol. 80, no. 3, pp. 328–338, 2010.

[28] G. P. Adam, *Voltage Source Converter: Modulation, Modelling, Control and Applications in Power Systems*, CreateSpace Independent Publishing Platform, Charleston, SC, USA, 1st edition, 2014.

[29] G. Beale, *Phase Lead Compensator Design Using Bode Plots*, George Mason University, Fairfax, Va, USA, 2010.

[30] G. Beale, *Lag-Lead Compensator Design Using Bode Plots*, George Mason University, Fairfax, Va, USA, 2003.

[31] G. Beale, *Phase Lag Compensator Design Using Bode Plots*, George Mason University, Fairfax, Va, USA, 2003.

[32] J. D. Irwin, *The Industrial Electronics Hand Book*, CRC Press & IEEE Press, USA, 1997, The Industrial Electronics Hand Book., USA, CRC Press IEEE Press,.

[33] P. V. Sreeraj, "Design and implementation of PID controller with lead compensator for thermal process," *International Journal of Computer Applications*, vol. 67, no. 1, pp. 26–31, 2013.

[34] M. R. Islam, S. Mekhilef, and R. Saidur, "Progress and recent trends of wind energy technology," *Renewable and Sustainable Energy Reviews*, vol. 21, pp. 456–468, 2013.

[35] S. Alsabbah, M. Aldhaifallah, and M. Al-Jarrah, "Design of multiregional supervisory fuzzy PID control of pH reactors," *Journal of Control Science and Engineering*, vol. 2015, Article ID 396879, 9 pages, 2015.

[36] M. A. Sahib, B. S. Ahmed, and M. Y. Potrus, "Application of combinatorial interaction design for DC servomotor PID controller tuning," *Journal of Control Science and Engineering*, vol. 2014, Article ID 576868, 7 pages, 2014.

[37] W. Wang, N. Chen, L. Z. Zhu, and D. G. Xu, "Phase angle compensation control strategy for low voltage ride through of doubly-fed induction generator," in *Proceedings of the 1st International Conference on Sustainable Power Generation and Supply (SUPERGEN '09)*, pp. 1–7, Nanjing, China, April 2009.

[38] H. Geng and G. Yang, "Linear and nonlinear schemes applied to pitch control of wind turbines," *Scientific World Journal*, vol. 2014, Article ID 406382, 9 pages, 2014.

[39] E. Kamala, A. Aitouchea, R. Ghorbanib, and M. Bayart, "Robust nonlinear control of wind energy conversion systems," *International Journal of Electrical Power & Energy Systems*, vol. 44, pp. 202–209, 2013.

[40] I. E. Davidson, H. Muashekele, and N. Mukapuli, "Benguela community/UNAM wind power demonstration project experiences in implementation," *Journal of Energy and Power Engineering*, vol. 8, pp. 1067–1072, 2014.

[41] A. Rolán, Á. Luna, G. Vázquez, D. Aguilar, and G. Azevedo, "Modeling of a variable speed wind turbine with a permanent magnet synchronous generator," in *Proceedings of the IEEE International Symposium on Industrial Electronics (ISIE '09)*, pp. 734–739, Seoul, Korea, July 2009.

[42] P. Kundur, *Power System Stability and Control*, McGraw-Hill Education, New York, Ny, USA, 1st edition, 1994.

[43] S. K. Bisoyi, R. K. Jarial, and R. A. Gupta, "Modeling and control of variable speed wind turbine equipped with PMSG," *International Journal of Emerging Technologies in Computational and Applied Sciences*, vol. 6, pp. 56–62, 2013.

[44] N. S. Nise, *Control Systems Engineering*, John Wiley & Sons, New York, NY, USA, 3rd edition, 2000.

[45] A. Kumar, *Control System*, McGraw Hill, New Delhi, India, 2006.

[46] O. Gomis-Bellmunt, A. Junyent-Ferré, A. Sumper, and J. Bergas-Jané, "Control of a wind farm based on synchronous generators with a central HVDC-VSC converter," *IEEE Transactions on Power Systems*, vol. 26, no. 3, pp. 1632–1640, 2011.

[47] C. B. C. Blake, *GBT or MOSFET: Choose Wisely*, International Rectifier, 2010.

[48] S. Keshavarz, *Design and evaluation of an active rectifier for a 4.1MW off-shore wind turbine [M.S. thesis]*, Department of Energy and Environment, Chambers University of Technology, Goteborg, Sweden, 2011.

[49] T. W. Shire, *VSC-HVDC based network reinforcement [M.S. thesis]*, Department of High-Voltage Components and Power Systems, Delft University of Technology, 2009.

22

A Rule Based Energy Management System of Experimental Battery/Supercapacitor Hybrid Energy Storage System for Electric Vehicles

Qiao Zhang, Weiwen Deng, Sumin Zhang, and Jian Wu

State Key Laboratory of Automotive Simulation and Control, Jilin University, Changchun 130022, China

Correspondence should be addressed to Sumin Zhang; zhangsumin@jlu.edu.cn

Academic Editor: Ahmed M. Massoud

In this paper, a simple and efficient rule based energy management system for battery and supercapacitor hybrid energy storage system (HESS) used in electric vehicles is presented. The objective of the proposed energy management system is to focus on exploiting the supercapacitor characteristics and on increasing the battery lifetime and system efficiency. The role of the energy management system is to yield battery reference current, which is subsequently used by the controller of the DC/DC converter. First, a current controller is designed to realize load current distribution between battery and supercapacitor. Then a voltage controller is designed to ensure the supercapacitor SOC to fluctuate within a preset reasonable variation range. Finally, a commercial experimental platform is developed to verify the proposed control strategy. In addition, the energy efficiency and the cost analysis of the hybrid system are carried out based on the experimental results to explore the most cost-effective tradeoff.

1. Introduction

Electric vehicles are considered as one of the most promising transportation tools for addressing issues faced by automotive industry worldwide on energy and environment [1–4]. Technologies employed for all kinds of electric vehicles are various, but their performances are largely dependent on the characteristics of adopted energy storage system (ESS) [5, 6]. Of all the ESSs, battery is one of the most widely used in kinds of electric vehicles, which has been an emerging area for ensuring a successful application of electric vehicles. However, battery alone as a power source has yet faced some challenges for practical engineering application, such as higher energy efficiency, smaller voltage drops, larger vehicle acceleration or deceleration rates, and better uphill climbing performance. Although high power battery can be made available, it is very bulky yet cost prohibitive [7, 8].

A possible solution to the battery dilemma is to add onboard supercapacitor to form HESS with the purpose that battery and supercapacitor function as a complementary role based on their individual dynamic characteristics [9, 10]. Battery has high energy density and relatively low cost per watt-hour but low specific power and short lifetime, while supercapacitor preserves high power density and long cycle life but relatively low energy density and high cost per watt-hour [11–14]. Consequently, a combination of these two types of ESSs will yield an equivalent ESS with both high energy density and high power density, where energy is stored in the battery and power is supplied by the supercapacitor. In this way, sudden peak current in battery, which can be resulting in a large reduction in lifetime [15, 16], can be avoided effectively. Besides, electric vehicle range can also be extended because of the high utilization of the regenerative breaking [17].

In order to encourage the development of battery and supercapacitor HESS, mounting research efforts have been devoted to improving the HESS performance from both low level topology structure [11, 18–20] and high level energy management [21–28]. In the literature, there are a number of hybridization topology structures where the supercapacitor has been implemented in combination with battery systems. Based on these topology structures, various energy management control strategies for battery and supercapacitor

HESS are also developed in the literature. These strategies can be broadly classified into rule based and optimization-based distribution. Rule based controllers are referred to as deterministic rule, learning network, and fuzzy logical controller. In [21], an ADVISOR-based battery and supercapacitor HESS was developed; the power distribution rule between battery and supercapacitor was determined based on an experiment approach. In [22], an integrated rule based metaheuristic approach was presented, and simulation results have proved the effectiveness of this multilevel EMS fulfilling the requested performance with better source usage and much lower installed capacities. In [23], a fuzzy logical controller based on particle swarm optimization algorithm was presented to achieve optimization power flow distribution between battery and supercapacitor. Reference [24] adopted the idea of neural networks and developed an efficient energy management system for hybrid electric vehicles with closed form approaches.

Optimization-based methods consider local and global optimization problems by using past driving cycle information to determine the power distribution mode among power sources. Reference [25] formulated an optimization problem by concerning the minimization of the magnitude/fluctuation of the current flowing in and out of the battery and the energy loss of the supercapacitor. An MPC (model predictive control) controller for a hybrid battery-supercapacitor power source was proposed and experimentally verified in [26]. The designed MPC controller enables the battery to share the low frequency power components and the high frequency power components are allocated to the supercapacitor. In [27], a wavelet transform-based power management strategy was proposed, in which load power is decomposed into different frequency components to the FC (fuel cell), battery, and supercapacitor, compatible with their respective dynamic characteristics. Reference [28] adopted a first-order low-pass filter, which was designed for frequency decomposition along with analysis of components responses under real world driving cycles. These strategies would increase battery lifetime attributing to effective frequency separation consideration.

In this paper, a rule based energy management system is developed for a semiactive hybrid system. The objective of the proposed energy management system is to focus on exploiting the supercapacitor characteristics and increasing the battery lifetime and system efficiency. The main contributions of this paper are summarized as follows. First, a simple but effective rule based energy management system is developed for a semiactive hybrid system. The main role of the energy management system is to yield battery reference current according to energy levels of the power sources and load demand level. Then two classical PI control loops are proposed to adjust the load distribution between battery and supercapacitor. The described energy management system enables the battery to share low frequency and smooth load current, and the corresponding high frequency and peak current are distributed to the supercapacitor. This system prevents an overstress on the battery while maintaining a good voltage balance of the supercapacitor. This is an important advantage to ensure the vehicle starts a new cycle if high acceleration is required. Second, an experimental platform of the hybrid system is developed to verify the described energy management system using two standard drive cycles for an electric vehicle. Finally, preliminary but significant energy efficiency and the cost analysis of the hybrid system are carried out based on the experimental results to explore the most cost-effective tradeoff.

The rest of this paper is organized as follows: Section 2 describes the dynamic characteristics of the battery and supercapacitor. Section 3 analyzes the hybrid topology structure. Section 4 describes the energy management control strategy. Section 5 verifies the proposed control strategy based on a commercial experimental platform. Finally, the paper is concluded in Section 6.

2. Characteristics of Battery and Supercapacitor as Power Sources

Batteries used in electric vehicle usually have the characteristics of high energy density and relatively low power density. The latter is a limitation to improve vehicle dynamic performance. High power capability requires the resistance of the battery to be low. Hence, knowledge of the resistance of a battery is critical to the ability to assess its power capability. In addition, the charging and discharge internal resistances of battery are usually different, which is dependent on charging and discharge operating conditions, including amplitude and frequency of charging and discharge of battery. For lithium-ion battery, the internal resistance could increase when high frequency loads are exerted. The ampere-hour capacity is affected by the discharging current rate and is modeled by the Peukert equation [13]. The charging and discharging internal resistances are nonlinear functions of current and state of charge (SOC).

Recently, another energy storage device, the supercapacitor, has attracted much interest because of its high power density (300 W/kg–5000 W/kg) and relatively low energy density. There are presently commercially available carbon/carbon supercapacitor devices (single cells and modules) from several companies, such as Maxwell, Ness, EPCOS, Nippon Chemi-Con, and Power Systems. All these companies market large devices with capacitance of 1000–5000 F. These devices are suitable for high power vehicle applications [29].

As described above, the difference between battery and supercapacitor can provide the complementary advantage to meet power and energy demand. From the perspective of power characteristics, the charging and discharging efficiency of system is usually considered as the main factor. By comparison, only one-half of the energy at the peak power from the battery is in the form of electrical energy to the load, and the other one-half is dissipated within the battery as heat in the ESR. That is to say the efficiency of batteries is around 50%. For supercapacitor, the peak power is usually for a 95% efficient discharge, in which only 5% of the energy from the device is dissipated as heat in the ESR. For a corresponding high-efficiency discharge, batteries would have a much lower power capability. Furthermore, the main drawback of the batteries is a slow-charging time, limited by a charging current; in contrast, the supercapacitor may be charged in

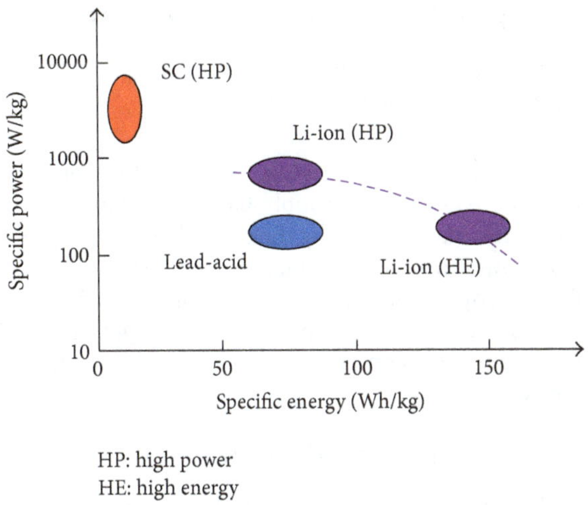

FIGURE 1: Comparison of battery and supercapacitor.

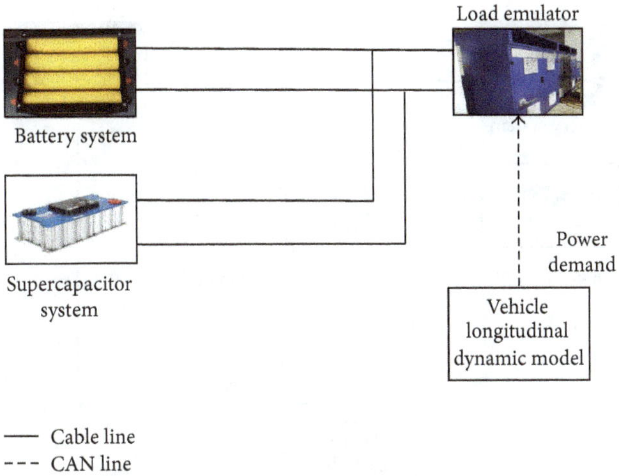

FIGURE 2: A passive topological structure.

a short time depending on a high-charging current (power) available from the main source [30].

Figure 1 shows the comparison of battery and supercapacitor in terms of power and energy. Even though it is true that a battery has the largest energy density (meaning more energy is stored per weight than other technologies), it is important to consider the availability of that energy. This is the traditional advantage of supercapacitor. With a time constant of less than 0.1 s, energy can be taken from a supercapacitor at a very high rate [31]. Electrochemical reactions in battery and double-layer effects in supercapacitor will yield different acceleration and deceleration transient behaviors of voltage and current, which are described as different time constants in dynamic equivalent circuit models.

3. Hybrid Topology Structure

Hybrid energy storage systems are formed in some typical topologies in order to manage the power flow between battery and supercapacitor, each of which has its own properties. In this section, the advantages and disadvantages of four commonly used hybrid structures for electric vehicles are described.

Figure 2 shows a passive topological structure, in which both battery and supercapacitor are directly connected in parallel. Although it is simple and easy to be realized in HESS, the power distribution is inherently limited by their internal resistance since the voltage of both battery and supercapacitor is the same [25].

Figure 3 shows one of the semiactive topological structures, in which the DC/DC converter is connected to the voltage side of the battery, and the battery voltage can be boosted up such that the battery pack can be made smaller to reduce its weight and volume. Further, with this structure, the battery voltage can be controlled more effectively [32].

Figure 4 shows another type of the semiactive topological structures, in which the voltage of supercapacitor is controlled by a bidirectional DC/DC converter, so that the supercapacitor can be boosted up to meet the driving

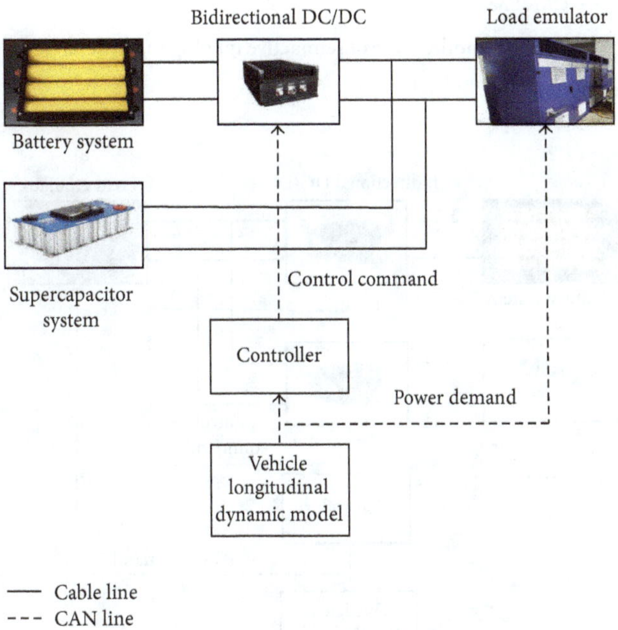

FIGURE 3: A semiactive topological structure.

power demand for electric vehicles. Likewise, it can also be reduced to a lower level for the purpose of energy recovery via regenerative braking. This would enable supercapacitor to operate in a wider voltage range and to curb voltage fluctuation and peak current damage to battery. Even if it was not controlled by a bidirectional DC/DC converter, the battery would still work in a high effective range [11].

A fully active topological structure is shown in Figure 5. In this structure, two bidirectional DC/DC converters are used such that the hybrid power system is decoupled between battery and supercapacitor. Therefore, both power sources can be controlled via each individual DC/DC converter independently. This structure can be more flexible, stable, and efficient for voltage control and power distribution between

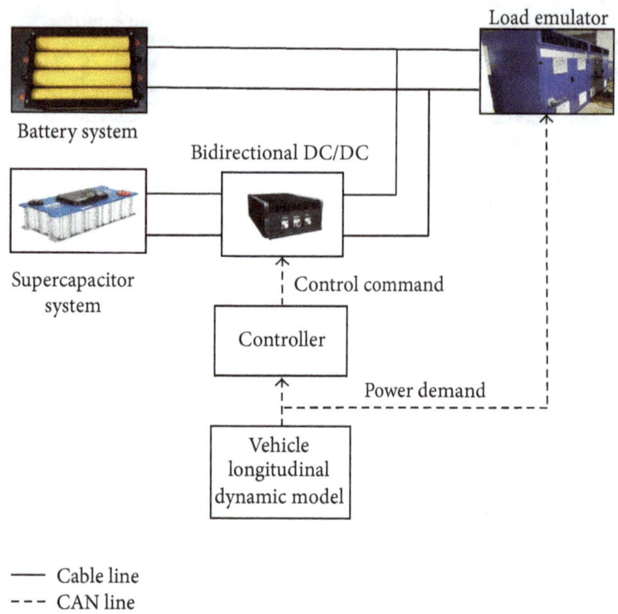

FIGURE 4: Another type of semiactive topological structures.

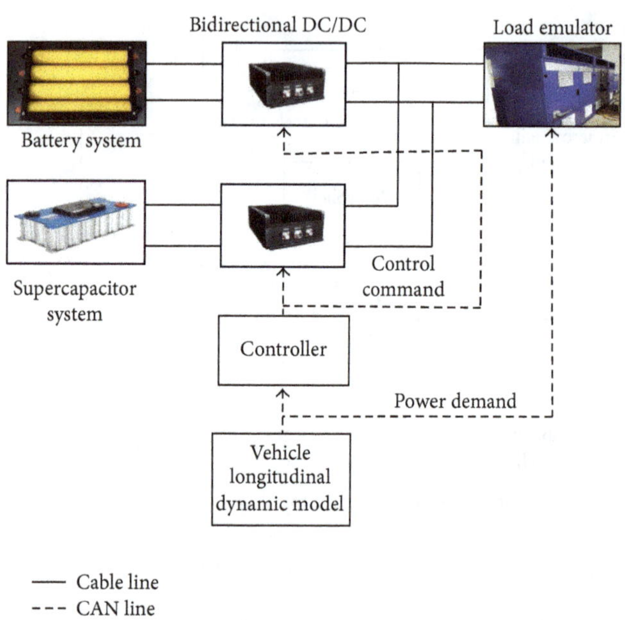

FIGURE 5: A fully active topological structure.

battery and supercapacitor. It can also reduce the size and weight of the hybrid energy storage system [19].

As a brief summary, the passive hybrid system is simple in structure and more cost-effective, but the fully active hybrid system offers the best performance. Therefore, a semiactive hybrid system is often a good tradeoff among them in terms of the performance, the structure complexity, and the cost-effectiveness.

4. Energy Management Control Strategy

The presented energy management system control frame in this research is illustrated in Figure 6.

4.1. Energy Management System. The function of energy management system is to supply battery reference current, which is subsequently used by the controller.

The power demand calculation of hybrid system can be obtained by considering vehicle dynamics, and the total power demand consisted of rolling resistance power, aerodynamic drag power, slope resistance power, and acceleration resistance power:

$$P_{\mathrm{veh}} = P_{\mathrm{roll}} + P_{\mathrm{aer}} + P_{\mathrm{slope}} + P_{\mathrm{acc}}, \tag{1}$$

where the rolling resistance power is described as

$$P_{\mathrm{roll}} = \frac{u_{\mathrm{veh}}}{\eta} \frac{Mgf\cos(\alpha)}{3600}. \tag{2}$$

The aerodynamic drag power is described as

$$P_{\mathrm{aer}} = \frac{1}{\eta} \frac{C_{\mathrm{aer}} A_{\mathrm{aer}}}{76140} u_{\mathrm{veh}}^{3}. \tag{3}$$

The slope resistance power is described as

$$P_{\mathrm{slope}} = \frac{u_{\mathrm{veh}}}{\eta} \frac{Mg\sin(\alpha)}{3600}. \tag{4}$$

The acceleration resistance power is described as

$$P_{\mathrm{acc}} = \frac{u_{\mathrm{veh}}}{\eta} \frac{\delta M}{3600} \frac{du_{\mathrm{veh}}}{dt}. \tag{5}$$

The total current demand can be calculated as

$$I_{\mathrm{load}} = \frac{P_{\mathrm{veh}}}{U_{\mathrm{bus}}}, \tag{6}$$

where U_{bus} is the bus voltage, M is the vehicle mass, u_{veh} is the vehicle velocity, g is gravity constant, f is the rolling resistance coefficient, α is the road slope angle, C_{aer} is the aerodynamic drag coefficient of the vehicle, A_{aer} is the frontal area of the vehicle, and η is the drive efficiency.

State of charge (SOC) is traditionally used to indicate the residual electricity of the battery; its definition is usually written by the equation

$$\mathrm{SOC}_{\mathrm{bat}} = \mathrm{SOC}_0 - k_{\mathrm{ch}} \cdot k_{\mathrm{dis}} \cdot \int \varepsilon \cdot \frac{I_{\mathrm{bat}} dt}{C_{\mathrm{bat}}}, \tag{7}$$

where SOC_0 describes the initial value of $\mathrm{SOC}_{\mathrm{bat}}$; k_{ch} and k_{dis} describe the influence coefficients on the current integration from charging current ($I_L < 0$) and discharging current ($I_L > 0$), respectively; if the battery is charging, $k_{\mathrm{dis}} = 1$, and if the battery is discharging, $k_{\mathrm{ch}} = 1$. C_{bat} describes the nominal capacity of battery; ε is the coulomb efficiency (including charging efficiency $\varepsilon_{\mathrm{ch}}$ and discharging efficiency $\varepsilon_{\mathrm{dis}}$).

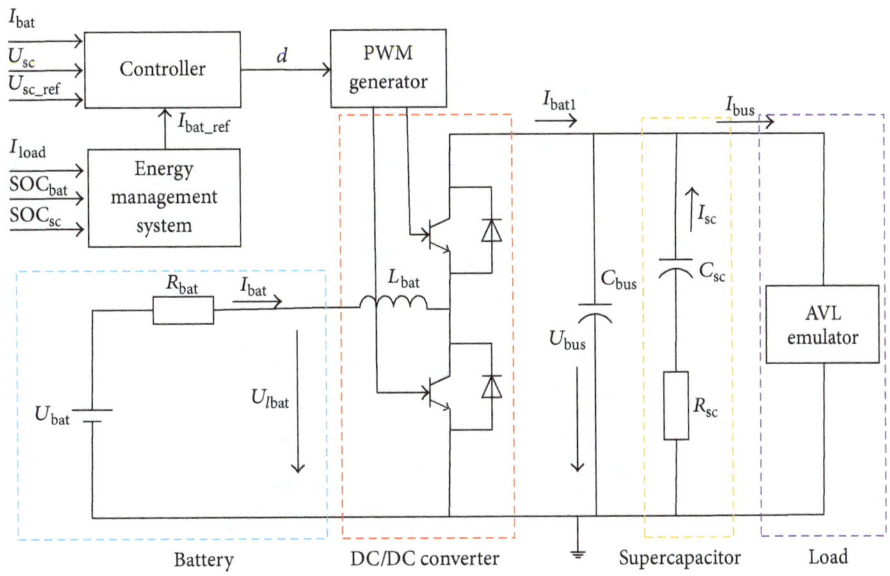

FIGURE 6: Control frame of the HESS.

In order to indicate the residual electricity of supercapacitor, the state of charge (SOC) of the supercapacitor is used to describe a percentage of the rated energy capacity, which depends on the terminal output voltage and is defined as in the equation

$$SOC_{SC} = \frac{(U_{LSC} - U_{cmin})}{(U_{cmax} - U_{cmin})}, \quad (8)$$

where U_{LSC} is supercapacitor load voltage and U_{cmax} and U_{cmin} are the maximum and minimum terminal voltage, respectively.

The proposed energy management system is a rule based and power-balancing strategy. The strategy is realized by a series of simple control logical rules. The main advantage of the proposed strategy is to protect battery from the high dynamics in current demand without overdischarging or overcharging the supercapacitor. Consequently, both battery lifetime and energy efficiency are increased.

The flowchart of DC/DC converter control mode strategy is shown in Figure 7. The control mode depends on the symbol of the load current demand. The positive load current represents the fact that the vehicle is driving. In this situation, battery or supercapacitor must supply the requirement driving current to meet vehicle driving demand. Therefore, the DC/DC converter need be switched to buck mode. However, the DC/DC control mode also depends on the charge and discharge relationships between battery and supercapacitor. Even if the load current is positive, the DC/DC converter is also switched to boost mode when the supercapacitor charges battery. In this way, the supercapacitor SOC can be adjusted to the expected variation range quickly and thus can ensure the HESS to meet the load current demand without overcharge or overdischarge of the battery. As a result, the battery operation condition can be smoothed greatly and battery lifetime is increased as well.

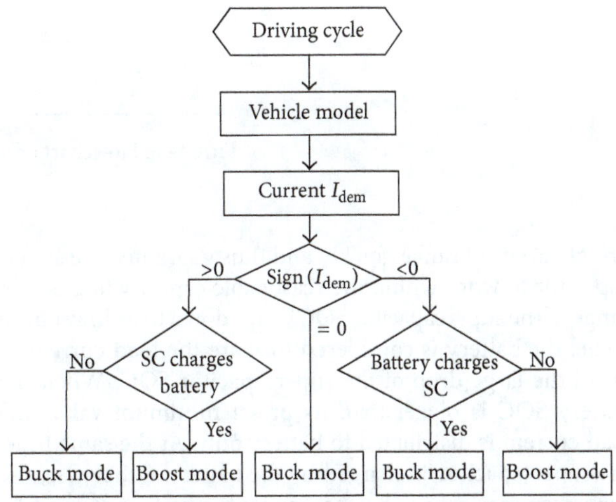

FIGURE 7: Flowchart of the DC/DC converter control mode strategy.

The flowchart of driving mode control strategy is shown in Figure 8. This decision-making flowchart considers as the first decision a comparison of the real supercapacitor SOC and the preset supercapacitor SOC variation range. If the real supercapacitor SOC is located in the preset supercapacitor SOC variation range, then the load current is distributed to supercapacitor only and the battery will not supply any load current. The purpose of this arrangement is to protect battery from the frequent charge and discharge process and increase battery lifetime. When the preset supercapacitor SOC variation range is broken, the battery is considered to balance the supercapacitor SOC or share the load current. In this situation, if the supercapacitor SOC exceeds its upper limit value, then the load current is distributed to supercapacitor only. Besides, the supercapacitor is charged by the battery. In this way, the supercapacitor SOC can be decreased to the

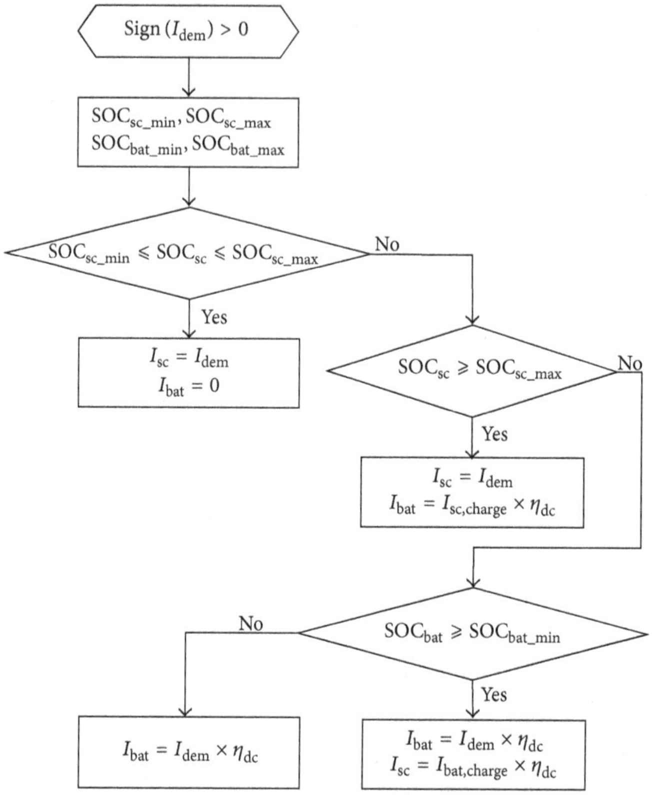

FIGURE 8: Flowchart of driving condition control mode.

preset variation range quickly and thus guarantee the supercapacitor to work within the reasonable capacity fluctuation range. If the supercapacitor SOC drops down to its lower limit value, the battery is considered to share the load current to avoid the large drop of the supercapacitor SOC. When the battery SOC is bigger than its preset minimum value, the load current is distributed to battery only. At the same time, the supercapacitor is charged by the battery. Otherwise, the supercapacitor cannot be charged by the battery. It must be noted that the described rules above mainly include five work modes for the battery and supercapacitor: the battery work only, the supercapacitor work only, the battery charges to the supercapacitor, the supercapacitor charges to the battery, and the battery and supercapacitor working together. In fact, when the supercapacitor exceeds the preset supercapacitor SOC variation range, the load current is distributed to battery and supercapacitor jointly. In this process, the charge is also carried out simultaneously.

The flowchart of idle speed mode control strategy is shown in Figure 9. Generally, the SOC of the supercapacitor is controlled within a certain reasonable fluctuation range. This is to ensure that electric vehicle is able to run even if a high acceleration or deceleration is required without overstressing the battery. Therefore, the idle speed mode control strategy is only to ensure the supercapacitor SOC to be controlled within preset fluctuation range. When the supercapacitor SOC is below the preset lower limit value, the supercapacitor is charged by the battery. On the contrary, the supercapacitor is discharged. Otherwise, no operation is carried out.

The flowchart of braking mode control strategy is shown in Figure 10. Similar to the driving mode control strategy, the battery current is firstly set to zero; this is very important to protect the battery from big current burst during the transient process. Then the only decision in this flowchart depends on the SOC of the supercapacitor and on the preset variation range. When the supercapacitor SOC is below the preset lower limit value, then the supercapacitor can absorb the current from the regenerative breaking. At the same time, the lacking energy is supplied by the battery charge. When the supercapacitor SOC exceeds the preset higher limit value, the regenerative breaking current is absorbed by battery. At the same time, the part energy is delivered to battery from the supercapacitor. Otherwise, the regenerative breaking current is absorbed by the supercapacitor only.

4.2. Controller. The input variable of controller is the battery reference current, and the output variable is the controller command. In order to realize control objective, a classical PI controller is adopted in this research. In this described configuration, the battery is connected to the DC/DC converter but the supercapacitor is directly connected to the bus without DC/DC converter. The current relations can be written by

$$C_{bus} \cdot \frac{dU_{bus}}{dt} = I_{bat1} + I_{SC} - I_{bus}. \quad (9)$$

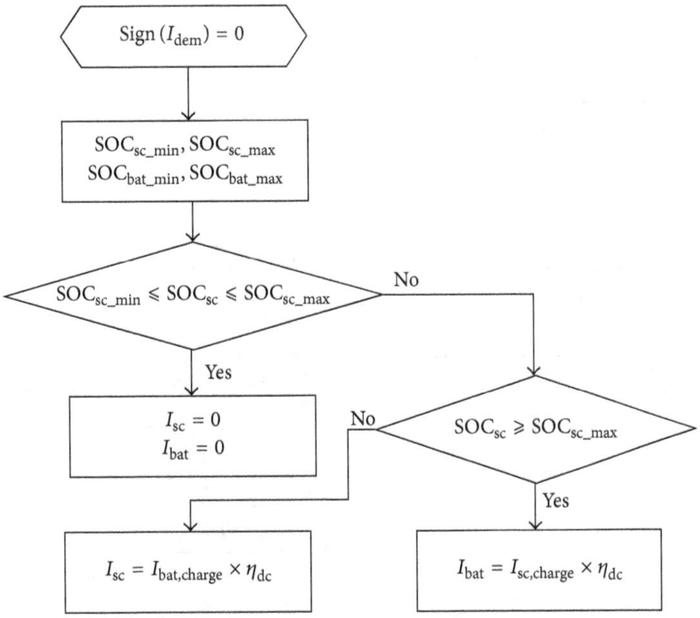

FIGURE 9: Flowchart of idle speed condition control mode.

The equivalent model of the DC/DC converter connected to the battery can be described as

$$L_{bat}\frac{dI_{bat}}{dt} + R_{bat}I_{bat} = U_{Lbat} - d \cdot U_{bus}. \qquad (10)$$

Equation (10) is a nonlinear first-order equation; by using the Laplace transform, we can obtain that

$$I_{bat}(s) = \frac{U_{Lbat}(s)}{L_{bat} \cdot s + R_{bat}} - \frac{d(s)U_{bus}}{L_{bat} \cdot s + R_{bat}}. \qquad (11)$$

The control loop of the battery current is described in Figure 11. The input is the DC/DC converter control command; the output is the battery current. In order to balance the hybrid system without overdischarging the supercapacitor, a supercapacitor voltage compensation loop has been implemented, which is shown in Figure 12.

5. Experimental Results and System Analysis

5.1. Experimental Results. In order to validate the proposed energy management control strategy, a commercial experimental platform is constructed. The whole experimental platform mainly includes two parts: hardware power system and software control system, which are shown in Figures 13 and 14.

The supercapacitor pack adopted in this experimental platform is the Maxwell/BCAP3000 type, rated 3000 F, 2.7 V having the parameters given in Table 1. The battery pack for the HESS is ternary lithium battery, which is considered as the next generation battery used in electric vehicle. The specific parameters of the battery pack are listed in Table 2. The presented DC/DC is a bidirectional DC/DC converter, by which both the driving current and the braking current can be controlled for the battery pack. The main parameters of the DC/DC converter are listed in Table 3. The Electric Control Unit (ECU) is a dSPACE-based MicroAutoBox (DS 1401). Two CAN controllers in the MicroAutoBox are adopted for the load current calculation and control algorithm calculation, respectively.

TABLE 1: Parameters of the supercapacitor pack.

Items	Specifications
Nominal voltage	240 V
Nominal capacity	55 F
Number of cells	88
Maximum continuous power	30 kW/13 s
Pack mass	45 kg ± 5%
Maximum operating temperature	+65°C
Minimum operating temperature	−40°C
Maximum storage temperature	+70°C
Minimum storage temperature	−40°C
Communication type	CAN2.0B, J1939
Leakage current	5.3 mA
Safe level	IP65
Vibration	IEC 16750
Lifetime	25°C, ≥10 years
Initial 48 V module resistance	6.3 mΩ
Shock	SAE J2464

The experiment was carried out to test the control strategy based on two driving cycles, that is, the USA Urban Dynamometer Driving Schedule (UDDS) and the New European Driving Cycle (NEDC). Simulation results and comparisons between the batteries only power system and the HESS system for UDDS driving cycle are shown in Figures 15–21.

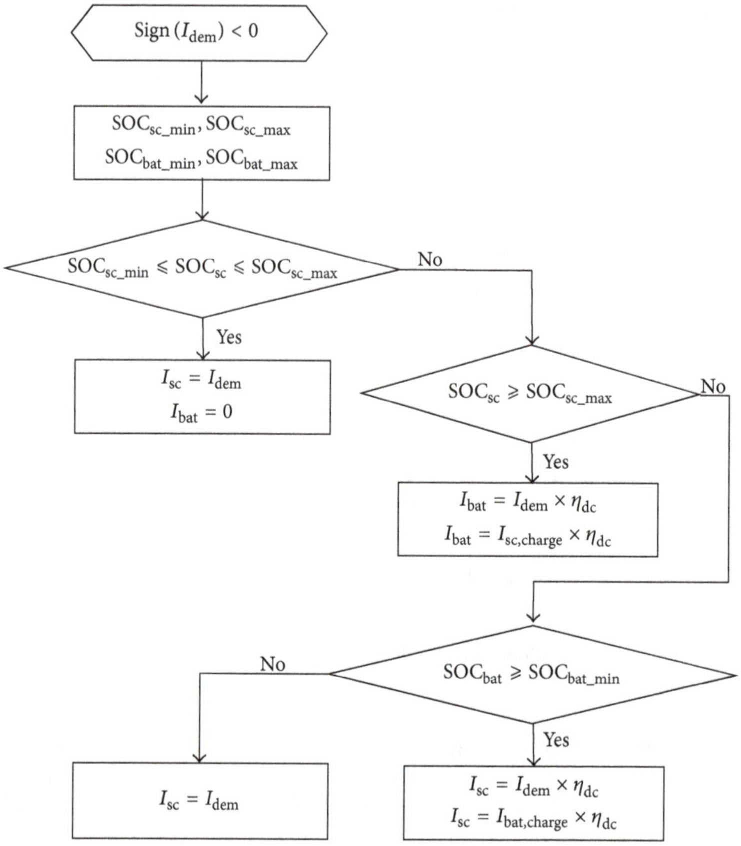

FIGURE 10: Flowchart of braking condition control mode.

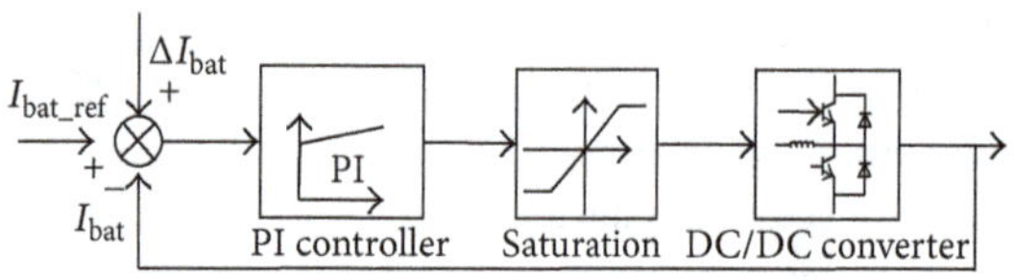

FIGURE 11: Battery current control loop.

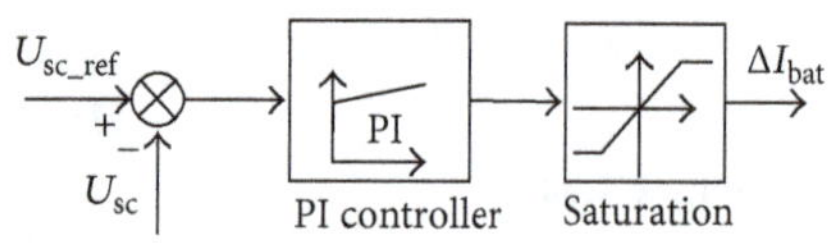

FIGURE 12: Supercapacitor voltage regulation loop.

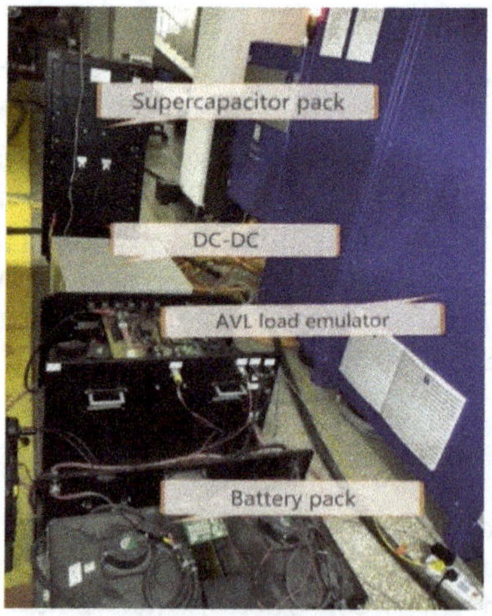

FIGURE 13: Hardware power system.

The charging and discharging currents of battery system are compared in Figure 16. Since the supercapacitor pack can absorb the regenerative braking energy quickly and supply a burst current demand, thus the impact of big charging and discharging current on the battery pack is avoided. It can be observed that the current of battery system for the HESS is mainly maintained in range from −20 A to 20 A, which means that depth of discharge (DOD) of the battery pack is less than 0.33 C, which is beneficial to extending battery lifetime because the number of cycles to failure increases exponentially as DOD decreases.

The evolutions of the battery voltage are compared in Figure 17. It can be obviously observed that large voltage drop of the HESS can be avoided compared to that of the battery only system; namely, a good voltage stabilization performance can be guaranteed for the battery system. It can be seen that the battery voltage of the HESS is maintained within the range from 279 V to 287 V, and the corresponding voltage difference is 8 V. For the battery pack with 72 series

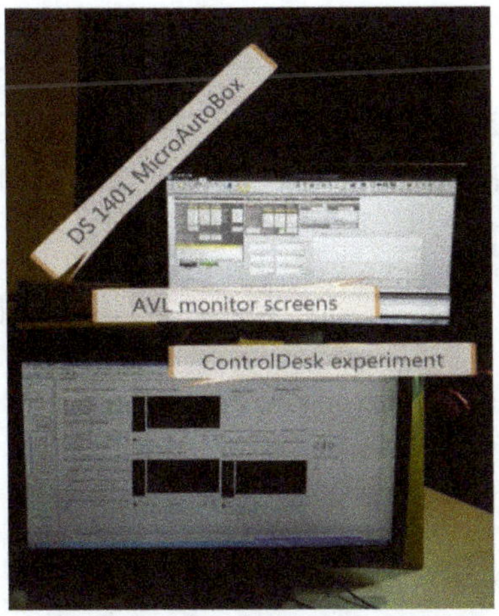

FIGURE 14: Software control system.

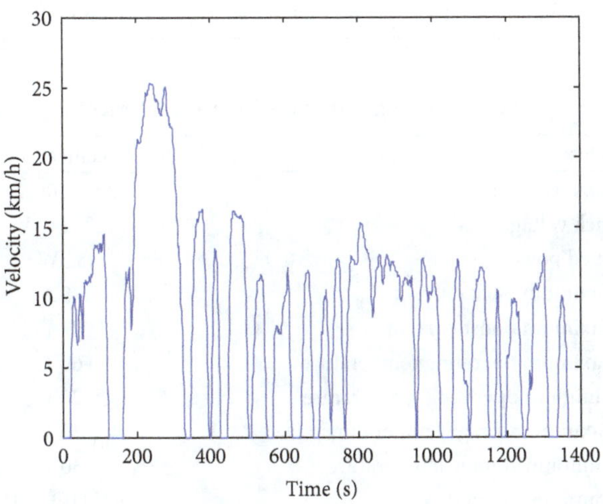

FIGURE 15: UDDS driving cycle.

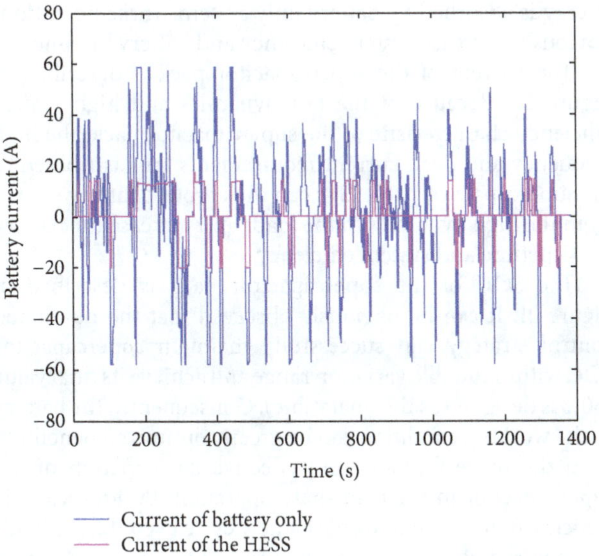

FIGURE 16: Comparison of the battery current curves.

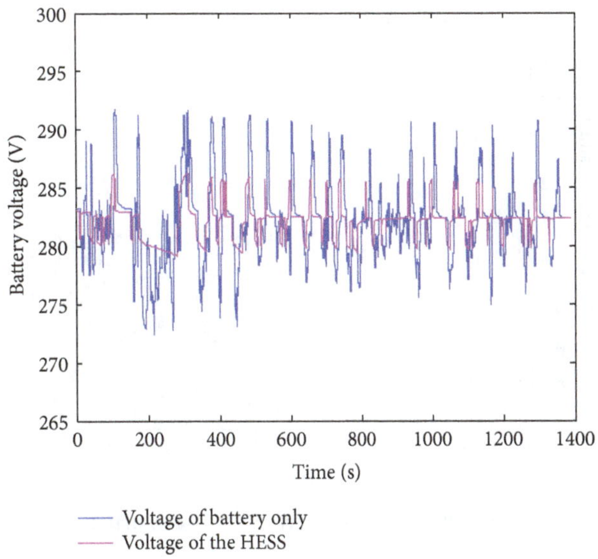

FIGURE 17: Comparison of the battery voltage curves.

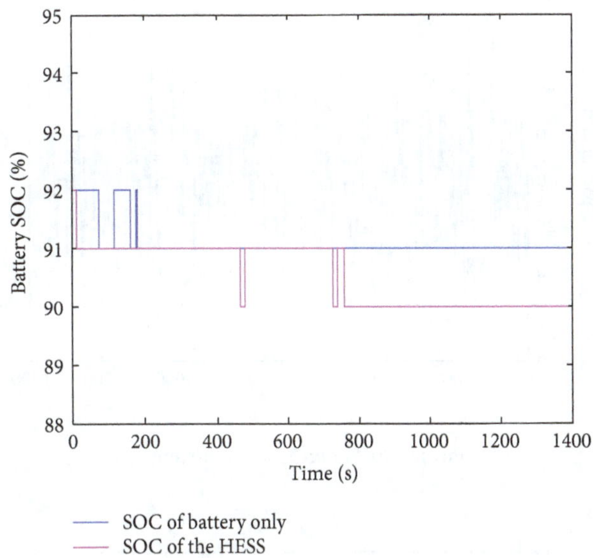

FIGURE 18: Comparison of the battery SOC curves.

of battery cells, a maximum 0.11 V voltage drop compared with a maximum 0.28 V voltage drop for battery only system is roughly estimated. Therefore, it is obvious that the battery system is operated in much smaller voltage fluctuation range and the potential battery cell balancing problem can be avoided to prevent individual cell voltages drift from time to time, which leads to rapid decreases of the total pack capacity, or even complete system failure.

The comparison of the battery SOC is shown in Figure 18. Since the supercapacitor pack absorbs the braking energy actively and efficiently and affords the additional peak power to meet the vehicle driving power requirement, the SOC of the battery pack is smoothed, which can be better found in Figure 25. By comparison, the benefit to electric vehicle range extension seems to be limited. This is because more braking

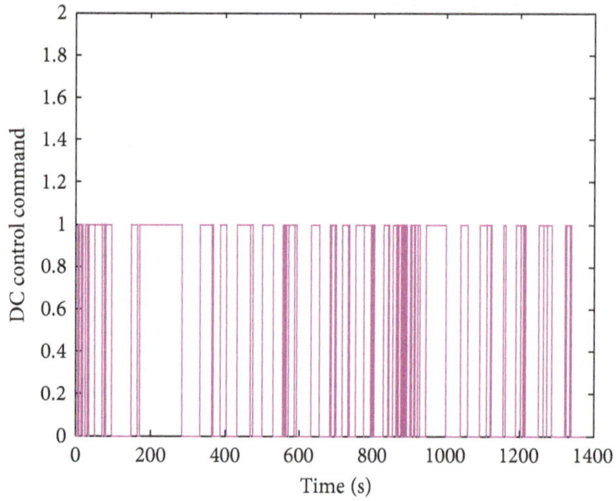

FIGURE 19: DC/DC converter control command.

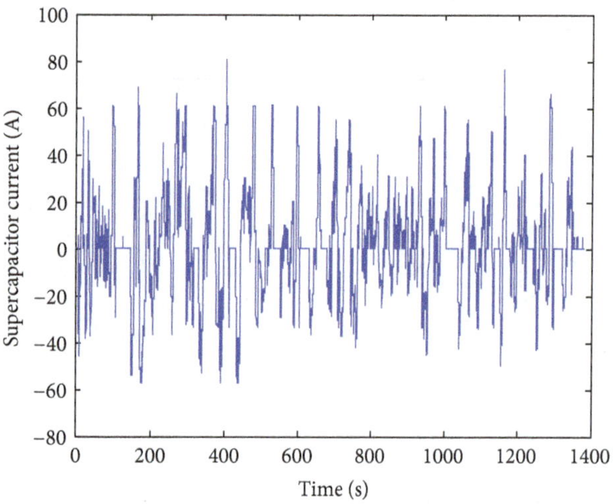

FIGURE 20: Supercapacitor current.

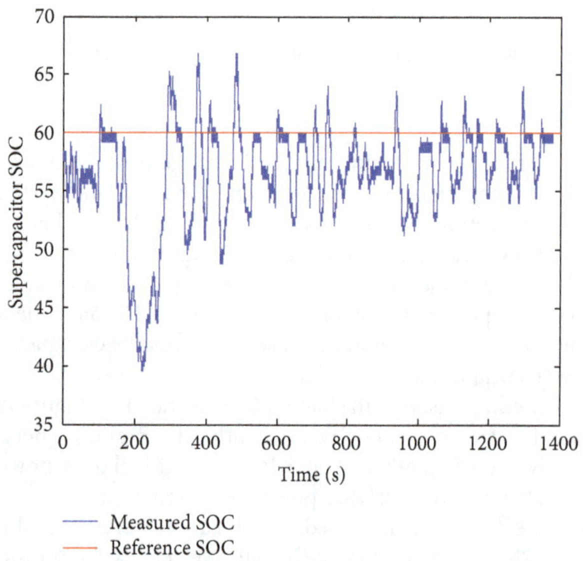

FIGURE 21: Supercapacitor SOC.

TABLE 2: Parameters of the battery pack.

Items	Specifications
Nominal voltage	280 V
Nominal capacity	60 Ah
Number of cells	74
Maximum continuous power	30 kW/13 s
Pack mass	280 kg ± 5%
Maximum charging temperature	+45°C
Minimum charging temperature	0°C
Maximum discharging temperature	+40°C
Minimum discharging temperature	−20°C
Maximum storage temperature	+45°C
Minimum storage temperature	−20°C
Lifetime	25°C, ≥1500 times
Communication type	CAN2.0B, J1939
Charging time	1.5 hours
Initial 50 V module resistance	20 mΩ
Shock	SAE J2464

TABLE 3: Parameters of the DC/DC converter.

Items	Specifications
Boost voltage	200–400 V
Buck voltage	120–240 V
Rated power	15 kW
Maximum buck current	125 A
Maximum boost current	75 A
Maximum operating temperature	+60°C
Minimum operating temperature	−20°C
Maximum storage temperature	+70°C
Minimum storage temperature	−30°C
Communication type	CAN2.0B, J1939
Ripple coefficient	≤1%

energy is absorbed by battery only system. These operations obviously decrease system efficiency and battery lifetime.

The current of the supercapacitor pack is described in Figure 20. Because of the fast dynamics and high system efficiency characteristic of the supercapacitor pack, the high frequency and peak current requirements are distributed to the supercapacitor pack. This can thus protect battery system from the high dynamics in the loads and increase the battery pack lifetime and system efficiency.

The SOC of the supercapacitor pack is described in Figure 21. It can be obviously observed that the developed control strategy can successfully maintain supercapacitor SOC within suitable variation range and achieve its final value (60% is designed as the final value). Consequently, the battery pack's working condition can be greatly optimized benefiting from the more frequent and effective participation of the supercapacitor in the load share operation. Besides, electric vehicle can be ensured to start a new cycle even if large loads are required given that the supercapacitor pack has enough energy and space to satisfy loads.

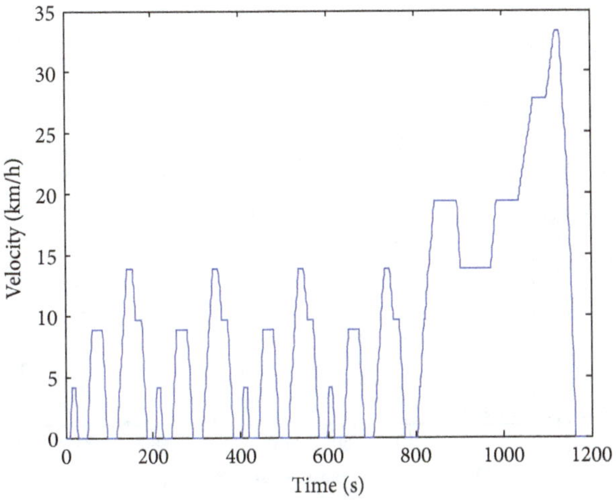

FIGURE 22: NEDC driving cycles.

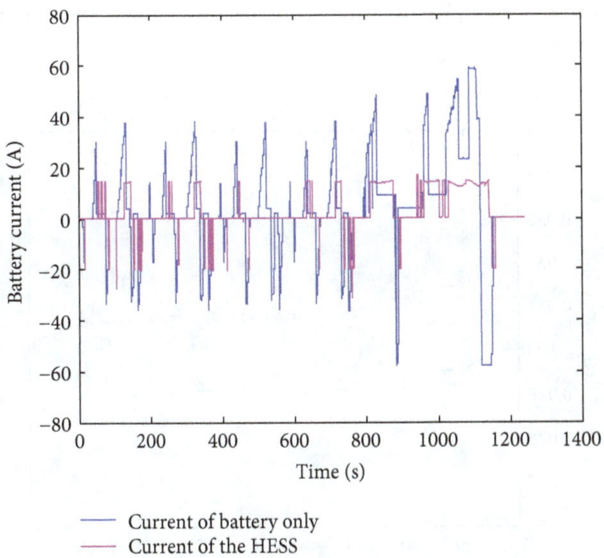

FIGURE 23: Comparison of the battery current curves.

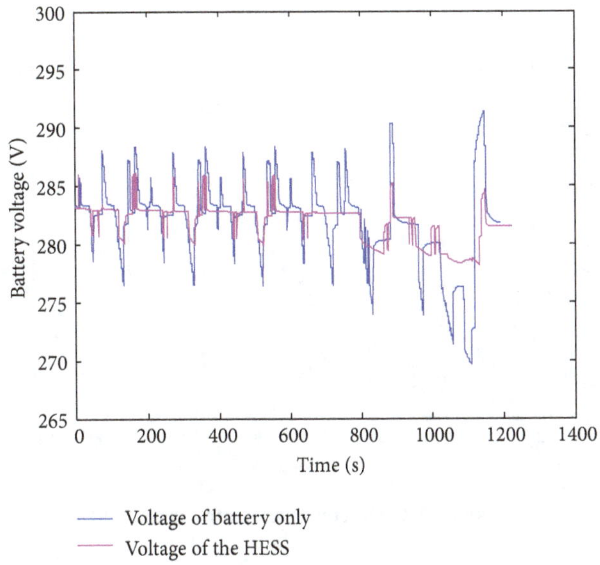

FIGURE 24: Comparison of the battery voltage curves.

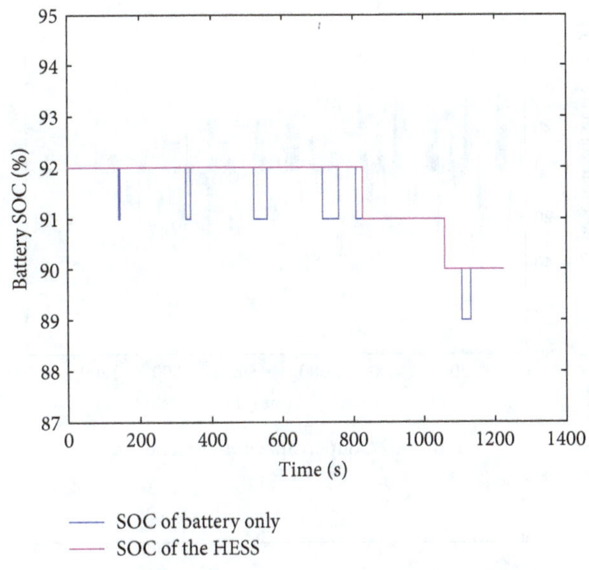

FIGURE 25: Comparison of the battery SOC curves.

Similar simulation was also carried out for the NEDC driving cycle, as shown in Figures 22–28. Again, in this case, the advantages of the proposed energy management control strategy proved to be effective in achieving battery lifetime and system efficiency.

5.2. Efficiency Analysis of the Hybrid System. In this subsection, the energy loss is analyzed to evaluate the effectiveness of the DC/DC converter used in the HESS. Since the energy efficiency has a big relevance with the resistances of the battery and supercapacitor packs and the efficiency of the DC/DC converter, thus the resistance test for the battery and supercapacitor packs and the efficiency of the DC/DC converter test are firstly carried out. The results are plotted in Figures 29–31.

The energy loss comparisons of two semiactive topological structures described in Figures 3 and 4 and battery only system for UDDS driving cycle are shown in Figure 32. It can be observed that the energy loss of the semiactive topological structures described in Figure 4 is about 400 kJ, and the energy loss in the semiactive topological structures described in Figure 3 is about 250 kJ. Therefore, the energy efficiency of the semiactive topological structures described in Figure 3 is higher than that of the semiactive topological structures described in Figure 4. This is because the supercapacitor pack is adjusted by the DC/DC converter to satisfy the load current frequently, consequently resulting in more energy loss from the DC/DC converter. Therefore, the increased range largely depends on the energy efficiency of the DC/DC converter. To clarify the issue for future DC/DC converter development in the HESS, the energy losses of the components in two semiactive topological structures are described in Figures 33 and 34. It can be observed that the energy loss of the HESS is

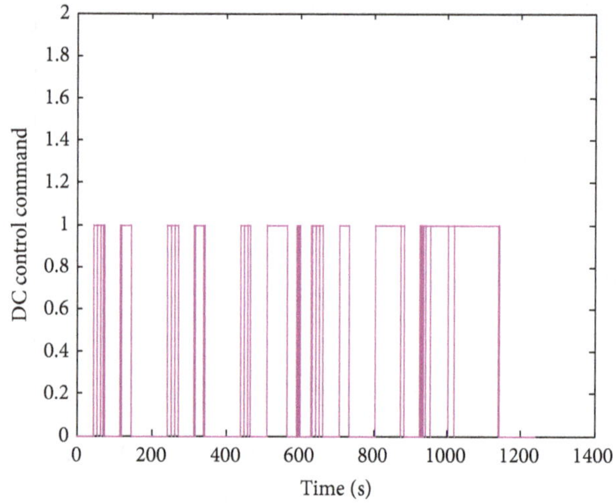

FIGURE 26: DC/DC converter control command.

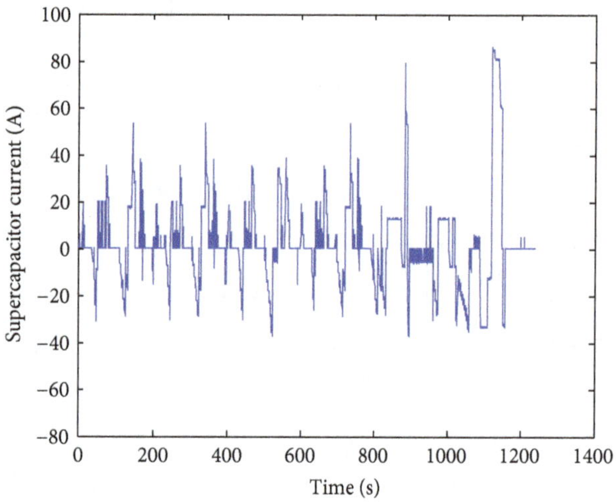

FIGURE 27: Supercapacitor current.

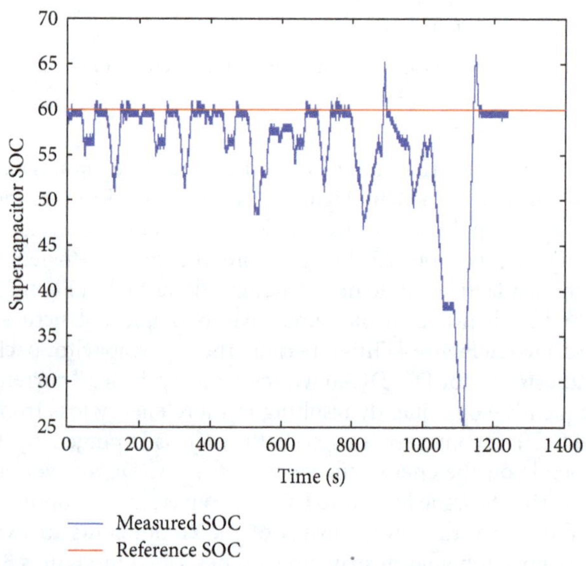

FIGURE 28: Supercapacitor SOC.

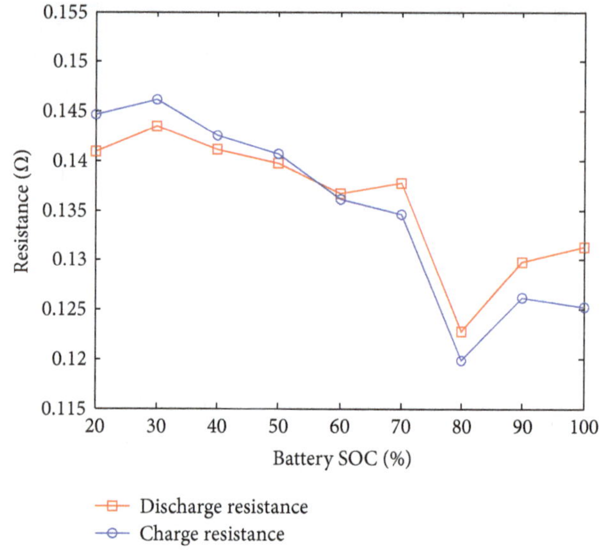

FIGURE 29: Charging/discharging internal resistances of the battery pack.

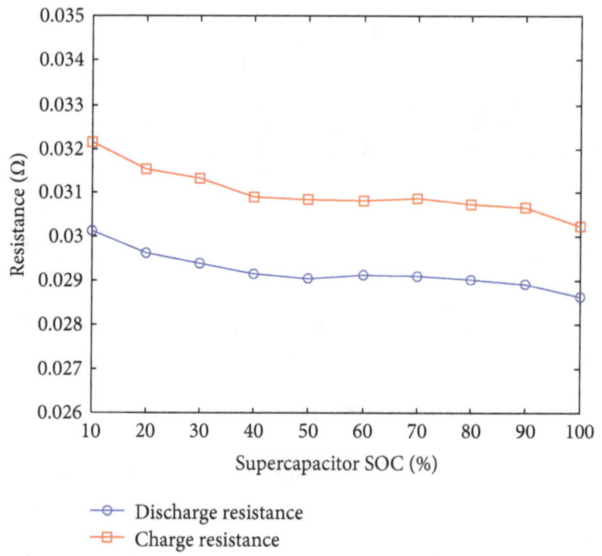

FIGURE 30: Charging/discharging internal resistances of the supercapacitor pack.

mainly from the energy loss of the DC/DC converter. Besides, the energy loss of the battery only system is higher than the total energy loss of the battery and supercapacitor pack in the HESS; it is thus suggested that the efficiency of the DC/DC converter needs to be increased to one certain limit value, which can effectively compensate for the energy loss difference between the HESS and the battery only system. Similar results can be found in Figures 35–37. According to the experiment results and theoretical analysis, based on the developed energy management strategy and the semiactive topological structure described in Figure 3, the DC/DC converter at least has 97% conversion efficiency to make the HESS energy effective compared to the battery only system.

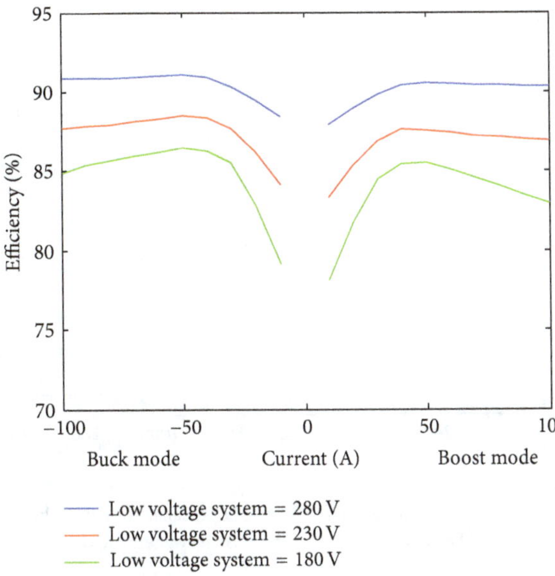

FIGURE 31: Efficiency map of the DC/DC converter.

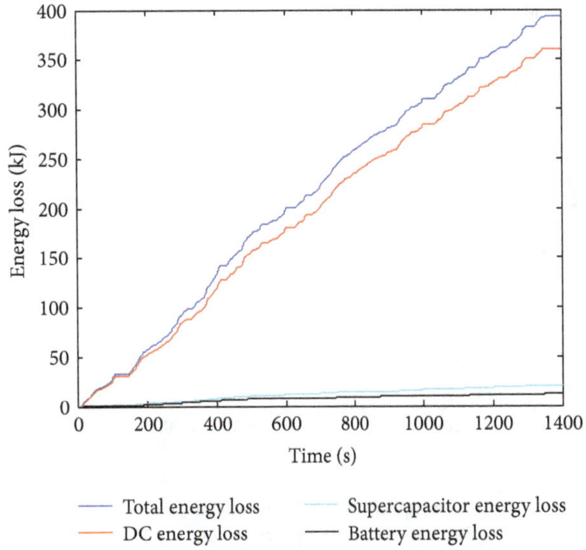

FIGURE 33: Energy loss of the semiactive topological structure described in Figure 4 for UDDS driving cycle.

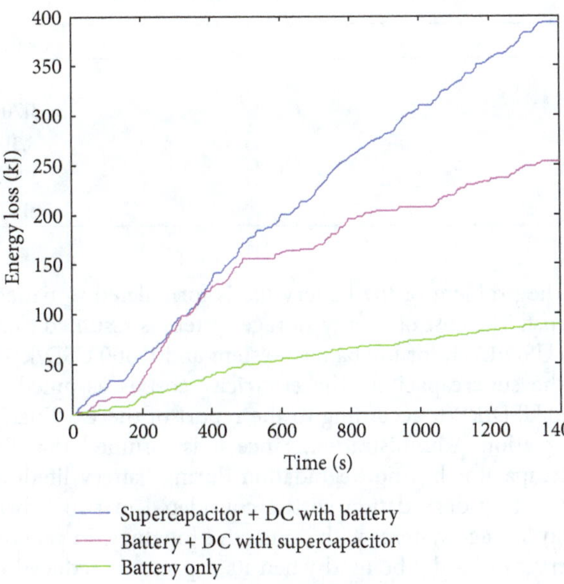

FIGURE 32: Energy loss comparisons of two semiactive topological structures and battery only system for UDDS driving cycle.

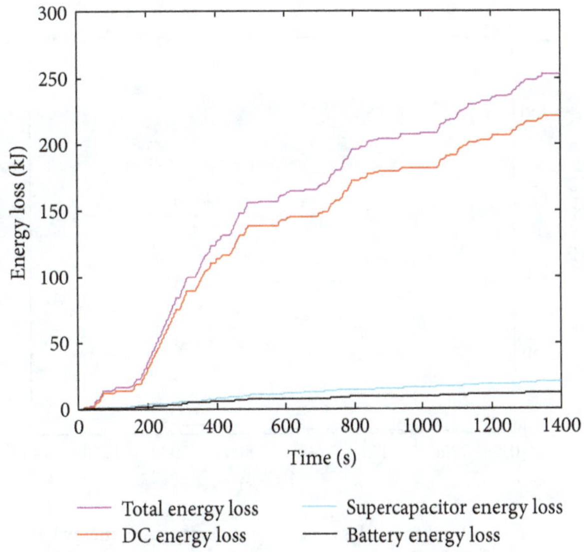

FIGURE 34: Energy loss of the semiactive topological structure described in Figure 3 for UDDS driving cycle.

5.3. Cost Analysis of the Hybrid System. In this section, the cost model of hybrid system is established. The model mainly consists of battery life cost and system electricity cost.

The battery life model is developed in order to analyze the impact of terrain inaccuracy on battery life. Since supercapacitor has much longer life cycle compared to that of battery, it is assumed that the supercapacitor has no degradation during the battery lifetime. The model on battery capacity dynamic degradation adopted in this research is a semiempirical life model [33]. The model includes four parameters, namely, time, temperature, depth of charge, and discharge rate. The variations of these parameters will influence battery lifetime directly. The formula of battery life model is given by

$$Q_{\text{loss}} = B \cdot e^{-((E_a + B \cdot C_{\text{rate}})/(R \cdot T_{\text{bat}}))} \left(A_h\right)^z, \qquad (12)$$

where Q_{loss} is the battery capacity loss which ranges from 0 to 1. B is the preexponential factor, E_a is the activation energy (J mol^{-1}), R is the gas constant (J (mol^{-1} k)$^{-1}$), T is the battery absolute temperature (K), A_h is the Ah-throughput, which is expressed as A_h, z is the power law factor, C_{rate} is the battery discharge rate, and B is the compensation factor of C_{rate}. The original formula is developed based on $LiFePO_4$ battery test results. For the consideration battery studied in this paper, a correction coefficient can be considered to predict battery

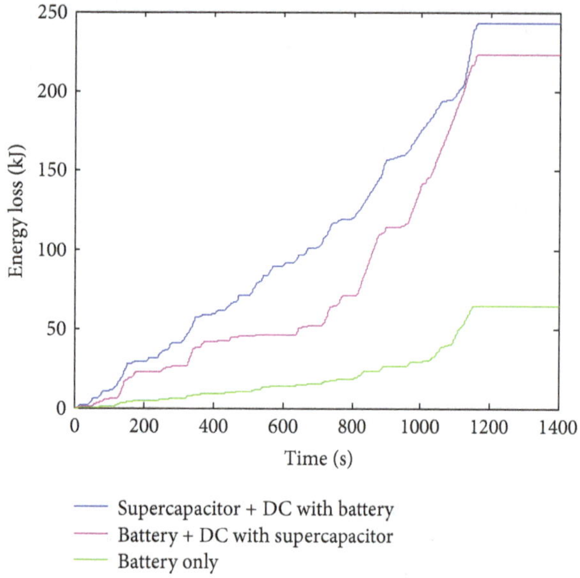

FIGURE 35: Energy loss comparisons of two semiactive topological structures and battery only system for NEDC driving cycle.

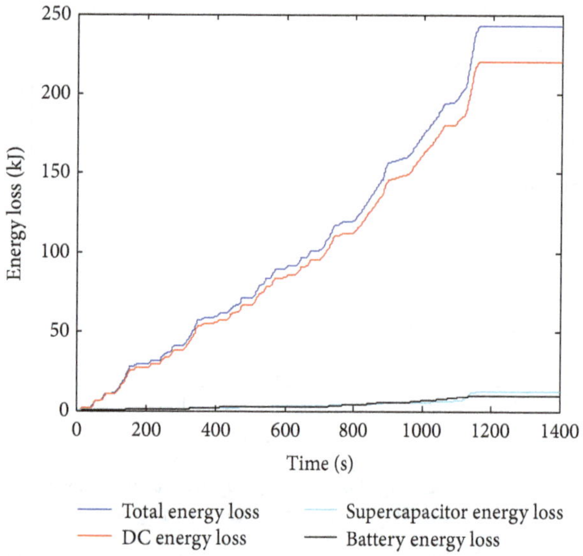

FIGURE 36: Energy loss of the semiactive topological structure described in Figure 4 for NEDC driving cycle.

lifetime more accurately. Then the formula can be rewritten as

$$Q_{\text{loss}} = \beta \cdot B \cdot e^{-((E_a + B \cdot C_{\text{rate}})/(R \cdot T_{\text{bat}}))} (A_h)^z, \tag{13}$$

where β is the correction coefficient: $\beta = 1.7 \times 10^{-4}$. Other parameters used in this formula are listed in Table 4. The Ah-throughput A_h is defined as

$$A_h = \frac{1}{3600} \int_{t_0}^{t_f} |I_{\text{bat}}| \, dt, \tag{14}$$

where t_0 is the initial time of a driving cycle and t_f is the final time of a driving cycle.

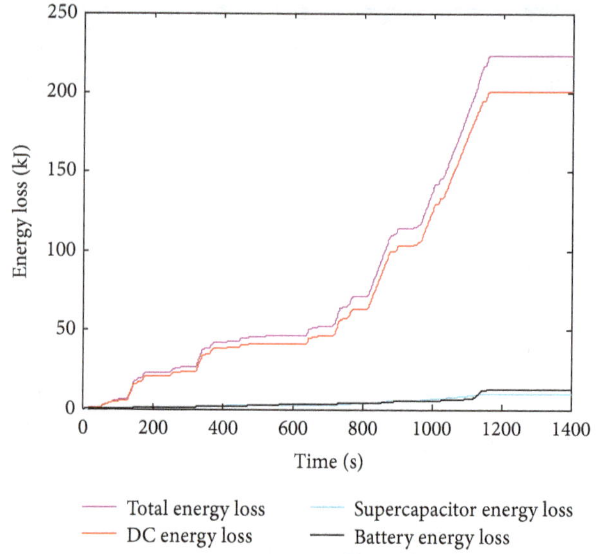

FIGURE 37: Energy loss of the semiactive topological structure described in Figure 3 for NEDC driving cycle.

TABLE 4: Parameters of battery life model.

Item	Value
B	30330
E_a	31700
R	8.314
B	370.3
z	0.55

The problem of the battery life is formulated as battery life cost. The cost of energy storage system is assumed to be 1600 USD/kWh for the battery system and 15000 USD/kWh for the supercapacitor. The electricity cost is assumed to be 0.1 USD/kWh according to the report of the US Energy Information Administration. Since it is assumed that the supercapacitor has no degradation during battery lifetime, only battery degradation cost is considered in two hybrid energy storage systems with semiactive topology. In general, battery can hardly be used when its capacity is reduced to 80% of its initial value. Therefore, the cost description of the battery life and the electricity can be given by

$$\begin{aligned} &\text{Cost}_{\text{bat,loss}}(t) \\ &\quad = 24.768 \\ &\qquad \times \int_0^t \frac{|I_{\text{bat}}|}{3600} dt \exp - \left(\frac{31700 - 370.3 C_{\text{rate}}}{8.314 T_{\text{bat}}} \right), \\ &\text{Cost}_{\text{ele}}(t) = \frac{0.1}{3600} \int_0^T [P_{\text{SC}}(t) + P_{\text{bat}}(t)]. \end{aligned} \tag{15}$$

Note that the electricity cost can be influenced by the resistance losses for both battery and supercapacitor and efficiency loss for the DC/DC converter. In this work, the resistances of battery and supercapacitor and the efficiency of DC converter are simplified as a fixed value. Thus the total

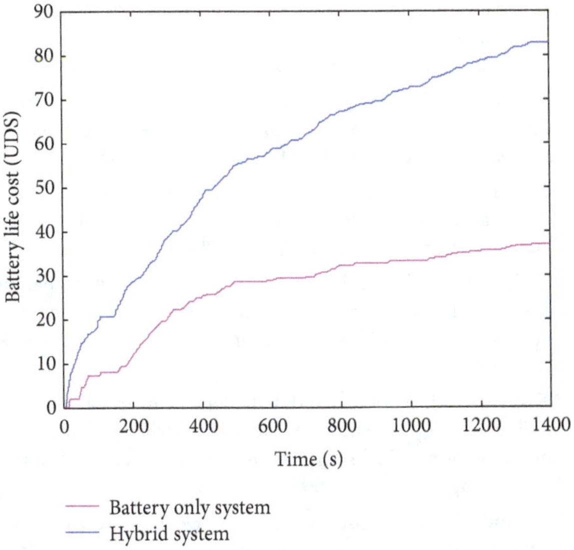

FIGURE 38: Comparison of the battery life cost based on UDDS driving cycle.

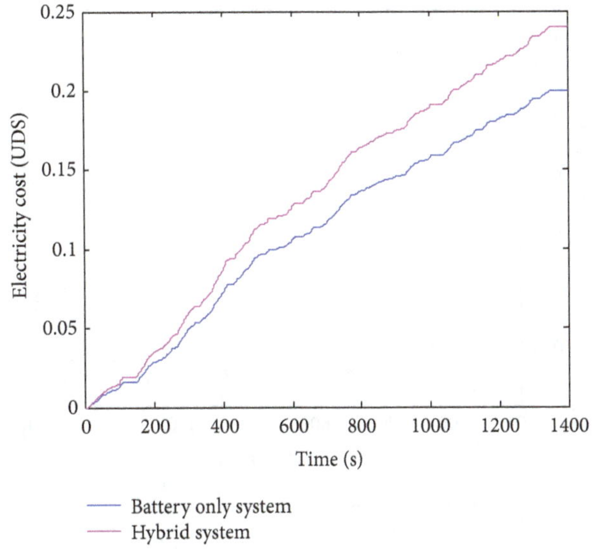

FIGURE 39: Comparison of the electricity cost based on UDDS driving cycle.

cost comprising battery life and electricity cost can be written as

$$\mathrm{Cost}(t) = \mathrm{Cost}_{\mathrm{life}}(t) + \mathrm{Cost}_{\mathrm{ele}}(t). \quad (16)$$

The comparison results of the battery life cost and the system electricity cost based on UDDS driving cycle and NEDC driving cycle are given in Figures 38–41. From Figure 38, it has been demonstrated clearly that hybrid system has an absolute advantage in improving battery life compared with the battery only system. However, the system electricity cost is increased because of the energy loss of the supercapacitor and the DC/DC converter. By comparison, the battery life cost is obviously higher than the system electricity cost. Therefore, to pursue the maximum benefit of the hybrid system, the battery life cost should be considered as the main factor in the total cost. In the future, when the battery is very cheap, the system electricity cost may be dominant in the total cost. By then, an appropriate balance between the battery life cost and the system electricity cost need be considered. From Figures 40 and 41, similar conclusions can be obtained.

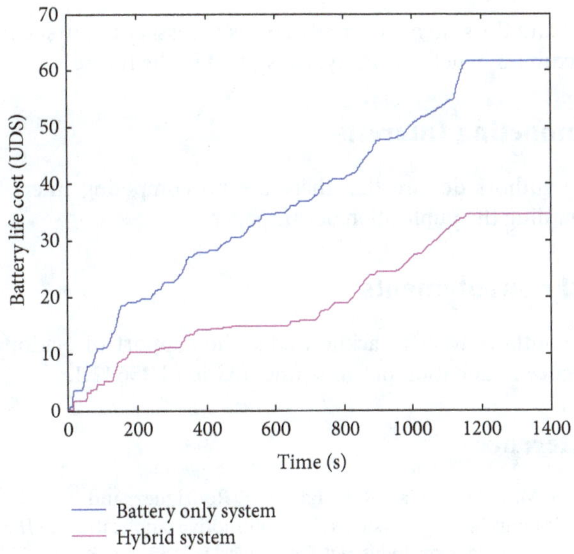

FIGURE 40: Comparison of the battery life cost based on NEDC driving cycle.

6. Conclusion

In this paper, a rule based energy management system is developed for the battery and supercapacitor HESS. The objective of the proposed system is to focus on exploiting the supercapacitor characteristics and on increasing the battery lifetime and system efficiency. Two controllers of the DC/DC converter are designed and integrated to achieve this purpose. Firstly, a current controller is designed to realize load current distribution between battery and supercapacitor. Then a voltage controller is designed to ensure the supercapacitor SOC fluctuate within a preset reasonable variation range.

Experiment results have shown that the system enables the battery to share the low frequency load current, which would be very helpful to increase battery lifetime. Correspondingly, the high frequency load current is distributed to the supercapacitor pack. Efficiency analysis has revealed that the semiactive topological structure described in Figure 3 has a higher energy efficiency compared with the semiactive topological structure described in Figure 4. The increased range depends on the energy efficiency of the DC/DC converter largely. Besides, the DC/DC converter at least has 97% conversion efficiency to make the HESS energy effective compared to the battery only system. The preliminary cost analysis of hybrid system has demonstrated that hybrid system can increase battery lifetime obviously compared with battery only system. At the same time, the analysis also highlights that an appropriate balance between the battery life

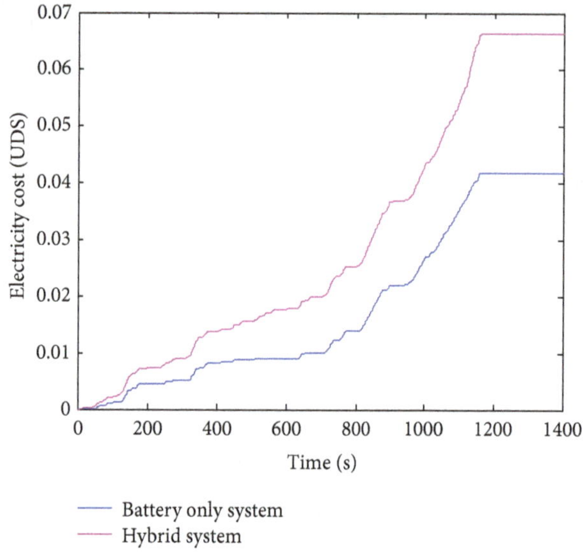

FIGURE 41: Comparison of the electricity cost based on NEDC driving cycle.

cost and the system electricity cost is necessary to pursue the maximum benefit of the hybrid system in the future.

Competing Interests

The authors declare that there are no competing interests regarding the publication of this paper.

Acknowledgments

The authors wish to acknowledge the support of National Science Foundation of China under Grant U1564211.

References

[1] S. M. Lukic, J. Cao, R. C. Bansal, F. Rodriguez, and A. Emadi, "Energy storage systems for automotive applications," *IEEE Transactions on Industrial Electronics*, vol. 55, no. 6, pp. 2258-2267, 2008.

[2] H.-W. He, R. Xiong, and Y.-H. Chang, "Dynamic modeling and simulation on a hybrid power system for electric vehicle applications," *Energies*, vol. 3, no. 11, pp. 1821–1830, 2010.

[3] R. F. Nelson, "Power requirements for batteries in hybrid electric vehicles," *Journal of Power Sources*, vol. 91, no. 1, pp. 2-26, 2000.

[4] K. T. Chau and C. C. Chan, "Emerging energy-efficient technologies for hybrid electric vehicles," *Proceedings of the IEEE*, vol. 95, no. 4, pp. 821–835, 2007.

[5] H. Rahimi-Eichi, U. Ojha, F. Baronti, and M.-Y. Chow, "Battery management system: an overview of its application in the smart grid and electric vehicles," *IEEE Industrial Electronics Magazine*, vol. 7, no. 2, pp. 4–16, 2013.

[6] A. Khaligh and Z. Li, "Battery, ultracapacitor, fuel cell, and hybrid energy storage systems for electric, hybrid electric, fuel cell, and plug-in hybrid electric vehicles: state of the art," *IEEE Transactions on Vehicular Technology*, vol. 59, no. 6, pp. 2806-2814, 2010.

[7] S. F. Tie and C. W. Tan, "A review of energy sources and energy management system in electric vehicles," *Renewable and Sustainable Energy Reviews*, vol. 20, pp. 82–102, 2013.

[8] G. Ren, G. Ma, and N. Cong, "Review of electrical energy storage system for vehicular applications," *Renewable and Sustainable Energy Reviews*, vol. 41, pp. 225–236, 2015.

[9] S. Pay and Y. Baghzouz, "Effectiveness of battery-supercapacitor combination in electric vehicles," in *Proceedings of the IEEE Bologna PowerTech Conference*, pp. 728–733, June 2003.

[10] R. M. Schupbach and J. C. Balda, "The role of ultracapacitors in an energy storage unit for vehicle power management," in *Proceedings of the IEEE Vehicle Technology Conference*, pp. 3236-3240, 2003.

[11] M. Ortúzar, J. Moreno, and J. Dixon, "Ultracapacitor-based auxiliary energy system for an electric vehicle: implementation and evaluation," *IEEE Transactions on Industrial Electronics*, vol. 54, no. 4, pp. 2147–2156, 2007.

[12] G. Guidi, T. M. Undeland, and Y. Hori, "Effectiveness of supercapacitors as power-assist in pure EV using a sodium-nickel chloride battery as main energy storage," in *Proceedings of the 24th International Battery, Hybrid and Fuel Cell Electric Vehicle Symposium and Exhibition*, pp. 2190–2198, May 2009.

[13] L. Gao, R. A. Dougal, and S. Liu, "Power enhancement of an actively controlled battery/ultracapacitor hybrid," *IEEE Transactions on Power Electronics*, vol. 20, no. 1, pp. 236–243, 2005.

[14] R. Carter, A. Cruden, and P. J. Hall, "Optimizing for efficiency or battery life in a battery/supercapacitor electric vehicle," *IEEE Transactions on Vehicular Technology*, vol. 61, no. 4, pp. 1526-1533, 2012.

[15] P. Ruetschi, "Aging mechanisms and service life of lead-acid batteries," *Journal of Power Sources*, vol. 127, no. 1-2, pp. 33–44, 2004.

[16] P. Lailler, F. Zaninotto, S. Nivet et al., "Study of the softening of the positive active-mass in valve-regulated lead-acid batteries for electric-vehicle applications," *Journal of Power Sources*, vol. 78, no. 1, pp. 204–213, 1999.

[17] N. Omar, M. Daowd, O. Hegazy, P. V. D. Bossche, T. Coosemans, and J. V. Mierlo, "Electrical double-layer capacitors in hybrid topologies—assessment and evaluation of their performance," *Energies*, vol. 5, no. 11, pp. 4533–4568, 2012.

[18] F. Ju, Q. Zhang, W. Deng, and J. Li, "Review of structures and control of battery-supercapacitor hybrid energy storage system for electric vehicles," in *Proceedings of the IEEE International Conference on Automation Science and Engineering (CASE '14)*, pp. 143–148, IEEE, Taipei, Taiwan, August 2014.

[19] A. Kuperman and I. Aharon, "Battery-ultracapacitor hybrids for pulsed current loads: a review," *Renewable and Sustainable Energy Reviews*, vol. 15, no. 2, pp. 981–992, 2011.

[20] O. C. Onar and A. Khaligh, "A novel integrated magnetic structure based DC/DC converter for hybrid battery/ultracapacitor energy storage systems," *IEEE Transactions on Smart Grid*, vol. 3, no. 1, pp. 296–307, 2012.

[21] A. C. Baisden and A. Emadi, "ADVISOR-based model of a battery and an ultra-capacitor energy source for hybrid electric vehicles," *IEEE Transactions on Vehicular Technology*, vol. 53, no. 1, pp. 199–205, 2004.

[22] J. P. Trovão, P. G. Pereirinha, H. M. Jorge, and C. H. Antunes, "A multi-level energy management system for multi-source electric vehicles—an integrated rule-based meta-heuristic approach," *Applied Energy*, vol. 105, pp. 304–318, 2013.

[23] Z. Chenghui, S. Qingsheng, C. Naxin, and L. Wuhua, "Particle swarm optimization for energy management fuzzy controller design in dual-source electric vehicle," in *Proceedings of the IEEE 38th Annual Power Electronics Specialists Conference (PESC '07)*, pp. 1405–1410, June 2007.

[24] Y. Ates, O. Erdinc, M. Uzunoglu, and B. Vural, "Energy management of an FC/UC hybrid vehicular power system using a combined neural network-wavelet transform based strategy," *International Journal of Hydrogen Energy*, vol. 35, no. 2, pp. 774–783, 2010.

[25] M.-E. Choi, S.-W. Kim, and S.-W. Seo, "Energy management optimization in a battery/supercapacitor hybrid energy storage system," *IEEE Transactions on Smart Grid*, vol. 3, no. 1, pp. 463–472, 2012.

[26] B. Hredzak, V. G. Agelidis, and M. Jang, "A model predictive control system for a hybrid battery-ultracapacitor power source," *IEEE Transactions on Power Electronics*, vol. 29, no. 3, pp. 1469–1479, 2014.

[27] X. Zhang, C. C. Mi, A. Masrur, and D. Daniszewski, "Wavelet-transform-based power management of hybrid vehicles with multiple on-board energy sources including fuel cell, battery and ultracapacitor," *Journal of Power Sources*, vol. 185, no. 2, pp. 1533–1543, 2008.

[28] Y. Kim, T.-K. Lee, and Z. Filipi, "Frequency domain power distribution strategy for series hybrid electric vehicles," *SAE International Journal of Alternative Powertrains*, vol. 1, no. 1, pp. 208–218, 2012.

[29] W. Gao, "Performance comparison of a fuel cell-battery hybrid powertrain and a fuel cell-ultracapacitor hybrid powertrain," *IEEE Transactions on Vehicular Technology*, vol. 54, no. 3, pp. 846–855, 2005.

[30] J. P. Zheng, T. R. Jow, and M. S. Ding, "Hybrid power sources for pulsed current applications," *IEEE Transactions on Aerospace and Electronic Systems*, vol. 37, no. 1, pp. 288–292, 2001.

[31] P. Thounthong and S. Rael, "The benefits of hybridization," *IEEE Industrial Electronics Magazine*, vol. 6, pp. 69–76, 2008.

[32] A. Khaligh and Z. Li, "Battery, ultracapacitor, fuel cell, and hybrid energy storage systems for electric, hybrid electric, fuel cell, and plug-in hybrid electric vehicles: State of the art," *IEEE Transactions on Vehicular Technology*, vol. 59, no. 6, pp. 2806–2814, 2010.

[33] J. Wang, P. Liu, J. Hicks-Garner et al., "Cycle-life model for graphite-$LiFePO_4$ cells," *Journal of Power Sources*, vol. 196, no. 8, pp. 3942–3948, 2011.

Permissions

All chapters in this book were first published in JCSE, by Hindawi Publishing Corporation; hereby published with permission under the Creative Commons Attribution License or equivalent. Every chapter published in this book has been scrutinized by our experts. Their significance has been extensively debated. The topics covered herein carry significant findings which will fuel the growth of the discipline. They may even be implemented as practical applications or may be referred to as a beginning point for another development.

The contributors of this book come from diverse backgrounds, making this book a truly international effort. This book will bring forth new frontiers with its revolutionizing research information and detailed analysis of the nascent developments around the world.

We would like to thank all the contributing authors for lending their expertise to make the book truly unique. They have played a crucial role in the development of this book. Without their invaluable contributions this book wouldn't have been possible. They have made vital efforts to compile up to date information on the varied aspects of this subject to make this book a valuable addition to the collection of many professionals and students.

This book was conceptualized with the vision of imparting up-to-date information and advanced data in this field. To ensure the same, a matchless editorial board was set up. Every individual on the board went through rigorous rounds of assessment to prove their worth. After which they invested a large part of their time researching and compiling the most relevant data for our readers.

The editorial board has been involved in producing this book since its inception. They have spent rigorous hours researching and exploring the diverse topics which have resulted in the successful publishing of this book. They have passed on their knowledge of decades through this book. To expedite this challenging task, the publisher supported the team at every step. A small team of assistant editors was also appointed to further simplify the editing procedure and attain best results for the readers.

Apart from the editorial board, the designing team has also invested a significant amount of their time in understanding the subject and creating the most relevant covers. They scrutinized every image to scout for the most suitable representation of the subject and create an appropriate cover for the book.

The publishing team has been an ardent support to the editorial, designing and production team. Their endless efforts to recruit the best for this project, has resulted in the accomplishment of this book. They are a veteran in the field of academics and their pool of knowledge is as vast as their experience in printing. Their expertise and guidance has proved useful at every step. Their uncompromising quality standards have made this book an exceptional effort. Their encouragement from time to time has been an inspiration for everyone.

The publisher and the editorial board hope that this book will prove to be a valuable piece of knowledge for researchers, students, practitioners and scholars across the globe.

List of Contributors

Haibo Mu, Yubo Song and Linzhong Liu
School of Traffic and Transportation, Lanzhou Jiaotong University, Lanzhou, Gansu 730070, China

Paul Oke and Sing Kiong Nguang
Department of Electrical and Computer Engineering, The University of Auckland, Auckland 1010, New Zealand

Wentai Qu
School of Information Science and Engineering, Ningbo Institute of Technology, Zhejiang University, Ningbo Higher Educational Zone, China

Yonghua Zhou
School of Electronic and Information Engineering, Beijing Jiaotong University, Beijing, China

Zhihui Wang
School of Electronic and Information Engineering, Beijing Jiaotong University, Beijing, China
Beijing Research and Development Center, Zhongxing Telecommunication Equipment Corporation, Shenzhen, China

Deng Liu
School of Electronic and Information Engineering, Beijing Jiaotong University, Beijing, China
China Railway Communication & Signaling (CRCS) Survey and Design Co., Ltd., Beijing, China

Yong-Hong Lan and Lei-Zhou
College of Information Engineering, Xiangtan University, Xiangtan, Hunan 411105, China

Feng Gao and Chengkai Ye
School of Automotive Engineering, Chongqing University, Chongqing 400044, China

Zilong Wang
China Automotive Technology and Research Center, Tianjin 300300, China

Xu Li
Chongqing Changàn Automobile Corporation Limited, Chongqing 401120, China

Funing Yang and Yanyang Wang
The Key Laboratory of Automotive Engineering, Xihua University, Chengdu 610039, China

Yunxia Li, Yufeng Lu, Huilai Sun and Peng Yan
School of Mechanical & Automotive Engineering, Qilu University of Technology (Shandong Academy of Sciences), Jinan, China

Zhaodi Pei, Xiaoxu Zhao and Huimei Yuan
College of Information Engineering, Capital Normal University, Beijing 100048, China

Zhen Peng
Information Management Department, Beijing Institute of Petrochemical Technology, Beijing 102617, China

Lifeng Wu
College of Information Engineering, Capital Normal University, Beijing 100048, China
Beijing Engineering Research Center of High Reliable Embedded System, Capital Normal University, Beijing 100048, China
Beijing Advanced Innovation Center for Imaging Technology, Capital Normal University, Beijing 100048, China

Mingyu Fu and Duansong Wang
Harbin Engineering University, Harbin, Heilongjiang, China

Cyuansi Shih and Chengyu Cao
Department of Mechanical Engineering, The University of Connecticut, Storrs, CT 06269, USA

Zengcai Wang, Yazhou Qi, Guoxin Zhang and Lei Zhao
School of Mechanical Engineering, Shandong University, Jinan, China
Key Laboratory of High Efficiency and Clean Mechanical Manufacture (Shandong University), Ministry of Education, Jinan, China

Changfan Zhang, Jianhua Liu, Jing He and Guangwei Liu
College of Electrical and Information Engineering, Hunan University of Technology, Zhuzhou 412000, China

Xiang Cheng
College of Electrical and Information Engineering, Hunan University of Technology, Zhuzhou 412000, China
CHNV Technology Co., Ltd., Dongguan 523000, China

Darong Huang, Zhenping Deng and Bo Mi
Institute of Information Science and Engineering, Chongqing Jiaotong University, Chongqing 400074, China

Qi Wu
Department of Computer Science, Guangdong Police College, Guangzhou 510230, China

Yuyan Guo
School of Automation, Northwestern Polytechnical University, Xi'an 710072, China

Amanda Danielle O. da S. Dantas and Carlos Eduardo T. Dórea
Department of Automation and Computing, CT, Federal University of Rio Grande do Norte, 59078-970 Natal, RN, Brazil

André Felipe O. de A. Dantas, João Tiago L. S. Campos and Domingos L. de Almeida Neto
Master of Engineering in Oil and Gas, Potiguar University, 59054-180 Natal, Brazil

Jiemei Zhao
School of Mathematics and Computer Science, Wuhan Polytechnic University, Wuhan 430023, China

Lina Jin
School of Computer and Communication Engineering, Liaoning Shihua University, Fushun, China

Shuanghe Yu and Dongxu Ren
College of Marine Electrical Engineering, Dalian Maritime University, Dalian, China

Mingyu Fu, Shuang Gao and Chenglong Wang
College of Automation, Harbin Engineering University, Harbin 150001, China

Dang-Bo Du, Jian-Xun Zhang, Zhi-Jie Zhou, Xiao-Sheng Si and Chang-Hua Hu
High-Tech Institute of Xi'an, Xi'an, Shaanxi 710025, China

Index

www.ingramcontent.com/pod-product-compliance
Lightning Source LLC
LaVergne TN
LVHW080904210926
848485LV00004B/587